Design and Analysis of Lean Production Systems

Ronald G. Askin
and
Jeffrey B. Goldberg
The University of Arizona

John Wiley & Sons, Inc.

Acquisitions Editor   *Wayne Anderson*
Marketing Manager   *Katherine Hepburn*
Production Editor   *Patricia McFadden*
Design Manager   *Madelyn Lesure*

This book is printed on acid-free paper. ∞

Library of Congress Cataloging-in-Publication Data
Askin, Ronald G.
Design and analysis of lean productions systems / by Ronald G. Askin and Jeffrey B. Goldberg.
p. cm.

ISBN 0-471-11593-2
1. Production management. 2. Costs, Industrial. I. Goldberg, Jeffrey B. II. Title.

TS155 .A693 2001
658.5--dc21

2001045360

10 9 8 7 6 5 4 3 2 1

To science, engineering, and industry,
     for building such a wondrous world;
To family and friends,
     for making my world wonderful.

Ron Askin

This book is dedicated to my wife Donna, my children Jessica, Emily, Ben and Brandon, and my mother Bettemae, who love me despite my faults and who gave me the time and space to complete this work.

Jeff Goldberg

Preface

Manufacturing companies are in the business of making, selling, and delivering goods. Considering resource availability, technological feasibility, and customer desires, a corporate strategy guides the design, manufacture, distribution, and service of products. Marketing provides input on what customers want. Engineering converts the data into product designs, and process planners convert those designs into instructions detailing how to manufacture the item. Manufacturing actually produces the products. Supporting these activities are many service functions such as logistics, which procures parts and manages distribution of finished products to customers. This is an ongoing process as products and technologies pass through their normal life cycles. Production planners are responsible for planning the production quantities and timing for each product and for controlling the material, equipment, human, and information resources used in production on a recurring basis.

Understanding the basics of production systems for planning, scheduling, and controlling production quantities and timing has long been a fundamental curricular goal in both industrial engineering and management programs. In business schools, this topic is typically included as part of a broader course in operations management. Industrial engineering programs, on the other hand, tend to delve more deeply into the analytical aspects of production systems, often devoting an entire course to this topic. The related operations management topics of facility layout and location, materials handling, and quality engineering are each presented in their own course. There are many excellent operations management texts designed for undergraduate business students but few texts designed for upper division industrial engineering students and quantitative business programs. Moreover, recent trends such as globalization, lean manufacturing, and integrated supply chain modeling have created new issues and models calling for a new generation of educational resources. This, then, is the motivation for this text. Our objective is to present a current view of production system design and modeling at a level appropriate for upper division undergraduates and beginning graduate students in industrial engineering. The book should also be of interest to graduate students in business schools and to practicing production planners and managers looking for an explanation of new production philosophies. The text fits between the nontechnical conceptual descriptions found in popular publications and the mathematically rigorous presentations found in archival research journals. We present a mix of traditional and modern models, philosophies, and tools. In discussing production systems, we attempt to go beyond the basic *what* questions and provide guidance on *how, when,* and *why* as well. While we emphasize planning models, at times we delve into more hands-on engineering topics such as methods for setup time reduction and fool-proofing of production operations. We attempt to integrate modern philosophies such as time-based competition, lean manufacturing, and supply chain integration into the presentation.

The flow of the book follows the standard production system decision hierarchy. After two introductory chapters, we present three chapters covering the characterization of market demand, long-range strategic planning and partnering, and medium-term aggregate planning. This is followed by a series of chapters describing alternative production-control systems. Separate chapters cover traditional inventory theory, pull control systems, and MRP push systems. A more advanced chapter on multistage planning models concludes this section. This chapter is intended for graduate students. The final section addresses lower-level shop-floor design and operation. We begin with an integration chapter on lean manufacturing concepts. This is followed by a more traditional chapter on shop-scheduling. The final chapter discusses general shop-floor control issues and several special environments. The following describes the book's outline in more detail.

Chapter 1 introduces the student to the business firm. It provides a view of how the various functional areas fit together to conduct the firm's business, explains the scope and role of the production system within the firm, and presents the usefulness and limitations of accounting data as a source for model parameters.

Chapter 2 provides a framework for future study. It provides important definitions and functions and components of production systems. We introduce the basic tradeoffs among inventory, setup cost, and shortages. Many of the basic principles and laws that govern the behavior of production systems, from Little's Law to learning curves, are introduced and illustrated in the context of production operations.

Chapter 3 covers the process of identifying market opportunities. We concentrate on mathematical models for forecasting demand, but we also cover nonquantitative methods and the design and monitoring of forecasting systems. This is a stand-alone chapter and can be skipped without loss of continuity. It is included at this early stage in the text to reinforce the notion that customer demand both empowers and constrains production.

Chapter 4 covers long-range strategic planning and the importance of coordinating information across the supply chain. It discusses key factors for specifying manufacturing strategy, and problems, such as the bull-whip effect, that can occur when information is not freely exchanged. Globalization has led to large-scale, long-term production system design problems that incorporate transportation and cultural considerations. We present a discussion of these issues and an introduction to building decision models in this environment.

Chapter 5 returns to the more traditional set of medium-term aggregate planning models. We emphasize the use of network models where appropriate and the general applicability of linear programming. We discuss the difficulty in explicitly modeling setup considerations. The chapter concludes with a discussion of disaggregation of medium-term family production schedules into a time-phased master production schedule for end items.

Chapter 6 covers reorder point inventory models. The EOQ is explored and extended for time-based criteria, internal manufacturing, constraints, price breaks, and stochastic demand. The chapter presents multiple product and newsvendor problems. The chapter then addresses dynamic demand with uncapacitated and capacitated problems, and illustrates affiliated managerial topics such as ABC classification systems for prioritizing items and exchange curves.

Chapter 7 begins with an explanation of the basic requirements of a production authorization mechanism and poses several alternative approaches, including pull and push. The chapter then emphasizes pull control strategies. It describes various kanban systems and discusses the impact of variance and dynamic behavior. It includes an evaluation and description of CONWIP systems.

Chapter 8 covers push systems, namely, materials requirements planning. It covers the mechanics of the system, along with the advantages and practical limitations. We show how a conceptually simple idea can become complex and break down quickly in a random, continuous environment. The chapter also addresses extensions of MRP to include capacity requirements planning and enterprise integration.

Chapter 9 presents more advanced models for multi-item, multistage planning. It introduces decomposition strategies and modifications to cost structures. We model coordination of batch sizes between stages and competition for capacity. This chapter is intentionally, and probably necessarily, more mathematically sophisticated than the other chapters. As such, it is intended for graduate students.

Chapter 10 discusses recent developments and consolidation of ideas into a lean manufacturing paradigm for eliminating waste and rationalizing procedures in manufacturing. It promotes fool-proofing of processes, setup time reduction, efficient facility layout, reduction of variability, and statistical process control. It also discusses the role of preventive maintenance and the usefulness of flow-charting processes.

Chapter 11 presents shop-floor sequencing and scheduling tools. We progress from single-machine to two-machine to flow-shop to job-shop procedures and discuss makespan, flowtime, and tardiness objectives. The shifting bottleneck procedure is shown to integrate previous modeling approaches. We conclude with an acknowledgment that these traditional, quantitative scheduling approaches often make simplistic assumptions. This leads to a discussion of knowledge-based approaches that attempt to be more realistic.

Chapter 12 ends the text with a collection of relevant scheduling topics. The chapter begins with hierarchical vs. heterarchical control and requirements for manufacturing execution systems, followed by special topics for flow systems, including line-balancing, lot-streaming, and re-entrant flow scheduling. It also covers the importance of tool management and issues that arise in flexible manufacturing systems.

The text may be used in several ways. There is more than enough material for a one-semester course at either the undergraduate or the beginning graduate level. We assume the reader has an understanding of basic operations research and statistics. (An Appendix summarizing linear programming is provided for those students needing a refresher. However, the text emphasizes system design and modeling concepts; it is not a primer on solving mathematical models.) For a first course in production systems for upper division industrial engineering students, we recommend covering the major topics in Chapters 1 through 8 and 10 through 12. At the instructor's discretion, some of the more quantitative topics in the later sections of Chapters 3 through 7 and 11 may be omitted without a loss of continuity. Indeed, it is in this manner that the authors have piloted the text. For undergraduates, we strongly recommend that the topics in Chapter 10 be covered, either in this course or in a related course. However, we would not consider covering topics such as the bottleneck scheduling heuristic in Chapter 11. Many undergraduates will find the bullwhip effect and multifacility models in Chapter 4 rather challenging as well. Several minicase studies are provided in the end of chapter exercises, and we do recommend using these or similar exercises for in-class discussions or group projects.

For graduate courses we recommend including the more sophisticated planning models in the later sections of Chapters 4 through 6 and 11 and in the multistage models of Chapter 9. The instructor may even wish to augment Chapter 4 of the text with the latest readings on supply chain management system design and supply contracts and Chapter 9 with advanced results on multiechelon inventory theory. To do so, however, will require the prerequisite of a solid operations research background. Chapters 10 and 12 may be omitted for students with a background in industrial engineering. Chapter 11 may also be omitted for graduate students if a separate course in scheduling theory will be taken.

Practicing production managers may choose topics based on their specific needs and background. An engineer with responsibility for operations may find that reading the text helps explain the aggregate planning and production-control information systems with which they currently interact. This background will allow them to select pieces from the text for implementing improvements while understanding the ramifications throughout the rest of the system. For the less technically oriented production manager, an understanding of the concepts and underlying principles of lean manufacturing can be gained without digesting all of the mathematical models. For within-plant operations, a review of Chapters 2, 7, and 10 will provide a relatively thorough yet readable description of how to create a lean production system. Details appropriate to the particular site can then be gleaned from study of the relevant topics in Chapters 6, 8, 11, and 12.

The reader may notice some redundancy between chapters. Certain key concepts reappear to reinforce their importance and applicability. Certain equations are reproduced to increase independence between chapters and permit flexibility in how one chooses to sequence the topics.

A number of colleagues and students have provided helpful comments on earlier versions of this manuscript. In particular we acknowledge the assistance of Alessandro Agnetis, Frank Ciarallo, Johann Demmel, Michael Deisenroth, Maged Dessouky, Erin Fitzpatrick, Marvin Gonzalez, Catherine Harmonosky, Bryan Harris, Chun-Yee Lee, and Jennifer Mello for their constructive suggestions.

We would like to thank our instructors, who by their own high quality research and teaching, taught us how to practice and teach industrial engineering. Jeff benefited from many individuals at Cornell and Michigan, but would particularly like to thank Jack Muckstadt, Bill Maxwell, Jim Bean, and Dick Wilson. Also, he would like to thank the many students he has taught over the past 16 years. They have suffered through the experiments, the projects, and other brilliant ideas that sometimes worked and oftentimes didn't. The true measure of a person is not fame or wealth, but is the positive impact, that they have on others. He hopes that he has made a positive, lasting impact and added value to your lives. Ron would like to especially thank those instructors at Lehigh and Georgia Tech that shaped his professional interests and thought processes. Lynwood Johnson, Doug Montgomery, and Wally Richardson stand out among many positive influences.

Ron Askin
Jeff Goldberg

Contents

Chapter 1

The Industrial Enterprise

$\mathbf{A}$utomobiles, telephones, furniture, clothes, and games. These and all the other products we use in our daily lives are designed, manufactured, and delivered with the intention of making a profit for the producer by enhancing the quality of life of the customer. To achieve this result, entrepreneurs must arrange to bring together the necessary financial, technological, information, and human resources to design, implement, and operate the complete manufacturing system. In this text, we concentrate on one aspect of that total manufacturing spectrum, the production planning and control system. For the most part, we will assume the products have already been designed, and the technologies to be used to make these products have been selected. We will focus on the questions of what to make, how much to make, and when to make it. The associated issues of acquiring raw materials, selecting the most efficient production methods, and deciding how to distribute products (get them to the customers) will also be discussed.

Production system design and operation involve managing production resources to meet customer demand. Managing the system requires the development and execution of production schedules detailing how to use resources to convert raw materials into finished products. In addition to employees, machines, tools, and materials, relevant resources include utilities, information, and established procedures. The physical items used and produced by the system are referred to as **inventory.** Inventories are created during the production process and held in the form of raw materials, component parts, subassemblies, and finished products. The scope of the production system includes long-term **planning,** medium-term production **scheduling,** and short-term **control** (dispatching) decisions. The acquisition of new capacity and development of productive employees require planning months in advance of actual usage of these resources. The interface between long-term planning and medium-term scheduling involves setting production goals for each product over the next several weeks. With these objectives set, short-term schedules must detail the sequence of activities for each worker and machine resource for the current day or week. We may be able to develop a plan for these activities, but the stochastic nature of the world as evidenced by machine breakdowns, late deliveries, defective parts, and changes in customer orders requires that we be able to react quickly to changing conditions. If we plan wisely, we will have the flexibility to adapt to the random events of the everyday world.

The reader should keep in mind that many steps are involved in converting raw materials into delivered products. We must typically fabricate many component parts, assemble these parts into the final product, and deliver the product to the point of sale. Fabrication and assembly may occur at a single location or may involve facilities spread around the world. Some companies choose to keep as many activities "in-house" as possible, whereas other companies are little more than production managers, subcontracting out most activities. The first approach allows tighter control, but the latter approach allows groups to focus on what they do best and permits the company to remain flexible and constantly seek the best supplier for each activity. The latter situation has become increasingly more common for complex products as companies seek to cope in a dynamic world.

Throughout the system we must ensure conformance to product quality requirements, and execute the necessary administrative support functions. This becomes increasingly complex when dealing with multiple cultures, measurement systems, languages, and standards. Figure 1.1 illustrates the overall production process from "ore to door" for a large product such as an automobile or computer. At each stage of this process, we incur cost and add value to the product. However, the valid criteria deal not with the individual steps, but with the total cost of the system and state of the final de-

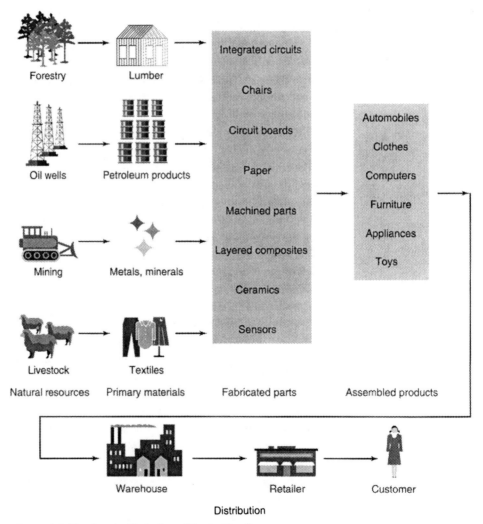

Figure 1.1 The Supply Chain from "Ore to Door"

livered product. Thus, we will endeavor to maintain a system-wide perspective on the effects of our planning and control decisions.

The industrial firm maintains its existence by providing customers with products whose societal value exceeds their cost of production plus delivery. The product may be a physical good or service.[1] In describing the industrial firm, we will look at the measures that determine the viability of the firm as indicated by its products and/or services, the necessary functional areas of the firm, and the specific activities in the product design, manufacture, and delivery process.

1.1   MEASURES OF COMPETITIVENESS

The industrial firm competes in multiple dimensions. When car shopping you will probably buy from the dealer that has the car you want at the lowest **price.** However, you very likely first limited your choice to a few select models, colors, and optional configurations. Thus, **product features,** both cosmetic and functional, are important. Likewise, you probably stayed clear of any models or dealers with poor **quality** or **service** reputations. And, of course, you restricted yourself to vehicles with **availability** within your time window. Firms compete in all these dimensions. Many firms today perform benchmark analyses of their performance relative to the best in their industry. Such analyses yield information on competitiveness and indicate requirements for success and opportunities for market penetration.

1.2   FUNCTIONAL AREAS OF THE FIRM

Industrial firms vary widely in the products they make and their internal philosophies toward employee development and technology acquisition, but they must all perform the same conceptual functions. Table 1.1 lists the major organizational units commonly found in industrial firms. It is best to view these units as functions instead of departments. Although traditionally each function may have been a separate department, an increasingly popular organizational structure creates employee reporting structures based on product line. Thus, each major product category, such as toasters, microwave ovens, and conventional ovens, would have its own unit to perform each function. This product-based organizational structure causes a greater identification with the product and facilitates cross-disciplinary teamwork. Unfortunately, many of the techniques needed by the individuals relate to their technical discipline and continuing technical education is best accomplished through that discipline. As a result, many companies have found it preferable to use a matrix structure. Individuals reside in an organizational unit based on their technical expertise such as accounting or maintenance, but they are normally on loan and temporarily report to a product-based team. Figure 1.2 illustrates this structure.

1.3   PRODUCT DESIGN, MANUFACTURE, AND DELIVERY

The primary mission of the firm relates to supplying goods and services to satisfy customer needs. Laws governing financial reporting, environmental impact, and product safety; the availability of capital, labor, and materials; customer preferences; and the state

---

[1]In this text, we will concentrate on physical products produced by manufacturing industries, but much of the discussion herein pertains to service industries as well.

Table 1.1   Major Functional Units of an Industrial Firm

Functional unit	Activities
Executive Committee	Strategic Planning
Design Engineering	Product Design
Product Engineering	Process Planning
	Test Design
Manufacturing	Production Planning
	Scheduling
	Materials Management
	Material Handling (incl. shipping)
	Tooling
Facilities	Building/Site Layout
	Maintenance
	Equipment Installation
	Utilities
Product Assurance	Quality Assurance
Research and Development	Product Development
	New Technology Evaluation and Implementation
Management Information Systems	Data Processing
	Report Generation
Procurement	Vendor Certification
	Purchasing
Finance	Budgeting
	Cash Flow Management
Accounting	Financial Reporting
	Managerial or Cost Accounting
	Accounts Payable
	Accounts Receivable
Marketing	Sales
	Order Entry
	Forecasting
	Customer Relations
	Advertising
Human Resources	Recruitment
	Labor Relations
	Wage Administration
	Employee Protection (Safety, Equal Opportunity)
	Training

of technological knowledge create constraints on how the firm accomplishes this mission. Corporate management strives to maximize profit by determining the optimal targets for the competitive factors discussed above subject to these constraints. The company must then design and implement systems and procedures to achieve these targets. This leads to the set of functional areas listed in Table 1.1. In addition, management must create an atmosphere conducive to success. Factors such as employee morale, organizational structure, and employee empowerment must be monitored to ensure their support for achieving corporate objectives.

Several functions must be executed to create and deliver products. From a chronological perspective, these functions include:

	Project Team (Leader)				
(Manager) Functional Department	(Sanjay) Body frame	(Herbert) Power train	(Kuei) Controls	. . .	. . .
(Richard) Mechanical Design	Bill Frank R.	Sam Tom	Samantha Joe		
(Barbara) Electrical Design	Gloria Frank T.	Ron Linda	 Ramesh		
(Fang Yu) Test	 Fred	Susan Kim	 Ralph		
(To be filled) Industrial Design	 Sherry	 Dave	 -		
(Henry) Marketing	 Rick		 -		
.					
.					

Figure 1.2 Matrix Organization Structure

1. Product, Process, Tooling, and Test Design
2. Manufacturing
3. Distribution
4. Product Support and Maintenance
5. Product and Process Disposal

These activities fall into the phases of product development, production, and usage. Figure 1.3 depicts the product life cycle and the involvement of major functional areas over time. The figure envisions a completely new product being brought to life. The process shown is somewhat simplified. In reality, design engineering stays involved throughout the product's life cycle issuing engineering changes to incorporate improvements in performance and cost reductions. In addition, most new designs represent minor upgrades or modifications of existing products. Eventually, the market becomes saturated, technological breakthroughs occur, and new creative concepts indicate it is time to develop a new generation of products.

1.3.1   Integrated Product, Process, Tooling, and Test Design

Integrated product design begins with collection of customer specification information. Products typically develop from either response to a customer request for quotation or marketing research. The first case occurs with military/government contracts and industrial tooling such as dies. The product idea exists, and the customer is in search of a producer. Alternatively, the producer's marketing department may be surveying potential customers to determine opportunities. Occasionally, the serendipitous discovery of a new technology or product creates a market. The 3M Company's revolutionary Scotchgard developed in this manner. A lab assistant accidentally spilled a chemical mix on a pair of

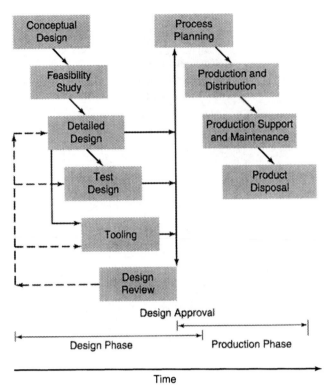

Figure 1.3 The Product Design and Production Process

canvas tennis shoes. The shoes then proved to be waterproof and stain-resistant. Most large companies maintain Research and Development groups charged with advancing basic science and finding commercial applications. Breakthroughs in electronics such as the transistor and wafer-fabrication processes have led to new products for computing and telecommunication that have transformed our daily lives.

After collecting customer desires, functional specifications are determined for the product and placed in the internal technical jargon of the firm. Breaking strength in psi and computer storage disk access time in milliseconds are examples. Knowledge on available technology is then synthesized to select the technological approach. Several initial concepts may be generated and compared. This is followed by detailed design and analysis along with formal reviews to accept or reject product designs. In today's rapidly changing technological world, the speed at which new products can be brought to the marketplace often determines the financial success of a company. Current approaches call for simultaneous or concurrent engineering of the product design, manufacturing processes, test equipment, and maintenance plan. Thus, a multidisciplinary team of designers works together to develop the product including its manufacture, test, and service plans. The product design as described in the product definition data must normally be approved by all relevant functional areas of the firm including marketing, purchasing, and finance in addition to the technical areas just noted. Production tooling and test equipment may be designed along with the product. Although this may seem wasteful because not all data necessary for tool and test design will be available until the product design is complete, the integration of these activities frequently reduces the number of iterations required to pass through the design and review cycle. Eliminating costly missteps and design changes in this manner allows the final product to hit the shelves quicker and with lower total cost.

Techniques such as the House of Quality (Clausing [1988]) can be used to guide the acquisition of data and setting of product specifications and to increase the chances of product success. A simplified house is illustrated in Figure 1.4. The **customer attributes** of importance are listed on the left of the house (each level of the house describes "what" the customer wants). A relative measure of importance should be assigned to each customer attribute. Next, the columns are the **engineering characteristics** that will indicate "how" we can measure in technical language whether the attributes are achieved. The elements in the central array indicate the degree of correspondence between each attribute and characteristic. The basement of the house contains an indication of the difficulty in improving each engineering characteristic, a summary of measures for the products of competitors, and our target levels. Product design usually involves breaking down the product into modules. Each module is assigned a subset of functions to perform (engineering characteristics to meet). The design of each module can be farmed out to the group with expertise in that area. Modules are then fit together with minimal interactions to satisfy all product requirements. For example, one group is responsible for the frame of your computer, another for the power supply. The memory chips and CPU are separate as are the modem and graphics board.

Process planning involves the transformation of part or product definition data (materials, geometry, topology, tolerances, demand volume) into detailed production instructions. Alting and Zhang (1989) summarize this process in ten steps, as shown in Table 1.2. Detailed manufacturing, assembly, test, and service instructions are then generated. An example of a process plan is shown in Figure 1.5.

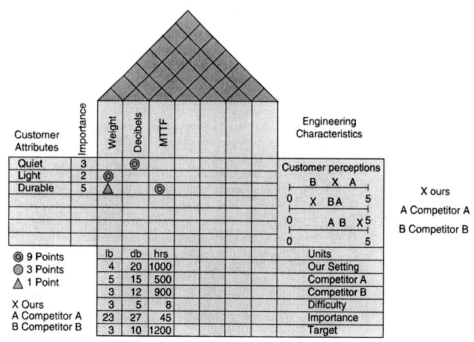

Figure 1.4 Sample House of Quality

Table 1.2 Creation of a Process Plan

1. Interpretation of product design data
2. Selection of machining processes
3. Selection of machine tools
4. Determination of fixtures and datum surfaces
5. Sequencing of operations
6. Selection of inspection devices
7. Determination of production tolerances
8. Determination of the proper cutting conditions
9. Calculation of standard times and costs
10. Generation of process sheets and NC code

A process plan consists of a sequence of **operations** for converting raw material into finished products. (Note that the finished product for one manufacturer may be a component part or bulk product, such as a coil of sheet metal or styrofoam pellets, that becomes the raw material for another manufacturer.) There are several types of operations, and it can be important to distinguish between these when collecting data for modeling the production facility. **Fabrication** involves transforming the raw material such as casting molten metal in a mold, drilling holes, or etching a pattern into the surface of a material. We will include required time-delay processes such as drying, aging, and cooling as a special type of fabrication operation. **Assembly** refers to the action of putting parts together to form a product. Many final products are produced on an assembly line where the various parts are connected by careful positioning, followed by soldering, fastening, or other joining action. We will also refer to **kitting** as an assembly-type operation. Kitting consists of picking from storage the parts needed to make a batch of product and loosely grouping these parts together in a package such as a bag or tote. Kitting frequently precedes assembly. **Separation** operations break a product or material into multiple components. These operations are common in the early stages of processing natural materials such as ore or oil and in the recycling of used products. It may be useful to identify **test** operations because these often have probabilistic outcomes with the subsequent routing of parts being dependent on the results of the test.

Operations possess several attributes in addition to operation type. **Setup** refers to the activity required to begin an operation such as adding tooling on a machine or checkout (running and testing an initial part). The **unit-processing time** indicates the time required per repetition of the operation. This may be a fixed time or a random variable. If parts are processed one at a time, then the batch-processing time is $S + Qt$—the sum of setup time (S) plus the product of batch size (Q) and unit-processing time (t). In some operations, units can be processed in parallel, thus reducing the processing time required per batch. Thus, the maximum simultaneous **processing quantity** represents another attribute. Process **yield** denotes the proportion of units that are successfully processed on average. The remaining units are either scrapped or sent for rework. Yield can be multidimensional in the sense that the process may produce product of varying quality. The output of the process may require grading with higher grades bringing higher prices. You have seen this practice when choosing the speed of your computer's microprocessor or choosing between regular and irregular clothes. Many chemical processes also exhibit this trait.

Operations are normally separated by a material handling activity. For the most part, it is sufficient to consider material handling actions as another operation type. The characteristics of that move are the number of units processed (moved) simultaneously and

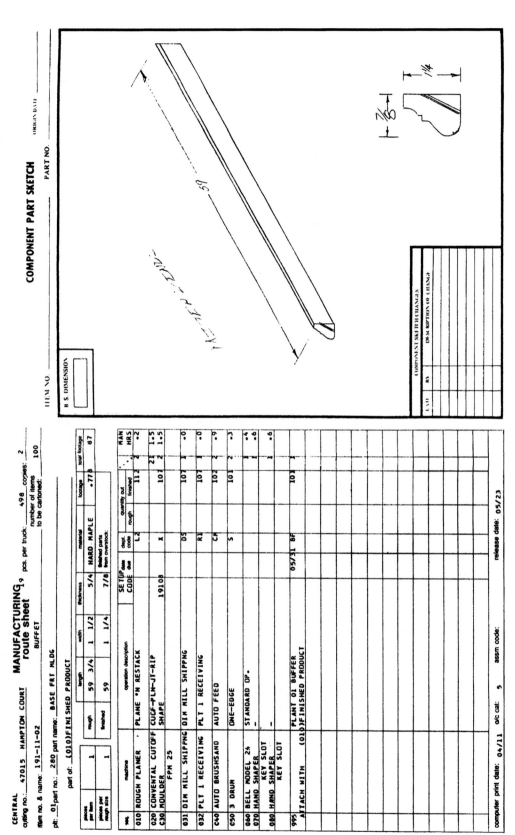

Figure 1.5 Sample Process Plan Route Sheet
(Courtesy of Stanley Furniture Co.)

9

the move time. We refer to the quantity of units moved per trip as the **unit load.** This may be a single item, a pallet, or tote bin full of material. The number of units processed in immediate succession on a machine is the **process batch.** The process batch normally consists of a single part type. Once the machine is set up, it produces the entire process batch before changing to a different part. The **transfer batch** is the number of units moved at the same time. It will often be the same as the unit load but may consist of several unit loads. The process batch then consists of one or more transfer batches.

If all jobs move through the same sequence of machines we refer to the facility as a **flow shop.** If each job has its own routing through the set of processors then we refer to the production facility as a **job shop.** When operations can be performed in any order, we refer to it as an **open shop.** Some facilities have **re-entrant flow** patterns where jobs repeatedly cycle through the same processors. Semiconductor manufacturing offers the most important example of this flow pattern as each layer of the device requires passing through many of the same processes of masking, photoresist, etching, etc.

The House of Quality can be used in a hierarchical manner for specifying materials and process plans. The engineering characteristics from the conceptual design stage, such as thermal conductivity, become the customer attributes at the detailed design stage. The engineering characteristics from detailed design become the attributes at the process planning stage.

1.3.2   Manufacturing

Manufacturing generates the value added by the industrial firm. It is here that raw materials pass through a series of operations to be converted into finished products for distribution to customers. The production system covered in this text has the responsibility to determine the "what," "how much," and "when" of manufacturing. The "how" is determined by process planning. The production system specifies the production rate and corresponding capacity requirements. It then coordinates the flow of materials through the facility. In addition, the production system must specify the levels of support services such as material purchases, inventory, and human resources needed to support this plan.

1.3.3   Distribution

Product value is realized only when the product is delivered to customers in useable condition. Distribution can be a prolonged and expensive activity. The cumulative cost of moving physical goods from the raw material through the finished product stage may account for as much as 50% of the product's cost. Final goods must often be kept in warehouses near final customers to reduce delivery time. If the final product is available in many variations, we may store generic assemblies and then customize upon receipt of an order to achieve a happy medium between excessive storage costs and instant availability. The combination of purchasing, manufacturing, and distribution is referred to as the **supply chain.**

1.3.4   Maintenance and Support

Rare is the corporate entity that would purchase a computer or duplicating machine without a warranty, on-site maintenance agreement, and, in most cases, on-line support. IBM can attribute much of its historical success to its strong customer support. Products are useless if they do not work and worse than useless if, once put into service, they cannot be counted on in an emergency. Documentation of how your newly purchased VCR operates may not be the most exciting reading, but you certainly expect to be able to find help when needed to ensure you can tape your favorite shows while on vacation.

Customers expect timely delivery of quality products. This requires maintenance of production equipment. The machines must be available for use when needed and must perform as expected. Preventive maintenance helps ensure that production plans can be met and repair activities help to avoid missing delivery schedules.

1.3.5   Disposal

In recent years, life-cycle management concepts have come to include product disposal and recycling. According to the 1994 General Motors Environmental Report, 94% of cars and trucks in the United States are recycled. Of those, 75% of the vehicle content is recycled (primarily metal). Some manufacturers will now supply customers with written feedback on the ultimate disposition of their vehicles. Environmental rules often require extensive documentation on the disposal of hazardous materials. Within the plant, the use of environmentally safe processes has become increasingly important. In addition to providing a better image, costs are often reduced and safety improved. Capture and reuse of unconsumed chemicals, elimination of unnecessary hazardous materials through better process control, different production methods, or use of cleaner alternatives are all valuable strategies. For instance, water-based solvents are now used in place of many pollutants.

1.4   BUSINESS PROCESSES

Business operations consist of a set of "processes." A process may be defined as a set of procedures designed to integrate people, knowledge, materials, equipment, energy, and information to accomplish a specific task. Examples of important business processes include:

- Design a product
- Enter a customer order
- Fill a customer order
- Manufacture a product
- Set long-term capacity plans
- Acquire raw materials
- Hire and train workers
- Bill and collect accounts receivable
- Record and reimburse accounts payable
- Develop a proposal for new business

To illustrate how a process can be documented and to familiarize the reader with the basic process of receiving and executing customer orders, Figure 1.6 presents an aggregate view of the customer order execution process. The figure assumes a producer with a business office, production facility and distribution center from which orders are shipped to customers. The three functions may be at the same or different geographical locations. The first major step is to receive the order and obtain agreement between the customer and supplier on price, delivery date, product quantity, and other attributes of the order. An order will often include multiple items. The production facility must break the order into

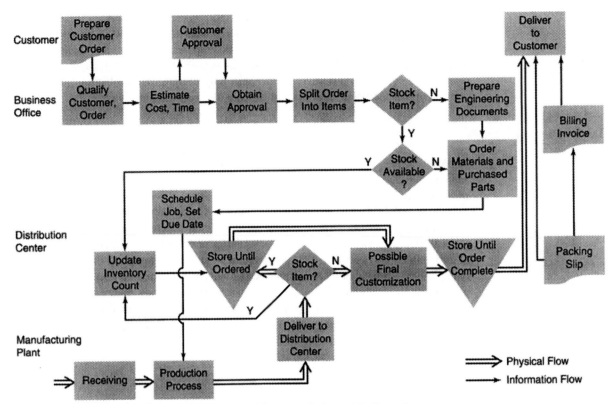

Figure 1.6 Overview of Customer Order and Delivery Process

line items and produce each. Some items may be standard and available off-the-shelf, whereas others may require custom production. Inventory status records must be maintained and production schedules coordinated to blend this order with those of other customers. Eventually, the items are collected and the completed order delivered. In some cases, the customer may be willing to accept partial shipments as items become available. In practice, the order execution process can become far more complex as we need to be able to implement engineering and customer change requests and react to changes in the yields and availability of production equipment.

There are three important characteristics for all processes.

1. Processes require resources. This is evident from the list of collected items in the above definition.

2. The process should be derived from the objectives of the firm. It should be possible to trace the justification for the process directly back to satisfaction of some objective of the firm. If the process is not needed for accomplishing the business objectives, it is extraneous. In light of the first characteristic, extraneous processes should be eliminated.

3. Each process must have a product **and** a customer. The customer dictates what the product should be and quality of the process is measured against customer satisfaction and delight. In practice, processes tend to have primary and secondary products and customers. In manufacturing and delivering a physical product such as an automobile or toaster, the primary product is the item itself and the primary

customer is the purchaser. However, the employees who made the product and spend much of their lives in the shop, and the service rep who must maintain that product in the field, are also customers.

In evaluating processes, there are several relatively universal measures: cost, quality, and time. Maximizing profit represents a major objective for most private sector entities. This can be accomplished through continually providing to customers products with high perceived value. This perceived value is derived from the functions performed by the product, the timeliness of the product's availability, the aesthetics associated with ownership and use of the product, the quality of its performance, and its cost. Minimizing internal cost to meet a given level of perceived customer value contributes to maximizing profit.

Quality represents several aspects of the overall evaluation process. Design quality refers to the degree of exquisiteness selected for the process (or product). This is closely associated with the market in which we choose to compete. Conformance quality refers to how well the targeted design quality is met. Design quality indicates what we plan to do; conformance quality indicates how well we do it. The quality of a $50,000 luxury automobile with headlights that are not precisely focused or high internal noise level is poor. This may be attributable to poor design intent or failure to conform to the design. Design for manufacturability bridges the gap between these two concepts to ensure that it is possible with existing human and technological capability to conform to the design.

Processes often cross organizational and functional boundaries. During conversion from raw material to finished product, the material may pass through several process-oriented departments such as receiving, cutting, milling, welding, and assembly. Designing a product involves marketing, design-engineering, test-engineering, manufacturing, purchasing, finance, and customer-support functions. Processes also are often interconnected. Order entry, purchasing, and production planning are highly interconnected processes. We cannot plan production without the parts being purchased, and we do not know what to produce without order entry.

Advances in information technology support significant improvement in the design of interconnected processes. The time to replenish shelves at the retail level and the cost of inventory can be substantially reduced if retail sales are immediately transmitted back to the order entry and production scheduling system. Electronic checkout systems and the international phone and data communication networks make this a feasible option. Information exchange is important within the production facility as well. When an item is scrapped because of a problem in the manufacturing process, the data should be relayed to related processes such as production scheduling.

1.5   ACCOUNTING SYSTEMS: ACQUISITION AND MANAGEMENT OF MODEL DATA

Accurate, timely, and complete data are essential for good decision making. The models we will be discussing in this text, and, indeed, many of the decisions made in production systems planning and control, depend on the availability of accurate cost data. Whether to use overtime, a second shift, or new labor and machines to meet increasing product demand; whether to make a component or buy it from a supplier; whether to automate an operation or use manual labor. Accounting systems accumulate much of the data necessary to model these problems and make decisions. As part of the process of financial reporting and support of managerial decisions, accounting systems track the cost of materials, labor, factory overhead, and administrative overhead. Labor cost is generally divided into **direct labor** and **indirect labor.** Direct labor includes work performed specifically on the part such as operating a machine while the part is being produced. Activities such as material handling and setup may also be charged to direct labor if the worker's responsibility is restricted to one product type. Usually, however, these are considered indirect labor and allocated to factory overhead.

Indirect labor also includes support activities such as maintenance, supervisors, and quality assurance. Indirect material costs would include general supplies, lubricants, and other materials consumed during operations that do not become part of a finished product. Other necessary costs that support operations such as utilities, insurance, equipment depreciation, and space rental also are lumped into factory overhead costs in traditional direct cost accounting systems.[2]

Partially completed products that are currently in the shop are referred to as work-in-process (WIP). Accounting systems then use the following relations to accumulate cost for each time period.

Total Manufacturing Cost = Direct Material Cost + Direct Labor Cost
                    + Factory Overhead

Cost of Goods Manufactured = Total Mfg. Cost + Initial WIP Value
                    − Ending WIP Value

Cost of Goods Sold = Cost of Goods Manufactured + Initial Finished Goods
                    − Final Finished Goods

Table 1.3 shows a generic cost of manufactured goods statement.

The objective is to eventually track costs to products. The traditional approach assigned direct material and labor costs to products and then apportioned overhead to products in proportion to their direct costs. As manufacturing systems became more automated

Table 1.3   Cost of Manufactured Goods, June, 1999

Direct Materials 5/31/99	$25,000	
Direct Materials Purchased	$12,800	
Direct Materials 6/30/99	($26,500)	
Cost of Direct Materials Used		$11,300
Direct Labor		$20,000
Indirect Labor	$8,500	
Indirect Materials	$4,000	
Utilities	$2,500	
Insurance	$800	
Rental Space	$3,000	
Depreciation	$9,600	
Misc.	$1,200	
Total Factory Overhead		$29,600
Work-in-Process 5/31/95	$53,000	
Work-in-Process 6/30/95	($52,600)	
Cost of WIP Used		$400
Cost of Goods Manufactured		$61,300

---

[2]Marketing and administrative offices costs form administrative overhead. These are kept separate from manufacturing costs but are clearly part of the overall business system.

and indirect costs became more dominant, this procedure proved unsatisfactory. **Activity-based costing** (ABC) attempts to first assign all costs to the generating activity. An assembly line, a manufacturing cell of several machines, and an inspection station could each be classified as an activity. The set of products using each activity is recorded. For each activity, a cost driver such as hours on the key machine is identified. The costs for the activity are then assigned to products based on their usage of the cost driver. Quality, equipment depreciation, maintenance, machine setup, and other costs can often be attributed to specific products. This can provide a more accurate picture of product costs. As part of this process, management should consider elimination of activities that are not essential for any product.

EXAMPLE 1.1

A publishing company sells a hardbound and softbound version of a textbook. Publishing requires two steps. First pages are produced and then collated into books and bound. Pages are produced identically for hard- and soft-cover versions. Direct labor is 2.5 hours per 100 books. Indirect cost for page making is $150,000 per year. The hardback covers are made and the books stitched together by an automated machine. Costs for this process are $500,000 per year, and the machine can produce 500,000 books per year. Soft covers are produced and glued by hand, taking 7.5 hours per 100 books. Indirect cost for this department is $100,000 per year (also 500,000 books). Labor costs $15 per hour and direct material costs are $4.00 per hardback, $2.00 per softcover book. Determine costs per 100 books for each version using both traditional and activity-based costing. Assume the facility is fully utilized, making 1,000,000 books per year.

SOLUTION

Direct costs are the same under both accounting methods. These costs are added for the page-making and -binding departments. The activity centers are page-making, hardcover-binding, and softcover-binding. The traditional approach allocates the $750,000 total annual overhead costs on the basis of direct labor cost. Thus, 80% of this cost is assigned to the soft-cover version because these take 10 direct labor hours per 100 compared with the 2.5 hours for hardback books. Clearly, this is a misleading estimate because page-making costs are the same for both versions, and the labor-intensive softcovers have lower actual overhead cost. This error is corrected by the activity-based estimate where indirect costs at each center are assigned to products. The full cost comparison is shown in Table 1.4. Direct labor and material are the same for both systems. Hardback books require 2.5 labor hours per hundred at the rate of $15 per hour. Softcover books require a total of

Table 1.4   Traditional vs. Activity-Based Costing

Traditional cost accounting		
	Hardback	Softcover
Direct Labor	$37.50	$150.00
Direct Material	$400.00	$200.00
Overhead (Direct Labor Basis)	$30.00	$120.00
Total Costs per 100	$467.50	$470.00
Activity-based costing		
Direct Labor	$37.50	$150.00
Direct Material	$400.00	$200.00
Page Dept. Overhead	$15.00	$15.00
Binding Dept. Overhead	$100.00	$20.00
Total Costs per 100	$552.50	$385.00

10 labor hours per hundred for page-making and -covering. The difference occurs in allocating the $750,000 per year of overhead. Because the softcover process uses 80% of the direct labor, it receives 80% of the overhead charge in traditional accounting. With ABC, the overhead costs are apportioned by work center to the products using that center. Page-making and -binding are separated. The full $500,000 overhead in automated hardcover binding is assigned to the hardcover books at the rate of $1 per book. Clearly, the two accounting approaches present different pictures of product cost. The traditional accounting approach may cause us to underprice hardback books and overprice softbacks. Both systems suffer from their inability to estimate cost gradients from attempting to increase or decrease the production rate.

Of course, some costs are still difficult to assign to specific parts. Also, costs are often random or merely approximations. Suppose, for instance, two identical robots are purchased. In the first case, the robot lasts five years, and product demand runs higher than expected at 20,000 units per year. Each unit of product must be charged 0.0001 of the robot's lifetime cost. On the other hand, if the robot only lasts four years and demand is only 5,000 per year, each unit will be charged for 0.0005 of the robot's cost. Is it fair to charge the product based on the random performance of the robot temporarily assigned to its production? Also, these actual costs are known only after the product and robot's life. At that point, it is too late to change plans. Partly for this reason, **standard costs** are often used instead of actual costs. The use of standard cost rates for resources removes the misleading effect of random events when analyzing cost reports. The standard costs reflect the norm for a particular activity. Variances in labor, material, and other costs are used to indicate the deviation between standard and actual realized costs.[3]

It is important to distinguish variable overhead from fixed overhead. Suppose we are considering producing a new part that was previously purchased. Power to run the machine and materials consumed during production such as cutting tool points are clearly variable with the level of production. Their cost, although probably part of overhead, should be included when estimating the cost of producing the part in-house. The salary of the supervisor of the product test department will probably not change and thus can be excluded from the analysis. The accounting system will have one way of compiling costs, yet the factors that are relevant in comparing options often depend on the scale, time frame, and set of alternatives. The experienced decision analyst must extract the appropriate cost data for the problem at hand.

Accounting systems vary widely depending on the type of business. Accumulating costs by process and assigning them to product types seem natural for continuous-flow and high-volume manufacturing. In lower-volume, higher-variety discrete parts manufacturing, the procedure can still be used. As before, costs are allocated to products based on the proportion of time they use each resource. Some businesses operate on a "cost plus" contract basis. This has often been true in the defense industry. In such cases, it may be necessary to directly accumulate costs by specific job or contract instead of by activity. Likewise, actual costs are needed instead of standard costs.

It is important that the builder of production system models understands the components of any accounting figure. In many cases, the data may not be precisely in the form desired for our planning models. We may have a different perspective on fixed vs. variable costs, and direct vs. indirect. Indirect labor may be a direct consequence of our production decisions and relate directly to production levels. If so, the costs of producing products should include this. Storing money in the form of inventory is a real opportunity

---

[3]The accounting use of the term variance refers to a deviation from the standard and differs from the concept of variance in statistics.

cost (more about this later) and a necessary part of operating production facilities, but the accounting system may not view cost collection in this manner. We must be cautious in using accounting data without knowing how it was defined and compiled. The cost of a machine hour may seem like a simple request. However, different accounting systems will report different figures. The figure could be standard or actual, may or may not include overhead allocated in one of several ways, and will be based on some assumed utilization when including variable cost.

Beyond this, our desired value will depend on the time frame for the model. In a long-range planning model, selling the machine and reducing the indirect labor force that supports that operation may be relevant costs. In a short-term routing decision, these options do not exist, and their fixed costs should be excluded. As another example, we may wish to know how much we would benefit if we could save an hour of the time required to change over a machine from producing one product type to another. A simple answer is the accountant's hourly cost basis for the machine (multiplied by the number of changeovers per time period). The actual answer is more complex. As we will see later, such time savings indicate that we should produce fewer units of a product at a time. This in turn will reduce WIP levels and response time to changes in customer demand. If this is a bottleneck machine, time saved in changeovers may allow us to produce more product across the entire factory leading to a significant increase in revenue. Unfortunately, accounting systems are not designed to track these secondary savings or to provide the value of improved customer satisfaction. The accounting system's estimate of the savings from improved changeover efficiency is only a lower bound on the actual benefit.

1.6   SUMMARY

The modern industrial firm strives to make a profit by timely delivery of high-quality, low-cost, functional products to customers. Marketing, engineering, and manufacturing work together to design products that will meet customer requirements. Business processes must be developed that receive information from and deliver products to customers while also supporting internal operations. In addition to coordinating the direct flow of materials from suppliers through manufacturing and on to distribution to customers, information must be collected to make sure products satisfy customer needs, meet governmental requirements and execute administrative functions. Employees and suppliers must be paid; equipment must be maintained.

To make appropriate decisions for the firm, it is imperative that accurate data be available and understood. Accounting systems accumulate data and aggregate data into information for financial reporting and managerial decision-making. It is important, therefore, that the accounting data accurately reflect the impact of decisions and that the definitions of accounting parameters are understood by decision-makers. Activity-based cost accounting systems attempt to provide an accurate portrayal of the factors generating all the costs of the firm. Nonetheless, we must be careful when using accounting figures to know precisely what is included and how the figures relate to our engineering choices. Improper classification of costs can distort management's understanding of the production process and lead to poor decision-making.

1.7   REFERENCES

ASKIN, R.G., & STANDRIDGE, C.R. *Modeling and analysis of manufacturing systems.* New York: John Wiley & Sons, Inc., 1993.

ALTING, L., & ZHANG, H. Computer-aided process planning—a state of the art survey. *International Journal of Production Research,* 1989.

COOPER, R., & KAPLAN, R.S. Measure costs right: make the right decisions. *Harvard Business Review,* 1988, 96–103.

DAVENPORT, T.H., & SHORT, J.E. The new industrial engineering: information technology and business process redesign. *Sloan Management Review,* 1990, 11–27.

DIETRICH, B.L. A taxonomy of discrete manufacturing systems. *Operations Research,* 1991, Vol. 39, No. 6, 886–902.

HAUSER, J.R., & CLAUSING, D. The house of quality. *Harvard Business Review,* 1988, 63–72.

KAPLAN, R.S. Measuring manufacturing performance: a new challenge for managerial accounting research. *Accounting Review,* 58(4), 1983, 686–704.

LEWIS, R.J. *Activity-based costing for marketing and manufacturing.* Westport, CT: Quorum Books, 1993.

WOMACK, J.P., JONES, D.T. & ROOS, D. *The machine that changed the world.* New York: MacMillan Publishing Co., 1990.

1.8 PROBLEMS

1.1. List the key resource types required by the firm to produce products.

1.2. Describe the key activities for each of the primary functional units of an industrial organization.

1.3. What are the primary measures of a firm's competitiveness? How do you think the relative importance of these will vary from one industry to another? Give examples.

1.4. Consider a color printer designed for the home PC market. List the key customer attributes. Define a set of engineering characteristics that could be used to specify product requirements.

1.5. What is the purpose of a process plan? List the types of information normally included in the plan.

1.6. A furniture shop produces chairs (among other products) with metal arms. The arms are formed from metal blanks in a three-machine cell. This is a make-to-order shop, i.e., the production floor makes a quantity of chairs of a specific type in response to a customer order. The blanks are transported in tubs that hold approximately 50 pieces. The typical order is for about 500 chairs. If tubs are moved in their entirety between machines in the cell, what constitutes the process batch, transfer batch, and unit load? Suppose full tubs are set at the beginning of the cell and empty tubs at the end. Units are paced one at a time between machines and completed units are placed in the "empty" tub. When the empty tub is full, it is moved to the chair assembly line. Now what constitutes the process batch, transfer batch, and unit load?

1.7. Define the major types of production operations.

1.8. List the important characteristics of a business process.

1.9. Visit a local fast food establishment. Observe the process of taking and filling a customer order. Construct a flow diagram of the process for a customer ordering a burger, fries, and shake and paying with a debit card.

1.10. Educational institutions produce educated students ready for employment and life-long learning. Consider the process of offering a course for academic credit as part of a degree program. Describe the resources involved, tie this process to the objectives of the firm, and define the product and customer(s) of this process.

1.11. Suppose you were considering purchasing a personal computer for use in your educational program. List your key features: price, performance, and delivery requirements.

1.12. Describe the differences between traditional cost accounting systems and activity-based costing. How is activity-based costing more appropriate for modern manufacturing systems?

1.13. Two products are produced in a factory. One thousand units per year of each product are manufactured. Both products have direct material cost of $10 per unit. Both products visit department A and department B in the plant. Department A is highly automated while B is largely manual operations. A unit of product one requires 0.2 labor hours in department A and 5 hours in department B. Product two takes 1 hour of labor in each department. Annual overhead costs are $500,000 and $100,000 respectively for departments A and B. Direct labor costs $8 per hour. Compute the cost per product for both traditional overhead allocation based on direct labor hours, and activity-based costing.

1.14. What problems can you see happening if the engineering department that finalizes product designs and changes does not communicate well with production planning and purchasing personnel?

Chapter 2

Introduction to the Production System and the Role of Inventory

The central theme of this text relates to designing the system to produce a given set of products and understanding how that system will perform under a given set of product demands and resource availability. By system design, we mean how physical resources (human and inanimate) are managed and information is used to produce products. We assume the products have already been designed and their process plans constructed. We hope that manufacturability has been factored into the product design and process planning decisions. In this chapter, we provide a basic description of production systems—what they are and how they operate. Because inventory plays a central role in the operation of a production system, we include an overview of inventory basics and how they relate to the production system.

2.1 THE PRODUCTION SYSTEM

The set of resources and procedures involved in converting raw material into products and delivering them to customers defines the production system. This is a rather broad definition, but from our perspective, the production and delivery of products are central to the firm. Functions have value only to the extent they enhance the firm's ability to do this profitably. In this section, we examine the objectives, information structure, and physical structure of production systems. Subsequent sections will cover the basic principles of production systems and background information on the production system models that will form the core of the remainder of this text.

2.1.1 System Objectives

A capitalistic entrepreneur might claim that the primary objective of the firm is to maximize profit. A behaviorist might view the firm as a living entity and view long-term survival as the primary objective. The humanist assigns the objective of providing jobs and goods for society. In reality, the three objectives support each other. A healthy firm maintains its long-term viability by providing goods and services upon which society, in the form of customers, places greater value than their cost. The excess revenue, or profit, motivates the owners to continue their investment in the company, thereby providing jobs. The production system has the responsibility for the planning and execution of the activities that use workers, energy, information, and equipment to convert raw materials into finished products. To be successful, the production system must deliver products with the desired functions, aesthetics, and quality to the customers at the right time and cost. This constitutes the objective of the production system.

2.1.2   System Components and Hierarchy

The command and control functions of industrial firms often follow a hierarchical structure. A typical hierarchy for an industrial firm is shown in Figure 2.1. The time frame and dollar value of decisions decrease as we move down the hierarchy from the corporate level to the equipment level. Each node in the figure resides at a horizontal level of the hierarchy. Above the node resides its "parent." Below it are its "children." In general, decisions are made at each node and passed down one level to children in the form of constraints and instructions. Current status and performance data are passed up the hierarchy from the "child" to the "parent" to facilitate the decision-making and guidance process.

Production-planning decisions are also typically made in a hierarchical manner. Figure 2.2 shows the stages in production system decision-making. The first column in the figure depicts the physical flow of material from raw material through delivered product. The third column lists the associated support functions and design activities that must precede production. The middle column shows the sequence of operational decisions for production-planning, scheduling, and control. These are listed in the hierarchical manner in which these decisions are frequently made. This text is primarily concerned with this middle column. Demand forecasts (see Chapter 3) define the opportunity for making profits and providing gainful employment by satisfying customers. At the strategic level, long-term decisions, such as capital investment in new facilities and development of the distribution network of regional warehouses, are decided (see Chapter 4). These plans range from one to five years, and decisions may take at least a year to implement. At the tactical level, aggregate production plans (see Chapter 5) are set. A typical aggregate plan states the levels of major product families to be produced monthly over the next year. It is here that we decide whether to change workforce levels, to schedule overtime, or to accumulate inventories. At the operational level, detailed scheduling of part batches is planned and their execution controlled (see Chapter 11). A significant portion of production-system planning covers the area between aggregate planning of product families and scheduling part batches. This includes the breakdown of family production quotas into a plan for timed order releases of part batches and the coordination of

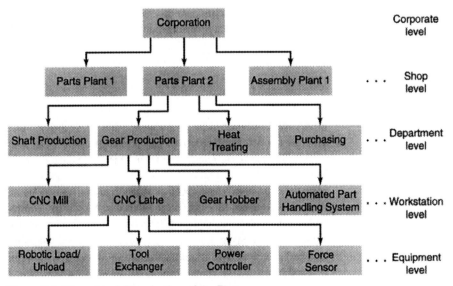

Figure 2.1 Hierarchical Organization of the Firm

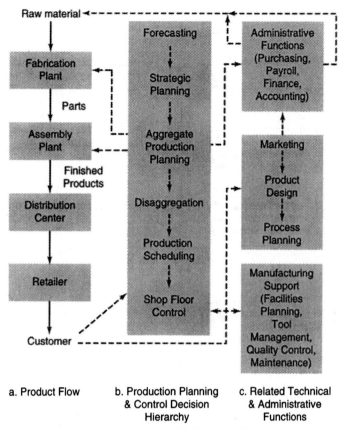

a. Product Flow   b. Production Planning   c. Related Technical
                     & Control Decision   & Administrative
                     Hierarchy   Functions

Figure 2.2 Production Activity and Information Flows

these orders through parts fabrication and assembly. Several approaches are described in Chapters 6 through 9. Figure 2.3 summarizes the inputs and outputs at each stage. Note that the outputs of the parent become the inputs to the child. Each of these main decision blocks is described in greater detail in the later chapters.

There are several reasons for hierarchical organization of production decisions. First, the sheer complexity of the entire system makes it impossible to jointly determine all plans simultaneously and in real time. There are too many decision variables, constraints, and interrelationships to build the monolithic model of the production system. Second, it is difficult to change plans already in motion. We may have ordered materials last week for use this week. Parts may have been made earlier this week for assembly at the end of the week. Subsequent changes in customer demand or new production cost data may mean that other plans would have been slightly more profitable. However, once these plans are set in motion, the cost or delay involved in making changes may be prohibitive. Thus, there is a natural lead time for each decision, and this varies by decision type. Building a new facility takes several years, buying a new machine takes several months, and setting up a machine to make a batch of parts takes several minutes. If a customer requests a delay on one of many orders currently in progress, we may change the batch of parts to be set up on the machine, but it is doubtful the decision to build a new plant will be affected. Decisions should only be made far enough in advance to allow the action selected to be executed. Delaying the decision until action must begin allows use of the most current data when making decisions. This increases the chance that we make the correct

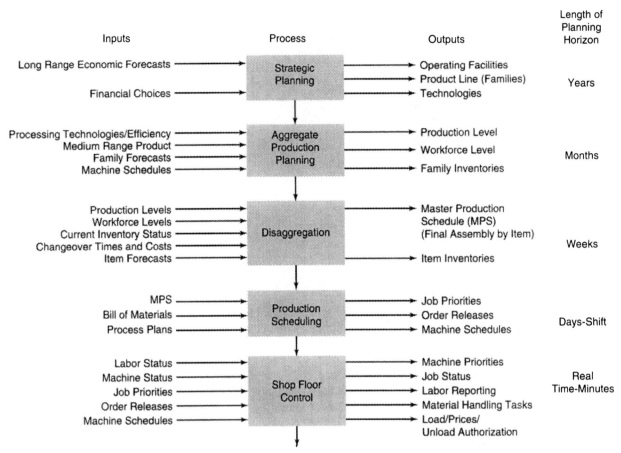

Figure 2.3 Production System Decision Hierarchy

decision. The machine setup decision can wait until it is necessary to obtain the tooling for the next batch of parts. If we made the decision too far in advance, priorities are likely to change before the action is taken. Third, most firms have a hierarchical structure for management—department foreman, shop supervisor, plant manager, corporate vice president. Each has different needs with respect to level of detail and frequency of information updates. It is natural to match the design of the production system to the organizational decision-making process. We need the right information to be conveyed to ensure compatibility of decisions, but the best decisions often result from local empowerment and control. The lathe operator needs to know the set of parts available for processing and their due dates. To have to pass information back to a corporate model detailing that one of the many lathes at the Pittsburgh plant is showing wear and will undergo two hours of preventive maintenance this evening, and then wait for a decision on what to do, would be inefficient. The operator should be able to judge if any pending orders can be safely produced this shift. To summarize, hierarchical structures allow coordination of decision actions among system entities while maintaining a minimal number of interacting connections. Although constrained by parent nodes to act in a manner that is complementary to other nodes, each entity is free to use its own local knowledge to execute in the most efficient manner and report its performance up the hierarchy to improve future system decisions.

2.1.3   Types of Production Systems

Production systems can be classified according to a number of factors. Some of the more important factors are listed in Table 2.1. We discuss each of these in the remainder of this section, emphasizing those distinctions that are important for building decision models. All these dimensions will play important roles in the models developed throughout this text.

2.1.3.1   Physical Organization Schemes

There are four basic types of manufacturing system organization: process, product, cellular, and fixed position. These are shown schematically in Figure 2.4, and a comparison is presented in Table 2.2. Product layouts are used for repetitive manufacturing where demand is large enough to economically justify the dedication of equipment to a single product. Production processes are laid out in a sequential manner so that product visits each area, one right after the other. Equipment is designed to have similar production rates at each stage. In a high-volume environment, product layouts are very efficient and typically have small throughput times. Process layouts are used when many different products are produced and there is a need for skilled expertise at production processes. This approach has the greatest flexibility to produce a variety of items but suffers from low equipment use and long throughput times. Whereas materials flow serially from one work station to the next in product layouts, arbitrary flows and handling loads must be accommodated in process layouts. Cellular organizations attempt to partition large facilities into smaller "cells" designed to produce a family of related parts or products. The relation may be because of similarity of machines and tooling (including fixtures) or usage in the same finished products. Each cell is a simpler minifactory, with short material moves and easier coordination. Product departments and cells are generally scheduled as a single entity. Each workstation is separately scheduled in a process configuration. Fixed-position layouts are used for large projects such as constructing a building. We will not concern ourselves with fixed-position layouts in this text. In practice, evolving product demands and process plans often result in incremental layout changes. When a new machine is purchased to help out the plant bottleneck, it may have to go where there is space instead of where it would best accommodate material flow during the next six months. The alternative may be frequent major relayouts that can be equally time-consuming and disruptive.

Table 2.1   Dimensions for Classifying Production Schemes

Dimensions	Levels
Physical Organization	Product vs. Process
Key Resources	Labor Intensive vs. Capital Intensive
Product Flow Control	Discrete Batches vs. Continuous Flow
Order Initiation	Make-to-Order vs. Make-to-Stock
Production Authorization	Push vs. Pull
Product Variety	Single vs. Multiple Products
Product Volume	Custom Jobs vs. Repetitive Mass Production
Fabrication/Assembly Structure	Single Production Stage vs. Multiple Stages
Time Horizon	Single Planning Period (Static) vs. Multiple Periods (Dynamic)

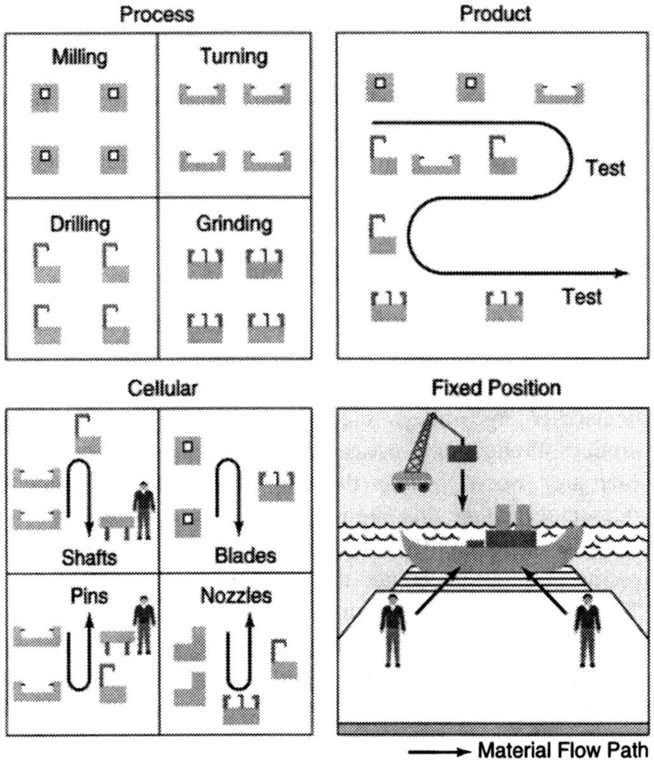

Figure 2.4 Layout Types

2.1.3.2 Key Resources

Labor and equipment are key resources in most industrial operations. Often one or the other of these is dominant. The high cost of furnaces in a steel mill and wafer fab equipment in a semiconductor manufacturing plant make them capital intensive. Such systems run multiple shifts, and adequate labor is employed to keep the process running. Increasing capacity takes a long time and is very expensive. Often, the majority of cost is fixed and thus it is expensive to have the equipment idle. Balancing production over time is essential. The expensive capital requirements limit the production rate. In general, we will label any resource that limits output as a **bottleneck.** Capacity of all other resources is maintained at a higher level to ensure optimal use of the bottleneck.

Assembly and light manufacturing tend to be labor-intensive. Equipment tends to be small and inexpensive. Capacity can be modified in a relatively short time by hiring and training workers (or through layoffs). Overtime and extra shifts can be added when extra capacity is needed. Thus, the options for reacting to fluctuations in demand are considerably different in the two environments.

Table 2.2 Basic Physical Layout Types

Characteristic	Product	Process	Cellular	Fixed position
Skill Level	Low	High	Mixed	High
Unit Production Cost	Low	High	Low	High
Equipment Utilization	High	Low	Moderate-High	Low-Moderate

2.1.3.3   Product Flow Control

Discrete part manufacturers (automobiles, furniture, consumer electronics, textiles, etc.) tend to make parts in batches. A batch may contain anywhere from a couple to several thousand identical parts. These parts move through the production system together. We may split the batch into subbatches for material handling purposes and to reduce processing time, but the batch essentially moves through the system together to be followed by another batch of a different part type. The information system will track the batch as a single shop order. This book will deal primarily with such batch operations.

Process industries tend to use continuous processing. Chemicals, foods, and pharmaceuticals are examples of products that typically are made using a continuous flow process. Product looks like a continuous mass as it flows through pipes or along conveyors. The individual parts are not distinguishable. There is some overlap between discrete part batch and process industries; for instance, aspirin may be made in distinct batches with different coatings or milligram potencies. Nonetheless, it is important to be cognizant of the fundamental differences. Process industries produce product either through continuous release of raw material to the production process or in very large batch quantities. Capacity is measured in generic units such as pounds per hour or batches per hour, usually with little opportunity to change the plant capacity in the short run. The entire batch is often being processed simultaneously at each step, and capacity is measured accordingly on the basis of the maximum batch size possible. If smaller batch sizes are used, capacity is reduced. This occurs because of processes that are paced by time and capacitated by volume. The product bakes in the oven for a fixed time, for example, whether the oven is full or half-empty. Process industries often have strict constraints on in-process inventories. Wherease discrete part industries can stack pallets of castings on the floor, process industries are characterized by limited capacity holding tanks for in-process material. This material is often fluid or otherwise in need of careful protection and thus requires specialized storage vessels. The number of different products held between production stages is limited by the number of storage vessels. The amount of each product stored is limited by the vessel capacity. There may also be interoperation time windows. A minimum time limit exists between operations, such as curing time, and if too much time elapses, the product may be contaminated and have to be scrapped. Such constraints rarely exist with discrete parts. Finally, products are often coupled in process industries. A single raw material will be split into several products at the next stage, representing different basic chemical mixes or grades of the same mix. Thus, whereas discrete parts usually aggregate with several components coming together to form a single finished product, in the process industries, we often see a single raw material split into multiple finished products. These differences create a need for different modeling techniques when analyzing production system plans.

2.1.3.4   Order Initiation

Order initiation actually refers to two dimensions. At the strategic level, we must decide whether we **make-to-stock** or **make-to-order.** The make-to-stock system sets target levels for the number of units of each product to keep on hand at all times. These units allow quick delivery to customers upon receipt of an order. This strategy makes sense when delivery response time is a key competitive factor and when a limited number of products are manufactured repeatedly, i.e., we have an idea what customers will want. Stocking allows us to schedule production in advance and to coordinate the delivery of raw materials with the production schedule. However, we require investment capital and storage space, and we run the risk of damage and obsolescence. Items that resemble "commodities" in the sense that many suppliers are available to the customer are generally stocked as well because immediate delivery is often expected. It is generally impractical to stock low-volume items with unpredictable demand and impossible to stock finished goods in a custom product environment. At the input side, critical spare parts that are difficult to acquire or for which there is a high cost of not having the item available when needed must be stocked. Essentially, a decision not to stock an item is a statement that the cost of making the item now and carrying it in inventory until it is demanded is more costly than incurring a shortage cost when demand occurs and delivering the item after some delay. The reader is asked to formalize this tradeoff in the end of chapter problems.

In make-to-order, we do not maintain inventories of finished products, but only produce items after they have been ordered. Make-to-order is appropriate when the production system can respond quickly to customer requests, products have a high degree of customization, or shelf life of products is short because of changing customer tastes or product spoilage.

Whether to stock an item is often a strategic decision. What services do we want to offer our customers, and what do our competitors offer? What financial and space constraints do we have? The stocking decision is part of the overall supply chain management system. This includes interactions with vendors and suppliers and interactions between sequential production stages within the firm. In many industries, many variations of final products exist, all based on a few assembly platforms. An automobile, for instance, may have many final configurations, but the product line has only a few choices for engine, instrument packages, audio systems, etc. Even your house thermostat may be basically the same as all others, but simply with a different cover style and retailer label. In such cases, it is good practice to store the assemblies or various high-level modules. These are then sent to final assembly based on customer orders. Total demand is often fairly predictable even if specific customers and model versions are not. By storing the high-level subassemblies, final assembly can be performed quickly providing a rapid delivery cycle time of seemingly customized products. We need not incur the high cost of storing a wide variety of finished goods that may never be ordered. In some cases, it is even possible for the warehouse to add the final label and minor customer details, allowing storage at regional warehouses. As an example, the same power units, disk drives, motherboards, software packages, modems, and monitors can be quickly assembled and packaged to form a variety of PC configurations. This approach is often called **assemble-to-order.**

2.1.3.5 Production Authorization

Production authorization refers to the tactical decision of when a worker or machine is allowed to start a task that is part of an open shop order. We will contrast two approaches: push and pull. Envision each work center as containing an input buffer of raw material, a processor, and an output buffer of completed parts as shown in Figure 2.5. In a push system, a high-level planning model will keep track of all orders and their status and send authorizations to workers/machines when they are to begin a job. We use the word "Push" because these orders, along with the required raw materials, are pushed into the work area's input buffer by this upper-level controller. The appearance of this material constitutes an authorization to work. In a "pull" system, workers are informed that it is time to perform a task when someone comes with an authorization to remove the finished product from the output buffer. For instance, suppose a machine presses side panels for several household appliances. The station is supposed to maintain a fixed number of each panel type in inventory. When parts are pulled out of the station's output buffer by a sub-

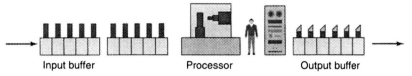

Input buffer   Processor   Output buffer

Figure 2.5 Schematic of a Workcenter

sequent assembly station, the station is automatically authorized to replace these parts. Detailed descriptions of these two approaches form the core of Chapters 7 and 8.

2.1.3.6   Product Volume and Variety

The combination of product volume and variety strongly impacts the choice of layout type and, therefore, the planning unit and decision hierarchy. Figure 2.6 shows the physical organization normally used for volume-variety combinations. If products use resources in a similar proportion, we can frequently aggregate these into a single product for planning purposes. If multiple products with different resource profiles are produced, then the products and all potential bottleneck processes must be explicitly considered in decision models. As product volume increases and variety decreases, dedicated process layouts and continuous flow manufacturing become more economical. High-volume products can justify specialized tooling and equipment.

2.1.3.7   Product Structure—Fabrication and Assembly

Most items are assembled from manufactured (fabricated) components. In complex products such as an automobile with 10,000 parts, many of the components may first be assembled into subassemblies (an engine, a radio, etc.) and then the subassemblies combined into products. Even with a simpler item such as a telephone, you may have component parts first being put into subassemblies, such as a soldered circuit card or handset, before final assembly. Each component part may require multiple fabrication processes, such as milling, drilling, reaming, and grinding. Thus, the production plan must account for the time-phased assignment of operations to workstations for each component and

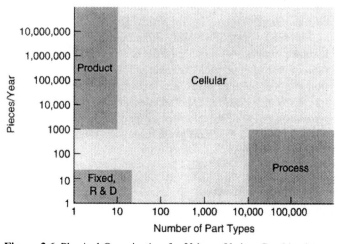

Figure 2.6 Physical Organization for Volume-Variety Combinations

subassembly to be used in final assembly. We must plan to have each component and subassembly arrive at the final assembly point at the right time and in the right sequence. Disruptions caused by machine breakdowns or defective items can directly impact priorities for related parts and thus the product structure must be included in the production model if we are to aim for the most efficient plans in a dynamic environment. Products with many parts lead to large planning models. When resource utilization can be high, lead times will vary and must be included in the model if we intend to coordinate the production of related parts. Likewise, if quality is not consistent, process yields must also be factored into the planning models.

2.1.3.8 Time Horizon (Static vs. Dynamic Environments)

Planning models are either static or dynamic. By static, we mean constant through time. Static models assume the same plan will be acceptable in each period, at least for the foreseeable future. Dynamic models explicitly consider changes in demand and resource availability to determine what should be done through time over a **planning horizon.** As such, dynamic models require additional data and greater effort to build and solve. Curiously, the choice of a planning horizon can have a significant impact on the outcome of a planning model and should be selected with care. Systems tend to have natural planning frequencies defined by the frequency at which plans seem to want to repeat. This natural frequency will become apparent as we progress through the text. The planning horizon should be an integer multiple of this natural cycle time.

2.2 THE ROLE OF INVENTORY

Inventory consists of the physical items moving through the production system. Inventory originates with shipment of raw materials and parts from the supplier and ends with delivery of the finished product to the customer. The cost of storing inventory accounts for a substantial proportion of manufacturing cost, often 20% or more. In this section, we describe what inventory is and why it is important. This leads to two key principles that we elaborate on later in the chapter. First, there is an optimal level of inventory, a level that allows production operations to continue in a smooth manner while minimizing the sum of storage, ordering, and shortage costs. Second, the primary determinant of the optimal inventory level is the cost to place an order for parts or to set up a production batch. Reducing this cost through process improvements can lead to substantial reductions in total cost and improvements in competitiveness.

Inventory costs are closely monitored by most firms. A common control measure is **inventory turns,** the number of times a year inventory is replenished as measured by the ratio of cost of goods sold to value of inventory in stock at any point in time. A $100 million per year company with ten inventory turns has an average of $10 million in inventory. In recent years, many companies have worked to improve their turns, sometimes with returns on the scale of an order of magnitude. Whereas values of 1 to 5 may have been common in years past, companies today are shooting for 50 or more turns a year to reach world-class status.

2.2.1 Inventory Definitions and Decisions

In forming plans for ordering inventory, we have two main decision variables. Determining values for these variables is the primary focus of Chapter 6. The first variable is the batch or order size, Q. The **batch size** is the number of units released to the shop floor to be produced. If we are ordering parts from an external vendor, we refer to Q as the **order size.** The second variable is the **reorder point,** r. The reorder point specifies the timing for placing a new order. It is stated in terms of the **inventory position** which is defined as:

$$Inventory\ Position = Inventory\ On\ Hand + On\ Order - Backorders$$

If inventory is being tracked in a continuous review manner, then when the inventory position drops to the reorder point, we order Q units. Units **on order** are those that have been ordered previously but not yet arrived. **Backorders** are items that have been promised to customers but not yet shipped because we are out of inventory. As soon as new units arrive, they are shipped out to customers to cancel out these backorders.

The time between placing an order and the availability of those items for use is called the **lead time,** τ. The lead time may be a fixed time interval or a random variable. Lead time includes the time to check inventory status, prepare and place the order, await receipt of the order, perform any quality checks, transport the items to their storage location, and update the inventory records. We use the term **cycle time** to indicate the time between producing or ordering successive batches of an item. The ratio of batch size (measured in *units*) to the demand rate (*units/time*) determines cycle time. The final concept to note before we proceed is the **safety stock.** Safety stock is the number of units of inventory we plan to have on hand when an order arrives. The safety stock is used to guard against unpredictable high demand during the lead time. If demand and lead time are fixed constants, we do not require safety stock. We simply plan to have the new units arrive just in time as the inventory level reaches zero. In practice, randomness means that we must have some planned extra units on hand to cover variability. Safety stock, *SS,* is computed as the excess of reorder point over lead time demand, i.e., $SS = r - \tau D$ where D is the mean demand rate.

2.2.2   Types of Inventory

It will be useful to distinguish between the different types of inventory.

1. Raw materials are kept to ensure availability for production. Inexpensive raw materials that are essential to the production process are often kept in large quantities on site. This practice guards against shortages of essential materials that might occur owing to plant shutdowns, labor disputes, or market demand. Maintaining a stock of these items ensures a supply and reduces the length of the required planning horizon. If we adopted a system whereby these materials were ordered precisely to meet the timing and quantity determined by our production plan, then our forecast and production planning horizons would have to be increased by the length of time required to obtain replenishments of these materials. We would also have very little flexibility to rapidly adjust plans if demand changes. The time quoted to potential customers for delivery would increase by the time required to obtain these materials. Spare parts that might be needed to repair essential equipment are generally stored as well.

2. Finished goods are completed products that are awaiting shipment to customers. This includes goods in warehouses, on shelves at retail outlets (if we still own the product), and even goods that have just undergone manufacture and are sitting on loading docks to be shipped.

3. Work-in-Process or WIP inventory consists of the batches of semifinished products currently in production. This includes batches of material and purchased parts from the time they are released to the shop floor until they enter finished goods status. The WIP inventory may be in queue awaiting the availability of their next workstation,[1] being loaded or processed on a machine, or being moved between workstations. In addition, WIP may exist because of parts being held in temporary storage pending approval from the customer or instructions from engineering due to a design change, or a decision on disposition resulting from a quality problem, or arrival of other parts/materials to be mated with these parts in the next production operation. In some plants, parts are first fabricated and then kept in a controlled storage facility until needed for final assembly. If the final assembly operation belongs to the system currently being considered, then we consider these parts in controlled storage as part of WIP. If they are waiting to be shipped to another facility that is not directly included in the decision model being constructed, then we refer to these as finished parts.

4. Pipeline inventories consist of goods in transit between facilities. This includes raw materials being delivered to the plant and finished goods being shipped to the warehouse or customer. In the current global marketplace, pipeline inventory can be substantial. It includes, for instance, all the foreign parts being shipped to an assembly plant in the United States and the final products being shipped overseas for sale. The key approaches to reducing pipeline inventories include finding local suppliers and producing (or finishing production) closer to the final customer. The schematic diagram in Figure 2.7 illustrates a typical fabrication/assembly system and the stages of pipeline inventory. Note that every unit of product sold goes through this process.

Companies keep track of inventory on hand, on order, and allocated to planned orders. Figure 2.8 shows a typical record. Each manufactured item is given a row in the Table. Column one gives units on hand. Column two reports units on order and scheduled to be

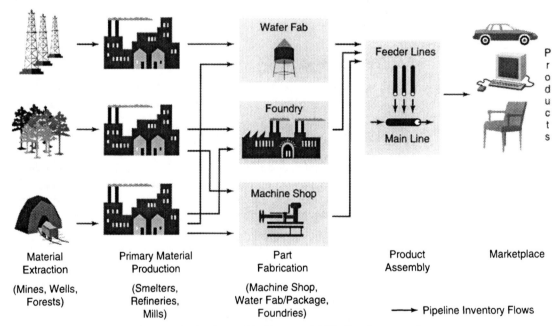

Material Extraction	Primary Material Production	Part Fabrication	Product Assembly	Marketplace
(Mines, Wells, Forests)	(Smelters, Refineries, Mills)	(Machine Shop, Water Fab/Package, Foundries)		

⟶ Pipeline Inventory Flows

Figure 2.7 A Fabrication/Assembly System with Pipeline Inventory Stages

---

[1]We use workstation in a generic sense. A workstation could be an automated or manually operated machine, workbench, drying facility, test station, or other operation location defined in the process plan.

Inventory Status Report 11/26/96				
Item	On-Hand	On-Order	Assigned	Units Unallocated
PZ-9643B	1,043		625	418
BR-1406Q		650	700	−50
PZ-9637B	185			185
PZ-1045C	496		150	346
BS-1110C	23	400	190	233

Figure 2.8 Sample Inventory Status Report

received. Column three records the number of units that have been allocated to forth-coming production orders. These are orders for products that will use this item in its production. The final column lists the number of unallocated units that are free to be assigned to new orders.

2.2.3 Justification of Inventory

Inventory will always exist. The transit pipeline places a lower limit on inventory. Moreover, the world is random. Limits on the effectiveness of preventive maintenance create a need to carry spare parts. Competitive pressures to supply common products quicker than they can be produced imply that we must keep finished goods inventory near our customers.

Other justifications exist for inventory as well. Price breaks are common when large quantities of materials and parts are purchased. Although we do not advocate the general practice of keeping large inventories, it is clearly economical in some instances to order large quantities to save on costs—both the costs of placing an order and the cost per item. When we order materials or produce products periodically in large batches, the supply starts large and is gradually drawn down. This cyclical activity is referred to as **cycle inventory.**

At the aggregate planning level, we may use the strategy of storing inventory in periods of low demand and consuming these stocks in periods of high demand to allow us to smooth our production rate. This may be more economical than hiring workers, using overtime, or maintaining excess equipment capacity to meet seasonal demand spikes. Figure 2.9 illustrates the relationship between production rate and inventory level when we have seasonal demand. Random demand may also justify carrying inventory. To guard against the unpredictable period-to-period variation in demand, we may carry a safety stock of finished goods. The safety stock is the minimum planned inventory level. The actual level will go below the safety stock setting when, by randomness, demand is large in a period. Having this safety stock allows us to meet demand without making the customer wait for delivery.

2.2.4 Inventory Costs and Tradeoffs

Holding inventory is costly. Some of the components of holding cost are direct costs and easy to tabulate. Others are indirect costs and often difficult to quantify. In constructing economic models for choosing the optimal levels of inventory, we will generally trade off

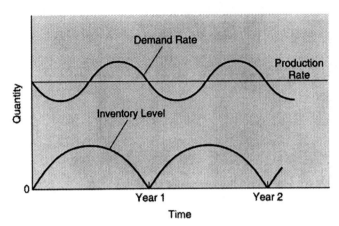

$$\text{Inventory Level at Time } t = \int_0^t (\text{Production Rate} - \text{Demand Rate})\,dt$$

Figure 2.9 Inventory Accumulation and Depletion for a Constant Production Rate with Seasonal Demand

the costs caused by 1) ordering or set up of machines, 2) investing and storing the goods, and 3) shortages (not having inventory available when needed).

2.2.4.1 Ordering and Setup Costs

When parts are purchased from suppliers, a fixed ordering cost can be associated with each replenishment. This includes the cost of first identifying the need to order. We must then execute the order. This may require extracting from the vendor database the appropriate list of suppliers and contacting each to determine their current cost and availability of parts. We then prepare the paperwork and place the order. The delivery cost may include a fixed component, such as the cost of sending a truck for pickup and delivery. When the order is received, a sample of the parts may go through a receiving inspection to determine quality. The order must then be transported to its place of use or stored until needed. The paperwork alone to track the movement of the batch of goods through each step can be substantial.

For parts produced in-house, we must still determine the need for replenishment and prepare the authorization to produce the parts. We must check the status of raw material stores and possibly place an order. Route sheets with instructions for each stage of the production process must be created. Routing data may be stored in a database, but it must be checked for compatibility with the current shop status and engineering changes, and the routing instructions must be mated with the raw material and delivered to the production workers. At each workstation along the route, the machine may need to be set up. The operator will need to examine the instructions for setting up the machine, retrieve the necessary tooling, align it on the machine, and normally make and inspect at least one part to ensure compliance with standards before the batch is run. Efforts to reduce this setup time have dominated industry in recent years, and we will see why setup is so important as we proceed through this chapter and throughout the text. If the machine is a "bottleneck" in the sense that lost production time reduces our ability to make products that could be sold at a profit, then the lost profit during machine setup time should be included as well. Whether the parts are ordered from suppliers or set up and produced internally, we denote the fixed cost per replenishment as A.

2.2.4.2   Inventory Carrying Costs

Carrying inventory incurs a variety of costs. Inventory takes up space that must be rented, heated, and cooled. Depending on the storage protocol, inventory may have to be moved occasionally because it is blocking access to other goods. We must keep track of where the inventory is located; thus, we have to construct and maintain an information system to track material. (The alternative of having to look for it whenever it is needed and never knowing what is available is probably more expensive.) Taxes are often based on the value of inventory as well as insurance costs. As we keep goods in inventory, some will be lost or damaged, others will perish (food items in particular) or become obsolete as new products are brought to market and customer tastes change. A major factor is the cost of capital invested in inventory. That investment could be earning an economic return if invested elsewhere. Thus, the company's cost of funds should be included in the determination of holding costs.

Precise determination of holding costs is difficult. We will follow the standard accounting approach of approximating these inventory carrying costs by charging an annual percentage of the average value of goods in inventory. We will call this the holding cost rate and denote it by h. For our approach, $h = iC$, with C being the cost of an item measured in $/item, and i being the rate *per year*. Typical values for i are 0.2 to 0.5 per year. Full holding costs are found by multiplying h by the average inventory level measured in *items*. Thus, if an item costs $100, the holding cost rate is 0.5 per year and we average 10 units on-hand, the annual carrying cost is: $Cost/yr = 100 \cdot (0.5) \cdot (10) = \$500/yr$.

A value for the holding cost rate will likely be available from the accounting system because it is used routinely in financial planning, reporting, and managerial decision-making. The appropriate item cost to use may be situation-dependent. We saw in Chapter 1 how different accounting systems may lead to different cost figures. The appropriate cost will include all purchasing, manufacturing, and transportation expenses required to get the item to its current condition and location.

2.2.4.3   Shortage Costs

When a customer demands an out of stock item, the customer will decide either to wait for delivery or to cancel the order. Within the plant, if a material is unavailable to start a production run, the work center may lack work, schedules for assembly and production of complementary parts may have to be modified, and completion of products may be delayed, resulting in late deliveries or lost sales. In general, we categorize the results of shortages and the accompanying costs into the two categories of **backorders** and **lost sales.** For backorders, we incur the cost of trying to expedite production of the item. This may result in using inefficient setup sequences on machines. Production schedules are modified, and material handlers may have to find and retrieve parts unexpectedly. Overtime may be required to produce the items quickly. We may also need to notify and offer incentives to appease customers. Customers may also decide to look elsewhere next time. Backorder costs are dependent on the situation, but a common modeling approach is to divide the costs into a fixed component, π, for each unit or order backordered and a variable cost, $\hat{\pi}$, per time unit late. Lost sales result in a loss of contribution to profit and overhead equal to the difference per unit between selling price and variable production cost. (Production cost should include warranty and delivery charges.) We will denote this cost as π. In both cases, future lost sales because of disgruntled customers should be included in the cost, along with customer interaction costs such as telephone calls, notification letters, etc. Unfortunately, many of the costs of shortages are intangible or difficult to estimate. As a result, companies will often define a desired service rate instead. For instance, the manager may declare the goal of filling 98% of customer orders within 2 days using stock on hand.

2.2.4.4 The Tradeoffs: Inventory vs. Setup and Shortages

For a fixed system configuration consisting of a set of production facilities, suppliers, customers, and information systems, there is a tradeoff between carrying inventory in the form of large batch sizes and increasing the number of setups or orders. Likewise, a tradeoff exists between carrying inventory in the form of safety stock and incurring shortages. The basic forms of these tradeoffs are illustrated in Figures 2.10a and 2.10b, respectively. Figure 2.10a shows the tradeoff between cycle inventory and setup. Let D be the annual demand rate and Q be the selected production batch quantity produced each time we set up or order. Now, as we increase the production batch size, average inventory goes up. Thus, inventory cost increases linearly with Q:

$$\text{Inventory Cost/Time} = \frac{h \cdot Q}{2}.$$

But, as shown in the Figure, we initiate fewer orders and, thus, the associated fixed costs are decline exponentially with Q:

$$\text{Setup Cost/Time} = \frac{A \cdot D}{Q}.$$

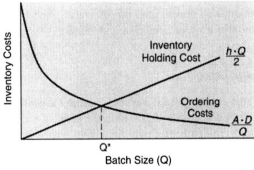

a. Inventory Costs vs. Batch Size

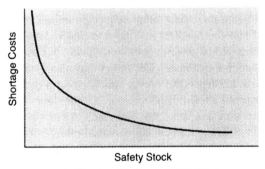

b. Shortage Costs vs. Safety Stock

Figure 2.10 Inventory Tradeoffs

Combining costs, the minimum total cost is achieved at the:

$$\text{Economic Order Quantity} \equiv Q^* = \sqrt{\frac{2 \cdot A \cdot D}{h}}. \qquad (2.1)$$

This result describes the most basic tradeoff in inventory planning. It will form the starting point for our discussion of detailed planning in Chapter 6.

Next, consider Figure 2.10b. As we hold more safety stock, we are better protected against randomness. Shortage costs go down, but we have to pay a larger average on-hand inventory cost. A quantitative analysis of these tradeoffs and decision models for selecting the optimal location on these curves is presented in Chapter 6. We will also see how information can be used to improve the tradeoff options.

2.3   THE ROLE OF INFORMATION

Information forms the basis for operating modern production systems. Gone are the days when a box of 3×5 index cards serves as the standard information system. Today, we have the potential to allow the staging area for a machine to be the truck in transit delivering product from the supplier directly to the machine where it will be processed. The Global Positioning System can tell us exactly where the material is at this moment. The Intelligent Vehicle Highway System can look at congestion and tell us when the truck should arrive. Our Vendor Certification programs assure us that the material is top quality, and our real-time scheduling system selects the right sequence of material to be processed on each machine based on this information. Sensors and on-line inspection systems can collect real-time data on process status and be integrated into the Statistical Process Control system. The system in turn communicates with support functions to schedule tool changes and regular maintenance, fixing problems before they result in the production of defective items.

Information collected by and transmitted throughout the system can serve as a substitute for safety capacity and inventory. As demand projections improve through the sharing of retail demand information back through the supply chain of producers and raw material suppliers, we move the entire cost vs. inventory tradeoff curve. The value of information shared in this manner is seen in Figure 2.11. Information allows the firm to use its resources more efficiently, thereby moving its tradeoff curve from A to B in the Figure.

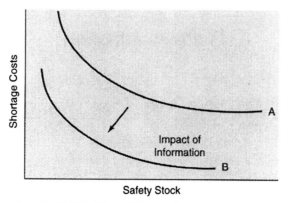

Figure 2.11 The Value of Information

The manner in which information is distributed to work centers to authorize production forms a key descriptor of the production system, as discussed above. Figure 2.12 illustrates several types of systems. In the MRP approach, information is stored centrally and distributed to each work center in batches as needed. Both the base stock and just-in-time approaches pass along information immediately causing the work centers to react to the most recent customer action. The systems differ in that the base stock model informs all earlier stages simultaneously, whereas the just-in-time model passes information only from each stage to its immediate preceding upstream stage. Each of these systems are addressed in detail in subsequent chapters.

Production authorization rules drive operations. These rules rely on accurate due date, inventory level, and machine/worker status data. With so much data now available, the current challenge is to determine what data are important and then to develop the tools to turn the data into useful information and distribute it to the individuals and systems that can use it effectively.

Advances in computing hardware and software have significantly lowered the cost to collect, distribute, and process data. This opens up new alternatives for decision architectures. With the limited availability of communication links and data processing capability in the past, hierarchical structures were necessary. Today, heterarchical control systems of autonomous agents are becoming feasible. Each resource, such as a ma-

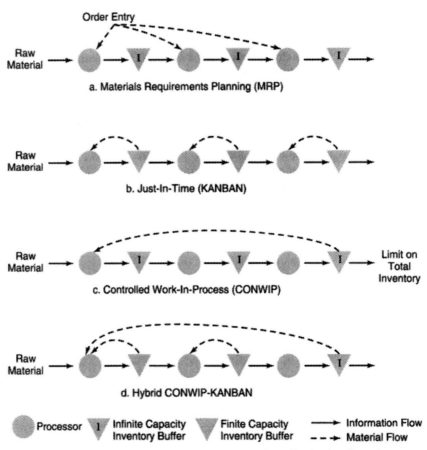

Figure 2.12 Information Flow for Various Systems (Three Production Stages)

chine center, becomes an intelligent agent, capable of making decisions and acting autonomously. The machine tracks its status and listens for opportunities. The machine agent can broadcast its availability to jobs and other machines, bidding for work in the process. Based on closeness and bids, the job agent accepts bids, contracts for more operations, and directs its routing through the best sequence of processing machines available at the time needed.

2.4 PRINCIPLES OF PRODUCTION SYSTEMS

Several guiding principles contain important lessons for understanding the performance of production systems. These principles either result from physical laws or have become accepted based on widespread recognition of their usefulness. Production system design and operating plans must take these principles into account if they are to be successful. We discuss several of the important principles in this section.

2.4.1 Learning Curves: Decreasing Marginal Costs

The first prototype of a product takes longer and costs more to produce than the second unit, the second costs more than the third, and so on. We learn by doing, and this learning allows for continuous improvement by identifying and eliminating problems and reducing variability. Experience indicates that each time we double the number of units or repetitions of an activity, the marginal cost or time per unit reduces by the same proportional amount. Mathematically, this can be expressed as:

$$Y(n) = Kn^{-a} \qquad (2.2)$$

where n is the number of the unit produced, $Y(n)$ is the cost (or time) to produce that n^{th} unit, and K and a are constants. Setting $n = 1$, we see that K is the cost to make the first unit. The parameter a specifies the learning rate. Figure 2.13 illustrates several typical learning rates and the corresponding a values.

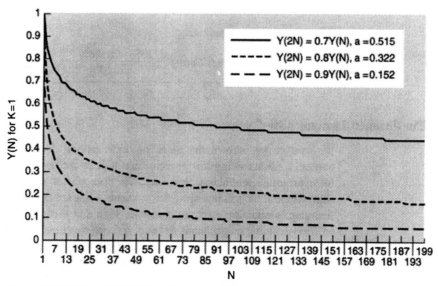

Figure 2.13 Relative Y(N) Values for Assorted Learning Rates

We can find the cumulative cost of producing N units from $\sum_{n=1}^{N} Y(n)$. This sum can be approximated by integrating $Y(n)$ over n from $n = 0.5$ to $n = N + 0.5$. (This approximation uses the integral from $n - \frac{1}{2}$ to $n + \frac{1}{2}$ to estimate the discrete cost of item n. We find the total cost of N items to be:

$$\int_{0.5}^{N+0.5} Y(n)dn = \frac{K[(N + 0.5)^{1-a} - 0.5^{1-a}]}{1 - a} \tag{2.3}$$

and the average cost per unit for N units is:

$$\text{Average Unit Cost} = \frac{K[(N + 0.5)^{1-a} - 0.5^{1-a}]}{N(1 - a)} \tag{2.4}$$

Expression (2.2) implies a constant learning rate. At any production level N, if we double production, we have the relative cost:

$$\frac{Y(2n)}{Y(n)} = \frac{K(2n)^{-a}}{Kn^{-a}} = 2^{-a}.$$

A typical learning curve would have $2^{-a} = 0.9$ or $a = 0.152$, but the learning rate will vary from one situation to another.

EXAMPLE 2.1

The cost of the first two prototypes of a new emergency braking system for an automobile totals $15,000. Experience has shown such products to have a 90% learning rate. Estimate the cost per unit of the 10,000th unit and the average cost of the first 10,000 units.

SOLUTION

For a 90% learning rate, $a = 0.152$. We can use the average cost expression and knowledge that the average cost of the first two units is $7,500 to find K.

$$K = \frac{N \cdot (1 - a) \cdot (\text{Ave. Cost})}{(N + 0.5)^{1-a} - 0.5^{1-a}} = \frac{2 \cdot (0.848) \cdot (7,500)}{2.5^{0.848} - 0.5^{0.848}} = \$7,855.$$

With knowledge of K and a, we can compute:

$$Y(10,000) = K \cdot (10,000)^{-a} = (7,855) \cdot (10,000)^{-0.152} = \$1,937.$$

Likewise, from Equation (2.4):

$$\text{Average Cost} = \frac{(7,855) \cdot (10,000)^{0.848} - 0.5^{0.848}}{(10,000) \cdot (0.848)} = \$2,284.$$

2.4.2 The Product Demand Life Cycle

Experience has shown that most products exhibit similarly shaped life-cycle demand curves. Understanding this phenomenon aids in the forecasting of demand and selection of appropriate production processes. The four phases are shown in Figure 2.14. During Introduction, the product struggles to create a market and receive recognition. Because of learning, production costs are generally high and profitability low. Product designs may still be evolving as problems develop and improvement opportunities are discovered. As product recognition increases in the marketplace, a larger proportion of individuals develop a desire for the product and sales boom, leading to Rapid Growth. Production costs are dropping during this period, and the hope is that the sales level reaches a plateau at which profits are achieved. Improvements to the product design and production processes

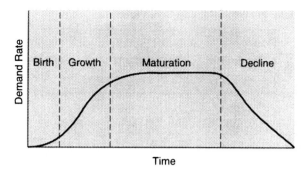

Figure 2.14 Product Life Cycle

are reducing cost and improving quality at this stage. Eventually, the market is saturated, and sales level off to replacement units and sales resulting from a dynamic demographic makeup of society (new teens wanting jeans). Maturation may be brief, as in the consumer electronics industry, or long, as in breakfast cereals and brand name household cleaners. In this phase, product designs and process plans are relatively settled. Automated manufacturing may become feasible for replacing manual tasks and inefficient general purpose equipment. Flexible automated manufacturing systems have become popular because they decrease the production volume and life-cycle time needed to reach this transition point in manufacturing. Finally, the market moves on, and sales enter a period of Decline. The length of the life cycle varies from six months for some electronics to many years for some staple items.

2.4.3   Production Setups Drive Operations for Low- and Medium-Volume Products

The cost and time to set up production facilities to manufacture a specific product drive much of what happens on the shop floor. When the time and cost to change from producing one product to another are large, then naturally we are limited to infrequent changovers. This results in the accumulation of large batches of material and finished goods. This inventory of goods requires space and management. Sufficient supplies of finished goods must be stockpiled to fill demand until the next production run for a product. With high-volume production, this is not a problem. Production processes can be dedicated to a single product and operated at a rate to match demand. However, market trends in recent years are forcing companies to produce a greater variety of products with lower individual demand rates.

After World War II, Taiichi Ohno, Toyota's chief production engineer, began attacking the problem of how Japanese automobile manufacturers could competitively produce thousands of vehicles per year, while in the United States production levels were in the millions. For one, the large presses that form car bodies require careful alignment of heavy dies when changing parts. Toyota could not afford to invest in as many presses and invested instead in methods for reducing changeover times. Gradually, a methodology developed that allowed dies to be changed in three minutes by machine operators instead of requiring a set up specialist an entire day. The huge inventory levels required by large production runs were no longer needed. Figure 2.15 shows the effect of long production runs on cycle inventory levels. A machine is set up. It produces a batch of identical parts and then is set up for the next part type. This process continues until all parts assigned to that machine have been produced and then the cycle repeats. The longer a setup takes, the longer we must produce each product to make up for the setup cost and lost machine time.

a. One Setup Per Week, Constant Consumption

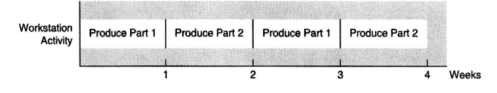

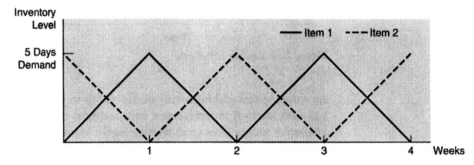

B. One Setup Per Day, Constant Consumption

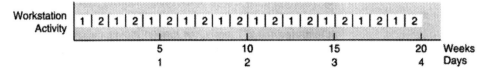

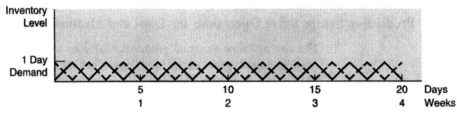

Figure 2.15 Impact of Run Lengths on Inventory Levels: A Workcenter with Two Products

As the length of a product run increases, so does the average inventory level, storage space requirement, and, therefore, separation between adjoining processes. Likewise, the time between producing and using parts increases. This negatively impacts quality because parts may become damaged or obsolete before use, and equipment problems exist longer before detection. This leads to the next principle.

2.4.4   Inventory Is Needed to Support Production

It has been popular in recent years to claim a goal of *zero inventory*. However, some inventory, in the form of materials and finished products, is necessary to meet the needs of internal and external customers in a timely manner. Moreover, it is economically practical to maintain some level of work-in-process (WIP) inventory to facilitate scheduling of production.[2] The theory of queues describes the relationship between WIP inventory lev-

---

[2] We will also discuss later the strategic use of inventory to allow for smoothing production levels when demand varies over time.

els and the production rate of the shop. This relationship is illustrated in Figure 2.16a. Variability in processing times and job arrival rates requires that some inventory is available to reduce the starvation of machines and workers. (A resource is starved when it is available but must remain idle because there are no jobs available for it to produce.) We would like to operate somewhere around point A in the Figure. We can perhaps move along the curve to point B, but thereafter, as we move toward point C, we achieve very little additional output while requiring much larger inventory levels. Figure 2.16a shows a simplified, theoretical relation. The actual relationship may be more like that shown in Figure 2.16b. As extra WIP hits the shop, conditions become crowded, parts are lost or damaged, and space needs increase and therefore distances between operations increase, eventually leading to a drop in production. *A necessary condition for effective production management is that production managers are aware of the specific realization of this relationship for their operation.* We should also strive to constantly reduce the minimum inventory required to maintain a given production rate. This can be done through efforts to improve planning and scheduling procedures, improve the timeliness and accuracy of shop information, increase the reliability of processes, reduce setup times, and implement marketing strategies to reduce demand fluctuations. Above all, inventory should not be used to cover problems such as unreliable production processes. This is an extremely wasteful practice that unfortunately is all too common despite being expensive and ineffective. In the past, many manufacturers thought they could live with poor processes by

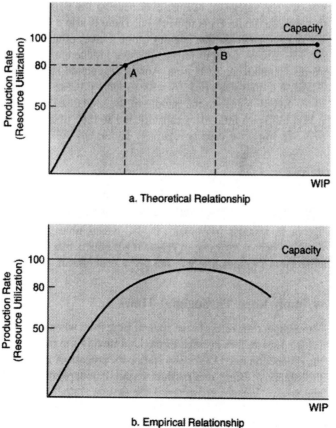

a. Theoretical Relationship

b. Empirical Relationship

Figure 2.16 Production Rate vs. Work-in-Process Inventory Level

carrying large inventories and hoping there would always be some acceptable parts in stock. Thus, there was little motivation to improve the processes. However, large inventories actually prevent the system from improving. If parts are not used immediately upon production, defects are not detected until much later when the entire batch has been ruined. Successful lean competitors operate with reliable processes, quick changeovers, low inventories, small space, low scrap and rework, and closer communication among adjacent processes.

2.4.5 Capacity Balancing

Production processes often involve multiple stages and require multiple resources. A coil of sheet metal will need to be slit and have holes punched, and then be pressed into shape to form the top of a stapler. It may need to be primed and painted. Eventually, it is added to other parts during assembly. Each of these stages requires different equipment. In general, it is desirable to have the capacity to produce the same number of staplers at each stage. Capacity here is measured by the number of staplers that the stage can make per time period (every day for instance). Clearly, total production per week is limited by the workstation with the smallest capacity. Extra capacity at any other workstation is wasted.

There are minor exceptions to this principle. We should normally maintain extra slack capacity for relatively inexpensive operations. To be efficient, we need to ensure that the expensive equipment can be kept busy when demand exists. Thus, if occasional breakdowns or equipment unavailability occurs, it may be economically justified to maintain extra capacity at the inexpensive operations that feed into the expensive operations. This way we can safeguard against ever causing the expensive machines to be idle because they are starved for parts to work on. There is also a relationship between the level of usage of equipment and the length of time it takes products to move through the production process. Excess capacity at inexpensive production stages may provide an economical means for reducing cycle time. Another exception results from the phenomenon that the output of a multistage process with variable processing times is maximized when capacity is slightly higher in the middle of the sequence. This "bowl phenomenon" implies it is best to have a little extra capacity and storage space near the middle of the production line. The impact is usually relatively minor, however.

2.4.6 Customer-Defined Value

Value is defined by what the customer wants and how much he or she will pay for it. Non-value added operations should be eliminated. This means unnecessary time spent in storage or in material movement. It also means unnecessary record-keeping and product-testing. There is no value in producing products that cannot be sold or in using high-tech production methods that will not prove economical.

2.4.7 Large Inventories Imply Long Throughput Times

Throughput time refers to the span of time from when a part enters a system until it leaves. Little's Law relates average throughput time (T) to the level of average inventory (I) and the production rate (X) for any stationary process. A process or system is *stationary* if the probability of being in a particular state is independent of time. The status of resources and sizes of queues characterize the system state. Stationary processes do not increase, decrease, cycle, or otherwise vary consistently over time. The law states that $I = XT$. The relationship holds at all levels. We could refer to an individual workstation, a department, a manufacturing plant, or the entire supply chain network. The result still applies as long as the inventory level and throughput time relate to the same system. In most cases, we will be concerned with a single manufacturing facility. At low-utilization levels, if we release more batches of material (inventory) into the shop, we are likely to see an increase in the production rate with only minor increases in lead time as was shown in Figure 2.16. However, as the production rate approaches the capacity of one or more workstations, additional releases of material will only result in increased lead times and a congested shop. Typically, a significant threshold occurs at approximately 80% of capacity. Beyond this level of production, increased inventory yields only minor increases in output unless very careful scheduling and release strategies are used to balance workloads at each workstation and maintain an even flow of jobs through the production facility.

2.4.8   Throughput Time $\geq$ Cumulative Move and Processing Time for a Transfer Batch

Previously, we defined the transfer batch as the quantity of parts transported together to the next destination. If this quantity is larger than one unit, then each unit in the transfer batch waits while the others are being processed. The result is obvious but meaningful. Because each unit must go through the production system, the sum of the times when it must be at each workstation or in transit between workstations is the minimum possible throughput time. Additional time in queue will normally be present as well. If we want to reduce throughput time, we must do one of the following: eliminate unnecessary processing steps, reduce waiting time for the load to be started once it arrives at a workstation, reduce time the load spends waiting to be moved, run processors at a faster rate, speed up the material handler, or reduce the number of units in the transfer batch. The choice of the appropriate transfer batch size is discussed in detail later in this text.

2.5   PRODUCTION SYSTEM MODELS

Testing out new ideas on full-scale systems is expensive, time-consuming, complex, and unnecessary. Instead, we build models to help us visualize and examine the aspects of a system in which we are interested. Models allow us to learn about the system and test various system designs. Models of production systems will allow us to test out in advance the impact of planning and control decisions and therefore avoid making errant decisions and minimize the disruption to the real process.

2.5.1   Definition of a Model

A model is a simplified, artificial representation of reality constructed to facilitate off-line study of a real object or system. There are many types of models, some of which we have already seen. Our information flow diagram in Figure 2.12 is a model used to represent how the system performs under different paradigms. We may also use geographical models to indicate the locations of production facilities and the distribution network for transporting materials from the raw material stage through delivery of finished products to customers. We can use philosophical (or conceptual) models to contemplate how the system should be configured. Small-scale physical models are often used to view prototypes of products and manufacturing systems and to detect potential interference problems. For the most part, however, we will use mathematical models. A set of mathematical equations will be constructed that describe objectives such as profit maximization, constraints such as resource availability, and physical laws such as conservation of mass. If these expressions are viewed as functions of the decision variables such as production quantities, we can explore the effect of potential planning and control decisions. In system design and improvement projects, we have found it useful to delineate the decision variables, objectives, and constraints. This is an effective way to summarize the synthesis step of the engineering design method. This formulation of the problem data also conveniently leads into the building of mathematical models for the problem. The reader should also note that simulation models that logically trace the sequence of events resulting from planning decisions can also be valuable aids to the productions and operations planner.

2.5.2 Role of Models

Models play an important role in learning about production systems and guiding our improvement of system design and operation. However, we must always remember that the model is a simplified abstraction of some subset of the real system. Models should only include those aspects of the real system that are necessary to approximate behavior over the range of conditions and questions being considered. Different models are necessary for different questions. For instance, suppose we wish to plan a minor increase in production for the next week. To enable this change, we will increase overtime for a few workers. In such a case, we may be able to ignore product quality and just model cost as a linear function of the change. However, major, long-term structural changes may require new equipment, maintenance schedules, purchasing agreements, and shift schedules. Overtime may be limited, and workers may tire of the increased pace after awhile. All of these could require nonlinear models to estimate the cost and quality impact. We must be careful not to expect the model to imitate the real system exactly. Models are useful for determining the key factors and interactions that affect performance, not necessarily for defining precise outcomes. Real-world behavior is too complex and random to capture completely in a model.

There are four primary sources of differences between the results of a modeling project and the behavior of the real world. These are depicted in Figure 2.17. First, we have an imperfect view of the world because of our own limited knowledge and historical experiences. We see only a portion of the real world (system) that we view through a cloud of perception and measurement error. Thus, we could not construct an exact replica even if we so desired. Second, we extract the essence of our image of the world to construct a simplified model of reality. This is to keep the model understandable and manageable within the available time and cost resources. The best model will generally be the simplest model that adequately replicates reality. We will refer to this as the *parsimonious principle of model building*. The final model may be established through the iterative process of hypothesizing a model form, building the model, and then inputting historical data and comparing model and actual outcomes. Once we are satisfied with the model,

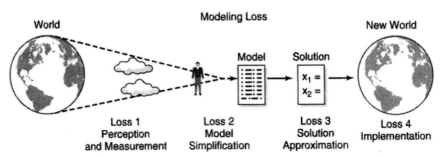

Figure 2.17 Modeling Loss

we exercise it. We may try to find an optimal solution, or we may just run the computer long enough to find a satisfactory outcome. In the latter case, we are only approximating the optimal solution and therefore have an obvious loss relative to the ideal solution. Our heuristic solution may vary from the optimal solution by 1, 5, or 10%. Even if we find the optimal solution to the model, to the extent that we only perform limited sensitivity analysis to understand the effect of each assumption and parameter, we deviate from full understanding and true optimal decision-making in an uncertain world. Finally, we must implement the solution. In doing so, we may have to change some decision variables to have the model accepted, the individuals we train to operate the system will not perform perfectly, and the ideal solution will vary over time as the environment evolves. Thus, there are both short- and long-term losses in actuating the solution. In summary, we have four losses in modeling—*perception, simplification, solution*, and *implementation*. Thus, as we build and manipulate models throughout the text, keep in mind that an optimal solution to a model need not be the best solution in practice.

Nevertheless, models have several important uses. Primary among these are the following:

1. Insight. Model-building includes verification and validation. Model-building is an iterative process of inferring the key determinants of system behavior, building the model with these factors, exercising the model on historical data, and comparing results to historical performance. The process continues until model output matches historical output closely enough for the model to be useful. Building the model requires identifying the important factors, collecting and analyzing data, and eliminating the causes of model inadequacy. This process provides insight into the factors that determine system and model behavior. Once the model is constructed, examining the output under new input conditions leads to even greater insight into system behavior. Eventually, the user may intrinsically understand the behavior well enough to obviate the need for executing the model to estimate output for given inputs.

2. Improving System Design. Models can predict system performance for various designs. Multiple design concepts can be subjected to the likely operating scenarios and performance compared. We can evaluate the effect of purchasing different machine types, building a network of regional warehouses, adding production capacity, or acquiring a more extensive data-collection system. New systems are expensive, and few systems are built today without extensive modeling and analysis of design alternatives. Models can help determine the best system design.

3. Improving System Operation. The model may be run under different possible operating scenarios to select the best. Various rules for assigning priority to orders, scheduling worker time, and setting target inventory levels can be tested across the possible levels of customer demand. Many possible operating plans can be tested by exercising the model and only the best implemented on the real system. Thus, the model allows us to test various decision rules for general use and to evaluate specific decisions for a given situation. In each case we can choose the best.

4. Rational Decision-Making. All too often, the most persistent or highest ranking individual wins out in group decision-making. That individual's personal agenda and biases may overrule the wisdom of the majority and evidence of the data. Models allow the group to center the discussion on scientifically testable hypotheses and to make data-based decisions. Individuals pushing specific solutions can be made to demonstrate their validity through the model. Defining the conditions and assumptions necessary for the model to yield a specific solution allows the discussion to center on the likelihood that a particular viewpoint is valid.

5. Communication. Most projects require the selling of the solution to other stakeholders. Visual models are useful in convincing others that decisions are appropriate. Animated simulation models in particular are useful for demonstrating the impact of production plans. Workers can be trained on the model to obtain desired skill levels and understanding of procedures. In some cases one model may be used to solve a problem and then another built to help sell and implement that solution.

2.6   A SYSTEMS PERSPECTIVE

The production system represents a key aspect of the firm. To be successful in designing and operating the system, we must maintain an integrated, global view of the entire supply chain from materials through product delivery. In addition, as we have seen there are other important activities in the firm as well. The production system must integrate and cooperate with these other systems, especially marketing, purchasing, quality assurance, accounting, design engineering, and manufacturing. We conclude this chapter with an illustration of what can happen if we fail to maintain a global perspective.

Consider a manufacturing plant that produces a single product type with random demand. The firm would like to produce to average demand, but feels it must keep a week's worth of product in inventory to serve as a buffer against random demand fluctuations. This extra week's worth of product is commonly referred to as safety stock. Thus, the firm plans to have two weeks' supply of finished product on hand at the start of each week; this represents the expected sales that week, plus safety stock. At the start of each week the inventory level is checked and the weekly production plan set to produce average weekly demand, plus any correction required to bring the safety stock back to the desired level. Average demand is estimated by demand in the previous week. Such rules are not uncommon in industry. Assuming average demand is uniformly distributed between 80 and 120 with an average of 100 units per week, let's see what will happen to inventory and production levels.

Performance of the system over ten weeks is listed in Table 2.3 and graphed in Figure 2.18. Demand values were generated as uniform random numbers between 80 and 120. The system began with a history of demand at 100 units and starting inventory and target inventory set to 200 units. We will denote demand in period t by D_t. P_t will be the production level selected in each period. I_t and I_t^* indicate the actual on-hand inventory

Table 2.3   Production System Activity over Time

Period	Starting inventory	Target inventory	Production	Demand
0	200.00	200.00	100.00	93.61
1	206.39	187.23	74.45	87.50
2	193.34	175.01	69.17	88.04
3	174.47	176.07	89.64	102.24
4	161.87	204.48	144.86	110.35
5	196.37	220.70	134.68	86.90
6	244.15	173.81	16.56	118.61
7	142.11	237.21	213.71	102.33
8	253.49	204.66	53.50	82.24
9	224.74	164.49	21.99	115.33
10	131.40	230.66	214.58	97.40

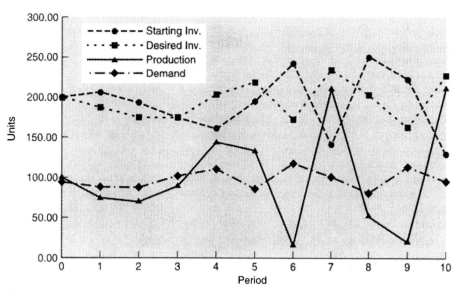

Figure 2.18 Inventory and Production Fluctuations over Time

and desired target inventory at the start of period t. The relationship between these variables is

$$I(t) = I(t - 1) + P(t - 1) - D(t - 1) \tag{2.5}$$

$$I^*(t) = 2 \cdot D(t - 1) \tag{2.6}$$

$$P(t) = I^*(t) - I(t) + D(t - 1) \tag{2.7}$$

The starting inventory in a period is equal to the amount we started with in the previous period plus what we produced minus sales. The desired inventory is always twice last week's demand since we expect to sell that amount again this week, plus keep safety stock. Production is set to adjust the deviation between actual and desired safety stock and to prepare to replace this week's expected demand. The most striking aspect of the table and figure is the large variability in production levels from week to week. This causes severe problems for the shop which must maintain excessive capacity to meet production plans in peak weeks. We also see that starting inventories vary more than desired inventories. These problems result from planning based on a poor understanding of system interactions. Safety stock should be used as a buffer to soften the impact of random demand, not to amplify its effect on production levels. The reader is asked in the end of chapter problems to chart the system if production is held constant at 100 units per week.

In this example, a misuse of marketing (demand) information coupled with a misunderstanding of the relationship among safety stock, inventory and production led to the instability in the system. Similar results can occur if the production system decisions are not properly linked with purchasing, quality, advertising, and engineering decisions.

2.7 SUMMARY

The production system encompasses the resources directly involved in the flow of material goods from acquisition of raw material through delivery of finished products. It is crucial that production system interact appropriately with the other functional areas of the firm.

Production systems can be classified based on their physical organization, product mix, key resources, sequence of processing operations, material flow control rules, and information system. Several principles have been found to guide production system design: 1) We should expect learning over time; 2) Products have a natural life cycle; 3) The time to setup machines is a key driver of performance; 4) An optimal level of inventory can be found for each system; 5) Capacity should be balanced across processes with excess capacity apportioned in accord with acquisition cost; 6) The production rate relates the throughput time to inventory level; and, 7) Activities should only exist to the extent they increase perceived customer value.

Models are important for gaining insight into how the system behaves. Models come in various types and are used in guiding system design and operation. Models can also be used to demonstrate system behavior to others. In building models it is important to obtain valid data; this requires good interpersonal skills and an understanding of how accounting systems accumulate costs.

In addition to the physical resources of people and machines, production systems run on policies, inventory and in-formation. In many ways, improving information flow can substitute for costly inventory without sacrificing performance or service to customers. Inventory expenses arise from the opportunity cost of investing money in inventory, transporting and caring for physical inventory stocks, preparing orders and machine setups, and guarding against shortages. An important strategic decision concerns whether products should be made to stock in anticipation of orders or only after orders are received. Generally, when many raw materials are used to make a few end products, then it is best to make-to-stock. If a relatively small number of raw materials are combined in many ways to create many possible end products, then we make-to-order. In between, if many different final products can be created from a relatively small number of modules, then we make modules-to-stock and then, for final products, we assemble-to-order.

2.8   REFERENCES

ASKIN, R.G., & STANDRIDGE, C.R. *Modeling and analysis of manufacturing systems.* New York: John Wiley & Sons, Inc., 1993.

ALTING, L. & ZHANG, H. Computer-aided process planning—a state of the art survey. *International Journal of Production Research,* 1989.

BONVIK, A.M., & GERSHWIN, S.B. Beyond kanban: creating and analyzing lean shop floor control policies. *1996 MSOM Conference Proceedings,* Dartmouth College, Hanover, NH, 1996, 4–51.

COOPER, R., & KAPLAN, R.S. Measure Costs Right: Make the Right Decisions. *Harvard Business Review,* 1988, 96–103.

DAVENPORT, T.H., & SHORT, J.E. The New Industrial Engineering: Information Technology and Business Process Redesign. *Sloan Management Review,* 1990, 11–27.

DIETRICH, B.L., A Taxonomy of Discrete Manufacturing Systems. *Operations Research,* 1991, Vol. 39, No. 6, 886–902.

HAUSER, J.R., & CLAUSING, D. The House of Quality. *Harvard Business Review,* 1988, 63–72.

KAPLAN, R.S. Measuring manufacturing performance: a new challenge for managerial accounting research. *Accounting Review,* 58(4), 1983, 686–704.

LEWIS, R.J. *Activity-based costing for marketing and manufacturing.* Westport, CT: Quorum Books, 1993.

LITTLE, J.D.C. A proof of the queueing formula L = λW. *Operations Research,* 9(3), 1961.

WOMACK, J.P., JONES, D.T. & ROOS, D. *The machine that changed the world.* New York: MacMillan Publishing Co., 1990.

2.9   PROBLEMS

2.1.   List the major types of inventory and the key factors that determine the optimal quantity of each type.

2.2.   Describe the typical decision levels in a production planning system and indicate the inputs and outputs at each level.

2.3.   Give two examples of how information can be used to improve service and reduce costs in a production system.

2.4.   What is a bottleneck and how does it affect the cost and output of a production system?

2.5.   What is the difference between *inventory position* and *on-hand inventory*?

2.6.   A product is produced in batches of 100 units. The machine requires 1 hour of setup time. Unit processing time is 10 minutes but the machine can process up to 5 units at a time. After processing, each unit must spend 2 hours on a cooling rack before it can be used. There is no limit to the number of units that can be cooled at a time. Assuming there is no additional waiting time for the machine, find the production time for batches of this product.

2.7.   Describe the various types of physical layouts used by production systems and for each give an example where it would be used.

2.8.   A customer enters a grocery store to do their weekly shopping. Wheaties is on their shopping list, but there are no boxes available on the store's shelves. List the possible actions the customer may take. For each action, give your estimate of the probability the customer chooses that action and the shortage cost incurred by the grocery store as a result.

2.9.   It costs a development lab $250.00 to fabricate a prototype part. The normal learning curve rate is 90 percent. Estimate the variable unit cost after 1,000 items are produced.

2.10. How would you explain to a marketing manager that even though your facility is not operating at 100% of its capacity, it may not be advisable to release more work to the shop floor in order to shorten production times and on-time delivery performance?

2.11. In touring a manufacturing plant what clues might you look for and what questions might you ask to judge how efficiently the plant is operating?

2.12. What are the factors that should be included in the inventory holding cost rate, setup cost, and shortage cost parameters? For each, describe how you would obtain the data necessary to estimate these values.

2.13. A popular buzz word in recent years has been "zero inventory." It is possible to achieve a level of zero inventory and still meet demand? Why or why not?

2.14. What is the relationship between the amount of inventory in the plant and the ability to meet customer demand and delivery due dates?

2.15. In a stable system inventory averages $10,000,000. The company sells $35,000,000 per year. What is the average time product spends in the system? How many inventory turns are there per year?

2.16. In each of the following cases indicate if the modeling loss is due to perception, simplification, solution, or implementation.

(a) Data indicates that actual costs are a nonlinear combination of several interacting factors. However to make the model easier to develop and solve a linear approximation of costs is used.

(b) A model has 100 million potential solutions. The analyst decides this would require too much computational effort and therefore only searches over the 1 million solutions deemed most likely to be optimal.

(c) After playing with a simulation model, an analyst determines a set of ten rules that determine optimal behavior in all cases. However, she also notes that using only a subset of three of these rules provides the capability to obtain a good solution in all situations likely to occur. The analyst then decides to train line workers in when to apply these three separate rules.

(d) In building a model of global logistics, an analyst is unaware that certain countries require that all products sold in that country must pay import duties unless at least 50% of the parts are manufactured locally.

2.17. Describe the major uses of models.

2.18. Suppose you had just solved an important company problem using a sophisticated algorithm you learned in your last year of school. How would you go about convincing your MBA degreed boss, your engineering coworker, and the high school educated shop workers that your solution was optimal?

2.19. The production supervisor regulates output by keeping a fixed number of jobs active on the shop floor at all times. If a new machine is purchased that will double the production rate, what will this do to the throughput time for jobs? If all jobs have due dates set by the customer, how should the acquisition of the new machine affect release dates? (Note: The due date states when the job must be delivered. The release date is when the job is started into production.)

2.20. Using the same demand values given in Table 2.3, track the inventory levels for the ten week period if production is fixed at 100 units for each period. Explain why the inventory levels also behave more smoothly with this rule.

2.21. Using the same demand values given in Table 2.3, track production and actual inventory levels for the ten week period if Target Inventory is always set to 200 units. Contrast the behavior of this system with the original.

2.22. Due to the complicated process required to set up a machine, a company has decided to produce a particular product only once a month. Demand occurs at a continuous and constant rate of 1,000 units per month. The company also keeps 50 units available as safety stock, just in case production is delayed. Sketch the level of cycle inventory and safety stock over the month. What is the average inventory level?

2.23. A batch of material must visit five workstations in series to be converted into finished product. Each batch makes 100 units. Setup time at each workstation is 1 hour. Variable processing time is 1.2 minutes per unit. After completing an operation the batch has to wait an average of 1 hour for the material handler to move it to the next machine. Assuming there is no time lost waiting for machines, find the time it takes for a batch of material to go through the system.

2.24. Repeat the previous problem assuming units are moved between operations in transfer batches of size 10 and that operations are close together so no time is lost waiting for material handling.

Chapter 3

Market Characterization

The firm strives to produce products or supply services that customers will purchase. Accordingly, determining what customers want to buy is a critical activity for success. While efficiency (doing things right) is desirable, effectiveness (doing the right things) is essential. Many millions of dollars of products are trashed or banished to remote warehouses every year because the company produced products customers were not interested in purchasing.

In this chapter, we provide several techniques for acquiring information on what customers want. Our emphasis is on techniques for forecasting continuing demand for existing products. We present techniques for short-term forecasting that are based purely on historical trends and more sophisticated techniques for modeling medium-term demand. However, the reader must be cautioned. In dynamic, chaotic environments, forecasts will have limited accuracy. Even when we make probability statements and discuss confidence intervals for customer demand, we will **necessarily be assuming** that the world will continue to operate in a manner consistent with our historical perspective. The real world is not always so accommodating. We will take a scientific approach to describing forecasting methods. By acquiring a thorough understanding of the marketplace, accurate historical data, and careful model building and monitoring, we can hopefully produce useful forecasts. However, mankind has yet to develop the ability to see into the future and, thus, we must always be aware that our estimates will usually be slightly in error and sometimes grossly mistaken.

3.1 FORECASTING SYSTEMS

Forecasting systems must meet the demand for data to support both long-term and short-term decision-making. Effective strategic planning at the corporate level requires knowledge of potential demand for product families far into the future. Strategic decisions include selecting the product lines to pursue and acquiring the appropriate capacity to produce these families. Thus, technological forecasting is an essential aspect. We must predict which developing technologies will become economically viable and what volume and variety of products will result. In the mid 20th century, it was predicted that the upper limit on the market for computers was one per large organization (university or government agency). More recently, the world domination of the watch market shifted from Switzerland to Japan when one country recognized the potential of digital quartz technology and the other did not.

One of the most common techniques for forecasting demand is to examine past demand data and to extrapolate any trends forward. We would ideally like to know exactly how much of each product customers demanded in each time period in the recent past. This information can be difficult to obtain. Instead, we are likely to find records of shipments or customer orders. Shipments differ from demand because some orders are shipped late, some are shipped early, and some demand goes unfulfilled owing to delivery problems and thus never shows up in shipment data. Many plants also tend to register higher shipments at the end of the accounting period (usually a month) as a natural result of their internal procedures. Although this is not a desirable approach and usually causes problems for the production system, it is a political reality given the accounting and managerial accountability systems in practice in most firms. The amount of orders booked may also be misleading. Order data do not reflect the fact that one customer wanted delivery the next day and another gave a month's lead time for delivery. Orders may also fail to capture the potential demand that was lost because we could not promise the desired delivery date. Advertising efforts may distort measurement of the natural demand process. Nevertheless, we must forecast demand using the available data. In some cases, knowledge of lost sales or shipment delays may make it possible to adjust sales orders or shipments to obtain demand.

3.1.1   Purpose and Use of Forecasts

Forecasts of product demand are necessary to guide many of the decisions made by the firm. Characterization of market potential determines the new products to be designed and technologies to be acquired by the firm. Together, these determine needs for part and material supply contracts, financing plans, and capacity expansion.

Forecasts are also used to set production plans describing the timing and quantity for manufacturing products. The production plans then generate orders for the acquistion of raw materials and purchased parts. Thus, when we forecast, we must forecast the need for these materials at least far enough into the future that we can obtain the materials in time for use. The length of time into the future that we forecast is called the **forecast horizon.** We should distinguish between **forecasts** and **predictions.** Both are conjectures about the future. We will reserve the term forecast, however, for conjectures derived from data by applying formal, analytical (mathematical) procedures.

A forecast must do much more than provide a number for anticipated demands. A confidence interval describing the range of possible demands provides the decision-maker with the ability to prevent potentially detrimental consequences of ending up with too much unsaleable product. Likewise, the aggressive producer in a high markup, low variable production cost environment may wisely gamble that demand realization may be near its upper limit. This producer may be willing to risk having to scrap excess production to ensure that shortages do not occur. While we will concentrate on forecasting average demand, the reader should also keep in mind the importance of forecast error variability as well. Forecast errors are often unbiased, meaning the forecasts are correct on average. If this is true and the errors are normally distributed, standard statistical theory may be used to form a 95% confidence interval on demand by using the interval $(F_t - 2\sigma_F, F_t + 2\sigma_F)$, where F_t is the forecast for period t and σ_F is the standard deviation of the forecast error in period t.

3.1.1.1   Hierarchical Forecasting Systems

Forecasts are needed at several levels of the management hierarchy. Long-term forecasts are needed to support strategic planning. Strategic plans include our intentions for acquiring new technologies, changing capacity (labor, machine, and plant), and setting advertising plans. Manufacturing strategic analysis blends market opportunities with corporate expertise and financial resources to determine strategic plans for each major product group. At this level, we need the market potential for each product group such as home kitchen appliances. Factors such as population demographics and government forecasts of national economic conditions for the next several years may dominate. At the low end of the planning hierarchy, we will be ordering parts and scheduling machines. At that point, we will need to know how many brown two-slot, high-grade toasters will be sold next month.

Forecasts can be produced bottom-up or top-down. In a bottom-up system, each end item is treated as a different product and forecasted directly. Thus, we would forecast the number of red model 26 toasters and white model 36 toasters ordered by customers in each upcoming period. To order sheet metal for toaster bodies, we would aggregate sales over all models. In top-down forecasting, we might forecast the number of toasters to be sold each month in the next year. Separately, we would forecast the percentage of toaster sales that will be white vs. brown and the proportion of toasters that will be model 26 vs. 36. By applying these proportions, we can infer the expected demand for each possible end item. For instance, suppose we forecast a market of 1000 toasters, of which 40% will be white and 75% will be model 26. If the color and model selections are independent, we can then expect to sell 1000(0.40)(0.75) = 300 white model 26 toasters. The top-down approach is more efficient in terms of required model development, but the bottom-up approach does not make any assumptions regarding independence of occurrence of features. Of course, we could estimate the four proportions of white model 26, white model 36, brown 26, and brown 36, but this begins to resemble then the bottom-up approach. In most cases, it is easier to obtain an accurate forecast of the number of toasters than of the number of toasters with each combination of parameters.

Adjustments are often made to data before a quantitative analysis of the series is attempted. Months have different numbers of working days, thus, it may be necessary to normalize values by dividing by the number of working days in the period. The time series becomes demand per day during the month instead of demand per month. Likewise, special advertising promotions, labor disputes during contract negotiations, and special purchase before an impending rate increase may all affect data. The data should be adjusted to remove these assignable causal factors before applying most of the techniques discussed in this chapter.

3.1.1.2 Model Building

The process of developing a forecasting procedure follows the procedure shown in Figure 3.1.[1] The first step requires defining the target variables to be forecasted. As discussed above, this may be individual items or product families. In addition, we must select the appropriate units for the variable. Dollars of sales may be relevant in a low inflation period but could be misleading over the long term. Labor-intensive operations often use labor hours; however, this measure may also be misleading because of productivity improvements and increasing automation. Capital intensive facilities often use measures such as pounds of product. Quite often in manufacturing environments, a natural unit of production will exist.

The second step requires collection of relevant data. For a new product, this will involve data from similar existing products, historical start-up results, market surveys, and expert predictions. For existing products, we may only need recent sales figures for this

---

[1]This process closely resembles the standard engineering method of Problem Definition, Data Synthesis, Data Analysis, Candidate Solution Generation, Solution Evaluation, Recommendations, and Implementation/ Monitoring.

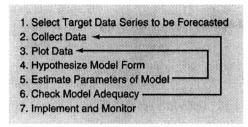

Figure 3.1 Model Building Process

item. If a more sophisticated process exists, econometric data such as interest rate forecasts or housing starts may be desired. This is particularly true if aggregate, long-term forecasts are desired. Once collected, the data are summarized. Graphs depicting the relevant data values plotted against time are helpful at this step.

Having studied the available information, the analyst must hypothesize a form for the forecasting model. A basic model, for instance, might state that demand in period t, D_t, is composed of a long-term mean demand, B_t, plus a seasonal component, s_t, plus a random error ($D_t = B_t + s_t + \varepsilon_t$). The following sections will detail a variety of model forms for different situations. We then try to fit the parameters of these models using the historical data to assess the validity of the hypothesized model forms. To test the validity of the model, we must compare the forecasts from this model with actual demand. A common practice is to split the data. Part of the data is used to estimate values for the model parameters (such as the mean B_t and seasonal components s_t). The model is then used to forecast values for the other historical periods for which actual demand is known. The model's forecasts can then be compared with the actual historical values. If the model performs sufficiently well, we accept it and implement it. We will continue to monitor its performance in use, however. If the model does not adequately predict the past, we certainly cannot expect it to do well in the future. Thus, we need to hypothesize a new model form and repeat the procedure.

3.1.1.3   Checking Model Adequacy

Several measures of forecast accuracy are used. Let e_t be the difference between the observed demand value and our forecast for period t i.e., $e_t = D_t - F_t$. The first measure relates to the unbiasedness of the forecasts. We would like these values to average 0. The average forecast error, as measured over T historical data points, is given by $\bar{e} = T^{-1}\sum_{t=1}^{T} e_t$. The magnitude of the forecast errors can be measured by the mean absolute deviation (MAD). Alternatively we could use the sample variance of the forecast errors, also known as forecast mean squared error (MSE). These are computed as:

$$MAD = \frac{\sum_{t=1}^{T}\left|e_t - \bar{e}\right|}{T - 1}, \text{ and } MSE = \frac{\sum_{t=1}^{T}(e_t - \bar{e})^2}{T - 1} \tag{3.1}$$

If the forecasts are unbiased, the $\bar{e}$ term can be ignored in the MAD and MSE expressions and T used in the denominator instead of $T - 1$. As an example, suppose our forecast model had predicted demands of 53, 53, and 53 for the past three periods, but actual demands were 55, 51, and 56. The forecast errors are then $e_1 = 55 - 53 = 2$, $e_2 = 51 - 53 = -2$, and $e_3 = 56 - 53 = 3$. The average forecast error is thus $\bar{e} = \dfrac{2 - 2 + 3}{3} = 1$. The $MAD = \dfrac{|2 - 1| + |-2 - 1| + |3 - 1|}{2} = 3$. The MSE is $MSE = \dfrac{(2 - 1)^2 + (-2 - 1)^2 + (3 - 1)^2}{2} = 7$. The standard deviation of the forecast errors is estimated by both $1.25 \cdot MAD$ and $MSE^{1/2}$. These yield the estimates $\tilde{\sigma}_e = 1.25(3) = 3.75$ and $\hat{\sigma}_e = 2.65$.

3.2 TIME SERIES EXTRAPOLATION FOR SHORT-TERM FORECASTING

Extrapolation of past trends in the data provides a simple approach to forecast over a relatively short forecast horizon. This approach is based on the assumption that whatever phenomenon was causing demand to vary from period to period in the recent past will continue. Demand values are assumed to be a function of time. And although we may not see time as an actual factor that causes demand to change, the historical evidence indicates a correlation between time and demand, and we can exploit this relationship to forecast future demand.

We begin with a stable process. From there we look at a general model that will allow modeling of growth, decline, and seasonality factors.

3.2.1 Simple Exponential Smoothing

The most basic model of demand states that the demand in time period t consists of a constant mean plus some random error term. In statistical notation, we write this as:

$$D_t = \mu + \varepsilon_t \tag{3.2}$$

where D_t is the demand in period t, μ is the mean (expected) demand in any period, and ε_t is a random error component with constant mean 0 and variance σ^2. Figure 3.2 illustrates such a process. Demand in each period varies randomly about the mean. This would be an appropriate model for a part during the stable maturation portion of its life cycle. At any time t, we have knowledge of demand from time 0 to t. In most cases, we desire a forecast of demand next period, that is, in period $t + 1$. We will denote this one period ahead forecast as F_{t+1}. If we wish to forecast demand in period $t + \tau, \tau > 1$, we will use the notation $F_{t, t+\tau}$ for the forecasted value.

In forecasting future demand values, we would like to use all past data. However, in case the demand process has shifted through time, it might be advisable to put more weight on the most recent data. We would also like the method to be computationally quick and not require a lot of data storage because we may have to forecast time series for many items. Simple exponential smoothing provides a means to accomplish all these goals. At any period t, we estimate the mean demand by a weighted average of all past data. The weights are selected such that the importance of any historical demand value decays exponentially towards 0 as that value ages. Algebraically, we obtain the simple expression:

$$S_t = \alpha D_t + (1 - \alpha)S_{t-1} \tag{3.3}$$

where S_t is the estimated mean demand including all data up through period t. The parameter α defines the weighting function for data. We can think of α as the proportion of weight we want to give the most recent observation relative to the average of all the pre-

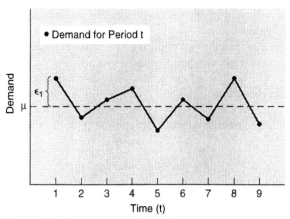

Figure 3.2 Constant Mean Demand Process

vious data. By this reasoning, we should specify α in the range $(0,1)$ although this is not mathematically necessary. A common rule of thumb suggests using $0.001 < \alpha \leq 0.3$. In general, the more faith we have in recent data, the larger we should set α. If observed deviations in demand tend to indicate a permanent change in the demand process, then α should be large. If these deviations are likely to be a one-time random perturbation, then α should be small.

By successive back substitution in Equation (3.3), we see that:

$$S_t = \alpha D_t + (1 - \alpha)[\alpha D_{t-1} + (1 - \alpha)S_{t-2}]$$
$$S_t = \alpha D_t + \alpha(1 - \alpha)D_{t-1} + (1 - \alpha)^2 [\alpha D_{t-2} + (1 - \alpha)S_{t-3}]$$
$$S_t = \alpha D_t + \alpha(1 - \alpha)D_{t-1} + \alpha(1 - \alpha)^2 D_{t-2} + (1 - \alpha)^3 [\ldots])$$
$$S_t = \alpha \sum_{k=0}^{\infty}(1 - \alpha)^k D_{t-k}$$

Thus, a demand value from k periods in the past has weight $\alpha(1 - \alpha)^k$. (Note that for $0 < \alpha < 1, \sum_{k=0}^{\infty}(1 - \alpha)^k = \dfrac{1}{1 - (1 - \alpha)} = \alpha^{-1}$, and thus the sum of all the weights is one.)

To estimate future demand, we need a **forecasting equation.** For a constant mean process, the forecasting equation is simply:

$$F_{t,t+\tau} = S_t, \qquad \tau = 1,2,\ldots$$

EXAMPLE 3.1

A mature product is forecasted using simple exponential smoothing. The current time is week 32, and the exponentially smoothed average is $S_{32} = 56.8$ cartons of product ordered per week. Demand in week 33 turns out to be 58 cartons. Using $\alpha = 0.2$, update the smoothed average and forecast for weeks 34 and 35.

SOLUTION

First use the updating formula to obtain $S_{33} = 0.2(58) + 0.8(56.8) = 57.04$.
The forecasting equation can then be applied to obtain

$$F_{33,34} = F_{33,35} = S_{33} = 57.04.$$

Once this process is started, it is easy to continue. However, we need a way to find the initial S_0. If historical data are available, it is customary to use the average of that data.

If not, we will need to rely on the judgmental forecasting methods described later in this chapter to get us started.

Simple exponential smoothing tracks a constant mean demand process very efficiently. However, what happens when demand is increasing or decreasing? The process will slowly adjust but will lag behind the actual mean demand. The average age of the data can be found from the weights as:

$$\text{Average Age of Data in } S_t = \sum_{i=0}^{\infty} i\alpha(1 - \alpha)^i = \frac{1}{\alpha} \text{ periods.}$$

Thus, if $\alpha = 0.2$, the smoothed average lags the current demand process by $\frac{1}{\alpha} = 5$ time periods. If we have a new product and demand is increasing by 10 units per period, then S_t will typically be 50 units too small, and our forecasts will be even farther off because demand is increasing each period and forecasts made at any time t are the same for all future periods. Clearly, we need to be able to model more complex demand patterns. In the next section, we describe a generalization of the exponential smoothing concept for modeling processes that have a trend and/or seasonal variation.

3.2.2   Seasonal Forecasting Methods (Winter's)

Consider a product with a demand history as shown in Figure 3.3. Demand is growing over time but also seems to vary according to a predictable seasonal variation pattern. Many items exhibit recurring weekly, monthly, or yearly patterns. A bakery may sell mainly breads early in the week but a number of cakes late in the week. Demand for lawn mowers increases in the spring while children's toys take off each fall. For such items we might consider the model:

$$D_t = (B_0 + t \cdot G_0)s_t + \varepsilon_t. \tag{3.4}$$

where B_0 is the **base** component measured at time 0, G_0 the **trend** component measured at time 0, and s_t a multiplicative **seasonal** adjustment factor for period t. The base component gives the deseasonalized mean of the process at a specific point in time. The growth component indicates how much the deseasonalized mean of the process is changing from one period to the next. Let N be the the number of periods in a season, i.e., 7 for the days of a week or 12 for the months of a year. We have a single base and trend component at any point in time, but we will need a seasonal factor for each period in a season. As before, ε_t is a random error component for period t with mean 0 and constant variance, and τ is the forecast lead time.

To be able to apply the model to a set of historical data and forecast future demand, we need to be able to 1) update the parameters each period as new information becomes available; 2) implement an expression for forecasting; and 3) initialize the model parameters B_0, G_0, and s_t.

3.2.2.1   Updating Equations

Let's look first at updating the parameters. Because expected demand changes each period, we need to identify our time origin carefully when specifying the base and seasonal components (the trend is assumed to always be the same). The standard procedure is to let time 0 be the present. Thus, we update our reference point for time each period as a new demand value is observed. The updating equations both add the new information from

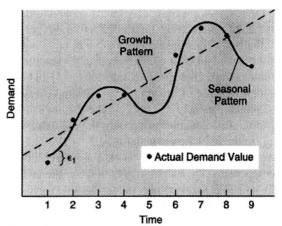

Figure 3.3 Demand Process with Trend plus Seasonality

the most recent demand observation and correct for the rolling time origin. Figure 3.4 illustrates the updating of the origin each period. Initially, January is the origin, and the base has a value of 100. However, the trend is +10. Assuming there are no seasonal factors and the February demand is observed to be 110 as expected, the trend would remain 10, but the base would be updated to 110 and the time origin becomes February.

The updating equations are:

$$B_t = \alpha_1\left(\frac{D_t}{s_{t-N}}\right) + (1 - \alpha_1)(B_{t-1} + G_{t-1}) \tag{3.5}$$

$$G_t = \alpha_2(B_t - B_{t-1}) + (1 - \alpha_2)G_{t-1} \tag{3.6}$$

$$s_t = \alpha_3\left(\frac{D_t}{B_t}\right) + (1 - \alpha_3)s_{t-N} \tag{3.7}$$

Each of the updating equations has the format α(New estimate) + $(1 - \alpha)$(Old estimate). The α values indicate the weight for the estimate based on the most recent demand

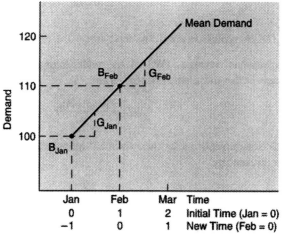

Figure 3.4 Updating the Time Origin

value while the remaining weight, $(1 - \alpha)$, is applied to the estimate based on the previous data, updated to the current time point as necessary. When considering the current time, $t = 0$, the basic equation becomes $D_t = B_t \cdot s_t + \varepsilon_t$. With the recognition that our best guess for the random error component is 0, the resulting relation $D_t \approx B_t \cdot s_t$ explains the new estimate term in the updating equations (3.5) and (3.7) for the base and seasonal factors. The new estimate of growth is the estimated change in the base between last period and the present, namely $B_t - B_{t-1}$. Likewise, to estimate the base for the updated time origin in the first equation using only the old data, we need to add a growth component, G_{t-1}, to the old base estimate, B_{t-1}.

We have added a subscript to α in case a different weighting is desired for each of the model components. The base and growth factor are updated each period; thus, our old rule of $0.001 < \alpha \le 0.3$ seems appropriate. Because each seasonal component is updated only once every N periods, we might want a higher weighting factor here.

3.2.2.2 Forecasting Equation

The forecasting equation is derived directly from the model form. Given the current parameter values, we can forecast τ periods into the future by:

$$F_{t,t+\tau} = (B_t + \tau \cdot G_t) \cdot s_{t+\tau-N} \text{ for } \tau \le N. \tag{3.8}$$

The restriction on τ is just because our seasonal factors only go back N periods. We can easily forecast further into the future, we just need to recycle through the seasonal factors to ensure that we use the correct factor for the period of the season that we are forecasting. For instance, whether we are forecasting for March of the coming year or several years hence, we use the seasonal factor that was last updated in the most recent March period for which we have data.

3.2.2.3 Initialization

All that remains is a method for initializing the parameters. We need initial base, growth, and seasonal factors to begin the updating process. We will assume we have two full seasons of historical data. The current time is 0; thus that data has a negative time index. Let V_i be the average of season i historical demands. Thus, $V_1 = N^{-1} \cdot \sum_{t=1-2N}^{-N} D_t$ and $V_2 = N^{-1} \cdot \sum_{t=0}^{1-N} D_t$. A typical heuristic method for estimating the growth factor is to use the changes in seasonal averages. (We ask the reader to explore the validity of this method in the problems at the end of the chapter.) Thus:

$$G_0 = \frac{V_2 - V_1}{N}. \tag{3.9}$$

The base at the current time is then obtained by updating from the midpoint of the past year to the present by:

$$B_0 = V_2 + \frac{N-1}{2} \cdot G_0. \tag{3.10}$$

The seasonal components give the ratio of demand to the base. From Equation 3.8, we can use the estimation rule:

$$s_{-t} = \frac{D_{-t}}{B_0 - t \cdot G_0} \text{ for } t = 0, \dots, N - 1.$$ (3.11)

If desired, multiple years of historical data could be used to estimate the appropriate seasonal factor and these values averaged. In accord with our model, the seasonal components should be normalized to vary about one, i.e., $\sum_{t=0}^{N-1} s_{-t} = N$. (This normalization is accomplished by multiplying each seasonal component by $N \sum_{t=0}^{N-1} s_{-t}$.)

EXAMPLE 3.2

Monthly sales of beer for a distributor are listed in Table 3.1 and shown in Figure 3.5. Forecast for January and February 1993 using Winter's model.

SOLUTION

The first step is to estimate initial values for the model parameters. To provide a time frame, we will let the current time, December 1992, be time 0. The averages for the two years of historical

Table 3.1 Monthly Beer Sales

Sales					
Period	Sales	Seasonal factor	Ave. sea. factor	Normalized factor	Time period
Jan-91	16.28	0.9676			−23
Feb-91	15.17	0.9016			−22
Mar-91	16.3	0.9687			−21
Apr-91	17.23	1.0239			−20
May-91	18.9	1.1230			−19
June-91	19.17	1.1390			−18
Jul-91	19.88	1.1811			−17
Aug-91	18.63	1.1068			−16
Sep-91	16.12	0.9576			−15
Oct-91	16.66	0.9896			−14
Nov-91	14.4	0.8553			−13
Dec-91	13.64	0.8102			−12
Jan-92	15.65	0.9295	0.9485	0.9474	−11
Feb-92	16.1	0.9562	0.9289	0.9278	−10
Mar-92	18.06	1.0725	1.0206	1.0194	−9
Apr-92	18	1.0689	1.0464	1.0451	−8
May-92	18.9	1.1222	1.1226	1.1213	−7
Jun-92	18.95	1.1251	1.1321	1.1307	−6
Jul-92	18.35	1.0895	1.1353	1.1339	−5
Aug-92	17.57	1.0431	1.0749	1.0737	−4
Sep-92	15.7	0.9320	0.9448	0.9437	−3
Oct-92	16.15	0.9587	0.9742	0.9730	−2
Nov-92	14.43	0.8565	0.8559	0.8549	−1
Dec-92	14.32	0.8499	0.8300	0.8291	0
				12.0142288	

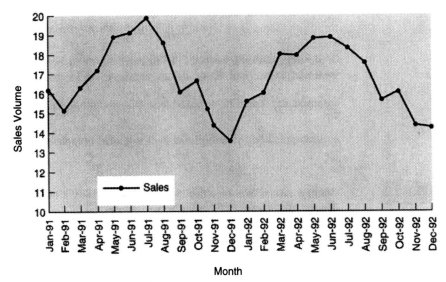

Figure 3.5 Monthly Beer Sales

data are $V_1 = \dfrac{\displaystyle\sum_{t=-23}^{-12} D_t}{12} = \dfrac{16.28 + \ldots + 13.64}{12} = 16.865$ for 1991 and $V_2 = 16.848$ for 1992.

Our estimate of the growth trend is then $G_0 = \dfrac{V_2 - V_1}{12} = \dfrac{16.848 - 16.865}{12} = -0.001$ per period.

To find the current base, we set $B_0 = V_2 + \left(\dfrac{12-1}{2}\right) \cdot G_0 = 16.848 + (5.5)(-0.001) = 16.8425$, which corrects the base from the middle of 1992 to December 1992. To estimate the seasonal factors, we need to divide each historical demand value by its corrected base plus growth. For December 1992, $s_0 = \dfrac{D_0}{B_0} = \dfrac{14.32}{16.8425} = 0.850$. For November 1992, $s_{-1} = \dfrac{D_{-1}}{B_0 - G_0} = \dfrac{14.43}{16.8425 - (-0.001)} = 0.857$. We continue in this manner to estimate the seasonal factor for each period. These are shown in the third column of Table 3.1. The fourth column shows the average over the two years for each month. The final column contains the normalized values (multiply each column four value by 12 and divide by the sum of the seasonal averages). We are now ready to forecast. For January and February 1993 our forecasts are $F_1 = (B_0 + G_0 \cdot 1)s_{-11} = (16.848 - 0.001) \cdot 0.9474 = 15.96$ and $F_2 = (B_0 + G_0 \cdot 2)s_{-10} = (16.848 - 0.002) \cdot 0.9278 = 15.63$.

EXAMPLE 3.3

Let's suppose it is now a month later and actual January demand was 15.36. Update the model parameters.

SOLUTION

Using Equations (3.5) through (3.7),

$$B_1 = 0.1\left(\dfrac{D_1}{s_{-11}}\right) + 0.9(B_0 + G_0) = 0.1\left(\dfrac{15.36}{0.9474}\right) + 0.9(16.848 - 0.001) = 16.784$$

$$G_1 = 0.1(B_1 - B_0) + 0.9G_0 = 0.1(16.784 - 16.848) + 0.9(-0.001) = -0.0073$$

and

$$s_1 = 0.2\left(\dfrac{D_1}{B_1}\right) + 0.8s_{-11} = 0.2\left(\dfrac{15.36}{16.784}\right) + 0.8(0.9474) = 0.9410.$$

Winter's seasonal model is simple to administer yet has performed well in tests, even in comparison to more sophisticated approaches. The seasonal components do not assume any specific pattern and, thus, are flexible for many scenarios. The growth component allows the model to adapt fairly rapidly to changes in the average demand level. If growth and seasonal components are not present in the data, these components will automatically remain around their neutral levels of 0 for growth and 1 for seasonality. Leaving these terms in the model will inflate the variance of the forecast error slightly but will not seriously degrade forecast performance in most cases.

3.2.3   Tracking Signals and Monitoring of Forecasts

Once a model that adequately tracks the historical data is developed, the model can be implemented on a rolling horizon basis to forecast future demand. By rolling horizon, we mean that for a period we first use the model to forecast for τ periods into the future. Then, when the next period's actual demand is observed, we use this new information to update the model parameters as shown in the last section. Once again, we forecast for the next τ periods.

It is important to track the performance of this process over time to ensure accurate forecasts. One common tracking signal is given by the ratio of the smoothed forecast error to the smoothed absolute forecast error. Each period we use the current forecast error $e_t = D_t - F_{t-1,t}$ to update the smoothed forecast error:

$$E_t = \gamma e_t + (1 - \gamma)E_{t-1} \tag{3.12}$$

where γ is another smoothing constant normally set at approximately 0.2. On average, the forecast errors should be randomly distributed about zero. Thus, E_t should be close to zero. E_t will increase in magnitude if we constantly forecast low or high. If the forecast errors are all above or below zero, this indicates a bias in the forecast. For instance, if the process has an increasing trend, then a simple exponential smoothing forecast model will always lag behind the actual demand and underestimate demand.

We also need to compute the smoothed absolute forecast error:

$$\Delta_t = \gamma |e_t| + (1 - \gamma)\Delta_{t-1}. \tag{3.13}$$

The smoothed absolute forecast error Δ_t estimates the mean absolute deviation (MAD) of the forecast errors. Because all absolute values are non-negative, Δ_t must also be non-negative and gives an estimate of the scale of forecast errors. It can be shown that the size of Δ_t depends on the standard deviation of the forecast errors. In fact, for normally distributed random error components:

$$E(\Delta_t) \approx 0.8\sigma_e. \tag{3.14}$$

To determine whether E_t is too large, i.e., significantly different from zero in a statistical sense, we compute the tracking signal:

$$Y_t = \frac{E_t}{\Delta_t} \tag{3.15}$$

each period. We must have $Y_t \leq 1$, and we only approach the equality when all forecast errors have the same sign, an obvious example of consistent forecast bias. To determine a range of reasonable values for the tracking signal, we need to know its statistical variance under the assumption that forecast errors have $E(e_t) = 0$. By repeated back substitution, we can show that Equation (3.12) can be written $E_t = \gamma \sum_{n=0}^{\infty} (1 - \gamma)^n e_{t-n}$ as $t \to \infty$. If the forecast errors have constant variance σ_e^2, then:

$$V(E_t) = \gamma^2 \cdot \sigma_e^2 \cdot \sum_{n=0}^{\infty} (1 - \gamma)^{2n} = \frac{\gamma^2 \cdot \sigma_e^2}{1 - (1 - \gamma)^2} = \frac{\gamma \sigma_e^2}{2 - \gamma}.$$

Accordingly, recalling Equation (3.14), the standard deviation of E_t is approximated by

$$\sigma_E = \sigma_e \cdot \sqrt{\frac{\gamma}{2 - \gamma}} \approx 1.25 \cdot \Delta \cdot \sqrt{\frac{\gamma}{2 - \gamma}}.$$

Our tracking signal automatically corrects for the scale of the errors by dividing by Δ_t. Thus, if we want a k standard deviation confidence interval at approximately zero for Y_t, we can use the interval

$$-1.25 \cdot k \cdot \sqrt{\frac{\gamma}{2 - \gamma}} \leq \frac{E_t}{\Delta_t} \leq 1.25 \cdot k \cdot \sqrt{\frac{\gamma}{2 - \gamma}}$$

for the acceptable range of the tracking signal. Values outside this range indicate a bias in the forecast errors. As an example, if the null hypothesis is that bias does not exist, a two standard deviation test with $\gamma = 0.2$ would reject this hypothesis when $\left| \dfrac{E_t}{\Delta_t} \right| >$

$$\frac{1.25(0.2)^{1/2}(2)}{(1.8)^{1/2}} = 0.83.$$

The derivation of the acceptable interval for the tracking signal made the assumption of uncorrelated forecast errors in deriving the variance of E_t. Unfortunately, this is not true. Forecast errors will tend to be correlated because the forecasts rely on the same historical data. Brown (1967) investigated this correlation and found that for simple exponential smoothing, the standard deviation of the smoothed error could be approximated by:

$$\sigma_E \approx 0.55 \cdot \sqrt{\gamma} \cdot \Delta.$$

This would yield the acceptable range:

$$-0.55 \cdot k \cdot \sqrt{\gamma} \leq \frac{E_t}{\Delta_t} \leq 0.55 \cdot k \cdot \sqrt{\gamma} \tag{3.16}$$

EXAMPLE 3.4

An exponential smoothing forecast model has generated the forecast errors -1.5, 0.5, 0.3, -0.7, -0.2, -0.4, and -0.6 over the last seven periods. From experience, we know that the forecast error variance is 0.81 when the process is under control. Test to determine whether the forecasts are unbiased. Use a smoothing coefficient of $\gamma = 0.2$.

SOLUTION

We need to select a value for k. We will let $k = 3$. Using Equation (3.16), the critical region then becomes $\left| \dfrac{E_t}{\Delta_t} \right| > (0.55) \cdot (3) \cdot \sqrt{0.2} = 0.74$. We initialize $E_0 = 0$ and $\Delta_0 = 0.8(\sigma_e^2)^{1/2} = 0.8(0.9) = 0.72$. Updating, $E_1 = 0.2(-1.5) + 0 = -0.3$ and $\Delta_1 = 0.2|-1.5| + 0.8(0.72) = 0.876$. The ratio is thus $\left| \dfrac{E_1}{\Delta_1} \right| = \left| \dfrac{-0.3}{0.876} \right| = 0.34$. Because the ratio is less than the critical value of 0.74, we conclude that there is no evidence that the forecast model is biased. Continuing with the other data, we find the results in Table 3.2. The tracking signal does not indicate any significant problem with the model.

Table 3.2 Tracking Signal Results

Period	Forecast error	Smoothed error	Smoothed abs error	Tracking signal
0		0.00	0.72	
1	−1.5	−0.30	0.88	0.34
2	0.5	−0.14	0.80	0.17
3	0.3	−0.05	0.70	0.07
4	−0.7	−0.18	0.70	0.26
5	−0.2	−0.19	0.60	0.31
6	−0.4	−0.23	0.56	0.41
7	−0.6	−0.30	0.57	0.53

3.2.4 Adaptive Control

Exponential smoothing methods assume a specific model form. If the underlying process changes, for example, if demand suddenly shifts to a new level or goes through a temporary growth period when using a simple exponential smoothing model, the forecasts will lag behind demand. In this case, we would like to use a larger value for the smoothing constant until the process mean catches up to the new level of demand. One approach is to use the tracking signal as the smoothing constant. We set $\alpha_t = \left| \dfrac{E_t}{\Delta_t} \right|$. As forecast errors grow, the smoothing constant will automatically be increased to place more weight on the recent data. This will increase the response rate to a change in the process. Suppose demand has been stable at the level of 100 units per period and then suddenly starts increasing at the rate of 10 units per period. Figure 3.6 illustrates the impact of α when simple exponential smoothing is chosen as a forecasting model. The figure graphs actual demand and the corresponding S_t values over time for various values of α. Recall that the forecast will lag actual demand by $1/\alpha$ periods. A small α of 0.02 will cause the forecast to lag well behind the actual demand. Setting $\alpha = 0.2$ will do better. The $1/\alpha = 5$ period average age of the data will produce an S_t value that is $5(10) = 50$ units behind actual demand. The tracking signal will note that we are constantly underestimating demand once the growth pattern begins. The smoothed error will rapidly approach the smoothed absolute error. This will make the tracking signal, and therefore α, approach 1. Thus, the S_t will lag demand by only one period. A similar analysis is shown in Figure 3.7 for the case that demand suddenly shifts to a new level. At time 6 in the Figure, average demand jumps from the level of 100 units per period to 150. The Figure shows the rate at which simple exponential smoothing will adjust for the three choices of α.

While adaptive control smoothing may seem appealing, remember that large forecast errors may indicate an incorrect model due to a structural change in the demand pattern. Adaptive control is not a panacea for covering model misspecification. If the model form is wrong, it would be preferable to build a correct model instead of continuing to play catch-up all the time.

3.2.5 Time Series Decomposition

We have discussed time series as if they are composed of trend, season, and random factors superimposed on the current base level. A classical decomposition approach is commonly used for economic time series that may also contain a business cycle factor. The demand

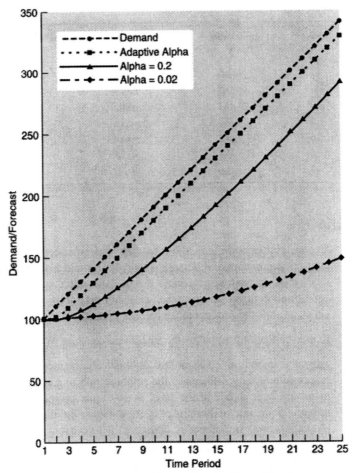

Figure 3.6 Simple Exponential Smoothing for Linear Trend

process is viewed as a function $D_t = f(S_t, G_t, C_t, \varepsilon_t)$, where S_t represents the seasonal component, G_t the growth (trend) component, C_t a possible cyclical component, and ε_t the random error. Both multiplicative ($D_t = S_t \cdot G_t \cdot C_t \cdot \varepsilon_t$) and additive models ($D_t = B_t + S_t + t \cdot G_t + C_t + \varepsilon_t$) are used, but the multiplicative is more common. We will use this model for demonstration purposes. The process can be made rather sophisticated, but the basic idea is to decompose the demand data by applying the following steps:

1. For each period t in the historical data, compute the moving average, $\overline{D}_t$, of the data over the past complete season (e.g., the average of 12 consecutive periods for monthly data). The moving average for a period should be computed using data centered about the target period t. Thus, for monthly data, we should average the period's demand with that for the 5.5 months preceding and after period t. Of course, this means that we will not be able to compute the average for the half season at the start and end of the historical data. The moving average for any time period includes one instance of each seasonal component, thus, these should cancel out. Likewise, the data should include some large and some small random disturbances that approximately cancel out. The resultant average should, therefore, represent the trend and cycle components. Algebraically, $\overline{D}_t \approx G_t \cdot C_t$.

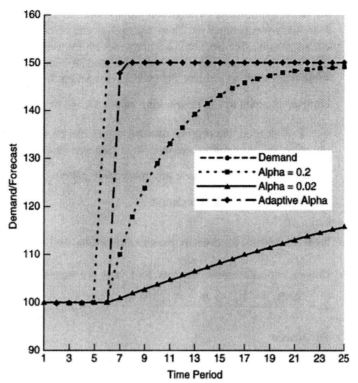

Figure 3.7 Simple Exponential Smoothing for a Change in Level

2. Divide each demand value by its season's moving average. $\dfrac{D_t}{\overline{D}_t} \approx \dfrac{S_t \cdot G_t \cdot C_t \cdot \varepsilon_t}{G_t \cdot C_t} =$ $S_t \cdot \varepsilon_t$, an estimate of the seasonality ratio (multiplicative factor) and random component for each period.

3. If multiple seasons of historical data exist, we combine estimates of seasonality for each period. For each period of the season, group all the seasonality ratios for that period together and take the average. This reduces the randomness and provides an improved seasonality ratio for each period of the season. Because deviations in the seasonal component from a base of 1 indicate seasonal effects, it is customary to normalize these components so that they average 1.0. This provides the seasonality estimates $\hat{S}_t$.

4. Deseasonalize the original D_t data by dividing each demand value by its period's seasonality ratio. The resultant deseasonalized demand values include trend, cycle, and random components. Thus, $D_t^s = \dfrac{D_t}{\hat{S}_t} \approx G_t \cdot C_t \cdot \varepsilon_t$.

5. Use a 3- to 5-period moving average of these values to reduce the effect of randomness. The moving average should be evenly spaced before and after the period. Thus, for a three-period average, the fourth demand value would be replaced by the average of the third, fourth, and fifth deseasonalized values. In this manner, we obtain $\overline{D}_t^s = \dfrac{D_{t-1}^s + D_t^s + D_{t+1}^s}{3} \approx G_t \cdot C_t$.

6. Hypothesize a basic growth trend model form such as $G_t = a + bt$, where a and b are unknown parameters. These parameters can be estimated from the data. Simple linear regression, (see Section 3.3.1) provides an easy method for this. Each moving average trend that contains cycle, growth, and random components can then be divided by the fitted trend value to estimate the cyclical component if we assume a neutral random error component equal to 1. The resulting values are $\hat{C}_t = \dfrac{\overline{D}_t^s}{\hat{G}_t}$.

7. If desired, the original demand values can be divided by the period's trend x cycle value and the seasonal component to estimate the random disturbance for that period. These values should not have any predictable pattern. $\hat{\varepsilon}_t = \dfrac{D_t}{\hat{C}_t \cdot \hat{G}_t \cdot \hat{S}_t}$.

We illustrate with an example.

EXAMPLE 3.5

Table 3.3 contains the quarterly domestic demand for steel wire products. Decompose this series.

SOLUTION

Column 3 contains moving averages. Each value is computed using the expression

$$\overline{D}_t = \frac{0.5D_{t-2} + D_{t-1} + D_t + D_{t+1} + 0.5D_{t+2}}{4}.$$

For instance:

$$\overline{D}_3 = \frac{0.5D_1 + D_2 + D_3 + D_4 + 0.5D_5}{4} = \frac{0.5(285) + (308) + 273 + 242 + 0.5(270)}{4} = 275.125$$

Table 3.3   Decomposition of Steel Wire Demand

Period	Demand	Moving average	Seasonal estimate	Deseasonalized demand	3 Month moving ave.	Cyclical factor	Random factor
Sp. 1988	285			276.96793			
Su. 1988	308			280.102	277.683	1.028	1.01
Fall 1988	273	275.125	0.992	275.981	276.799	1.039	1.00
Win. 1988	242	270.625	0.894	274.314	270.895	1.032	1.01
Sp. 1989	270	263.75	1.024	262.391	265.903	1.028	0.99
Su. 1989	287	255.75	1.122	261.004	255.001	1.000	1.02
Fall 1989	239	249.25	0.959	241.609	247.641	0.986	0.98
Win. 1989	212	242.875	0.873	240.308	240.976	0.974	1.00
Sp. 1990	248	237.125	1.046	241.011	238.650	0.980	1.01
Su. 1990	258	232.5	1.110	234.631	233.355	0.973	1.01
Fall 1990	222	225.375	0.985	224.424	225.564	0.955	0.99
Win. 1990	192	217.5	0.883	217.638	215.705	0.929	1.01
Sp. 1991	211	214.375	0.984	205.053	211.226	0.924	0.97
Su. 1991	232	215	1.079	210.986	213.825	0.951	0.99
Fall 1991	223	218.875	1.019	225.435	219.531	0.994	1.03
Win. 1991	196	223.375	0.877	222.172	226.300	1.042	0.98
Sp. 1992	238	224.375	1.061	231.293	224.212	1.051	1.03
Su. 1992	241	222	1.086	219.171	224.962	1.073	0.97
Fall 1992	222			224.424	215.121	1.045	1.04
Win. 1992	178			201.768			

For a season with an odd number of periods, the partial weighting of the beginning and endpoints would not be required. Column 4 contains the ratio of the moving average values to the original demand. Following step 2, we average these to compute the seasonal factors for each quarter. Averaging the fall 1988, 1989, 1990, and 1991 values, we obtain 0.9888. Likewise, winter, spring, and summer factors average to 0.8818, 1.0286, and 1.0991, respectively. These four seasonal measures sum to 3.9983. We normalize these about 1 by multiplying by 4/3.9983 to obtain the fall through summer seasonal factors 0.9892, 0.8822, 1.0290, and 1.0996. The seasonal factors are used to obtain the deseasonalized demands listed in column 5 by dividing each column 2 demand value by the factor for its seasonal period. Randomness is then removed by taking a three-month moving average as shown in column 6. Next, we apply step 6 and fit a simple linear regression model to the data in column 6. Letting period 1 correspond to spring 1988, the fitted trend model is $G_t = 277.68 - 3.7822t$. Thus, a fitted (seasonalized and trended) estimate for Spring 1988 would be $F_{sp88} = [277.68 - (3.7822) \cdot 1] \cdot 1.0290 = 281.84$. Now let's look ahead five years to spring 1993. This would be period 21 but would use the same Spring seasonal factor. To forecast for spring 1993, $F_{sp93} = [277.68 - (3.7822) \cdot 21] \cdot 1.0290 = 204$. These estimates do not include a cyclical factor. From the Table, we can see evidence of a four-year cycle that could be factored into our estimate if we so desired. Figure 3.8 plots the original demand values, the deseasonalized values from column 5 of the Table, and the fitted values of the forecasting model.

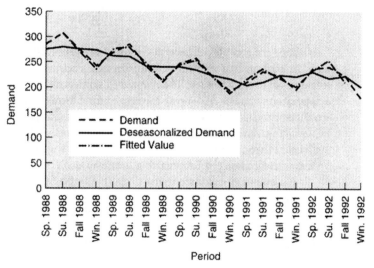

Figure 3.8 Steel Wire Product Demand

3.2.6   Box-Jenkins Time Series Models

Box and Jenkins (1970) popularized a general class of time series forecasting known as ARIMA (p, d, q) models. This stands for auto-regressive, integrated, moving average models, with p autoregressive terms, d differences, and q moving average terms. Autoregressive terms are lagged (earlier) values of demand itself. For instance, if the process is relatively constant, last period's demand, D_{t-1}, might be a good estimate of D_t. To remove randomness and allow modeling of polynomial growth patterns, several lagged terms, i.e., $D_{t-1}, D_{t-2}, \dots , D_{t-p}$, might be included in the model. It is often possible to replace a large number of autoregressive terms with a single or small number of *moving average* terms. These moving average terms are lagged values of the random error terms. A single moving average term model, called MA(1), could be written $D_t = \mu_t + \varepsilon_t - \theta_1 \varepsilon_{t-1}$. Nonstationary time series can often be transformed into stationary series by differencing the values, i.e., by replacing the original series by the differenced series $\tilde{D}_t = D_t - D_{t-1}$.

In general, we can build models with any number of autoregressive moving average and difference terms. The customary notation is to let B represent the time backshift operator. Thus, $B^k D_t = D_{t-k}$. An ARIMA(1,1,1) model would then be of the form:

$$(1 - \varphi_1 B)(1 - B)D_t = (1 - \theta_1 B)\varepsilon_t \text{ or } D_t = (1 - \varphi_1)D_{t-1} - \varphi_1 D_{t-2} + \varepsilon_t - \theta_1 \varepsilon_{t-1} \quad (3.17)$$

In addition, if the data is seasonal, a multiplicative set of ARIMA terms could be added to the base model. The seasonal factors are similar in form but perform backshifting (lags) in multiples of N periods.

Although these Box-Jenkins models seem to offer a wealth of flexibility, this results in the need to use more sophisticated techniques to determine the model form. The set of lagged autocorrelations and partial autocorrelations are the primary factors for determining model form. Autocorrelations are the correlation coefficients for values k periods apart in the data series. The partial autocorrelations are the correlations between demand values r periods apart with the effect of the intermediate demand values removed.

To estimate the autocorrelation coefficient for a lag of k periods, we compute:

$$r_k = \frac{(T-k)^{-1} \sum_{t=k+1}^{T} (D_t - \overline{D})(D_{t-k} - \overline{D})}{T^{-1} \sum_{t=1}^{T} (D_t - \overline{D})^2}. \quad (3.18)$$

In building the model, differences are first taken until the data appears stationary. In general, the number of significant autocorrelation coefficients indicates the number p of autoregressive terms needed. The number of significant partial autocorrelations indicates the appropriate number q of moving average terms. However, sample autocorrelation and partial autocorrelation coefficients can be misleading unless a long series of historical data is available during which time the underlying process held the same process. Because of statistical error, a set of historical data may be fit well by several model forms and provide no clear choice for forecasting future values.

The empirical evidence suggests that the flexibility provided by this general class fails to outperform simpler autoregressive and Winter's style models in most cases. Also, the nonlinear numerical estimation techniques required to fit ARIMA models come with all the computational and numerical accuracy problems of most nonlinear routines. As such, we will not discuss this class further. The interested reader should consult the end of chapter references.

3.3 MEDIUM-TERM EXTRAPOLATIVE AND CAUSAL MODELS

The previous section described models designed for short-term forecasting. Empirical models extrapolated the historical time trends into the future. The model form was determined by statistical correlation between the observed demand pattern and a simple function of time. In this section, we look at more sophisticated empirical models and also begin to examine causal models. Causal models attempt to explain behavior through a cause and effect relationship between variables. This relationship is then exploited to forecast one variable as a function of another variable.

3.3.1 Linear Regression Methods

In regression modeling, we model demand (or any variable of interest) as a mathematical function of a set of other variables whose values are known in practice. The variable we are trying to predict is sometimes called the dependent variable, and the others, the independent or predictor variables. In many cases, linear models of the form:

$$D_t = \beta_0 + \beta_1 X_1 + \beta_2 X_2 + \dots + \beta_k X_k + \varepsilon_t \tag{3.19}$$

can provide good estimates of the response variable where $X_1, \dots, X_k$ are k predictor variables, $\beta_0, \dots, \beta_k$ are parameters that we need to estimate to calibrate the relationship between the set of predictors and the response, and the ε_t are random disturbance components indicating the unexplainable variation seen in the demand series. The ε_t should be without any apparent structure. The standard model assumptions are that $E(\varepsilon_t) = 0$, $V(\varepsilon_t) = \sigma^2 \; \forall t$, and

$$Cov(\varepsilon_{t1}, \varepsilon_{t2}) = 0 \text{ for } t1 \neq t2.$$

The independent X_i variables may causally affect demand. For instance, in forecasting the demand for residential door locks, we could use the number of residential building permits issued in the previous month. In other cases, we can use variables that just seem to have a statistical correlation to the response variable. If sales are on an increasing growth trend, "time" may be a good predictor variable, because both increase each month. Autoregressive or lagged values of demand may play a causal role in forecasting demand or just provide a good base for forecasting the future. If the product life is known, such as a pair of running shoes wearing out in three months or spark plugs needing replacement once a year, then lagged values of demand may indicate new replacement sales. Word of mouth may also be a prime marketing tool. If each satisfied customer extols the virtues of the product to close friends and relatives, we may be able to model the average number of such spin-off sales per item as a function of time after sale. Consider a furniture or appliance manufacturer. The first sales of a new product model after a trade show may be for use in showrooms at the retail outlets. Each of these may then lead to some predictable profile of additional sales during the coming months. In this case, showroom sales in previous periods may be important variables to include in the model.

3.3.1.1   Parameter Estimation

Suppose we have examined a set of historical data and hypothesized a model form. The hypothesized form can be obtained by plotting the dependent variable against the predictor variables and looking for relationships. Figure 3.9 indicates hypotheses that may be drawn from such plots. Using the historical data, we then estimate the β_i values. The most common approach is through the method of *least squares*. In least squares, we find the β_i values that minimize the sum of the squared errors between the observed demand values and those predicted by the resultant model. In Figure 3.9, this would be the sum of the squares of the vertical distances between the plotted data points and the value of the fitted model at these points. To estimate the parameters, we need the hypothesized model form, a set of historical values for demand, and corresponding values of the predictor variables for each period. The observed value of the ith predictor variable at time t is denoted X_{ti}. The least squares problem is then:

$$\text{Minimize } L(\beta) = \sum_{t=1}^{T} (D_t - \beta_0 - \beta_1 X_{t1} - \dots - \beta_k X_{tk})^2 \tag{3.20}$$

The least squares function L is convex in the unknown parameters β_i. From calculus, we know, therefore, that the problem can be solved by taking the derivative of L with respect to each of the β_i, setting these expressions equal to 0, and solving for the solution to these equations. Differentiating with respect to β_i, we obtain:

$$\frac{\delta L(B)}{\delta B_i} = -2 \sum_{t=1}^{T} X_{ti} (D_t - \beta_0 - \beta_1 X_{t1} - \dots - \beta_k X_{tk})$$

$$= -2 \left(\sum_{t=1}^{T} D_t X_{ti} - T\beta_0 - \beta_1 \sum_{t=1}^{T} X_{t1} X_{ti} - \dots - \beta_k \sum_{t=1}^{T} X_{ti} X_{tk} \right)$$

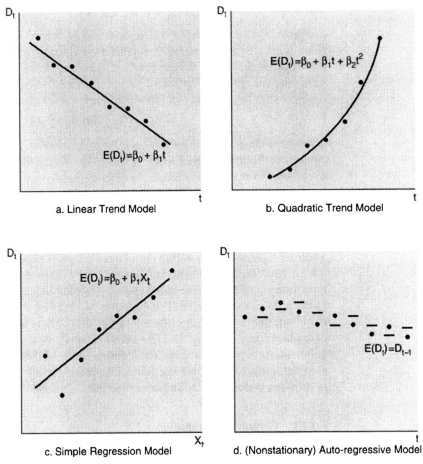

a. Linear Trend Model

b. Quadratic Trend Model

c. Simple Regression Model

d. (Nonstationary) Auto-regressive Model

Figure 3.9 Hypothesizing Model Form

It is easier to represent this result using matrix notation. We define the response vector $\underline{D}$ of demands, the vector of unknown parameters $\underline{\beta}$, the matrix of predictor variables X, and the vector of random disturbances $\underline{\varepsilon}$ as:

$$\underline{D} = \begin{bmatrix} D_1 \\ D_2 \\ \vdots \\ D_T \end{bmatrix}, \underline{\beta} = \begin{bmatrix} \beta_0 \\ \beta_1 \\ \beta_2 \\ \vdots \\ \beta_k \end{bmatrix}, X = \begin{bmatrix} 1 & X_{11} & X_{12} & \cdots & X_{1k} \\ 1 & X_{21} & X_{22} & \cdots & X_{2k} \\ \vdots & \vdots & \vdots & \ddots & \vdots \\ 1 & X_{T1} & X_{T2} & \cdots & X_{Tk} \end{bmatrix}, \text{ and } \underline{\varepsilon} = \begin{bmatrix} \varepsilon_1 \\ \varepsilon_2 \\ \vdots \\ \varepsilon_T \end{bmatrix}.$$

Let X^t be the transpose of X. The first-order conditions for minimizing L are then:

$$\frac{\partial L(\underline{B})}{\partial \underline{\beta}} = -2(X^t \underline{D} - X^t X \underline{\beta}) = 0 \tag{3.21}$$

Rearranging terms in (3.20), we obtain the $k + 1$ least squares normal equations:

$$(X^t X)\underline{\hat{\beta}} = X^t \underline{D} \tag{3.22}$$

Rearranging these equations, we find the parameter estimates:

$$\underline{\hat{\beta}} = (X^t X)^{-1} X^t \underline{D}. \tag{3.23}$$

Note that we use the $\hat{\beta}$ notation to indicate estimated values of the parameters $\beta_j, j = 1, \ldots , k$. Because we know the values in the X matrix, the expression for $\hat{\beta}$ is simply a linear combination (weighted sum) of the historical demand values. The ith row and j^{th} column term in $X^t X$ is just the cross product of terms in columns i and j of X. Likewise, the elements in $X^t \underline{D}$ are the cross products of the columns of X with demand vector $\underline{D}$.

Least squares is a very powerful technique. If the model form and the assumptions on the error components ε_t are correct, least squares yields the unbiased estimators (they are correct on average) with the smallest variance among all possible linear combinations of the observed response variable values. These are called *BLUE* (Best Linear Unbiased Estimators).

EXAMPLE 3.6

Table 3.4 contains data on shipments of news print for the past 14 years. The data are plotted in Figure 3.10. Using time as a predictor variable, fit the model $D_t = \beta_0 + \beta_1 t + \varepsilon_t$. Let the year 1980 correspond to *time = 1*.

SOLUTION

The second column of Table 3.4 becomes the vector $\underline{D}$. The corresponding X matrix is:

$$X = \begin{bmatrix} 1 & 1 \\ 1 & 2 \\ 1 & 3 \\ 1 & 4 \\ 1 & 5 \\ 1 & 6 \\ 1 & 7 \\ 1 & 8 \\ 1 & 9 \\ 1 & 10 \\ 1 & 11 \\ 1 & 12 \\ 1 & 13 \\ 1 & 14 \end{bmatrix}$$

Table 3.4 News print Shipments

Year	Shipments	Fitted value	Residual
1980	4234.	4252.5	−18.5
1981	4735.	4417.3	317.7
1982	4526.	4582.2	−56.2
1983	4674.	4747.0	−73.0
1984	5065.	4911.8	153.2
1985	4927.	5076.7	−149.7
1986	5115.	5241.5	−126.5
1987	5310.	5406.3	−96.3
1988	5415.	5571.2	−156.2
1989	5515.	5736.0	−221.0
1990	6007.	5900.8	106.1
1991	6152.	6065.7	86.3
1992	6464.	6230.5	233.5
1993	6396.	6395.3	0.6

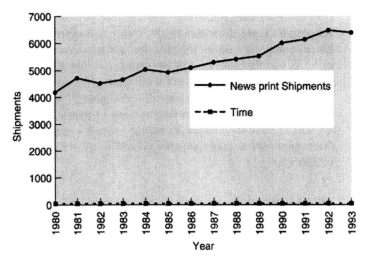

Figure 3.10 News print Shipments

Solving the least squares normal equations (3.22), we obtain:

$$\hat{\underline{\beta}} = \begin{bmatrix} \hat{\beta}_0 \\ \hat{\beta}_1 \end{bmatrix} = \begin{bmatrix} 4087.68 \\ 164.83 \end{bmatrix}.$$

Thus, $\hat{D}_t = 4087.68 + 164.83t$. Setting $t = 1$, we have a predicted value for 1980 of $\hat{D}_{1980} = 4087.68 + 164.83(1) = 4252.51$.

3.3.1.2 Model Adequacy

After estimating parameters for the hypothesized model form, it is essential to check for the adequacy of the model. To be adequate, the model must produce forecasts with acceptable errors. Forecast errors should be small and randomly distributed about 0. Checks for model adequacy begin by looking at the differences between the observed demand values and those estimated by the fitted model. The differences, $r_t = D_t - \hat{D}_t$ where $\hat{D}_t = \hat{\beta}_0 + \hat{\beta}_1 X_{t1} + \ldots + \hat{\beta}_k X_{tk}$, are called the residuals. The residuals can be thought of as the estimates of the assumedly random ε_t. Thus, the residuals should average 0 and have small variance independent of time. We leave it as an exercise to show that the residuals from least squares estimates will always sum to 0. However, we can compute their sample variance as an estimate of σ_ε^2 and look for nonrandom patterns. Large variances, obvious outliers (abnormally large residuals), or trends in the residuals over time indicate a problem with the model. In estimating the variance, we must keep in mind that each of the $k + 1$ least squares normal equations places a restriction on the residuals. Each restriction, such as forcing the residuals to sum to 0, costs a degree of freedom in the data. Thus, instead of having T independent residuals, we only have $T - (k + 1)$. To compute the sample variance of the residuals, we should divide the sum of the squared residuals by this $T - (k + 1)$. This produces the estimator:

$$\hat{\sigma}_\varepsilon^2 = \frac{\sum_{t=1}^{T} r_t^2}{T - k - 1}$$

which is sometimes called mean square for error.

In addition to checking the size of the residuals, it is useful to test to determine whether the model is helpful. If we assume the ε_t are normally distributed, we can test the hypothesis H_0: $\beta_1 = \beta_2 = \cdots = \beta_k = 0$ against the alternative that at least one parameter is nonzero, i.e., at least one predictor variable helps to forecast demand. This test is performed by comparing the amount of variation in demand data that is explainable by the predictor variables to that attributable to the random errors. If H_0 is true, the ratio of these two components of total demand variability form an F distributed random variable with k and $T - k - 1$ degrees of freedom. Thus, we can test formally by comparing the ratio

$$ F_0 = \frac{k^{-1} \sum_{t=1}^{T}(\hat{D}_t - \overline{D})^2}{\hat{\sigma}_\varepsilon^2} . $$

to the critical value from an F distribution. In practice, provided $T - k - 1$ is not too small, we can conclude that at least one predictor variable is significant if $F_0 \geq 4$. The numerator of this ratio is called mean square for regression. Figure 3.11 illustrates the partitioning for each data point into a portion explained by the regression model and an error component. The F statistic computes the ratio of the sums of squares of these two components. It is standard practice to summarize these results in an analysis of variance table of the form:

Source of Variation	Sum of squares	Degrees of freedom	Mean square	F
Regression	$\sum_{t=1}^{T}(\hat{D}_t - \overline{D})^2$	k	$\sum_{t=1}^{T}(\hat{D}_t - \overline{D})^2/k$	$F_0 = \dfrac{MS_{Regression}}{MS_{Error}}$
Error	$\sum_{t=1}^{T} r_t^2$	$T - k - 1$	$\sum_{t=1}^{T} r_t^2/(T - k - 1)$	
Total	$\sum_{t=1}^{T}(D_t - \overline{D})^2$	$T - 1$		

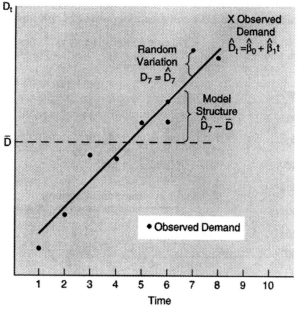

Figure 3.11 Partitioning of Data Variation

If the model is to be used for forecasting, then it seems reasonable to test it on future data. This is difficult because the future has not yet occurred. We can get around this minor difficulty by splitting the data into two groups. The first, and normally larger, group is used to estimate the model parameters via Equation (3.22). The second set of data is not used in estimation. Instead, once we estimate the model parameters, we use the model to forecast demand values for the second set. We then compare the deviations between the model forecasts and actual values. A reasonable measure is the square root of the averaged squared forecast error.

One popular twist on data-splitting is to try to forecast each of the T values in the historical data set by using a model built from the other $T - 1$ historical values. The root mean squared forecast error here, sometimes called the PRESS statistic, can be used to compare the performance of two candidate models.

EXAMPLE 3.7

Examine the adequacy of the model fit in Example 3.6.

SOLUTION

Residuals and fitted values are shown in Table 3.4. As an example, consider the 1993 shipments of 6396. Our fitted value is $\hat{D}_{14} = 4087.68 + 164.83(14) = 6395.3$. The corresponding residual is $D_{14} - \hat{D}_{14} = 0.7$. By summing the squared differences between the fitted values and average demand, we obtain the regression sum of squares. The sum of the squared residuals provides error sum of squares. We can fill out the entire analysis of variance table to obtain:

Source of variation	Sum of squares	Degrees of freedom	Mean square	F
Regression	6181154.	1	6181154.	226.6
Error	327357.	12	27279.8	
Total	6508511.	13		

The large F value indicates that time is a useful predictor. The original data are plotted vs. the fitted line in Figure 3.12. The residuals are plotted in Figure 3.13. These plots indicate a mild concern with a tendency to have negative residuals in the middle of the data and positive residuals on either end. This could indicate some nonlinear relationship. However, the tendency is only borderline significant. Without more data, we cannot be sure. For our purposes, the model seems to provide a reasonable estimate. We can then forecast for period 15 (1994) by:

$$\hat{D}_{15} = 4087.68 + 164.83(15) = 6560.1.$$

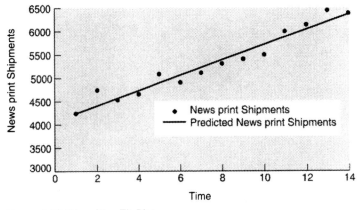

Figure 3.12 Time Line Fit Plot

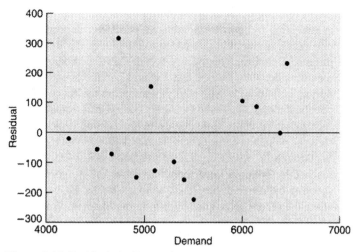

Figure 3.13 Residuals vs. Demand

3.3.1.3   Model Building

Regression modeling requires picking the best set of predictor variables and their proper form in the model. Model building is the process of selecting the equation form for parameter estimation. Consider, for instance, the growth data in Figure 3.14. Two possible models are the quadratic expression $D_t = \beta_0 + \beta_1 t + \beta_2 t^2 + \varepsilon_t$ and the exponential growth pattern $D_t = \beta_0 e^{\beta_1 t} \varepsilon_t$. (This latter model can be put in to the linear regression format by taking logs and using the form $\log D_t = \log \beta_0 + \beta_1 t + \log \varepsilon_t$. Here we use log D_t as the response variable, and the model's intercept actually estimates $\log \beta_0$.) Which model is best? The first approach may be to attempt to identify a scientific law or principle that would support a model form. If no relevant scientific principle is known, we could check the adequacy of each model and choose the model that produces the smallest residuals.

One problem with regression occurs when the predictor variables are correlated. It is difficult in this case to determine which of the predictors contains the best information

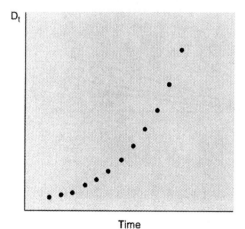

Figure 3.14 Nonlinear Growth Model

for forecasting demand. Even more frustrating is the fact that the effect of each predictor as measured by its corresponding β_i parameter depends on the model form selected. As new predictor variables are added to the model, the $\hat{\beta}_i$ values for the existing variables in the model change. This occurs because several predictors contain the same information regarding demand, and the least squares parameter estimation procedure will look for the best combination of explanatory power from the set of predictor variables provided. This phenomenon is illustrated in Figure 3.15. Least squares takes the multiple of each predictor variable that will get closest to the demand vector. Geometrically, least squares projects the demand vector $\underline{D}$ onto the $k + 1$ dimensional space spanned by the intercept, and X_1 through X_k variables. $\hat{\beta}$ is chosen such that $X\hat{\beta}$ is the closest point to $\underline{D}$ that can be reached by taking a linear combination of the $\underline{X_i}$ vectors. In the figure, the axes correspond to the observations. Thus, to locate vector $\underline{X_i}$, we move X_{1i} along the first axis, X_{2i} along the second axis, and so forth. In our example, if the model only has X_1, then it will set $\hat{\beta}_1 \approx 1$. However, if X_2 is also added to the model, we can get closer to $\underline{D}$ by using $\hat{\beta}_1 \approx -1$, $\hat{\beta}_2 \approx 2$. This significant change in $\hat{\beta}_1$ is unsettling. It occurs because X_1 and X_2 are so close. This is referred to as *multicollinearity* of the predictor variables. More sophisticated statistical methods exist for trying to find a better estimate of both parameters in such cases (see Montgomery and Peck [1982], for example), but it is simplest to just eliminate one of the variables from the model.

The general rule of thumb is to select the model with as few variables as possible that adequately fits demand. One simple interpretation of this principle is to select the model that produces the smallest estimate of σ_ε^2, i.e., the smallest mean square for error. This can mean trying many possible combinations, but modern computers make this an easy task. A number of other suggestions can be found in the references listed at the end of the chapter.

As a final note, at times we may be interested in testing whether a specific predictor variable is helpful. The best approach is to fit two models: one with the variable and one without. Other variables may be left in the model both times. If the variable under study is significant, it should increase the regression sum of squares enough that the loss of one degree of freedom is not significant. This is tested by checking the ratio:

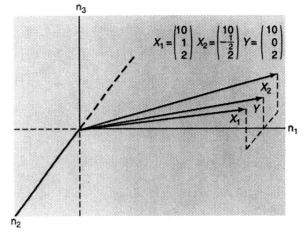

$$X_1 = \begin{pmatrix} 10 \\ 1 \\ 2 \end{pmatrix} \quad X_2 = \begin{pmatrix} 10 \\ -\frac{1}{2} \\ \frac{5}{2} \end{pmatrix} \quad Y = \begin{pmatrix} 10 \\ 0 \\ 2 \end{pmatrix}$$

Figure 3.15 Multicollinearity

$$F_0 = \frac{\text{Regression Sum of Squares with Variable} - \text{Regression Sum of Squares without Variable}}{\text{Mean Square for Error}}$$

This is an F test with one and $T - k - 1$ degrees of freedom. The larger the F_0 statistic, the more likely the variable adds significantly to our forecast. This test can be used to iteratively build bigger models by only adding new variables to the model if they significantly reduce the residuals as indicated by this test.

EXAMPLE 3.8

Suppose we added a quadratic time effect to the model of Example 3.7.

SOLUTION

Fitting the model $D_t = \beta_0 + \beta_1 t + \beta_2 t^2 + \varepsilon_t$ yields $\hat{D}_t = 4304.25 + 83.62t + 5.414t^2$ with the analysis of variance table:

Source of variation	Sum of squares	Degrees of freedom	Mean square	F
Regression	6,266,513.	2	3,133,257.	142.4
Error	241,997.6	11	21,999.8	
Total	6,508,511.	13		

The residuals are shown in Figure 3.16. The mean square for error has decreased, and the residuals show less of a curved trend. If we divide the incremental regression sum of squares by the mean square for error, we obtain:

$$F_0 = \frac{6,266,513 - 6,181,154}{21,999.8} = 3.9.$$

This is a marginally significant result (for an F test with 1 and 11 degrees of freedom and a 5% significance level). Thus, we may conclude that although the quadratic term tends to help slightly, it is not crucial. The analyst must decide whether the extra effort to maintain the quadratic model is worthwhile.

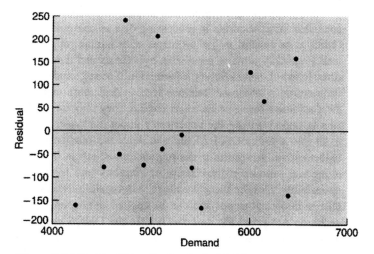

Figure 3.16 Residual Plots for Quadratic Model

3.3.2 Econometric Models

Aggregate forecasts such as the volume of U.S. consumer goods to be sold or the market for structural steel over the next year are often performed by organizations that specialize in econometric forecasts. Using the principles of economics, econometricians build large systems of equations that relate many variables together. The individual equations resemble multiple regression models, except the system of equations are often such that the independent variables are related between equations. As an example, suppose we were in the home refrigerator market. We might have a model of the form:

Refrigerator Sales$_t$ = f(Housing starts$_{t-1}$, Price$_t$, Advertising$_t$, Consumer interest rate$_t$).

If we knew the values of the four independent variables, (housing starts, price, advertising, and consumer interest rate), we could predict sales. Housing starts are affected by GNP and interest rates. Price may be a function of material costs, advertising, labor costs, and sales. Labor costs may depend on inflation, which also effects interest rates and GNP. We see that all the equations are interrelated, and parameter estimation must be done jointly across equations. These systems of equations require complex, nonlinear techniques for estimating parameters and may grow very large. It is also difficult to obtain accurate, timely data. The federal government reports such values but constantly revises them for months after issue. Several private forecasting services build and maintain large econometric forecasting systems. These organizations make forecasts and other related analyses available to major corporations on a fee basis.

3.4 JUDGMENTAL FORECASTING

Historical data cannot predict structural changes in the marketplace, plans to change advertising budgets, the commercialization of new technologies, or pending introduction of competing products. However, managers may be quite attuned to such possibilities. Thus, expert judgment offers the best means for obtaining accurate forecasts in some environments. This is particularly true for medium- to long-term forecasting where the likelihood of significant changes in the competitive environment is high. However, we should also keep in mind that humans are easily swayed by biases and are not as consistent or as accurate in processing data as computers. Wheelwright and Makridakis (1985) note several major problems with human judgmental forecasts. First, humans tend to quickly jump to a possible hypothesis and then search for supporting evidence, often ignoring contradictory information. Second, studies have shown that as additional information is obtained, humans tend to feel more confident in their projections, but their accuracy does not increase. Indeed, they may unjustly use the additional information to underestimate the uncertainty about the future. Third, the desire of individuals to fit into a group can lead them to defer too readily to the beliefs of the majority or leader and fail to scrutinize the suggestions of others. The fact that the entire group goes along can produce a false sense of certainty and lessen the feeling of individual responsibility. Despite these warnings, judgmental methods can provide insight not available in historical trends and can be applied in new environments where historical time series do not exist. We describe briefly the Delphi procedure for obtaining consensus from a group of experts and the use of customer surveys to acquire forecasting data. Other judgmental techniques have also been proposed. An overview is provided in Wheelwright and Makridakis (1985).

3.4.1 The Delphi Method

The Delphi method is a common approach for eliciting knowledge from individuals to find the range of likely outcomes concerning a factor of interest. The methodology has been widely used to address questions such as what new technologies will be important in the next ten years, when a new process will become commercially viable, or the size of the market for a new product. A set of informed "experts" is selected to participate in the decision process. Each expert is asked to fill out a form responding to the specific question(s) of interest. Depending on the question, the response could be a written forecast or list of expected events. Brief justifications for the opinions may be included. These responses are compiled by the facilitator and then redistributed to the experts. Responses are listed anonymously to ensure that all individuals will be given an equal hearing and to prevent domination by a senior executive or intimidating individual. This revised form should indicate which responses adhered to the general consensus and which were divergent from the general view. After considering the input from the other experts, the participants are asked to reconsider their own forecast and repeat the input procedure. Anyone still holding to one of the discrepant views should provide a brief explanation for that stand. This exchange of values and justifications may continue for several iterations in hope of obtaining a consensus or at least until the decision-maker is confident of having received the relevant wisdom of these experts. The hope is that all relevant information exists in the expert pool and, given access to all this information in the form of forecasts and explanations, the majority of the experts will be able to properly weigh the evidence and to come to an accurate consensus.

Several warnings should be noted about the Delphi technique. Results depend strongly on the wording of the questions and the specific set of experts selected. Thus, it is critical that both these issues be carefully selected. Also, it is not guaranteed that the set of opinions will converge to a unanimous scenario or even a clear consensus. Although this may appear frustrating, it is perhaps an asset. The future is not knowable. We should make robust decisions that will allow profitability and continuity no matter which of the highly plausible futures occurs. On the other hand, there is a tendency for people to want to fit in and conform. Thus, the "groupthink" mentality can take over, causing participants to adopt a viewpoint that agrees with others in the group. Independence of personal judgments is then lost. In a hierarchical firm, many participants may tend to automatically agree with the opinions of the participant at the top. This is particularly prevalent in firms where errors of judgment are punished but rewards for creative successes are small.

3.4.2 Market Surveys

Market surveys are a valuable tool for forecasting demand. Statistical techniques can be used to determine customer attitudes and the potential market for products. You have probably been asked for your opinion either while shopping in a mall or over the telephone. Surveys are generally one of two types. Potential customers are randomly selected and asked for their attitudes about the product or their level of optimism concerning the future direction of the economy. Results are tabulated and used to infer estimates of future demand. The second type of survey has the producer calling existing customers to obtain their plans for purchases in the near future. This approach is more direct; however, potential new customers may be overlooked. For new products, companies often contact their lead customers. These are major customers who tend to adopt new technology quickly. Interviews with these major customers can indicate whether the product is likely to be widely accepted and what product requirements are necessary for the new item to be successful.

3.5 COMBINING FORECASTS

When multiple sources of information or independent forecasts are available, it is not necessary to choose among them. In fact, the best choice is often to use a weighted combination of these forecasts. Suppose, for instance, we have an extrapolative forecast of sales, F_t^1, and a judgmental forecast, F_t^2, that includes extra knowledge on the part of management. Assume further that these two approaches have performed independently in the past (with respect to forecast errors), and their variances are σ_1^2 and σ_2^2, respectively. If we use the forecast:

$$F_t = \theta \cdot F_t^1 + (1 - \theta) \cdot F_t^2 \qquad (3.24)$$

then the forecast error variance is:

$$V(e_t) = \theta^2 \sigma_1^2 + (1 - \theta)^2 \sigma_2^2. \qquad (3.25)$$

For proper choice of θ, $0 < \theta < 1$, the combined forecast will have a smaller forecast error variance than either original forecast. By differentiating the variance expression (3.20) with respect to θ and setting the result equal to zero, we can show that the forecast error variance is minimized by choice of:

$$\theta^* = \frac{\sigma_2^2}{\sigma_1^2 + \sigma_2^2}. \qquad (3.26)$$

EXAMPLE 3.9

A company has used senior marketing managers to forecast sales. Historically, they have been provided unbiased forecasts with an error variance of 128. A young staff member has proposed an extrapolative time series model. On historical data, the model has had an error variance of 73. Tracking signals on both methods indicate they are unbiased. A scatterplot of the forecast errors indicates they are not strongly correlated. Combine the forecasts and estimate the forecast error variance for the combined method. If the managers forecast 1,245 and the model forecasts 1,290 for next period, find a better combined forecast.

SOLUTION

We find the optimal weighting factor by $\theta^* = \dfrac{73}{128 + 73} = 0.363$. The forecast error then has

$$V(e_t) = \theta^2 \cdot \sigma_1^2 + (1 - \theta)^2 \cdot \sigma_2^2 = (0.363)^2 \cdot (128) + (0.637)^2 \cdot (73) = 46.5.$$

The combined forecast for next period is $F_1 = 0.363 \cdot (1,245) + 0.637 \cdot (1,290) = 1,273.7$.

3.6 SUMMARY

Market opportunities are the driving force behind production decisions. These opportunities are compiled in the form of demand forecasts. Forecasting systems are designed to provide the input needed for planning production. These systems can take a top-down or bottom-up approach to forecasting demand for product families and individual items.

A variety of quantitative and judgmental methods exist for forecasting demand. For new products and highly volatile environments, judgmental methods are often best. Important methods here include market surveys of current and potential customers and iterative discussions among experts such as the Delphi method.

Future demand of existing products can often be forecast with quantitative models. These models vary from simple methods that attempt to determine basic growth and seasonal patterns to complex, multiequation, causal, econometric models.

In all cases, the methods attempt to separate the structure of the demand process from random fluctuations. The simple models assume past time trends will continue and simply extrapolate past data into the future. Simple exponential smoothing and Winter's multiplicative model with base, growth, and seasonal factors offer simple methods that work reasonably well in many cases. They are often used when many different time series must be forecasted. Regression models are used to relate demand to correlated variables that are observable and either knowable or predictable in the future. Econometric models build complex models to describe the causal interrelationships between financial and market data. In all cases, it is important both to validate the model initially and to track its performance through time to ensure accuracy of forecasts. Tracking signals can provide confidence that the forecasts are unbiased and that the error variance is not increasing.

3.7 REFERENCES

Box, G.E.P., & Jenkins, G. *Time series analysis, forecasting and control.* San Francisco, CA: Holden-Day, Inc., 1970.

Brown, R.G. *Decision rules for inventory management.* New York: Holt, Rinehart, and Winston, Inc., 1967.

Brown, R.G. *Smoothing, forecasting and prediction of discrete time series.* Englewood Cliffs, NJ: Prentice-Hall, Inc., 1962.

Montgomery, D.C., & Johnson, L.A. *Forecasting and time series analysis.* New York: McGraw-Hill, 1976.

Montgomery, D.C., & Peck, E. *Regression analysis.* New York: John Wiley & Sons, 1982.

Makridakis, S.G., & Wheelwright, S.C. *The handbook of forecasting.* New York: John Wiley & Sons, 1985.

Makridakis, S.G., Andersen, E., Parzen, E., Carbone, R., Fildes, R., Hibon, M., Lewandowski, R., & Winkler, R. *The forecasting accuracy of major time series methods.* New York: John Wiley and Sons, 1984.

Trigg, D.W., & Leach, A.G. Exponential Smoothing with an Adaptive Response Rate. *Operational Research Quarterly,* 18(1), 1967, 53–59.

3.8 PROBLEMS

3.1. Why is demand forecasting important? How are forecasts used?

3.2. What is the difference between top-down and bottom-up forecasting?

3.3. Explain the advantages and disadvantages of using a general regression model instead of simple exponential smoothing to forecast demand.

3.4. For checking model adequacy, why is the sum of absolute forecast errors a better measure than simply the sum of forecast errors?

3.5. Demand for the past four periods has been 104, 108, 105, and 99. Estimate the initial value of the simple exponential smoothing statistic by averaging the historical data. Update the statistic using the four periods of data. What is your forecast for period 5? Use a smoothing factor of 0.2.

3.6. A simple exponential smoothing model has a current mean of 156.8. Assuming demand for next period is 170, update the value of the smoothing statistic. Let $\alpha = 0.1$.

3.7. Reconsider the method used in section 3.3 to find initial estimates of the parameters in the seasonal model. In which of the following cases does this method provide good starting estimates of the growth factor:

a) seasonal components increase from period 1 to N;

b) seasonal components are randomly distributed about 1.0 for all periods of the season;

c) all seasonal components are the same.

3.8. Annual shipments of glass food containers for the past decade are given in Table 3.5. Plot these data, hypothesize an appropriate model form, estimate the parameters, and forecast shipments for 1993 and 1994.

3.9. Quarterly shipments of glass food containers for the past three years are given in Table 3.6. Plot these data, hypothesize an appropriate model form, estimate the parameters, and forecast for the next year.

Table 3.5 Annual Glass Beverage Shipments

Year	Shipments	Year	Shipments
1983	60,108.	1988	66,675.
1984	64,302.	1989	67,973.
1985	59,935.	1990	71,967.
1986	62,795.	1991	74,683.
1987	62,673.	1992	73,346.

Table 3.6 Quarterly Glass Container Demand

Year	First qtr.	Second qtr.	Third qtr.	Fourth qtr.
1990	17,446.	17,748.	18,971.	17,802.
1991	17,595.	19,360.	19,413.	18,315.
1992	17,785.	18,338.	19,140.	18,083.

3.10. Use the normal equations to show that the residuals from a least squares fit of the model $D_t = \beta_0 + \beta_1 X_{t1}$ will always sum to 0.

3.11. Using the news print shipments in Table 3.4, fit a first-order auto-regressive model of the form $D_t = \beta_0 + \beta_1 D_{t-1} + \varepsilon_t$.

3.12. Using the news print shipments in Table 3.4, fit an exponential smoothing model with smoothing parameters of 0.1. Compare your results with those of Problem 3.6.

3.13. Using the news print shipment data of Table 3.4, fit an exponential smoothing model with smoothing parameters set to 0.2. Is there a significant growth or seasonal trend component?

3.14. An electronics company has sold 145, 157, 124, 169, 179, 201, 197, and 230 units of a certain stereo system quarterly for the past two years. Plot the data and fit an appropriate model. Forecast sales for the next two quarters.

3.15. Use Winter's exponential smoothing model with intercept and slope components for the electronics data in problem 3.14. Compute the average one-step-ahead forecast error. Use a tracking signal to determine whether the model is in control. (Use 0.1 for smoothing parameters.)

3.16. Using Winter's seasonal trend model, forecast quarterly glass food container consumption for all of 1993 (see historical data in Table 3.6).

3.17. Quarterly sales (in thousands of units) of a product over the last two years have been 26, 29, 30, 27, 28, 32, 33, and 30. Plot the data and forecast for the next period using Winter's model.

3.18. In problem 17, assume sales for the ninth quarter turned out to be 32. Update your parameter estimates and forecast sales for period 10.

3.19. Consider the result in Section 3.2.1 for the average age of data in exponential smoothing. Suppose you use simple exponential smoothing when the true model follows a linear trend. Show that your forecast will tend to lag actual demand. Develop an expression for the expected forecast error (bias).

3.20. Consider the idea of high-definition TV. Using your judgment: 1) estimate the time until the first systems will be available commercially; 2) estimate the cost to a company to develop the necessary technology; and 3) sketch your estimate of the production cost of an average system for the first five years after development. Compare your results with those of two other students. Modify your estimates if your views have changed from this information exchange.

3.21. Apply the time series decomposition technique to the beer sales data in Table 3.1. Compare your results to those in Example 3.2.

3.22. U.S. production of hydrochloric acid over a twelve-year period is shown in Table 3.7. Plot the data. Hypothesize a model form. Estimate parameter values. Check adequacy of the model. Repeat until you are satisfied with your model, then forecast 1993 demand.

Table 3.7 Domestic Hydrochloric Acid Production

Year	Production
1981	2574.
1982	2450.
1983	2468.
1984	2693.
1985	2803.
1986	2392.
1987	2996.
1988	2640.
1989	3268.
1990	3140.
1991	3381.
1992	3566.

3.23. Collect data on enrollments at your school for the past ten years. Attempt to model this data. Does a model with base and trend appear adequate? Suggest predictor variables such as city or state population that may help forecast enrollments. Collect these data and build a new model using regression analysis.

3.24. Consider the model $D_t = \beta_0 + \varepsilon_t$. Derive a closed form expression for the least squares estimates of β_0.

3.25. Resolve the previous problem assuming that newer data is more valuable. To do this, weight the observations based on their age and minimize $L = \sum_{t=1}^{T} \gamma^{T-t}(D_t - \beta_0)^2$, where $0 < \gamma \leq 1$ is the geometrically decreasing weighting factor. Show the connection between your result and that from simple exponential smoothing.

3.26. Consider the model $D_t = \beta_0 + \beta_1 t + \varepsilon_t$. Derive an equation for the least squares estimates of the parameters using the T historical demand values.

3.27. Resolve for the least squares estimates in the previous problem assuming that newer data are more valuable. To do this, weigh the observations based on their age. Then, minimize $L = \sum_{t=1}^{T} \gamma^{T-t}(D_t - \beta_0 - \beta_1 t)^2$, where $0 < \gamma \leq 1$ is the geometrically decreasing weighting factor.

3.28. Show that the choice of θ given in section 3.5 for combining forecasts does indeed minimize forecast error variance for independent forecasts and that the combined forecast has smaller variance than either individual forecast.

3.29. Two forecasting models have been built to track a demand time series. These forecast error variances have proven to be 1.45 and 2.05, respectively. Determine the expression for the optimal combined forecast assuming the two forecasts are independent.

3.30. An experienced staff member has proven capable of forecasting demand with an error variance of 4.5. The company can also purchase an external forecast from an econometrics service. Tests on that model have shown the forecast error to have a variance of 5.6. Both forecasts appear to be unbiased, and the errors are uncorrelated. The methods forecast 210 and 214, respectively, for next period. Find an optimal combined forecast.

3.31. Suppose it is known that two forecasting models have a covariance between their forecast error variances of σ_{12}^2. Determine an expression for the optimal combined forecast.

3.32. Assuming the model assumptions are valid, find the variance of the fitted values for the historical data using a linear regression model. Second, find the variance of the one-step-ahead forecast errors. Using these results, add the assumption that the demand values are normally distributed about their mean and find an expression for a 95% confidence interval about next period's demand.

Chapter 4

Manufacturing Strategy and the Supply Chain

4.1 MANUFACTURING STRATEGY

If you don't know where you're headed, you can't tell if you're moving in the right direction. The company's strategic plan provides a view of where the company is headed and a road map showing the path it intends to take to get there. The strategic plan describes the desired positioning of the company in the marketplace and the manner in which it will strive to satisfy customers. Included are target measures for factors such as sales, market share, inventory levels, quality, delivery time, and profitability. The strategic manufacturing plan identifies product families that will be produced, production technologies that will be used to produce those products, and general policies that will be pursued for purchasing, production, and distribution.

Formed at the upper levels of the firm by senior management, the company's strategic plan defines the frame into which lower-level production system decisions must fit. The components of the plan defining corporate strategy for marketing, manufacturing, and distribution must fit together. One company may choose to concentrate on marketing with all manufacturing and distribution being subcontracted out to independent providers. Another company may concentrate on manufacturing and contract with independent salespeople and advertising agencies for marketing. Whatever the choice, the central corporate strategy should guide all lower-level strategic and tactical planning decisions. After setting the manufacturing strategy, the company must ensure that subsequent decisions reinforce that strategy. The location of production facilities, the design of the distribution network, and the inventory policies adopted must match the choice of geographical markets and speed of delivery. Employment, training, and benefits policies must match the decision regarding capital or labor-intensive production and the importance of a knowledgeable workforce. It would, for instance, be appropriate to build comfortable facilities in a scenic setting and offer opportunities for company-financed graduate education if the company decided that its competitive advantage relied on the creative ingenuity of its engineering staff. In general, this reliance on strategic decisions as a guiding framework should move down to the lowest level of planning, such as the salary and training schedule of an individual employee or the inventory policy for a specific raw material. In the remainder of this section, we detail the components of manufacturing strategy and how they fit together.

4.1.1 Dimensions of Manufacturing Strategy

Manufacturing strategy has several dimensions, including the range of product lines produced, the geographical and economic markets served, the technology base used, the core competencies emphasized, the degree of vertical integration pursued, and competitive positioning in the marketplace. Market opportunities, and unfilled niches in particular, offer an important guide to forming these plans. The location of the company's products on the spectrum, from product-based to technology-based, provides another clue. Product-based companies normally operate in highly competitive consumer markets and may be referred to as *market-pulled* companies. Exceptional service, low cost, and an increasing variety of choices and features over time become necessities for competitive success. The product is defined by the customer, and the producer must find the best technology for producing, delivering, and supporting the product. On the other hand, the *technology-push* companies are leaders in scientific research and offer unique products. As technological innovations develop, the company may develop new types of products for which to apply those technologies and be forced to convince the public of the value of those items. For instance, a decade ago, few people realized that a full life would require a pager and cell phone. Yet, instantaneous communication—to any one, at any time, from anywhere—will soon be an expectation. The development of products is impelled by technology. Automated location and direction systems, made possible by global positioning systems, will soon be standard for automobiles and hikers. Post-its, composite tennis racquets, and early microwave ovens are examples of technology-push products. Once the technology is developed, we often see a proliferation of product versions built on the same platform. For instance, note the variety of watches with the same quartz crystal movement, or automobiles with the same engine, brake system, or chassis design. Research in technology-driven companies tends to focus on basic science, whereas companies in the consumer goods market concentrate on industrial design activities, such as product aesthetics (look, feel, and ease of use).

4.1.1.1 Core Competencies

One aspect of manufacturing strategy relates to defining the **core competencies** of the company. Core competencies are best described as a "mission" or goal that will provide an advantage in the marketplace. We may be tempted to view a specialized technology protected by patents or an effective marketing network of sales representatives as a core competency. It is best, however, to define core competencies along more general lines that will help direct strategic planning decisions to maintain this competency over time. One approach uses the measures of competitiveness, which are discussed in Chapter 1: price, features, service, delivery, and quality. Motorola became widely recognized in the 1980s for its commitment to quality in its Six Sigma program. Feeling the need to compete with the perception of high-quality Japanese electronics, the Six Sigma program strived to reduce the percent of defective products produced to no more than 3 per million. A company that claims rapid delivery as a core competency would need to plan an efficient distribution system. The production planning and control system would need timely data on inventory levels and demand rates, as well as the ability to change production plans on short notice to incorporate updates in the data.

In addition, concepts such as a highly skilled workforce can constitute the core competency. Often, core competencies can be identified in a company's advertisement. For example, a telecommunications company will guarantee a low rate or clear signal.

McDonald's can consider assured availability of a consistent quality product as a core competency. Defining the core competency as being the technology leader will promote continual research and development. On the other hand, calling a current state-of-the-art manufacturing process a core competency may lead the company to rely too heavily on that technology instead of continuing to develop new and improved approaches to accomplish the same customer-valued objective. Soon, competitors will imitate, and then surpass, that technology, leaving the company without a market. Kuglin [1998] emphasizes that a core competency must:

- Facilitate access to markets;
- Be perceived by customers as adding significant value to products; and
- Be difficult to imitate, thus providing a barrier to competitors.

It is important to distinguish between the enterprise's current competencies and the core competencies necessary for market success. Surveys of suppliers and customers can be useful in defining the important market competencies.

4.1.1.2 Customer Markets and Distribution

Another aspect of strategy relates to the specific markets to serve. Markets in this sense are defined by the geographical location and the expectations of the customer. The world has become a global marketplace. Later in this chapter, we discuss the construction of global supply-production-distribution networks to efficiently serve geographical markets. We need to be concerned with product differences required for different cultures and local regulations on importing and recycling products, as well as the logistical details of actually delivering the finished product to the customer. Such details can affect where products should be produced and which countries actually offer profitable opportunities when all factors are considered. Customer expectations determine design, delivery, and service requirements for competitiveness. Markets can be segmented, based on the level of product features expected. This is sometimes referred to as quality-of-product design. Both $40,000 luxury sedans and $10,000 entry-level vehicles have markets around the world, but the customers are different individuals with different requirements. The more expensive vehicle would be expected to have more power, a quieter ride, and impeccable reliability. In addition, combined geographical cultural factors must be addressed. The $40,000 luxury sedan in northern Europe may not need as clean an engine or as efficient an air conditioning system as the $10,000 vehicle destined for the arid Southwestern United States. Market surveys and interviews with focus groups of potential customers can be helpful in identifying the functional and aesthetic expectations of customers in each combination of a geographical and economic submarket.

4.1.1.3 Vertical Integration

Many companies resort to **vertical integration** to ensure a reliable supply of materials. Vertical integration refers to a company owning two or more of the production stages—from basic raw materials through finished goods delivery. When Texaco drills an oil well, ships the crude to its own refinery, and then transports the gasoline through its pipeline to a neighborhood gas station owned by the company, that is considered vertical integration. Likewise, when Coca-Cola manufactures the syrup, then bottles and sells it through a company-owned vending machine, that, too, is vertical integration. The company can control the quality, price, and availability of raw materials, and can also assume responsibility for maintaining a market for the finished product. The vertically integrated company is less susceptible to price wars or competition that may squeeze out the profits at one level of the product hierarchy. Vertical integration also relates to the company's supply chain management strategy, the manner in which the company partners with suppliers and customers, and manages the flow of items and information from basic materials to delivered products. This topic will be examined in more detail later in this chapter.

4.1.1.4 The Level of Flexibility

The ease and extent to which a facility can adapt to change is referred to as **flexibility**. Flexibility comes in many varieties. Both external (to the firm) and internal variability create the need for flexibility. External factors include changing levels of product demand or changes in the proportional mix of products demanded. This includes changes in features, quality, and prices expected by customers to stay competitive. Internal factors include machine breakdowns and normal engineering design changes to improve product design and manufacturability. These occurrences generate the need for alternative ways to manufacture the product. Although many types of flexibility have been discussed in the literature (see for instance, Browne et al. [1984] and Sethi and Sethi [1990]), no standardized definition or measures have yet been adopted by industry. We can, however, divide the types into strategic and operational categories. Strategic flexibility offers the potential to adjust to unpredictable and dynamically changing market opportunities. Operational flexibility allows parts to be re-routed in the face of minor demand changes, i.e., machine breakdowns or temporary scheduling bottlenecks. Strategic flexibility encompasses *volume flexibility, expansion flexibility,* and *product flexibility.* Volume flexibility measures the ability of the production system to operate profitably at multiple levels of total demand. A high-volume flexibility system is one with a low break-even point.[1] Expansion flexibility measures the possibility of adding capacity to meet more significant increases in the level of total demand. Product flexibility relates to the variety of product types that can be produced by the system. Can new products be made without major disruption to operations and finances? This includes the ability to satisfy customer demands for new versions of existing products and generations of new products. Operational flexibility includes *part-mix flexibility, routing flexibility, operation flexibility, machine flexibility,* and *material handling flexibility.* The ability of the system to produce products in different proportions without major tooling changeovers or reconfiguration of equipment, measures part-mix flexibility. Flexible machines that can be used to make a variety of similar parts or be overhauled to produce new product families, increase part-mix flexibility. The presence of duplicate machines (sometimes referred to as parallel processors) enhances routing flexibility that is measured by the number of routes each product can take through the production facility. This is related to operation flexibility that describes the number of different process plans and processing sequences that can be used to manufacture a part. Options for using a sequence of special purpose machines in lieu of one machining center, or purchasing product modules instead of producing internally, impact operation flexibility. Machine flexibility supports operation flexibility. Although operation and routing flexibility focus on the product, machine flexibility focuses on the process. If a machine can be adjusted to perform quickly more than one operation, it has machine flexibility. Material handling flexibility refers to the system's ability to transport varying

---

[1]Break-even point refers to the level of demand at which the company can cover all costs. We assume the company will be profitable if demand exceeds the break-even point.

item and unit loads, the frequencies with which loads can be moved, and the ability to vary pick-up and delivery points. As well as allowing product mix and routing changes, a flexible material handling system allows frequent modification of the facility layout to adjust to changing market demands while maintaining efficiency.

The firm can acquire flexibility through several strategies. Purchasing excess capacity permits more rapid adaptation to demand changes. Excess capacity can be acquired by employing more workers or installing more equipment. Other strategies may be more cost effective, however. Cross-training workers and purchasing machines with flexible operation capabilities allow the company to quickly adjust to changing product mixes and machine breakdowns. In general, people tend to provide the most flexibility, but the flexibility of intelligent, automated systems is constantly improving. For instance, free-roaming automated guided vehicles and bar-coded transport containers and parts allow for flexible material handling. Information can also be a strategic factor in flexibility. Accurate market forecasts allow re-engineering of products in an effective time frame and the chance to develop efficient schedules for resource utilization. Direct reports of customer purchases allow production plans to be adjusted to meet market conditions. Knowledge of the current status of equipment allows rapid replanning of order releases and process routes to avoid congestion.

4.1.2   Supporting Decisions

The company makes a variety of planning decisions. Each is an opportunity to support its manufacturing strategy. In this section, we describe several important decision types.

4.1.2.1   Make versus Buy Decision

Deciding which parts to make and which to buy (purchase from vendors) is often an important decision for the firm. This decision relates to the choice of core competencies and the customer's values. Most companies are unwilling to relinquish control of key parts to vendors. Commodities, parts that are available from many vendors at competitive prices, and normally with quick delivery, are seldom produced by the manufacturer of the final product. It is often cheaper to purchase these items, and there is ordinarily little risk in doing so. On the other hand, parts that directly relate to the core competencies of the company are usually produced internally.

A simple economic trade-off model can sometimes help clarify the make versus buy decision. Suppose a part can be purchased from a vendor at a cost of c_1 per unit including relevant ordering and receiving costs. If we decide to make the item internally, we must invest C in a new plant and equipment, but we can make the units at a variable production cost of c_2 per unit. The corresponding cost curves, as a function of total production quantity X, are shown in Figure 4.1. Algebraically, we see that we should produce the item if:

$$C + c_2 \cdot X \leq c_1 \cdot X \tag{4.1}$$

or, we produce if the expected production quantity satisfies[2]:

$$X > \frac{C}{c_1 - c_2}. \tag{4.2}$$

---

[2]We assume $c_1 > c_2$, otherwise there is no economic justification for internal production. Other strategic factors may still favor internal production however.

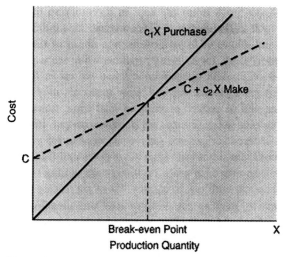

Figure 4.1 The Make versus Buy Tradeoff

EXAMPLE 4.1

A new bracket has been designed to correct a stability problem identified with a communication device when used in an environment subject to high vibration. The company has the machine capacity to produce the new bracket but estimates a present worth cost of $12,500 to obtain the tooling needed to produce the bracket. An additional worker would have to be hired to support internal production. The hiring and training cost is estimated at $1,000. Combining raw material and labor costs with estimated standard hours for production, the estimated internal variable production cost is $1.12 per unit. A vendor has quoted a price of $1.55 per unit. Forecasted demand is 10,000 units per year for the next two years. The company believes the vendor can meet the delivery and quality standard. Should the bracket be made internally or purchased?

SOLUTION

The breakeven point is found from Equation (4.2) as

$$X = \frac{12,500 + 1,000}{1.55 - 1.12} = 31,395.$$

Since lifetime demand is expected to be only 20,000 units, we should purchase the bracket.

Of course, this model is rather simplistic. A more complete model might include uncertainty in demand, price breaks if we buy in large quantities, supplier limitations, multiple technology and stair step capacity costs, nonlinear variable production costs, and risk-based decision criteria. We also need to worry about control issues. Can we really count on the vendor to deliver the requested number of units on time? Will product quality be dependable? Will the price remain stable? Could we decrease our production costs over time through process improvements? Can we obtain the necessary capital to buy the equipment needed to make this product? Will in-house production strain equipment and human resources? These and other questions should be addressed as part of the make versus buy decision.

4.1.2.2   Make-to-Stock versus Make-to-Order

Another basic decision concerns whether we should anticipate demand for a product and store it in inventory, or wait and produce it only after it has been ordered by a customer. Make-to-Stock items are kept in inventory so that they can be shipped to customers as soon as the customer order arrives. This quick delivery can be a competitive advantage and allows us to plan production ahead of time. On the other hand, we must incur inventory costs for carrying products while we wait for demand to occur, and we run the risk of product obsolescence when we err in forecasting demand. In general, high-volume products are stocked. Raw materials may also be stocked to save on ordering costs and to reduce production lead time. Some key spare parts may also be stockpiled to reduce repair times for plant equipment. Items not stockpiled are produced only when ordered (or ordered only when needed for planned production). This would not be practical for items with high setup costs and frequent small customer orders, but is often the best approach when customers are willing to wait for product delivery, and when orders occur with low frequency. These are termed Make-to-Order items. Although strategic factors, such as the delivery lead time demanded by customers, often dominate the justification for whether an item should be stocked, we can model the economic aspects of this decision.

Assumptions:

1. Unit variable costs and setup/ordering costs are constant and equal for the stocking and nonstocking cases.

2. The demand rate is constant and deterministic.

3. A backorder cost of π is incurred each time we need to order under no stocking. This includes goodwill loss due to slower delivery of the final product to the customer.

4. The cost of monitoring inventory levels under the stocking option costs C_s per time.

Let Q_{ns} be the average order quantity under the nonstocking option, D the annual demand rate for the item, and A the setup/ordering cost. If we do not stock the item, we will have $\dfrac{D}{Q_{ns}}$ orders per time period. The cost of ordering per period, including backorders, will be $\dfrac{(A + \pi) \cdot D}{Q_{ns}}$. If we do stock the item, we know from the EOQ model of Chapter 2 that the cost of setup/ordering, plus inventory holding cost, will be $\sqrt{2ADh}$ per time where h is the holding cost per unit, per time. Additionally, we have the inventory monitoring cost, C_s. Thus, our decision is to stock the item if:

$$\sqrt{2ADh} + C_s < \frac{(A + \pi) \cdot D}{Q_{ns}}.$$

Assuming the make-to-order cost exceeds C_s, we make-to-stock if:

$$h < \frac{\left(\dfrac{(A + \pi) \cdot D}{Q_{ns}} - C_s \right)^2}{2 \cdot A \cdot D} \tag{4.3}$$

Otherwise, we make-to-order.

EXAMPLE 4.2

A raw material normally used in batches of 30 units has annual demand of 150 units. The ordering cost is $145. The annualized cost of adding a part number to the automated inventory tracking system is estimated at $400. The cost per order of delayed product delivery is estimated at $100. The cost to hold a part for one year is $2.00. Should we stock the item?

SOLUTION

The condition of Equation (4.3) becomes

$$h < \frac{\left[\dfrac{(A + \pi) \cdot D}{Q_{ns}} - C_s\right]^2}{2 \cdot A \cdot D} = \frac{\left[\dfrac{(245)(150)}{30} - 400\right]^2}{2(145)(150)} = 15.65$$

Since the actual holding cost of \$2.00 is well under the threshold and the term in brackets is positive, we should stock the item. Using the EOQ, we place orders for:

$$Q = \sqrt{\frac{2AD}{h}} = \sqrt{\frac{2(145)(150)}{2}} = 147 \approx 150 \text{ units.}$$

4.1.2.3   Selection of Technologies and Equipment

A choice of processing technologies exists for most products. Spot-welding can be performed by manual labor or robots; parts can be moved by people pushing carts, drivers on forklifts, or automated guided vehicles. The level of capital investment versus manual labor used forms a strategic decision. Even within a level of capital investment, choices exist between intelligent, flexible technology, and special purpose processors. Do we purchase a computer-controlled machining center that can be programmed and tooled to produce many machining operations or design a machine to efficiently punch one particular sized hole in one particular material? Economics and the degree of certainty concerning future demand certainly play a key role here. But corporate image and recruiting are also important. The most talented employees may be attracted to companies that pursue new technological innovations. Customers may prefer the company with the high-tech image and showpiece factory.

Let's assume we have set the scale of the facility, and we now must select the technology to be used to produce the selected production volume. By technology, we may mean a particular machine, a process requiring several related machines that produce a feature or product, or an entire manufacturing facility. This will depend on the application being modeled. Let D_j be the selected production level of product j, $j = 1, ..., n$. If products require multiple operations, we will expand the list j to include each major operation of each product. Let i, $i = 1, ..., m$ be the technologies available for performing one or more of the operations required. Technologies may be specific types of machines or different strategies. Let C_i be the fixed cost per period, per unit of technology i that is acquired. Fixed cost refers to the amortized cost of purchasing and installing the machine, and renting the space it occupies. The variable cost of using a unit of technology i to produce product j, is v_{ij}. Variable costs include factors such as having a worker load and unload and, if necessary, operate the machine, and the cost of utilities, and other support functions while the machine is running. We will denote the productivity factors by a_{ij}, the number of units of product (operation) j that can be produced per unit of technology i. Our decision variables are:

Decision Variables:

   x_{ij} the number of units of technology i used to produce j; and

   y_i the integer number of units of technology i acquired.

A mathematical statement of the problem of selecting the optimal set of technologies for this plan is:

$$Minimize \sum_{i=1}^{m} C_i \cdot y_i + \sum_{i=1}^{m} \sum_{j=1}^{n} v_{ij} \cdot x_{ij} \qquad (4.4)$$

subject to:

$$\sum_{j=1}^{n} a_{ij} \cdot x_{ij} \geq D_j, \; \forall j \qquad (4.5)$$

$$\sum_{j=1}^{n} x_{ij} \leq y_i, \; \forall i \qquad (4.6)$$

$$y_i \; integer; \; x_{ij} \geq 0 \qquad (4.7)$$

The model will select the technologies to be acquired and how they will be used to produce the assorted products. The objective function (4.4) minimizes the sum of fixed acquisition plus variable processing costs. Constraints (4.5) ensure that the amount of technologies allocated to each product is sufficient to meet demand. Each $a_{ij} \cdot x_{ij}$ tells us the number of units of product j made using technology i. This is summed over all technologies to determine total production of j. For each technology type i, Equation (4.6) requires us to purchase the number of units of technology i that are used to produce all of the products. Finally, Equation (4.7) forces us to purchase an integral number of technology units and register the basic non-negativity restrictions on production.

The mathematical program (4.4) to (4.7) can be difficult to solve for problems with many products (operations) and technological choices. In most cases, the presence of integer variables, such as the y_i's, makes an optimization problem difficult. A variety of heuristic approaches exist for finding good solutions. For instance, we could ignore the integer restrictions on the y_i, solve the resulting linear problem, and then round up the y_i to the next largest integer. By relaxing the integrality restrictions, constraints (4.6) can be shown to satisfy $\sum_{j} x_{ij} = y_i$ for all i. This allows us to write the objective function in terms of just x_{ij} terms and the solution will be to set:

$$\begin{cases} x_{ij} = \dfrac{D_j}{a_{ij}}; & i = \arg \min_i \dfrac{(C_i + v_{ij}) \cdot D_j}{a_{ij}}. \\ x_{ij} = 0; & otherwise \end{cases} \qquad (4.8)$$

To explain (4.8), first note that if we use technology i for making all of product j, we will need $x_{ij} = \dfrac{D_j}{a_{ij}}$ units of i. The heuristic assumes we can buy only as much of technology i as necessary, including a fractional amount if desired. It then selects the cheapest technology for each product considering both variable and apportioned fixed costs. The reader should consult a general optimization text for suggestions on finding optimal solutions.

EXAMPLE 4.3

A facility is being constructed to produce two product families for sale in Eastern Europe. The choice of technologies is between a special purpose process for each family or purchasing some flexible capability that can produce either product family. Period demand is 100 units for family one and 200 units for family two. Additional data are as follows:

i	C_i	v_{i1}	v_{i2}	a_{i1}	a_{i2}
1 (Flexible)	$150	$100	$100	60	60
2 (Family 1)	$100	$100	infinite	75	0
3 (Family 2)	$100	infinite	$100	0	75

Select a good technological solution for manufacturing these product families.

SOLUTION

The two choices for product family 1 are the flexible process with cost:

$$\frac{(C_1 + v_{11})D_1}{a_{11}} = \frac{(150 + 100)(100)}{60} = \$416.7$$

or the dedicated process with cost:

$$\frac{(C_2 + v_{21})D_2}{a_{21}} = \frac{(200)(100)}{75} = \$266.7.$$

The dedicated process is better and we set:

$$x_{21} = \frac{D_1}{a_{21}} = \frac{100}{75} = 1.33.$$

The two choices for product family 2 are the flexible process with cost:

$$\frac{(C_1 + v_{12}) \cdot D_2}{a_{12}} = \$833.3$$

or the dedicated process with cost \$533.3. The dedicated process is better and we set $x_{32} = 2.7$. In actuality, we must buy integral amounts of capacity, however. Therefore, our real cost includes 2 units of the dedicated process for family 1 and 3 units of the dedicated process for family 2. The fixed cost is, therefore, $2(100) + 3(100) = 500$. The variable production cost for family 1 will be

$$\frac{v_{21} \cdot D_1}{a_{21}} = \frac{100(100)}{75} = \$133.3$$

and

$$\frac{v_{32} \cdot D_2}{a_{32}} = \frac{100(200)}{75} = \$266.7$$

for family 2. Thus, actual total cost will be

$$\$500 + \$133.3 + \$266.7 = \$900,$$

as we set $y_1 = 2$ and $y_2 = 3$.

We now have a good equipment acquisition plan, assuming the given data are correct. If, in reality, there is some uncertainty in the actual demands, it may be preferable to acquire some flexible capacity. Then, once we determine actual demand, we can allocate the flexible machines to the products as needed to meet demand. Another option we should consider is purchasing 1 unit of the dedicated technology for product family one, 2 units of dedicated capacity for product family two, and then purchasing flexible machines to manufacture the remaining demand for each family.

4.1.2.4   Justification of New Systems and Technologies

The competition does not stand still. If you want to be competitive, you must improve over time. Strategic planning and investment decisions must reflect this reality. Unfortunately, traditional discounted cash flow (DCF) techniques, such as net present value and rate of return analyses, are inappropriate for evaluating complex strategic investment decisions. The DCF approaches compare changes against a do-nothing alternative that assumes a static world. The traditional approaches overlook factors that cannot be quantified by dollars and have difficulty accounting for risk. It is very difficult to put a dollar value on improved customer appreciation, resulting from faster response to inquiries as we install a new order tracking system, or improved employee morale after being offered the opportunity to learn and use the newest process technology. Nevertheless, we should strive to include these intangibles into the justification of new information systems and processing technologies.

Kaplan [1986] notes that DCF approaches lead to nonoptimal facilities with obsolete process technology. Fortunately, alternative approaches exist. Uncertainty can be modeled by treating cash flows as random variables and estimating the distribution of net present value. The value of flexibility can then be evaluated by considering the distribution of possible customer demand, market price curves, discounting sales revenue, and operating expenses under each possible scenario. Intangibles such as the impact of a new system on customer appreciation and employee morale can be considered by defining qualitative benefits for each intangible factor under each alternative and computing a net present qualitative flow value. For instance, suppose we are using the interest rate "r" in our net present worth computations, and the planning horizon is T periods. Define Q_{tkj} as the qualitative flow value in period t for intangible attribute k, if investment alternative j is adopted. Intangible attributes include issues, such as the types of flexibility and compatibility with core competencies, and additional factors such as public relations, employee morale, and product manufacturability, discussed earlier in this chapter. Financial risk could also be treated as an intangible, if desired. Qualitative flow values can be thought of as intangible or surrogate revenue. Higher values imply a better score. The discounted net present qualitative flow (NPQF) value of investment alternative j with respect to attribute k is then:

$$NPQF_{kj} = \sum_{t=1}^{T} Q_{tkj} \cdot (1 + r)^{-t}. \tag{4.9}$$

Once the net present qualitative value is computed, multiobjective decision techniques can be used to compare alternatives based on monetary and intangible benefits. For comparison purposes, qualitative flows can be measured between $+1$ (ideal solution) and -1 (worst solution), and each attribute can be assigned a relative weight.

EXAMPLE 4.4

A company that prides itself on its quick delivery is considering which of two packaging systems to use—manual or automated. The automated system will cost $300,000 to purchase plus $50,000 per year to operate. The key attributes are cost, market position, and manufacturing flexibility. Once installed, the automated system is expected to provide a public relations benefit when trying to attract new customers because lower error rates and quicker delivery times can be promised. The manual system may be a liability when customer expectations change and our competitors adapt in the next several years. The proposed system will automatically identify parts and merge stored product into orders for shipping. Some training is required to perform the tasks manually. Workers have restricted capability and must be retrained as product lines change. The automated system is expected to be more flexible to product mix changes. Cash flows and qualitative rankings of the alternatives for the next three years are given in the following table. After some discussion, the company decided that each attribute can be considered of equal importance when measured between ± 1. A $0 cost system would be a $+1$, and a three-year present worth cost of $1 million would be the maximum possible investment and be rated as a -1.

Year	Attribute	Manual	Automated
1	Cost	$125,000	$350,000
2	Cost	$130,000	$50,000
3	Cost	$145,000	$50,000
1	Public Relations	0.0	$-.3$
2	Public Relations	-0.20	0.5
3	Public Relations	-0.5	0.4
1	Flexibility	0.3	0.4
2	Flexibility	0.2	0.5
3	Flexibility	0.2	0.5

The company uses a minimum attractive rate of return of 15% to evaluate investment opportunities. Which system should be selected based on total net present value? (Assume returns occur at the end of the year.)

SOLUTION

We begin by using Equation (4.9) to compute a net present value for each alternative on each attribute. For the cost attribute for the manual system we obtain:

$$NPV_{cost, \ manual} = (-125,000)(1.15)^{-1} + (-130,000)(1.15)^{-2} + (-145,000)(1.15)^{-3}) = -\$214,014.$$

Similarly,

$$NPV_{cost, \ automated} = (-350,000)(1.15)^{-1} + (-50,000)(1.15^{-2} + 1.15^{-3}) = -\$427,973.$$

For the other attributes,

$$NPQF_{public \ relations, \ manual} = 0 - .20(1.15)^{-2} - 0.5(1.15)^{-3} = -0.428.$$

Computing in a similar fashion yields,

$$NPQF_{public \ relations, \ automated} = 0.382, \ NPQF_{flexibility, \ manual} = 0.361,$$

and

$$NPQF_{flexibility, \ automated} = 0.750.$$

The net present values are next scaled to the $[-1, 1]$ interval. For public relations and flexibility, the maximum net present qualitative factor value would be:

$$|NPQF| \le 1(1.15^{-1} + 1.15^{-2} + 1.15^{-3}) = 1.587.$$

Normalizing the qualitative flows by dividing by 1.587, and then taking a sum of the three attributes we obtain:

$$NPQF^{normalized}_{public \ relations, \ manual} = -0.428/1.587 = -0.270,$$
$$NPQF^{normalized}_{public \ relations, \ automated} = 0.382/1.587 = 0.241,$$
$$NPQF^{normalized}_{flexibility, \ manual} = 0.361/1.587 = 0.227,$$

and

$$NPQF^{normalized}_{flexibility, \ automated} = 0.750/1.587 = 0.473.$$

To normalize the cash flows, we must divide by $1,000,000(1.587) = 1,587,000$. For the cost of the manual system, this yields

$$NPV^{normalized}_{cost, \ manual} = \frac{-214,014}{1,587,000} = -0.135.$$

Taking the weighted sum across attributes, we obtain:

$$NPV_{manual} = \frac{-0.270 + 0.227 + (-0.135)}{3} = -0.059;$$

and

$$NPV_{automated} = \frac{0.241 + 0.473 + (-0.270)}{3} = 0.148.$$

Although the automated system costs more, both initially and over the three-year horizon, when the intangibles are added in, the automated system has a positive present worth compared with the negative present worth of the manual system.

Other approaches exist for including qualitative factors and choosing between alternatives in multicriteria decision problems. The Analytical Hierarchy Process, or AHP,

(Saaty [1980], Zahedi [1986], and Golden et al. [1989]) has been widely studied and used for making such choices. In AHP, analysts judge the relative desirability of alternatives through pairwise comparisons. Key criteria are first divided into subcriteria to create a problem hierarchy. For instance, the criterion "Performance" might be subdivided into "accuracy," "repeatability," and "reliability." At each level pairwise comparisons are made to assess the relative importance of each subcriterion. Items on the same level should be of the same magnitude of importance, thus relative ratings are between 1 and 9. Finally, the decision alternatives are compared on each of the subcriteria. Weighting the relative evaluations of the decision alternatives by the relative weights found for the various subcriteria, a composite score is summed for each alternative. The AHP approach has the advantage of permitting consistency checks on pairwise comparisons to ensure rational comparisons and can automatically adjudicate between minor inconsistencies in comparisons. Sensitivity analysis can also be performed.

4.2 SUPPLY CHAIN MANAGEMENT CONCEPTS

From ore to door, materials have to be transported. Beginning with the extraction of basic elements and grains from the earth, materials, parts, and products must be packed, loaded, transported, unloaded, and unpacked at every stage of the journey as we convert raw materials into fabricated parts, then finished products, and, finally, delivered customer orders. The planning and execution of these activities encompass the function of *logistics*. The logistics function views the entire supply chain from raw materials to delivered products. Its task centers on coordinating the flow of materials and associated information.

Logistics activities typically account for 25% of the cost of manufacturing. Effective logistics requires recognition of the geographical location and functional capability of all potential material suppliers, production facilities, and customer markets, and the transportation options connecting those resources. The combination of material sources, production facilities, distribution locations, and transport options providing the most competitive option with respect to product features, cost, quality, and service to customers is selected. The set of resources selected may include partners external to the firm as well as internal departments. Given the time span and geographical distance involved in the logistics network, information management becomes critical. The information required to execute the logistics function must be shared across time, space, and partners. In this section we will examine how logistics has evolved in the modern global marketplace.

4.2.1 Global Logistics

The latter half of the twentieth century has been marked by an accelerating trend toward globalization. Disparate wage rates, shifting raw material sources, expanding overseas markets, government policies, and worldwide technical expertise have enticed firms to take advantage of international opportunities by creating global production and distribution networks. Opportunities exist for acquiring component parts and new technologies worldwide to improve production operations and to reduce transit costs. With markets expanding worldwide, it may no longer be desirable to manufacture and ship a product from a single factory. International markets have grown rapidly in recent years compared with the U.S. market. Consider, for instance, traditional U.S. companies, such as Coca Cola and Procter and Gamble. As of 1997, two-thirds of Coca-Cola's and half of Procter & Gamble's revenues came from overseas operations. Competition now requires that customized products be made in small quantities to satisfy local tastes, laws, and infrastructure, and supplied quickly to demand sources. Product designs, instruction sheets, and styling must match local culture. Try plugging your U.S.-made CD player into an European electric outlet or installing your VCR using instructions written in Japanese. Expanding environmental concerns in some countries now require recovery and reuse of products at the end of their life cycle. Whereas a few years ago, product designs switched from screws to glues to facilitate automated assembly, current designs must support product disassembly in some locales thus making the use of glue undesirable.

Today, it is common to see a company design, produce, and distribute products through a global network to provide the best customer service at the lowest price. The opportunities provided by such global operations often exceed the difficulties of dealing with diverse cultures, languages, governmental regulations, measurement/standards systems, and workforces. Inexpensive, instantaneous information exchange and improved transportation networks for physical goods facilitate globalization. It is not unusual for a product today to be designed, manufactured, and distributed on multiple continents. A new powered consumer item may have its power unit (engine) and controller designed in Europe, its body and circuitry designed in the United States, and its software designed in India. Raw materials may be shipped from Africa and South America to parts-fabrication plants in the United States and Europe. Parts are then shipped to Korea and Mexico for assembly, with finished products being marketed in Europe, Japan, and the United States. Material flow in this global logistics network is shown in Figure 4.2. Coordination throughout the entire logistical system must be planned and managed. In addition to the technical problems, at each stage, human communication problems must be overcome, customs duties paid, products packaged, shipped, and received, and environmental regulations met.

Multinational joint ventures have become commonplace with products often reflecting this collaboration. Both companies benefit by sharing technology and markets. North American and Asian companies now have joint ventures in the automotive, steel, and semiconductor industries. The mixing of parts in finished products has also become com-

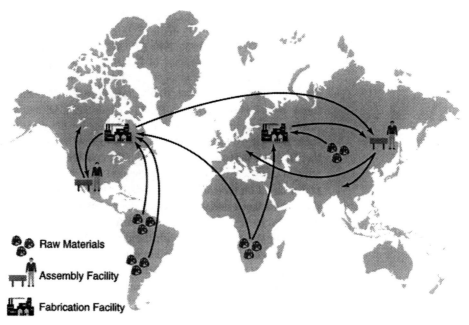

Figure 4.2 Global Material and Product Flows Example

monplace. It is difficult to identify a Toyota automobile assembled in the United States from Japanese, Mexican, European, and U.S. parts as a "Japanese" automobile. Although the general motivation of multinational firms would be to buy each part from the best supplier, many countries have established local *content requirements.* These requirements legislate that a minimum percentage of the value of products sold must be from local manufacturing operations. Otherwise, high import duties are imposed. Rules vary from country to country with respect to minimum percentages. The Philippines, for instance, requires 40% of value to be local, whereas Australia requires 85%. Components that would otherwise be purchased in the global market or imported from their own factories in other countries, must be purchased locally instead. In some cases, local suppliers are not even available. This forces the company to either produce components internally or invest in developing local suppliers. Later in this chapter, we will develop a model for assisting in dealing with such requirements.

Exchange rates among currencies of different countries play an important role in global logistics. Suppose a U.S. company selects a supplier in Brazil for a basic raw material such as bauxite or latex. Initially, the exchange rate has one U.S. dollar equal to 1.20 Brazilian reals. Suppose the dollar then rises against the real such that one dollar buys 1.30 reals. The Brazilian supplier quotes prices in reals. Let's assume the company purchases 1,000 reals worth of material per day. Originally it cost the firm

$$\frac{1,000 reals}{1.2\ reals/\$} = \$833.33$$

to purchase the material. However, now it only costs

$$\frac{1,000 reals}{1.3\ reals/\$} = \$769.23$$

per day. A similar phenomenon exists in trying to sell a product internationally. The product may need to be priced in the customer's local currency. Thus, the amount of dollar received by the U.S. parent company varies with the exchange rate. In general, exchange rates increase the variability of profit when doing business internationally. Company profits are subject to seemingly random shifts in exchange rates in addition to the normal variances in demand and production efficiencies.

There are positive aspects of the effect of exchange rates. In a multinational supply chain, companies can use changes in exchange rates to their advantage and incorporate this effect into their strategic plan. To illustrate, consider the manufacturer of surgical latex gloves. Latex can be purchased from Brazil or Indonesia. Gloves can be made either in the U.S. plant or in a facility in South Korea. Markets for the gloves exist in both the U.S. and Europe. Because exchange rates vary, the company will choose to purchase raw latex from either Brazil or Indonesia, depending on the country with the weakest (cheapest) currency relative to the dollar at that time. Production will shift between the United States and Korean plants based on their relative exchange rates. If the Korean won drops in value versus the dollar such that a dollar buys more wons, then labor and other expenses become cheaper in Korea and more production will be shifted to that plant. The opposite occurs if the won increases in value compared with the dollar. Marketing is also affected. As the U.S. dollar strengthens against the Eurodollar, each Eurodollar is worth less and the company will increase marketing efforts in the United States and reduce efforts in Europe. The strategy will be reversed as the dollar drops against the European currency.[3]

---

[3]These shifts in purchasing, production, and marketing will likely occur as exchange rates vary. The actual magnitude of the shifts will depend on other factors as well, such as transportation costs, production capacities at the plants, and the maximum size of each market region.

4.2.2   Logistics Information Systems

With the increase in geographical and vertical breadth of logistics systems and the trend toward outsourcing of component production and even final assembly and distribution, coordination between diverse facilities becomes crucial for efficient operation. Thus, the logistics information system becomes critical for keeping all players on the same page. The system must acquire, store, and transmit data, and compile it into useful information. Demands, deliveries, and stock levels must be coordinated across the vertical supply chain (from raw material suppliers to distribution warehouses) and geographical supply chain (international network of facilities, suppliers, and customers). Fortunately, modern communications makes this possible. Electronic data interchange (EDI) allows customer-order entry to occur instantaneously. This information can then be similarly distributed to all affected locations. Global positioning systems (GPS) can now be used to track all units, even those currently in transit, to update delivery schedules. In the past, a company would have to wait until parts were actually received and the corresponding data input manually into an inventory system before realizing that parts were available for use. Expected shipping dates or requested delivery dates were seldom accurate enough for scheduling production, and variations in delivered quantity and quality made the problem worse. Today, exact quantities can be tracked through automatic identification such as bar coding and remote radio frequency (RF) scanning, allowing these items to be tracked while en route. By combining this with a statistical process control vendor certification process, product quantity and quality can be dependably predicted, and operations can be scheduled using anticipated material receipts.

The expense of implementing the communications technology constitutes one down side to this process. Thus, the cost must be traded off with the gain in agility and customer responsiveness. Part of the expense comes from the need to design the information system and communications protocols for EDI. Not only the details of electronic data transmittal protocols, but also the definitions of data must be agreed to by all participants. For instance, what is a unit of product: a pallet of 144 boxes or a box of twelve items? What happens if a pallet is chosen, but some pallets are only half full? How do we know which specific items are included if many versions of a part are purchased? Do we track the green trim, size 10, March 01 shoe style separately from the black trim, size 10, December 00 version? We can add this detail if the supplier has a separate part number for the green and black items, but suppose we also need to know the production date to identify a part? This level of detail may not have been necessary when the system was first designed, and, therefore, not made part of the data exchange format. Part and order numbers will differ among vendors and customers and must be matched somehow. Computer hardware and software may be incompatible among companies or even among departments in the same organization. In addition, some companies may object to certain pieces of data being transmitted because of proprietary and security concerns. These issues must be addressed when designing the information system and communications protocols.

After investing in developing a logistics information system that connects multiple companies, the cost to change the partnering arrangements tends to lead to long-term relationships. Vertical company relationships for materials, part fabrication, product assembly, and distribution are fixed by the investment in the joint relationship. Each company in the chain has a stake in seeing its partners prosper and, therefore, negotiations may become more mutually supportive than confrontational. However, if one company controls the information system and market power, that company has the ability to manipulate the negotiations in its own self-interest and squeeze the partners. Finally, we note that the information system has strategic long-term uses. A wealth of detailed data is available. The data can be mined to evaluate performance, aid in solving problems, and identify opportunities. Data can be aggregated into long-range reports on performance and costs for upper management use in selecting long-term strategies and alliances. The availability of the Internet and related information sources also allows rapid benchmarking and exploration of other partnering relationships so that management can develop the optimal material and information network.

4.2.2.1 Variance Acceleration: The Bullwhip Effect

We can demonstrate the importance of coordination across the supply chain by tracking the effect of information myopia on production schedules. Without proper information exchange and planning coordination, variation at the final customer end of the supply chain can accelerate upstream in a chaotic manner. This is sometimes referred to as the *bullwhip effect*. Small changes in demand at the retail level cause major variations in orders for raw materials and component production.

Consider, first, the simple two-stage producer-retailer supply chain of Figure 4.3. (Of course, in practice there would probably be many retailers, but we aggregate these into a single entity for our model.) Suppose it takes the plant one week to produce an order and one week to ship it to the retailer, on average. Thus, orders placed at the start of week one will usually arrive at the start of week three. Initially, demand averages 100 units per week. The retailer keeps an extra week's demand in stock to guard against demand variation, possible quality problems, and lead-time variation. The plant does not see demand directly, it simply reacts to the orders placed by the retailer each week. Let's see what happens if demand jumps to 110 units in week three.

At the start of each week, units ordered two weeks ago are received, and the retailer places a new order. The desired starting inventory is two weeks' supply (expected weekly demand plus safety stock). After placing the order at the start of this week, we will not be able to impact on-hand inventory for three more weeks: 1 week until we order again + 2 weeks' lead time. Adding current on-hand inventory, plus units on order (those ordered last week that will be delivered next week) and the new order being placed, this amount must cover demand for this week, plus the next two weeks, plus a planned one-week safety stock. Thus, we place an order that includes units on hand, plus in-transit, plus on-order: equaling four weeks' supply. We can formalize this notion. Let $\overline{D}$ be the average demand rate per week, τ be the lead-time to receive orders, and let SS represent the desired safety stock. The desired on-hand inventory as we start a week is $I^* = \overline{D} + SS$ to handle expected demand for the week with a safety factor. To determine the order quantity, we use an "order-up-to R" policy[4] where we order up to the desired starting inventory, plus lead-time demand:

$$R = \overline{D} \cdot (\tau + 1) + SS = \overline{D} \cdot \tau + I^* \qquad (4.10)$$

The actual quantity ordered at the start of week t is:

$$Q_t = R - I_{t-1} - Q_{t-1} \qquad (4.11)$$

where I_{t-1} is the inventory carried over from period $t - 1$ including the new order arriving at time t. The order quantity equals the order-up-to level, minus the inventory on-hand and on-order.

The plant then produces exactly to order. Orders and inventory levels are shown in Table 4.1. The target "order-up-to" level is four weeks' demand. Orders in week three are

---

[4]We could denote this as an (R, T) policy with $T = 1$ week. Every T time units we order enough to bring the inventory position to the level R.

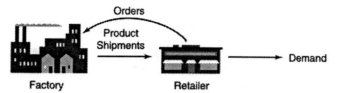

Figure 4.3 Basic Two Level Supply Chain

Table 4.1 Variance Propagation in a Two Level System

Week	Starting retail inventory	On order	Order-up-to target	Order quantity	Demand
1	200	100	400	100	100
2	200	100	400	100	100
3	200	100	400	100	110
4	190	100	440	150	100
5	190	150	400	60	100
6	240	60	400	100	100
7	200	100	400	100	100
8	200	100	400	100	100

placed before demand is realized and stay at 100 units. However, the next week we find that demand has jumped to 110 units, and inventory has dropped to 190. At this point, fear takes over. Instead of using the safety stock as the buffer it was intended to be, the typical reaction is to try to maintain the four-week "order-up-to" rule with weekly demand now set to 110. With 190 units on-hand (after receiving the order placed in period one) and 100 units on order, we conclude that an order for $4(110) - 190 - 100 = 150$ units is needed. A demand increase of 10% suddenly translated into a 50% increase in orders at the plant. The plant shifts into high gear this period to increase production. In actuality, we only needed to produce 10 extra units. This reality manifests itself in the next period when orders drop to 60 units as we correct for the excessive ordering of week four. Eventually, we settle down to our original pattern, but only after stressing the plant, probably with expensive overtime, hiring, or subcontracting, followed by undertime or layoffs.

You probably feel it was foolish to have overcompensated, and indeed, with exponential smoothing of demand forecasts, it is questionable that the target order-up-to point would actually go to 440 in week four. However, the forecast would have been raised based on the 110-unit demand, and even if the forecasted demand increased to only 105, we would still have a target of 420 units and, therefore, have ordered $420 - 290 = 130$ units. Thus, the variance acceleration effect exists; we are only quibbling about the magnitude. You may also think that over time, the plant would recognize this phenomenon and cease to over-react. That might be the case if demand was so well managed that these effects could be clearly witnessed. However, in practice, demand is constantly changing, and all the plant sees is variability in arriving orders. It is difficult for plant management to associate variability in orders with specific minor perturbations in customer demand, particularly when customer demand is not observed by plant personnel. As we see below, the problem worsens when the supply chain gets longer.

EXAMPLE 4.5

A supply chain consists of a raw material supplier, production facility, and retail outlet. The retailer places an order at the start of each week and receives the order three weeks later. The retailer attempts to keep its starting inventory level equal to 1.5 weeks of demand representing expected

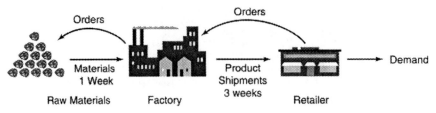

Figure 4.4 Three Stage Supply Chain

weekly sales plus safety stock of one-half-weeks average demand. Actual demand for the week is not known until after the order is placed. The retailer uses a two-week moving average of demand to estimate expected weekly sales. When the production factory receives the order, it immediately places an order for raw materials and begins producing the number of units ordered by the retailer. The raw materials arrive one week later. The factory also tries to keep 1.5 weeks of materials on-hand at the start of each period. Find the impact on the retailer if customer demand suddenly shifts from 100 to 110 units per week. Next, determine how this change in demand affects the factory. The system is sketched in Figure 4.4.

SOLUTION

The retailer sees a three-week lead time. Thus, following the model used in Eq. 4.10, its order-up-to rule is:

$$R_t^R = E(\textit{Lead Time Demand}) + \textit{Desired Starting Inventory} = 3\overline{D}_t^R + 1.5\overline{D}_t^R = 4.5\overline{D}_t^R.$$

The order size placed at the start of period t by the retailer will be

$$Q_t^R = 4.5\overline{D}_t^R - I_{t-1}^R - Q_{t-1}^R - Q_{t-2}^R,$$

which corrects the order-up-to level for actual starting inventory and units on order. Average demand will be given by the moving average.

$$\overline{D}_t^R = \frac{D_{t-1} + D_{t-2}}{2}.$$

Assuming demand increases to 110 units per week starting in week 2, Table 4.2 shows the progression of order and inventory quantities for the retailer. The table reflects status at the start of the week except for the final column that shows demand that is to occur during that week. The "on-order" column indicates the units ordered in the past two weeks. These have not yet arrived. As expected, the retailer over-reacts in week 3. However, order quantities stabilize in week 5 at 110 units, and the starting inventory finds its new base level at 165 units (1.5 times average weekly demand) in week 7. The order-up-to quantity stabilizes at 4.5 times the new average demand.

Table 4.2   Weekly Status Data for the Retailer

Period	Initial retail inventory	Quantity on order	Average demand	Order-up-to level	Retailer order quantity	Demand
1	150	200	100	450	100	100
2	150	200	100	450	100	110
3	140	200	105	472.5	132.5	110
4	130	232.5	110	495	132.5	110
5	120	265	110	495	110	110
6	142.5	242.5	110	495	110	110
7	165	220	110	495	110	110
8	165	220	110	495	110	110

Table 4.3   Weekly Status Data for the Factory

Period	Demand	Initial materials inventory	Average demand	Order-up-to level	Raw material order
1	100	150	100	250	100
2	100	150	100	250	100
3	132.5	117.5	116.3	290.6	173.1
4	132.5	158.1	132.5	331.2	173.1
5	110	221.3	116.3	290.6	69.3
6	110	180.6	110	275	94.4
7	110	165	110	275	110
8	110	165	110	275	110

Table 4.3 shows that because of the bullwhip effect, the impact is larger on the factory. The "demand" for the factory in Table 4.3 follows directly from the orders placed by the retailer in Table 4.2. Retail orders are received at the start of each week, and the factory starts that number of units into production. At the same time, the factory orders raw materials for replacement, using the rule

$$R_t^F = 2.5\overline{D}_t^F.$$

When week 3 demand hits 132.5 units, the factory is forced to delve into its safety stock to start production. This leaves 117.5 units in materials inventory. At the same time, average demand increases to

$$\overline{D}_3^F = \frac{D_2^F + D_3^F}{2} = \frac{100 + 132.5}{2} = 116.25$$

causing the order-up-to level to increase to

$$R_3^F = 2.5(116.25) = 290.6.$$

Therefore, factory orders for raw materials jump to

$$290.6 - 117.5 = 173.1$$

units in week 3. These 173.1 units arrive at the start of week 4, but meanwhile, a second straight order for 132.5 units from the retailer has raised the average demand estimate again and the order-up-to level increases to 331.2 units. Another order for 173.1 raw materials is placed, and it is not until demand drops off the next period that we realize we may have overproduced. With mean demand back to 116.1 units in week 5, and a starting inventory of 221.3 units raw material, orders are cut back significantly. Finally, in week 7 we return to a new steady state.

The bullwhip effect continues to amplify as the number of supply stages grows. A more complete multistage supply chain operating without coordination between stages is shown in Figure 4.5. Customers arrive at the retail outlets (simplified to be a single outlet in the Figure) and buy from the shelves. At the start of the week, the retailer places an order to replenish stock from the warehouse. It takes one week for the order to arrive. Orders are placed just before receiving the previous week's order. As before, the order quantity is based on the rule of attempting to end each week with an inventory of one week's supply. The one-week supply is planned as a safety stock in case demand is higher than expected or orders are late in arriving. The warehouse uses a similar rule to order replenishment from the factory each week. The delay is one week for receiving factory shipments. The factory keeps an inventory of finished products to meet orders from the ware-

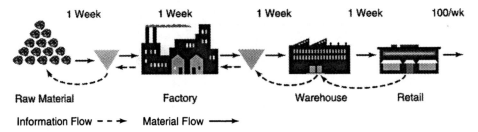

Figure 4.5 Hierarchical Supply System Without Integration

house and raw materials to support production. Production takes one week after materials are released to the shop floor. Products are added to inventory at the end of the week in which they are released. It takes one week to obtain raw materials. As before, assume demand averages 100 units per week. The target safety stock inventory level is 100 units at each stage except 200 at the factory output point. This accounting detail simply reflects the fact that factory production is added to inventory before the end of the week, whereas shipments are all received at the start of the following week. Solid arrows show material movement, and dotted arrows show order information. Let I_t^a represent the end of period t inventory at location a. Locations are denoted "R" for retailer, "W" for the warehouse, "FI" for the factory input, and "FO" for finished goods at the factory. Likewise, let Q_t^a represent orders placed by the facility a at the start of period t. Demand will be noted as D_t and factory production as X_t. Average values are indicated by a "bar" over the value and represent an average of historical data. We can quantify the status of each stage in the supply chain over time by the relations:

$$I_t^R = I_{t-1}^R + Q_t^R - D_t \qquad (4.12)$$

$$Q_t^R = D_{t-1} + [\overline{D}_{t-1} - I_{t-1}^R] + [\overline{D}_{t-1} - Q_{t-1}^R] \qquad (4.13)$$
$$= D_{t-1} + 2\overline{D}_{t-1} - [I_{t-1}^R + Q_{t-1}^R]$$

$$I_t^W = I_{t-1}^W + Q_{t-1}^W - Q_t^R \qquad (4.14)$$

$$Q_t^W = Q_{t-1}^R + 2 \cdot Q_{t-1}^R - [I_{t-1}^W + Q_{t-1}^W] \qquad (4.15)$$

$$I_t^{FO} = I_{t-1}^{FO} + X_t - Q_t^W \qquad (4.16)$$

$$X_t = Q_{t-1}^W + 2 \cdot Q_{t-1}^W - I_{t-1}^{FO} \qquad (4.17)$$

$$I_t^{FI} = I_{t-1}^{FI} + Q_{t-1}^{FI} - X_t \qquad (4.18)$$

$$Q_t^{FI} = X_{t-1} + 2 \cdot \overline{X}_{t-1} - [I_{t-1}^{FI} + Q_{t-1}^{FI}] \qquad (4.19)$$

The inventory equations (4.12), (4.14), (4.16), and (4.18) have the general form that inventory available at the end of period t equals inventory leftover from period $t - 1$, plus receipts in period t minus shipments in period t. Consider Equation (4.12). The retailer's inventory at the end of any period t (I_t^R) equals the retailers starting inventory left over from the previous period (I_{t-1}^R), plus the order that was placed last week and received this week (Q_{t-1}^R), minus this week's customer demand (D_t). These inventory balance equations will play a major role in subsequent chapters. The remaining ordering and production equations implement our replacement and inventory target goals. Storage locations place orders to replace units shipped out last week, plus an adjustment to bring the level of available inventory (inventory that is on-hand or on-order) equal to the average demand for two weeks. The two weeks represent our time exposure until we can next affect inventory levels. If we place an order at the start of week t, then this order must meet requirements for the next two periods because we cannot order again until period $t + 1$, and then we must wait another week for receipt of goods. Equation (4.13) records the weekly order quantity expression for the retailer. First, we estimate demand for this coming week by using the goods sold last week (D_{t-1}). Next, we try to correct for any deviation between desired and actual safety stock. Desired safety stock is one week's supply on-hand, plus one week in transit. Thus, we add a correction for the difference between average demand ($\overline{D}_{t-1}$) and on-hand inventory (I^R_{t-1}), plus a similar correction for the difference between average demand and quantity on order (Q^R_{t-1}).

Now, suppose in week six, demand exhibits a one-time increase from 100 to 110 units per week. Without coordinated planning and data exchange, the dynamic activity of the system is surprising. Inventory levels are shown in Figure 4.6. In week 6 demand goes up 10 units, causing retail inventory to drop 10 units. Warehouse inventory then drops a little more in week 7, followed by an even larger drop in factory product inventory in week 8 and a large drop in raw material inventory in week 9. Each stage jumps into action and over-reacts the following week. Eventually the system settles down. Figure 4.7 shows the corresponding level of orders and production each week. Effects are minor at the upper level of the supply chain (the retailer), but the effects of the change are multiplied as they filter down the chain. In week 7 the retailer increases the order size to the warehouse to 125 units. In addition to the new perceived level of 110 units per week, extra units are ordered to make up for the depleted inventory caused by the 10 additional units sold last week and to increase safety stock for the new perceived level of average demand. In our model, we use the average demand over the last four weeks as a measure of $\overline{D}_t$, thus

$$\overline{D}_t = (D_t + D_{t-1} + D_{t-2} + D_{t-3})/4.$$

A phenomenon similar in form, but with larger amplitude then happens in week 8 with the warehouse ordering 163 units from the factory. The multiplier is attributable to the warehouse now seeing a jump in orders of 25 units although demand actually increased by only 10 units per week. The factory, seeing a large jump in orders and worrying about its insufficient supply of goods, then produces 200 units in week 9, consuming all raw

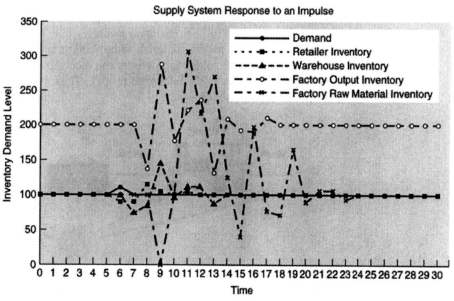

Figure 4.6 Inventory Levels without Coordination

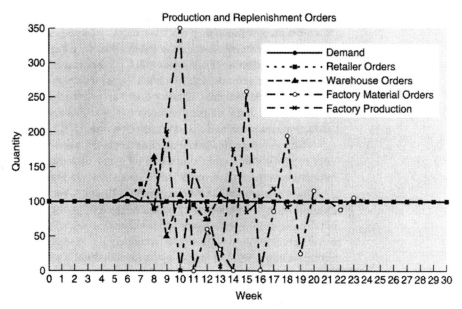

Figure 4.7 Production Levels and Orders without Coordination

materials. Imagine the disruption at the factory when it is suddenly asked to double production in one week. Of course, we do not really need 200 units every week, and the warehouse orders fewer units in week 9, causing factory production to drop sharply in week 10. The system does not stabilize until week 13. In the meantime, production and inventory levels vary widely. Examine the graphs in Figures 4.6 and 4.7. Note that as we move from the retailer, to the warehouse, and to the factory, the curves become affected one week later, and with a greater amplitude. The 10% increase in demand almost caused shortages at the factory despite having a full week's supply of safety stock at every level of the supply chain!

Now consider what happens with data exchange in an integrated, cooperative supply chain. Complete information is available to all production stages as shown in Figure 4.8. Our strategy will be to maintain inventory only at the retail level. Orders placed at the retailer will be for raw materials to be shipped to the plant. The next week these will fix production at the plant, followed by shipments to the warehouse the following week, and then shipments to the retailer to restock. Thus, customer demand in week 1 affects

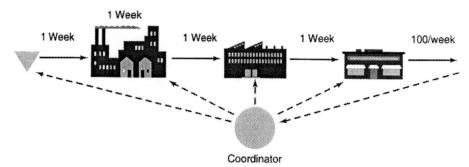

Figure 4.8 Supply Chain with Information Exchange and Coordination

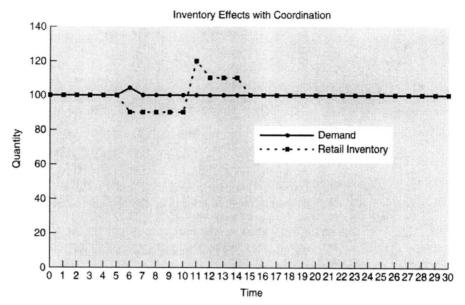

Figure 4.9 Inventory Levels with Information Exchange and Coordination

raw material orders for week 2, production in week 3, product arrivals at the warehouse at the end of week 4 (start of week 5), and product arrivals at the retailer at the start of week 6. The response to the same 110-unit demand in week 6 is shown for inventory levels in Figure 4.9 and orders/production in Figure 4.10. Note that the system still over-reacts at first because of the perception of the need for more inventory, but the overshoot is minor and dampens quickly. The ordering decisions for the coordinated system follow

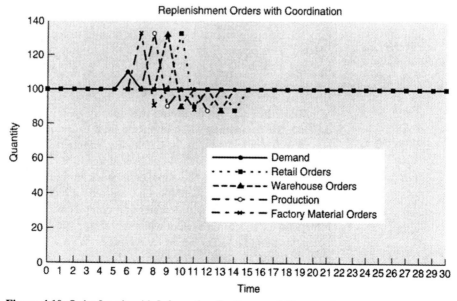

Figure 4.10 Order Levels with Information Exchange and Coordination

the simple rules:

$$Q_t^R = Q_{t-1}^W \tag{4.20}$$

$$Q_t^W = X_{t-1} \tag{4.21}$$

$$X_t = Q_{t-1}^F \tag{4.22}$$

$$Q_t^F = D_{t-1} + 5*\overline{D}_{t-1} - [I_{t-1}^R + Q_{t-1}^R + Q_t^R + Q_t^W + X_t] \tag{4.23}$$

Equation (4.23) orders raw materials to replace most recent demand plus an adjustment to compare expected usage over the five-week lead time to the actual pipeline inventory. Comparing Figures 4.6 and 4.9, the reduction in inventory cost is obvious for the coordinated strategy. Comparing Figures 4.7 and 4.10 shows the reduction in production volatility obtained through coordination. The operational and cost differences between the two systems, differing solely by the decision to coordinate, are extreme and demonstrate the current widespread interest in the area of supply chain coordination.

We have seen in this section how well-intentioned decisions can be misguided and lead to suboptimal, almost chaotic activity in the presence of uncertainty. Forecasting rules and safety stock policies tend to increase demand variability as we pass up through the supply chain. Other factors can cause similar reactions. For economic reasons, order quantities naturally tend to increase in size and decrease in frequency as we move from final products to raw materials. It almost never makes sense to order raw materials more frequently than they will be put into production. However, it may seem economical to save ordering costs by ordering enough materials to last through several production runs. If materials are cheap, we may reduce the risk of running out by ordering large quantities very infrequently, although the materials are used in smaller batches on a frequent basis. Thus, as we move from the final customer level toward raw materials, order quantities will naturally either stay the same or increase at each stage, and order frequencies will decrease. This process is discussed at length in chapter 9.

Occasional shortages of supplies present another source of variance acceleration. Material shortages, wherein customer orders temporarily exceed supply, lead to rationing. Suppliers ship each customer only a portion of the items ordered. Customers, deprived of materials, quickly realize they are receiving only partial shipments. Therefore, the company orders extra units expecting to receive only a fraction of the amount ordered. This practice, again used as a tool to minimize risk caused by uncertainty (in material availability), causes orders to increase at each stage as we move upstream toward the raw material. Once again, this activity accentuates itself. If all customers desiring a commodity increase their orders in expectation of a shortage, then the percent of orders filled will drop. Seeing this drop in materials received compared with those ordered, the companies will exaggerate their order quantities even more the next time, and the process feeds on itself. Eventually, suppliers increase capacity and begin filling orders completely. This leads to an oversupply of the item. Customers then cut back extensively on orders, and the supplier is stuck with excess capacity and no demand. A classic example of this phenomenon occurred with computer hard disk drives. In 1997, production capacity lagged demand and companies, such as Western Digital, put customers on an allocation system wherein orders were only partially filled. Consequently, computer manufacturers began to overorder. With a large backlog of orders, the company invested in additional capacity. However, as soon as shipments could match or even exceed the real demand rate, orders were cancelled. An inventory glut resulted, and prices declined. Burdened with investment costs and falling incomes, the company's stock price plummeted.

4.2.3 Product Design and Customization

Product design can complement logistics strategy. Competition in delivery speed, local content laws, and cultural differences encourages distributing final assembly among market areas. This becomes possible when product design allows for simple final assembly using parts supplied locally. A generic product frame is designed that forms the basis for an entire family of products. Minor changes are made when finishing the product to adapt it to a specific customer or regional market. This may be as simple as supplying a different electrical connector or inserting product information in different languages. However, product differentiation may also be more substantial, including different options and power supplies using the same product backbone. Product assembly involves combining the correct set of modules for the desired customer. The advantage comes from producing and storing a single generic base model that can be quickly customized to fit customer specifications. Thus, if the demand for various options changes or if demand shifts from one regional market to another, the producer can rapidly adjust with minimal cost. Final assembly may even be geographically integrated with distribution warehouses. Orders are received at the warehouses, and products are assembled at that site using the stored modules. This allows rapid delivery with minimal inventory. Billington [1994] describes an application of this strategy at the Hewlett-Packard Company. Deskjet Plus printers are maintained in a generic configuration at regional distribution centers. On order receipt, the center "localizes" the printer by adding the appropriate power supply. Lee and Tang [1997] further describe the general benefits of using standardized parts, modular product designs, and process plans that delay product differentiation.

4.2.4 Vendor Selection, Certification, and Contracting

A central component of supply-chain management involves managing relationships with suppliers. One approach, vertical integration, was discussed earlier in this chapter. However, many companies have found that it is better to concentrate on focused core competencies for internal production and to outsource materials and logistics. Parts vendors are available for selection from around the world. The chosen vendor(s) must be able to meet customer needs across the global manufacturing system. The choice of vendors must be based on a complete picture of the relationship. Table 4.4 lists the key factors to be considered. It lists cost as only one factor and often not the most important. The availability of rapid delivery of the desired quantity of items is essential, as is conformance to specifications. Factors such as the unit-load form or the package in which the product arrives can determine its compatibility with the receiving and material control systems.

The key is to develop a cost-effective system that guarantees delivery of materials as needed. Sole-sourcing offers one option. Instead of dealing with a large number of suppliers for a product and seeking the low-cost bidder each time a part is to be purchased, long-term contracts are negotiated with a single (or few) supplier(s) for each part. This symbiotic relationship allows the supplier and customer to communicate in greater detail to ensure that the material being supplied meets actual specifications. Employees often visit each other's plants, and it is not uncommon to see large manufacturers station employees at the supplier's plants to certify processes and help troubleshoot problems. The supplier, knowing that a long-term contract exists, is motivated to work closely with the customer. With sole-sourcing and long-term relationships, it becomes important to see that both companies survive. Thus, price negotiations tend to be designed to ensure that both companies make a fair profit. Contract negotiations cover a variety of factors. Essentially, the supplier and customer decide how to split profit and risk. The contract may specify

Table 4.4 Criteria for Vendor Selection

Category	Measures
Product Cost	Purchase
	Price Breaks
	Delivery Charges
	Tariffs and Import Duties
	Local Content Requirements/Penalties
Delivery	Delivery Frequency
	Minimum Batch Size
	Maximum Available Rate
	Package Form
	Communication Speed, Ease, and Accuracy
Quality	Completeness of Features
	Quality Level (Defect Rate)
	Reliability
Service	Warranty
	Liability Coverage
	Repair
Organization	Financial Stability
	Eagerness for Business and Leverage
	Labor Relations

price as a function of the quantity purchased or the price may be based on the customer's agreement to purchase at least a minimum quantity of units each period or over some specified time horizon. A return policy for units the customer does not consume (or sell) will typically be included in the contract. The specification of this policy indicates who will absorb the risk of demand variability. In the case that the parts are being purchased by a retailer for sale, the supplier may place limits on the selling price to protect other retailers and its own sales or demand for substitute items. Finally, the contract may stipulate an allocation rule detailing how the supplier's stock of items will be distributed among its retailers when shortages exist. Higher-paying retailers may receive preference. As discussed earlier in the chapter, proportional allocation rules lead to inflated orders being placed in times of shortage, which in turn can wreak havoc with production schedules and raw material orders.

Whereas long-term contracts have advantages for both partners (namely, secure supply and revenue) the growth of product data interchange, e-commerce, and automated quotation systems supports short-term, even one-time supply relationships. Only the future can tell which approach will dominate, but it is clear that suppliers will be required to provide continuously improving service in the areas of price, speed of delivery, flexibility of terms, and quality of goods.

4.2.5 Operational Decisions in Distribution Systems

Distribution of finished goods can account for a significant proportion of logistics costs. After production, parts are often sent to regional distribution centers and stored until needed by local wholesalers or retailers. These distribution centers can be very large, housing millions of square feet of floor space and servicing tens of thousands of customer orders or tens of trucks per day. Goods can be shipped by train, boat, truck, air, or pipeline. Many items today are shipped by intermodal transit. Products may be loaded into a container that can be handled by several means of transport, from the back of a truck to riding on a rail car. Starting at the factory, the filled container is driven or sent by rail to the nearest port. From there, it is automatically loaded onto a ship. After it arrives at the destination port, the container load is placed on the back of a cab and driven to the customer.

Rail transport is very inexpensive, at least in the United States, but it can be rather slow, and delays can occur. Rail transport is commonly used for shipping heavy and cheap commodities such as coal or grain. Trucks represent a good alternative for shorter transport of smaller loads. Trucks provide point-to-point service compared to the central depot constraints of trains and ships. Air transport is reserved for small, light items, generally with high cost, short life spans, or high urgency that are either traveling in small quantities or covering a distance that makes trucking impractical. Examples are mail, critical repair parts, medical supplies, and perishable items. Pipelines are inexpensive but normally only useful for special products such as crushed ore, oil, and gas.

Key decisions affecting distribution include the scheduling of production at each facility and the setting of inventory levels at each point. Scheduling decisions are driven by demand forecasts, production lead times, and facility capacities. Inventories are kept to smooth production during demand cycles, to guard against interruptions caused by breakdowns and strikes, and to reduce delivery times to customers by shipping from nearby warehouses. However, storing inventory is costly (this subject is explored in later chapters) and runs the risk of product deterioration and obsolescence. (It is not uncommon for a warehouse to contain large quantities of products not in demand while being void of those in demand.)

Cross-docking has become a major strategy for rapidly meeting demand with minimal inventory. In cross-docking, goods are never put into storage at distribution centers. Instead, goods are received from suppliers and manufacturing plants and immediately routed to the shipping bay for loading onto trucks. This process reduces handling and storage costs significantly but assumes that product acquisition and shipping can be closely coordinated. A shipment of products may arrive each day from several manufacturing plants. These products are unloaded and immediately allocated to customer orders, each order potentially requiring a small number of units from each truck that has arrived over the past day or two. The products arriving from the various suppliers are combined to meet individual customer orders, loaded on the truck, and shipped. Wal-Mart has become widely recognized for this practice. Wal-Mart collects point-of-sale data on customer demand and compiles the information into replenishment orders, which are electronically transmitted back to one of their approximately twenty distribution centers around the country. These centers then ship replenishment orders to stores twice a week. By restocking shelves more frequently than competitors, customers can be assured of availability of high-demand items while the company avoids the cost of centralized inventory and the wasted shelf space allocated to slow-moving items. Toyota uses cross-docking at its distribution center in Ontario, CA. This central location serves eleven regional distribution centers, which, in turn, serve 1,400 dealers. Where possible, arriving cases are bar coded with their destination and automatically sorted to the destination dock. When arriving cases must be broken down for shipment, receiving cases are scanned and placed on a sortation line. As the case passes the sorting station for specific distribution centers, an electronic message informs the worker of how many parts are to be removed from the case and repacked for shipment to that distribution center.

4.2.6   Locating Inventory

The location of component and finished-product inventories largely determine system cost and delivery time. The key issue, then, becomes where to store inventory in the system. The rule of thumb is to first try to eliminate the need for inventory. Adding capacity, enhancing communication, improving process and supplier quality, and improving forecasting models can all aid in this endeavor. When this fails, maintain inventory at the cheapest level that allows meeting demand on time. It is usually cheaper to store items upstream than downstream in the supply chain. If deliveries from suppliers are uncertain, raw material stocks are maintained. If plant capacity is unreliable, finished goods (from that facility) are stocked. Both unit cost and the amount of inventory needed to achieve a specific service level depend on the stage of the supply chain. This phenomenon is examined in the next section.

4.2.6.1   Basic Trade-Off: Centralization versus Responsiveness

Suppose we have defined our multistage supply chain. We have one or more manufacturing plants. Each plant serves several distribution centers. (Eight to ten of these spread throughout the United States can put us within a day's truck delivery of most domestic locations.) We may add a few more for our European and Asian markets. Each distribution center supplies a delivery warehouse in each of the major local markets in its region. Local markets may be defined as standard metropolitan areas of over 100,000 people or regions meeting specified sales volume. These local warehouses distribute frequently to the individual retail outlets in their immediate area. How do we determine the ordering policies and target safety stock levels at each location? The basic trade-off is between the added safety from pooling demand variability and storing inventory centrally versus the faster delivery available from local storage at the warehouses.

First, note that the distribution centers are the customers of the manufacturing plants, the local warehouses are the customers of the distribution centers, and the retail outlets are the customers of the local warehouses. Assuming that demands of the various customers are independent, we can provide the same service level of stock availability with less inventory if we keep the safety stock closer to the manufacturing plant. Safety stocks are typically set as some multiple of the standard deviation of demand. Suppose retailer i has weekly demand being independent between weeks and retailers. Denote standard deviations[5] of retail demands by σ_i, $i = 1, \ldots, N$. The safety factor is k, chosen to provide some desired level of service such as guaranteeing that 98% of demand can be filled from on-hand stock. If each retail outlet carries its own safety stock, the required amount of system safety stock is

$$SS_R = k \sum_{i=1}^{N} \sigma_i.$$

Suppose, instead, we keep all the safety stock at the warehouse. If demand exceeds expectations, then either the retail outlet places a special order to receive an extra shipment or the customer must wait while demand is backlogged until the next scheduled delivery. Each of these has a cost. However, if we centralize risk at the warehouse, we need only guard against the cumulative demand variability. The required centralized safety stock is:

$$SS_W = k \left(\sum_{i=1}^{n} \sigma_i^2 \right)^{1/2}.$$

Since, for

$$N > 1, \sum_{i=1}^{N} \sigma_i > \left(\sum_{i=1}^{N} \sigma_i^2 \right)^{1/2},$$

---

[5]In practice, the standard deviation is often a multiple of the mean demand rate. We will assume the demand is stationary so that we can use the constant factor σ.

we know $SS_W < SS_R$. If all $\sigma_i = \sigma$, then for equal levels of product availability, $SS_R = \sqrt{N} \cdot SS_W$. Thus, centralized safety can reduce inventory levels significantly when there are many customers, **if** we are willing to accept the extra time to fill some orders. In many markets, delivery is becoming a more important factor than minor price differences, and localized inventory storage makes sense despite the additional inventory required.

EXAMPLE 4.6

A warehouse delivers weekly to sixteen local retailers. Demand at each retailer is normally distributed with a mean of 150 units per week and a standard deviation of 10 units. Sales at the retailers are independent. If a retailer starts to run low on stock, a special delivery must be made during the week to avoid shortages. The system should have a 98% fill rate. Find the differences in safety stock between a safety stock at retailer and safety stock at warehouse policy.

SOLUTION

If safety stock is kept at the individual retailers, each requires enough to satisfy the 98th percentile of demand. For a normal distribution, this translates to $k = 2.1$. Thus, $SS = 2.1 \sigma = 2.1(10) = 21$ units. With sixteen retailers, this leads to a safety stock of $21(16) = 336$ units. At the warehouse level, total variance is $16(100) = 1600$, and the standard deviation of demand is 40. To obtain the same 98% fill rate, we need only $2.1(40) = 84$ units.

4.2.6.2 Distributing Inventory Through the Supply Chain

So far, we have seen that the proper use of information can reduce inventory needs and improve the stability of the system (the variance acceleration effect) and that safety stock can be reduced if we centralize its location. In discussing variance acceleration, we assumed the factory responded quickly to demand changes and each level of the supply chain shipped goods in direct response to orders from its customers. Products were effectively *pulled* through the system by customers. In many cases, factory production is fixed in the near-term to facilitate planning of workforce schedules, material purchases, and leveling of the load on work centers. If this is the case, we have the option of letting product be pulled through the system by customers or *pushing* it through the supply hierarchy based on projections of future needs. We can demonstrate these alternatives using the approaches of **distribution requirements planning** (DRP) for pulling and **allocation schemes** for pushing. To fully comprehend these approaches, it is helpful to first study the material in Chapter 5, which discusses aggregate production planning, and Chapter 8, which details materials requirements planning. However, we can briefly describe the two approaches.

In DRP, product is pulled through the supply chain. Retailers place orders with the local warehouses based on the retailer's demand forecasts. Subtracting inventory that is on-hand or already in the order pipeline from the projected demand, the warehouse then places orders with the regional distribution center. The distribution center nets its on-hand and on-order inventory against these orders and then releases orders for the remaining units to the factory. The factory operates based on an aggregate production plan formed by looking at current system inventory, demand forecasts, resource availability, and relevant costs. (The details of setting this aggregate production plan are discussed in the next chapter.) Excess product inventory is kept at the factory until ordered by the distribution centers. This corresponds to the system we assumed when discussing the variance acceleration phenomenon. End-customer demand effectively pulls product through. Allocation systems also schedule factory production on the basis of an aggregate plan. However, instead of storing excess inventory at the factory, inventory pushes through toward the retailers and away from the factory. As units are produced, they are sent to the distribution centers in proportion to the projected demand at each center. One allocation rule is to equalize the projected "run-out" time for each center. The run-out time is the point at which forecasted demand is expected to completely consume current inventory including units in transit and on-hand. For instance, suppose a distribution center has 500 units on hand and 250 in transit from the factory. The three warehouses served by the center have current inventory levels of 100, 125, and 300, respectively, with weekly demand rates of 150, 150, and 125. The system then has a total of $500 + 250 + 100 + 125 + 300 = 1275$ units available. Because system demand is $150 + 150 + 125 = 425$ units per week, we have enough inventory in the system to last $1275/425 =$ three weeks. Warehouse demand forms the basis for allocation. Based on demand and a three-week supply, the total gross allocation to the warehouses is $3(150) = 450$, $3(150) = 450$ and $3(125) = 375$, respectively. Subtracting the available inventories at each warehouse, the distribution center should plan on distributing $450 - 100 = 350$ units to warehouse 1, $450 - 125 = 325$ units to warehouse 2, and $375 - 300 = 75$ units to warehouse 3. Each warehouse will then distribute its inventory to its retailers using a similar rule.

EXAMPLE 4.7

Consider a three-stage supply chain for a product. The factory ships to a warehouse and direct to a local retailer (C). The warehouse serves two additional retailers (A and B). Current inventory levels and demand forecasts for the next ten weeks are shown in bold type in Table 4.5. The factory produces the product once every four weeks. The aggregate plan (see Chapter 5) schedules production for the next twelve weeks, calling for production of 950 units in weeks 1, 5, and 9. A week is re-

Table 4.5   Inventory and Planned Orders for Distribution Requirements Planning

Week	0	1	2	3	4	5	6	7	8	9	10
Retailer A (L. T. = 1)											
Demand Forecast		**100**	**100**	**100**	**100**	**100**	**100**	**100**	**100**	**100**	**100**
On-Hand Inventory	**175**	75									
Planned Order Release		25	100	100	100	100	100	100	100	100	
Retailer B (L. T. = 2)											
Demand Forecast		**70**	**70**	**70**	**70**	**70**	**80**	**80**	**80**	**80**	**80**
On-Hand Inventory	**94**	24	10*								
Planned Order Release	(56)	60	70	70	80	80	80	80	80		
Retailer C (L. T. = 0)											
Demand Forecast		**100**	**100**	**100**	**80**	**80**	**80**	**80**	**120**	**120**	**120**
On-Hand Inventory	**187**	87									
Planned Order Release			13	100	80	80	80	80	120	120	120
Warehouse (L. T. = 1)											
Expected Shipments		85	170	170	180	180	180	180	180	100	
On-Hand Inventory	**327**	242	72								
Planned Order Release			98	180	180	180	180	180	100		
Factory											
Expected Shipments			111	280	260	260	260	260	220	120	120
Projected On-Hand	**158**	158	997	717	457	197	887	627	407	287	1117
Planned Production		**950**				950				950	

*Includes an outstanding order for 56 units scheduled to arrive at the start of period 2.

Notes: On-Hand Inventory levels are measured at the end of the period. L.T. indicates lead time for stock replenishment.

quired to produce the product. The warehouse requires one week to process an order. Delivery to retailer A is very quick, but it takes an additional week to ship to retailer B. Determine the shipments to the warehouses using DRP.

SOLUTION

The coordination between levels of the supply chain is shown in Table 4.5. Input data, based on forecasts and current inventory status, are shown in bold type. The planning logic determines the other values. For the retailers, initial on-hand inventory is used to cover forecasted demand for as long as possible. Remaining demand forms a net requirement for more product. This is met by orders to the warehouse. The orders are sent early enough so that product will be received before it is demanded. The time offset between placing orders and actual demand is given by the lead time. Examine the data in the table for retailer A. The current time is the end of week 0. Retailer A currently has 175 units on hand. These will cover demand for week 1 plus all but 25 units in week 2. Because the retailer's lead time is one week, the retailer places an order for 25 units at the start of week 1 in anticipation of these arriving in time to satisfy customer demand in week 2. Thereafter, because the retailer has no extra inventory, orders match forecasted demand. The same process is used for retailers B and C recognizing their lead times.

Demand for the warehouse emanates from the orders released at retailers A and B. Thus, in week 1, the warehouse has an expected demand or shipment level of 85 units, 25 to retailer A and 60 to retailer B. This increases to 170 units in week 2. The factory derives expected shipments from the sum of the orders placed by retailer C and the warehouse. Retailer C wants 13 units in week 2 and the warehouse wants 98, yielding a demand level of 111 for the factory in week 2. The fundamental principle here is that every stage holds onto its inventory until orders are received.

We see that DRP will tend to store inventory upstream until it is demanded and pulled downstream. In our example, we assumed a higher level planning function specified production quantities at the factory. This may be necessary to ensure effective scheduling of production capacity. The factory's capacity may have to be allocated across multiple product lines, and in each period, we may have to match production levels with capacity. If we had let the orders from the warehouse and retailer C determine production quantities at the factory in the same manner we used retailer orders to determine warehouse orders, we might have created scheduling and capacity problems at the factory.

An additional caution should be noted. We ran the coordination plan assuming lead times were exact. If lead times are variable owing to possible shipping or transit delays from the factory, then we may have wanted to add some safety stock at the warehouses. Likewise, because of demand uncertainty we could have carried safety stock at the retailer outlets. These modifications can be easily incorporated in this scheme by several methods. Perhaps the easiest method is to artificially inflate the demand forecast in period 1 by the desired safety stock quantity. The modeling process will automatically plan to deliver this quantity to the retailer, and the extra units will remain in stock from period to period if demand occurs at the expected level. In a more sophisticated forecasting system, retailer-demand forecasts could be set equal to some upper percentage point on the cumulative demand forecast curve for each week. For instance, if demand for retailer A was normally distributed with mean of 100 and standard deviation of 10, then for a 95% fill rate in each period, we would use forecasted values of

$$F_{A1} = 100 + Z_{.05} \cdot (10) = 116.5$$

in period 1 and a cumulative forecast for the first two periods of

$$F_{A2} = 100 \cdot (2) + Z_{.05} \cdot (10^2 + 10^2)^{1/2} = 223.3.$$

These issues are presented in greater detail when we discuss material requirements planning in Chapter 8.

EXAMPLE 4.8

Repeat Example 4.7 using a runout-based allocation scheme.

SOLUTION

Inventory levels and allocation decisions are shown in Table 4.6. The factory plans to ship all production immediately. Likewise, planned shipments to retailers A and B match receipts at the warehouse. Allocation decisions are made to maximize the time before each "customer" will require replenishment, taking into account projected demand and lead times. Because the lead time from the warehouse to retailer B is one week more than that for retailer A, shipments from the warehouse must consider an additional week's demand at B over A.

Let's briefly examine how these decisions are made. First, on-hand inventory is allocated to forecasted demand for each retailer. The remaining unmet demand becomes a net requirement. For instance, retailer B has 94 units on hand and will receive 56 units at the start of period 2 (this planned receipt was used in the previous example and is restated here for consistency). These 150 units can cover demand for the first two periods plus 10 units in period 3. With a forecast of 70 units, this leaves a net requirement of 60 units in period 3, as shown in the Table. The net shipment requirements for the warehouse in any period t are given by the sum of the net requirements at retailer A in period $t + 1$ and retailer B in period $t + 2$. These temporal offsets reflect lead times. Once needs are determined, we can begin the allocation process. The warehouse has 327 units to divide between retailers A and B. This can cover all requirements in periods 2 and 3 for retailer A (125 units), and periods 3 and 4 for retailer B (130 units), and still leave 72 units. These 72 units are insufficient to cover the 100 units needed by A in period 4, plus the 70 units needed by B in period 5. However,

Table 4.6 Planned Inventory and Shipments for Allocation Scheme

Week	0	1	2	3	4	5	6	7	8	9	10
Retailer A											
Demand Forecast		100	100	100	100	100	100	100	100	100	100
On-Hand Inventory	175	75		42							
Net Requirement			25		58	100	100	100	100	100	100
Planned Receipt			167								
Retailer B											
Demand Forecast		70	70	70	70	70	80	80	80	80	80
On-Hand Inventory	94	24	10*		30						
Net Requirement				60		40	80	80	80	80	80
Planned Receipt			56	160							
Retailer C											
Demand Forecast		100	100	100	80	80	80	80	120	120	120
On-Hand Inventory	187	87	(17)	(221)	(141)	(61)					
Net Requirement			13	100	80	80	80	80	120	120	120
Planned Receipt		(30) ←	(304) ←								
Warehouse											
Net Shipment Reqts				98	180	180	180	180	180	100	
On-Hand Inventory	327										
Planned Receipt			→ (128)	→ (646)							
Planned Shipments		327	(128)	(646)							
Factory											
On-Hand Inventory	158										
Planned Production		950				950				950	
Planned Shipment		158	950				950				950

*includes anticipated receipt of 56 units

( ) after allocation of initial factory inventory and period 1 production batch they can cover $72/(100 + 70)$, or 42%, of these needs. Thus, we add $0.42(100) = 42$ units to retailer A's shipment and $0.42(70) = 30$ units to retailer B's shipment. This makes a shipment of $125 + 42 = 167$ units to retailer A and $130 + 30 = 160$ units to retailer B, as shown in Table 4.6. Other warehouse and factory allocations are made in the same manner as soon as product becomes available. Disbursement of the initial units and period 1 production from the factory to the warehouse and retailer C are shown in ( ) in the table. The warehouse would then immediately ship these to A and B.

4.2.6.3 Echelon versus Installation Accounting

The concept of item inventory becomes more complex for multistage systems. Consider the product structure shown in Figure 4.11. Each circle in the Figure represents a different part or set of operations in the production of a part. We will refer to these as production stages. For each stage j we have a quantity of I_j units of inventory on-hand. However, is I_3 really the number of units of stage 3 product available? Each unit of stage 1 inventory also contains a stage 3 product (plus a stage 2 part, and possibly additional work). Thus, we could claim that the stage 3 inventory is $I_1 + I_3$. This defines the difference between stage 3 installation stock, I_3, and echelon stock, $I_1 + I_3$. Installation stock at stage j equals only those units residing at stage j. Echelon stock includes installation stock, plus all those units residing in higher-level assemblies and products. We will denote the echelon stock as E_j and, in this case, $E_3 = I_1 + I_3$.

As with our definition of inventory, we can differentiate between installation and echelon holding cost. The installation holding cost rate fits our traditional view of $h_j = iC_j$ at stage j, where h_j represents the accumulation of holding cost per unit of average inventory per time, i is the inventory cost rate, and C_j is the value of a stage j item. Part of this C_j cost results from the incorporation of lower-level items in stage j product. For instance, stage 3 in Figure 4.11 includes a stage 4 and a stage 5 part. We define the echelon holding cost of an item at a stage to be the value added at that stage over and beyond its constituent components. We could represent the value of a stage j item as

$$C_j = \sum_{k \in IP(j)} C_k + v_j,$$

where $IP(j)$ is the set of immediate predecessors (components) of j, and v_j is the positive value added at stage j. For instance, as shown in Figure 4.12,

$$C_3 = C_4 + C_5 + v_3.$$

This value added results in incremental cost of holding a stage 3 product in inventory over and above the cost of storing one each of the stage 4 and stage 5 component items. Thus,

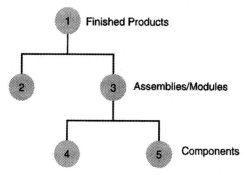

Figure 4.11 Sample Three Level, Five Stage Product Structure

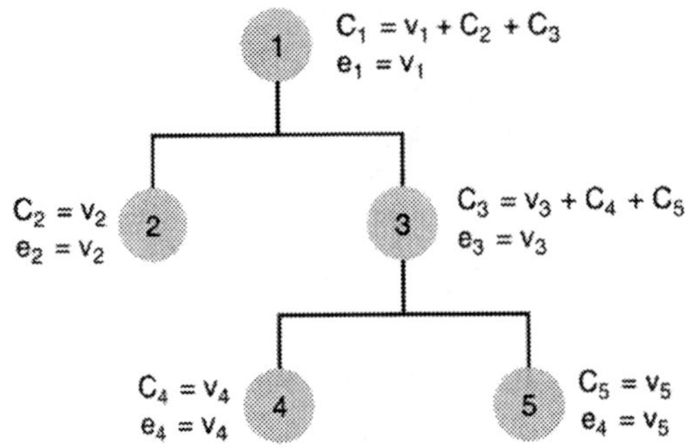

Figure 4.12 Item Echelon and Installation Cost Structure (Excluding Holding Cost Rate)

if our choice is whether to store a stage 3 item or keep the components in stock ready for assembly, the relevant incremental cost of assembling the product is attributable to v_3. This leads to the concept of echelon holding cost, e_j, as the holding cost resulting from the value added at a stage. Thus,

$$e_j = iv_j.$$

An important result comes from the fact that total inventory cost can be computed on the basis of installation stock and costs or echelon stock and costs. Thus, for a product hierarchy of N stages,

$$\text{Installation Holding Cost = Echelon Holding Cost, i.e., } \sum_{j=1}^{N} h_j \cdot I_j = \sum_{j=1}^{N} e_j \cdot E_j. \quad (4.24)$$

A formal proof of this result is presented in Chapter 9 where we elaborate on multiechelon planning models.

Recognizing the incremental impact of echelon costs for storing inventory in a multistage system allows us to justify keeping inventory close to the customer. This can be shown with the standard EOQ inventory model. Recall that the order size is given by

$$Q = \sqrt{\frac{2AD}{h}},$$

where A is the cost to place an order, D is the demand rate, and h is the relevant holding cost per period. Demand and ordering costs are unchanged by the echelon perspective. At the lowest level of the factory, holding and echelon costs are the same. The production batch size is the same for either model. However, as we move higher in the supply chain, to the distribution centers, warehouses, and retailers, echelon costs, which are determined by incremental costs, will be less than the full installation costs. If we use the smaller echelon holding cost for a stage in the EOQ expression, we will obtain a larger batch size. Because average inventory is equal to half the batch size, echelon costs will lead to higher inventory levels as we get closer to the final customer. From our conservation property of installation and echelon cost approaches, we can infer that the system-oriented echelon costing approach justifies holding inventory closer to the final customer.

EXAMPLE 4.9

Consider a two-stage system composed of a distribution center and local retailer system. Demand is 100 units per period, and ordering cost is $400 for the distribution center and $100 for the retailer. Installation holding costs are $1 and $1.25 per unit per period, respectively, for the center and retailer. Find the order sizes and average inventory levels for the installation and echelon cost approaches.

SOLUTION

Using the EOQ expression, order size for the distribution center for both echelon and installation costs is

$$Q_{DC} = \sqrt{\frac{2AD}{h}} = \sqrt{\frac{2(400)(100)}{1}} = 283.$$

Now, consider the retailer. With echelon costs of $e = 1.25 - 1.00 = 0.25$, the order size is

$$Q_R = \sqrt{\frac{2AD}{e}} = \sqrt{\frac{2(100)(100)}{0.25}} = 283.$$

The order sizes are the same. Because $Q_{DC} = Q_R$, the distribution center will not hold inventory but immediately pass it onto the retailer as shown in Figure 4.13a. Average inventory levels for the echelon model are

$$\bar{I}^e_{DC} = 0 \text{ and } \bar{I}^e_R = \frac{283}{2} \approx 142,$$

where "e" implies an echelon cost basis. With installation costs,

$$Q_R = \sqrt{\frac{2AD}{h}} = \sqrt{\frac{2(100)(100)}{1.25}} = 126.$$

In practice, we would probably increase this slightly to 142 units so that the retailer orders twice as often as the distribution center. The timing of these orders would be synchronized such that as soon as the DC receives an order, it passes half that order onto the retailer. Average inventory levels are $\bar{I}^h_{DC} = \bar{I}^h_R = 71$, where "h" indicates an installation cost basis. The corresponding inventory levels are shown in Figure 4.13b. Figures 4.13c and 4.13d provide the echelon inventory pictures relating to Figures 4.13a and 4.13b, respectively. Retailer inventories are the same. The echelon inventory at the DC is the sum of the installation inventories for the retailer and DC.

The natural question now arises: Which is better? Let's examine the actual costs. Total system costs include setup, plus holding costs for both sites. The costs are given by:

$$\text{System Cost} = \left[\frac{D \cdot A_{DC}}{Q_{DC}} + h_{DC} \cdot \bar{I}_{DC} \right] + \left[\frac{D \cdot A_R}{Q_R} + h_R \cdot \bar{I}_R \right]. \tag{4.25}$$

For the echelon cost solution:

$$\text{System Cost} = \left[\frac{(100)(400)}{283} + 1 \cdot (0) \right] + \left[\frac{(100)(100)}{283} + 1.25 \cdot \left(\frac{283}{2} \right) \right]$$

$$= \$141.34 + \$212.21 = \$353.55.$$

For the installation cost solution:

$$\text{System Cost} = \left[\frac{(100)(400)}{283} + 1(70.75) \right] + \left[\frac{(100)(100)}{141.5} + 1.25(70.75) \right]$$

$$= \$212.09 + \$159.11 = \$371.20.$$

From the system perspective, echelon based allocation is preferred. However, if the DC and retailer are different profit centers, they may not agree to work for the common good. The retailer, examining his/her own cost structure, realizes that it is best to order smaller batches if they are charged the $1.25 holding cost. The organizational structure, evaluation, and accounting system will ultimately determine the policy adopted.

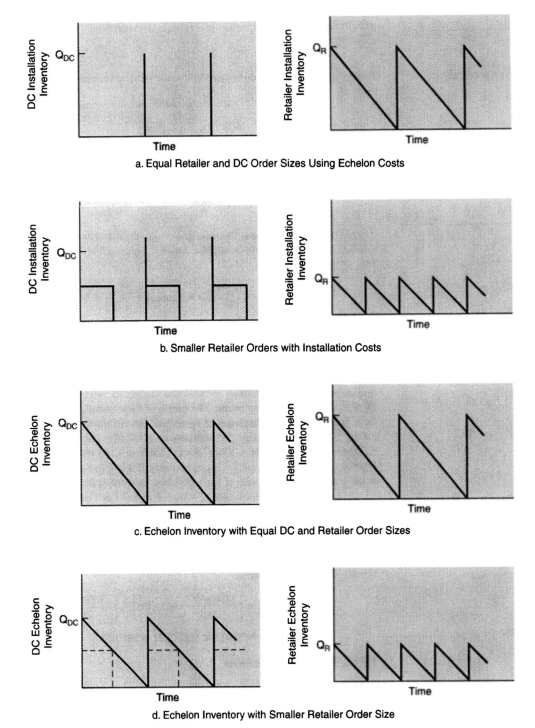

Figure 4.13 Illustration of Echelon and Installation Inventory

4.3   MULTIFACILITY LOCATION-ALLOCATION MODELS: DESIGNING THE LOGISTICS NETWORK

Strategic planning includes the sizing and location of production facilities and the allocation of product family production responsibility to those facilities. The objective is to minimize cost while meeting customer service objectives in all markets. A market refers to a combination of product and geographical region. A region could be a country or local region within a country (such as the southwestern United States). Typically, production responsibility will be assigned at the level of product families. A family may be a line of toasters, or motherboards for a line of PCs. Each family is assigned to one or more production facilities. Each facility must be allocated the necessary labor, tooling, and supporting resources to produce its assigned product families. This allocation problem has become more important in recent years as we have witnessed the globalization of manufacturing. It is not uncommon to have parts being assembled in one country that were fabricated on other continents. The finished product may then be shipped worldwide. Production and logistics costs can vary considerably based on whether we decide to produce our products in Europe, North America, the Far East, or in several of these locales. Tax rates, exchange rate uncertainties, and local content requirements can play a major role in such decisions as well. Cohen and Lee [1989] summarized a number of strategies that organizations may adopt for plant location and logistics. When markets are spread out, production capacity is flexible, distribution costs are high, and tax structures favor local production, a regional strategy may be used. A production facility is located in each geographical region to produce a wide variety of products that are then distributed locally. With less flexible capacity and strong economies of scale, the organization may adopt a product focus. In this case, plants are built in different regions, but each produces a limited line of products. Products are then distributed internationally, possibly globally, from the distribution center colocated at each production plant. Typically, raw materials will be obtained locally, and the availability of material suppliers may contribute to the decision of where to locate each product's facility. When economies of scale, market demand, and international cost structures warrant, a consolidation strategy may be adopted, whereby large-scale production occurs at a single plant. Usually, product development and a large distribution center are colocated at the site.

We will present a mathematical programming model, adapted from Munson and Rosenblatt [1997], for aiding in the facility location and distribution decision. To keep the model simple, we assume that production can be aggregated into a single product that is assembled from a set of components. We further assume that product demand in each market region is known and constant over time. Product will be shipped directly from production plants to the regional markets. We let the index j indicate the set of possible manufacturing (assembly) plant locations. Locations may indicate countries where we could build a plant, or they may represent one or more specific locations within each country; potential locations have been identified by considering labor supply, utility rates, transportation network, local environment, and related factors. A known set of possible suppliers exists for each component. This set may include the options of building our own component plants in some of the countries served. The decision variables for the model are:

Decision Variables:

$Y_j = 1$ if we utilize an existing facility or open a new production facility at location j and 0 otherwise;

S_{jl} = amount of product assembled at location j for shipment to customer l;

X_{ijk} = amount of component i purchased from supplier k and used for production in location j;

To choose our production system rationally, we need the following parameter data:

Technological Coefficients:

c_{ijk} = unit cost to purchase and deliver component i from supplier k to production location j;

D_l = demand per time period for customer location l;

F_j = fixed cost per period to open a production facility in location j;

K_j = capacity per period of the facility being considered at location j;

L_l = minimum local content of value added required by customer region l;

p_l = sale price (or value) of item sold in customer region l;

r_i = amount of component i needed to produce a unit of product;

s_{jl} = unit shipment plus import cost to serve customer l from a production plant in location j;

v_{jl} = value added (variable production cost) at plant location j for a product designed for market l.

(Note that the finished product may be customized for each market and, therefore, production cost may depend on the customer market l);

$V\{l\}$ = set of suppliers and sites in the same country as market area l.

$$\text{Minimize} \sum_j F_j \cdot Y_j + \sum_i \sum_j \sum_k c_{ijk} \cdot X_{ijk} + \sum_l \sum_j (v_{jl} + s_{jl}) \cdot S_{jl} \qquad (4.26)$$

subject to:

$$\sum_{j \in V\{l\}} \left(\sum_i \sum_{k \in V\{l\}} c_{ijk} \cdot X_{ijk} \right) \cdot \frac{S_{jl}}{\sum_u S_{ju}} + \sum_{j \in V\{l\}} v_{lj} \cdot S_{jl} \geq L_l \cdot D_l \cdot p_l; \ \forall l \qquad (4.27)$$

$$S_{jl} \leq Y_j \cdot K_j; \ \forall j,l \qquad (4.28)$$

$$\sum_j S_{jl} \leq K_j; \ \forall j \qquad (4.29)$$

$$\sum_k X_{ijk} = \sum_l r_i \cdot S_{jl}; \ \forall i, j \qquad (4.30)$$

$$\sum_j S_{jl} = D_l; \ \forall l \qquad (4.31)$$

$$Y_i \in \{0,1\}; \ X_{ijk} \geq 0; \ S_{jl} \geq 0 \qquad (4.32)$$

The objective function (4.26) minimizes the sum of fixed costs to use production facilities, the cost to obtain components, the cost to produce at the plant, and the cost to ship product to the final market destination. To introduce the basic model and keep it relatively simple, our objective function minimizes cost. Because tax rates vary by country, it would be preferable to track profits by country and to maximize total after tax profits. The reader is asked to make this extension in the end-of-chapter exercises. The constraints (4.27) enforce the local content regulations. The left hand side of the equation computes the value of components purchased in the country, plus valued added by production in the country. Total internal expenditures are proportioned, based on the percentage of production that is sold within the country. The right hand side determines the required local content value based on regulations and total local sales. We assume that markets share a one-to-one correspondence to governmental jurisdictions; otherwise, we may need to aggregate markets in these constraints. The precise form of this equation may vary depending on the specific form of the local regulations. Value, for instance, may be measured in terms of number of parts, value added, or selling price.[6] Equation (4.28) performs accounting functions for the model. The Equation set forbids production at a plant unless the fixed cost to build the facility is incurred. Plant capacity is enforced in Equation (4.29) by restricting total shipments from each production facility to be no more than the capacity of that facility. The necessity of acquiring a sufficient number of components of each type to allow production at the assembly plants is assured in Equation set (4.30). The r_i factor allows for differences in units between components and products. For example, there may be 10 square feet of sheet metal required to make a microwave oven or 10 transistors of a particular specification on each circuit card. Equation (4.31) ensures that demand will be met in each customer market. The left hand side of the equation sums up deliveries from all production plants to a single customer. Finally, the integer nature of plant construction and the non-negativity restrictions on production and shipments are satisfied by Equation (4.32).

The binary facility variables and nonlinear nature of the content restrictions (4.27) make the problem difficult to solve. Fortunately, the system design problem does not require real-time decisions. The analyst may specify a limited set of scenarios for the set of manufacturing plants to use and then solve the model for each scenario. For an existing network, the analyst may periodically consider the possibility of adding or deleting several new plant locations or suppliers. By assuming that plants will produce to near capacity and always serve the local market, we can remove the nonlinearity by replacing the

$$\frac{S_{jl}}{\sum_u S_{ju}} \text{ by } \frac{D_l}{K_j}$$

in Equation (4.27).

The model may be extended to include a variety of factors. As noted previously, the "product" could be divided into multiple product families. If demand and supply options vary over time, we could model time as a sequence of annual periods. A different strategy might be used for each year, and we might even consider storing inventory in periods of slack demand. The set of potential facilities indexed by j implies we have already specified the capacity for each site. However, we could include several different facility sizes in each country simply by including these as options with different fixed costs and capacities. Other types of local content rules could also be modeled. In the next chapter we describe dynamic system-wide production and distribution models in greater detail. Our objective here was primarily to present the issues involved in designing global logistics networks and to demonstrate the possibility of using mathematical models to aid in network design.

---

[6]As stated, this equation is nonlinear in the S_{jl} variables. We can linearize the model at the expense of defining more variables. The X_{ijk} variables could be expanded to X_{ijkl} to include an indication of where the component will ultimately be used.

4.4 SYNTHESIZING THE CONCEPTS

As noted by Tompkins [1998], the secret to success lies in leadership that synthesizes the core supply chain competencies. We must understand and build effective product design processes, operating systems, and distribution systems. This includes capturing the voice of the customer, catering to the whims of the best customers, and, where profitable, satisfying the needs of the smaller customers. We must select the proper combination of push and pull production and distribution systems[7] to meet delivery requirements with minimal inventory cost. This requires effective maintenance and quality assurance procedures and intelligent warehousing and distribution control plans to allow dependable and efficient product delivery. We must develop long-term partnerships with suppliers and customers with a willingness to share information, risks, and rewards. Then we must reap the benefits of these conditions by exchanging information and coordinating plans. Finally, effectively applying today's supply chain concepts will not suffice to ensure continued success in the future. The company must understand the process of change and harness its potential. If we don't improve, the competition will, and we will be left behind. Instead of moving from peak to valley, as the competition exceeds our improvements, we should anticipate technological and organizational breakthroughs and push the frontier forward, or at least be agile enough to follow closely behind.

4.5 SUMMARY

Markets today tend to be global, raising the opportunity for tremendous growth but, at the same time, increasing competition. Global competition requires establishment of an effective supply chain that spans the spectrum from materials purchases to product delivery, including the functions of manufacturing and customer service. The supply chain design must consider material availability, logistics costs, and local cultures, as well as market opportunities. Attention should be given to building mutually beneficial relationships with partners in the supply chain.

Manufacturing strategy plays an important role in deciding how a firm will position itself in the supply chain. Setting the strategy requires an understanding of the company's markets and competitive advantage opportunities. The core competencies vital for success are defined and protected to maximize competitiveness. The degree of vertical integration is selected, and necessary linkages with suppliers are formed. The company must decide how it can most efficiently add value. This typically includes choices concerning which products to make, where to make them, where to acquire the required materials, and where to distribute the finished products. As part of this process, the company decides the degree of flexibility to incorporate into operations, how to take advantage of information technologies most effectively, and where to stock inventory. Inventory may be pushed or pulled through the supply chain, and safety stock can be aggregated at distribution centers or distributed to retail sites. Mathematical programming models can assist in defining the network of suppliers, production facilities, and distribution facilities that minimizes total costs. Ultimately, the strategic plan must consider plants, processes, people, and products in forming the purchasing, manufacturing, distribution, and service plan.

Information exchange becomes critical for effective operation to minimize inventory and respond rapidly to customers. Without the free flow of information to all partners in the supply chain, random fluctuations in demand tend to accelerate up the supply chain, causing large fluctuations in orders and production levels. It is typically best to coordinate decisions at each level of the supply chain with a planning horizon at least as long as the cumulative lead time through the system. Distribution requirements planning and allocation schemes are used to pull or push inventory through the multiple stages of the supply chain.

Some inventory storage will always be necessary to guard against random events. The preferred location for safety stock depends heavily on the customer's expectation for delivery time, product cost structure, degree of randomness, and number of sites involved. Safety stock should be aggregated at the highest level of the hierarchy that still permits delivery to customers as needed without excessive cost or delay. An echelon cost model can help determine the optimal inventory strategy for the integrated supply system.

---

[7]Push and pull systems are addressed in Chapters 7 and 8.

4.6 REFERENCES

ARNTZEN, B.C., BROWN, G.G. HARRISON, T.P. & TRAFTON, L.L. Global supply chain management at digital equipment corporation. *Interfaces,* 1995, 25(1), 69–93.

BARD, J.F., & FEO, T.A. An algorithm for the manufacturing equipment selection problem. *IIE Transactions,* 1991, 23(1), 83–92.

BROWNE, J., DUBOIS, D. RATHMULL, K. SETHI, S.P. & STECKE, K.E. Classification of flexible manufacturing systems. *The FMS Magazine,* 1984, 2, 114–117.

CANADA, J.R., & SULLIVAN, W.G. *Economic and multiattribute evaluation of advanced manufacturing technologies.* Englewood Cliffs, NJ: Prentice-Hall, 1989.

COHEN, M.A., & LEE, H.L. Resource deployment analysis of global manufacturing and distribution networks. *Journal of manufacturing and operations management,* 1989, 2, 81–104.

DEMMEL. J.G., & ASKIN, R.G. A multiple objective decision model for the evaluation of advanced manufacturing system technologies. *Journal of Manufacturing Systems,* 1992, 11(3), 179–194.

DORNIER, P., ERNST, R. FENDER, M. & KOUVELIS, P. *Global operations and logistics.* New York: John Wiley & Sons, 1998.

GOLDEN, B.L. WASIL, E.A. & LEVY, D.E. The analytic hierarchy process: *Applications and studies,* Springer-Verlag, Berlin, 1989.

KAPLAN, R.S. Must CIM be justified by faith alone? *Harvard Business Review,* 1986, 64(2), 87–95.

KUGLIN, F.A. *Customer-centered supply chain management.* New York: American Management Association, 1998.

LEE, H.L., & BILLINGTON, C. The evolution of supply chain management models and practice at Hewlett-Packard. *Interfaces,* 1995, 25(5), 42–63.

LEE, H.L., PADMANABHAN, V., & WHANG, S. The bullwhip effect in supply chains. *Sloan Management Review,* 1997, 38(3), 93–102.

LEE, H. L., & TANG, C.S. "Modelling the costs and benefits of delayed product differentiation." *Management Science,* 1977, 43(1), 40–53.

MAGRETTA, J. Fast, global and entrepreneurial: supply chain management, Hong Kong style. *Harvard Business Review,* 1998, 76(5), 102–114.

MUNSON, C., & ROSENBLATT, M.J. The impact of local content rules on global sourcing decisions. *Production and Operations Management,* 1997, 6(3), 277–290.

NARASIMHAN, S., McLEAVEY, D.W., & BILLINGTON, P. *Production planning and inventory control.* Englewood Cliffs, NJ: Prentice-Hall, 1995.

SAATY, T.L. The analytic hierarchy process: planning, priority setting, resource allocation. McGraw-Hill, New York, 1980.

SETHI, A.K., & SETHI, S.P. Flexibility in manufacturing: a survey. *International Journal of Flexible Manufacturing Systems,* 1990, 2(3), 289–329.

TAYUR, S., GANESHAN, R., & MAGAZINE, M. *Quantitative models for supply chain management.* Boston: Kluwer Academic Publishers, 1999.

TOMPKINS, J.A., Time to rise above supply chain management. *Material Handling Engineering,* October 1998, SCF16–SCF 18.

VIDAL C.J., & GOETSCHALCKX, M. Strategic production distribution models: a critical review with emphasis on global supply chain models. *European Journal of Operational Research,* 1997, 98, 1–18.

WITT, C.E., Crossdocking. *Material Handling Engineering,* July 1998, 44–49.

ZADEHI, F. The analytic hierarchy process—a survey of the method and its applications, *Interfaces,* 1986, 16(4), 96–108.

4.7 PROBLEMS

4.1. What is the difference between a technology-push and a market-pull product? How should this impact the company's product design and marketing efforts?

4.2. Define the term *core competency* and explain its role in manufacturing strategy.

4.3. What are the competitive advantages of vertical integration?

4.4. List the different strategies available for achieving flexibility in manufacturing operations.

4.5. Describe the different strategies for organizing production facilities across a geographically dispersed supply chain and indicate the conditions under which each is appropriate.

4.6. A company can purchase a component for $0.35. However, the facility has extra capacity, and if they invest $5,000 in new tooling, the item can be made for an estimated $0.23 per unit. Find the break-even point for determining whether the component should be made or bought.

4.7. A common bracket can be purchased in large quantities for $0.67. The company can make the bracket at a variable cost of $0.41 by buying a machine for $33,000. Demand over the next two years (the anticipated life of the product and machine) is estimated to be normally distributed with mean 100,000 units and a standard deviation of 10,000 units. Would you recommend making or buying this bracket?

4.8. The purchasing manager for a job shop must decide whether to stock a particular material or just order it as needed for custom jobs. To stock the item, the inventory status would have to be checked once a week at a cost of $500 annually. The cost to place an order is estimated at $125. Irregular orders placed for specific jobs would require an additional $85 delivery charge. The company would have to place approximately twenty such orders per year. Demand for the $8 item is 2,300 units per year. The annual holding cost rate is 40%. Should the manager decide to stock the item?

4.9. A plant is being built to manufacture two product families—A and B. The facility must produce 100,000 units of each family per year. Three technologies are available. Technology one costs $125,000 per system per year, and a system can produce 50,000 units of product A, or 25,000 Bs (or any proportional combination) annually. Technology two costs $65,000 per year and can produce 45,000 of either product per year. A third technology exists that can only produce product B. The system could produce all 100,000 units for a cost of $180,000 per year.

a. Develop a mathematical programming formulation for the problem of selecting technologies.

b. Find a good heuristic solution to the problem by first relaxing the integer restriction on purchasing processes and then rounding up the resultant acquisition variables.

c. In addition to expected cost, what other factors should be considered?

4.10. An industrial engineer believes a new data collection and scheduling system will improve shop performance. The system will cost $150,000 to purchase, and will require a class II employee to operate. Class II employees have a compensation package of $46,000 per year. The current scheduling system is operated by two employees costing $31,000 each per year. The benefits of an improved schedule are difficult to estimate, but the engineer is confident that work-in-process inventory holding cost will be reduced at least 33% from the current cost of $80,000 per year. The reduction in inventory will also reduce congestion and response time to meet customer orders. Current delivery times are ten weeks. The plant manager has agreed that a 50% reduction in production flow time would be as important as incurring the cost of the system. System life is estimated at three years. The company uses a minimum attractive rate of return of 17%. Should the engineer recommend purchasing the system?

4.11. FACE Co. distributes cosmetics to department stores. FACE recognizes that fast delivery of a full range of cosmetic products accounts for its success. Currently, products are bought in large quantities and stored in regional warehouses. The annual cost of carrying inventory, including losses for unsold items, is $12,000,000. As an alternative, the company could purchase a manufacturing system for $3,000,000 and operate the system for $1,500,000 per year. Inventory costs would be eliminated and delivery time would be unaffected. However, the variety of products offered would be reduced by 50%. Assuming product variety is twice as important as cost and delivery time for the range of values that would result from these two alternatives, construct an AHP decision hierarchy to decide whether the system should be purchased. The company requires a payback period of two years, at most, for investments. The annual discount rate for the company is 20%.

4.12. A U.S. company has a network of manufacturing facilities around the world. How should changes in exchange rates affect its decisions about where to purchase raw materials and where to produce products?

4.13. A U.S. company that makes sheet rock enters into a long-range contract to buy 1,000,000 lbs. of gypsum each month from a mine in Mexico. Suppose the exchange rate for the Mexican peso changes from 10.20 pesos per dollar to 10.85 pesos per dollar. What effect will this have on the profits of the U.S. company?

4.14. Consider a two-stage plant-retailer supply chain. Lead time for the plant to fill retail orders is one week. The plant produces exactly to demand and does not keep inventory. Desired safety stock at the retailer is equal to half-week demand, and future demand is estimated by the average of the past two weeks. Find the impact on plant production (customer orders) if demand permanently increases from 100 to 110 units per period.

4.15. Suppose in the previous problem that the plant maintains a finished goods inventory with a target safety stock level of one week's demand based on a three-week moving average. Shipments are made from the plant's inventory to the retailer and take one week in transit. It also takes one week to produce items; therefore, whatever the plant produces this week is not available for shipping until next week.

a. State the order-up-to rule used by the retailer.

b. State the rule used by the plant to determine the production quantity each period.

c. Find the impact on plant production and inventory levels if demand permanently increases from 100 to 110 units per period.

4.16. Consider a two-stage plant-retailer supply chain. Lead time for orders is three weeks. Safety stock is equal to one week's demand, and future demand is estimated by the average of the past two weeks. Find the impact on plant production (customer orders) if demand increases from 100 units to 120 units for a single period and then returns to 100 units.

4.17. Consider a two-stage plant-retailer supply chain. Lead time for orders is two weeks. Safety stock is equal to one week's demand, and future demand is estimated by the average of the past two weeks. Find the impact on plant production (customer orders) if demand changes from 100 units a week to cycling between 90 units and 110 units per week.

4.18. Repeat the variance accelerations example in Section 4.2.2.1, assuming demand increases to 110 units in period six and then returns to 100 units per week in week 7 and thereafter. Use a spread sheet to track the effect on inventory levels and factory production.

4.19. A retailer places an order at the start of each week and receives the order two weeks later. The retailer attempts to keep its starting inventory level equal to 1.5 weeks of demand representing expected sales for that week, plus, half-week safety stock. (Actual demand is not known until after the order is placed.) The retailer uses a three-week moving average of demand to estimate mean demand. When the factory receives the order it immediately places an order for raw materials. The factory also immediately starts producing the number of units ordered by the retailer. Lead time in the factory is two weeks once production begins. The raw materials arrive one week after the order is placed. The factory tries to keep 1.5 weeks of raw materials in stock at the start of each week. Mean demand is based on a two-week moving average of retail orders.

a. Diagram the supply chain and state the order-up-to rule used by the retailer.

b. State the order-up-to rule used by the factory in ordering raw materials.

c. Using a spread sheet format, track performance of this system for eight weeks if demand suddenly shifts from 100 units per week to 110 units per week starting in period 3.

d. Suppose that, instead of placing an order for raw materials each week and then using the existing raw material inventory, the factory decided to eliminate the raw material safety stock. Each week an order would be placed for enough material to fill the retailer's order. When those materials arrived one week later, the factory would start its two-week production cycle. What effect would this strategy have on delivery lead time to the customer, and how would it affect inventory swings in reaction to demand changes?

4.20. Describe the key factors in selecting and certifying a vendor.

4.21. Describe the impact of cross-docking on inventory cost and delivery lead time in a multistage supply chain.

4.22. A distribution center serves twenty local outlets. Demand at the outlets is normally distributed with a total mean of 24,000 units per month. Individual outlet demands are independent and all have a standard deviation equal to 0.15 mean demand. The company has a target of filling 98% of demand from stock. What is the difference in required safety stock if this is kept at the distribution center instead of the outlets?

4.23. Repeat Example 4.7, assuming the production plan calls for producing 500 units of the product every other week, starting in week 1. Determine the shipping plan for the next ten weeks using distribution requirements planning.

4.24. Repeat the previous problem using a runout based allocation scheme. Fully allocate all factory shipments.

4.25. A final product is composed of two part As, one part B, and one part C. Each part C requires one part D and two part Es. Let I_i be the installation inventory level of item i. Using E_i for echelon stock, give the expression for each item's echelon stock in terms of installation inventory values.

4.26. Weber Manufacturing is trying to define its supply chain to serve demand in an area encompassing three countries. A production plant can be built in any one or more of the three countries. Monthly demands by customers in the three countries are 10,000 units in country A, 25,000 in country B, and 30,000 in country C. A plant with capacity 40,000 units per month (the optimal plant size for this product) can be built in any one of the three countries for an amortized monthly fixed cost of $20,000. Unit variable production costs in the three countries will be $1.25, $1.10, and $1.40, respectively. Potential raw material suppliers already exist in

countries A and B. The product sells for $4.50. Country A does not have a content law. Country B imposes a 50% tariff if less than 50% of the product's value is produced in country B. Country C imposes a 75% tariff if less than 25% of the product's value is produced in country C. The raw material is a basic commodity costing $1 per unit. Shipping costs per unit are as follows:

Raw Material Shipping Costs/Unit

From country	To country		
	A	B	C
A	$0	$0.15	$0.15
B	$0.15	$0	$0.20

Finished Product Shipping costs/Unit

From country	To country		
	A	B	C
A	$0	$0.25	$0.30
B	$0.25	$0	$0.35
C	$0.40	$0.35	$0

a. Formulate this supply chain design problem as a mathematical program. Define all decision variables.

b. Solve the formulation to determine the optimal supply system decisions.

4.27. Expand the formulation in part A of the previous problem to allow for the possibility of also constructing a raw material plant in country C. Define any necessary notation.

4.28. Suppose that in Problem 4.26, smaller plants could also be built with capacity of 20,000 units for an amortized monthly cost of $15,000 and a 10% increase in variable costs. How would this change your formulation?

4.29. Modify the formulation in Section 4.3 to track profits by country and to maximize total after-tax profits. In doing so, define notation for material purchase costs, tax rates, exchange rates, variable production costs, and selling price for each product. Note that transportation costs for materials and distribution costs for products must be charged to either the shipping plant or the receiving location. Discuss how varying tax rates between countries should affect the decision of how to charge these costs and set transfer prices between international facilities.

CASE STUDY 4.8 *Elba Electronics*

John James, vice president for Manufacturing of Elba Electronics, stared out the window past the autumn leaves floating down the Elba River. He was lost in thought and didn't notice the reflection of the setting sun as it glistened off the lake, nor the reddish-orange wash it painted over the surrounding mountains. He had just finished reading Rick Armstrong's recommendation to close part of the Elba facility and move all production of low-end components and circuit card manufacturing offshore. Rick was the chief financial officer of Elba, and made a strong case for significant cost saving from the move. Of course, Rick wasn't a native of Elba; he hadn't worked at the plant along with his childhood buddies for the past twenty-five years, and he wasn't on a first-name basis with virtually every worker in the plant. John, troubled by his conflicting emotions and loyalties, could not help but wonder whether the decision was truly so straightforward.

Elba Electronics opened its plant along the river banks in 1955. Three friends returning to the small town from the Korean War recognized the potential for supplying basic electrical components used in the production of radios. The town of 15,000 offered limited opportunities. The three friends decided to make use of the GI Bill. After earning degrees at the public university located outside the state capital forty miles away, they returned to Elba and opened the plant with support from the entire population. Early on, their catalog of parts expanded to serve the rapidly growing television industry. Except for a few lean years in the mid-1970s and 1980s, the company had prospered and grown steadily through the years, until recently. John's uncle Bill and his two cofounders had retired about ten years ago and sold the company to a large conglomerate. Elba's business was usually dominated by high-volume production of basic commodity electrical components. However, the founders had twice led the company out of recessions by investing their retained earnings in new production equipment and exclusive rights to revolutionary new product technologies. These high-tech components carried higher profit margins. Unfortunately, those businesses required constant investment to remain competitive. In the past ten years, the company's core business had returned to producing basic axial and radial components and the production of printed circuit boards. These businesses were more labor-intensive, but Elba Electronics had developed a strong reputation for high-quality and dependable delivery. Consequently, they were a preferred vendor for many of their customers. In recent years, however, profits were dropping. Overseas competitors, with lower labor costs, were cutting prices. Elba's customers were slowly migrating, and now most of their product was shipped to Mexico, Ireland, Korea, and Singapore. Elba was forced to absorb the increased shipping cost and to incur increased inventory cost by warehousing components overseas to ensure timely delivery. John felt trapped between his loyalty to the employees and the new company owners. He was hoping he could discern some weakness in the CFO's analysis. His thoughts focused on two questions:

1. What noneconomic factors deserve consideration, and how should they be measured and weighted?

2. Are there rational alternatives to the options of status quo and closing the Elba site? John turned back to his desk and picked up the memo to give it a second look.

To: B.R. Pullman, President, Elba Electronics
From: Rick Armstrong, chief financial officer
RE: Corporate Financial Planning Committee Viability Investigation

The FPC has completed its study of the long-term viability of Elba's current operating plan. It is our conclusion that continued operation of the Elba facility plant is unwarranted. Let me emphasize that the corporate members of the committee were most steadfast in this view and are in the process of drafting a memo to the corporate executive council recommending that the facility be closed by the end of the year with the operation being moved to Southeast Asia.

The committee's decision came down to one of economics. We currently employ 265 workers here in Elba, and recorded $37.6M in revenues last year. Revenues have been growing at the rate of 3% annually in recent years. The committee focused on the rapidly increasing distribution and warehousing costs that have led to erosion in our profit margin. A summary of current cash flows and the three-year trend appears as:

Cost category	Annual cost	3yr growth rate
Labor	$11.8M[1]	3%
Marketing	$2.0M	2%
Depreciation and Utilities	$3.3M	4%
Distribution	$11.1M	12%
Finance (Inventory)	$4.6M	10%
Materials	$4.6M	3%
Net Earnings	0.2M	−5%

[1]Labor cost is 35% administrative, 50% direct, and 15% indirect shop labor.

A preliminary study indicated that a move to Southeast Asia could probably be completed with a start-up cost of $6 million after accounting for the sale of local facilities and expected tax breaks. This includes transferring the production equipment. We currently have on the books a plant value of $12 million, plus, an additional $8 million in equipment. Once established offshore, labor costs will drop 60%, and distribution costs would decline 30%. This assumes that component manufacture and card production will each continue to account for approximately 50% of revenues. Although accounting for 50% of revenues, card manufacture requires 70% of our investment in equipment, 20% of the labor force, and 33% of the labor cost. While component distribution is almost exclusively to Southeast Asia at present, cards are shipped in roughly equal quantities to Phoenix, Arizona, Dallas, Texas, Limerick, Ireland, and Seoul, Korea. The key problems anticipated include: 1) a temporary cutback in production caused by the move and a training period while we bring the new plant up to speed; 2) loss of some key production and engineering staff who will not want to move overseas; and 3) temporary quality problems during startup, with a small permanent drop in quality resulting from the loss of an experienced and educated workforce.

I have asked all committee members to keep this recommendation confidential for the time being. I assume you will want to have the site council meet to prepare an action plan.

Chapter 5

Aggregate Planning

Once strategic plans are set, we know the families of products to be produced and the resource capacities at each manufacturing facility. As indicated in Figure 1.5, aggregate production planning forms the next step in the decision hierarchy. Demand forecasts and resource availability over time serve as inputs to the aggregate planning process. The aggregate plan then specifies the production quantity for each product family in each period, at each facility, over a moderate- to long-term planning horizon. The planning horizon is usually six to eighteen months for aggregate planning, with periods from one week to one month. A by-product of the aggregate plan is the planned amount of each key resource to be used over the horizon.

A facility may produce thousands of products, using hundreds of resources (labor grades, machine types, raw materials, etc.). To fully model all final products and resources would require too much detailed data and result in models that are too large to solve. This is unnecessary at this stage. Thus, aggregate units are used to describe production and resource usage. All items that share common setups (are produced together) and place proportionally similar demands on resources or have proportional demands can be aggregated. Items that have these characteristics are called families. For instance, suppose we make several models of ink-jet black and white printers. All the printers are produced on the same equipment by the same workers. We do not need to know the precise number of each model we will produce next month; we do, however, need to know how many labor and machine hours are required. Indeed, if we have estimates of the relative proportion of each model demanded, i.e., 10% of model 4150, 20% of model 4260, and so forth, then we can easily define a generic "printer" that is a weighted composite of all printer models. If multiple product families use the same resources, it may not even be necessary to include ink-jet black and white printers in our model. We may be able to use a single variable to represent all printers, including all black and white, color, ink-jet, and laser models. This would make sense if a single labor grade and machine group produced all these models and formed the constraining factor on capacity. However, if the laser printing subassembly formed a major-capacity bottleneck, we would need to model the laser printer family separately.

Figure 5.1 shows an example of the aggregation process for products and resources. With regard to "units," aggregate plans are often developed in terms of production hours, pounds, or dollars. For instance, suppose ovens required two hours of manual labor on average. We could measure oven production by letting our decision variable be labor hours allocated to ovens and then dividing by 2 whenever we wanted the actual number of ovens to be planned. Pounds of production make sense as a unit of measure for some capacity-oriented processes, particularly those that produce basic

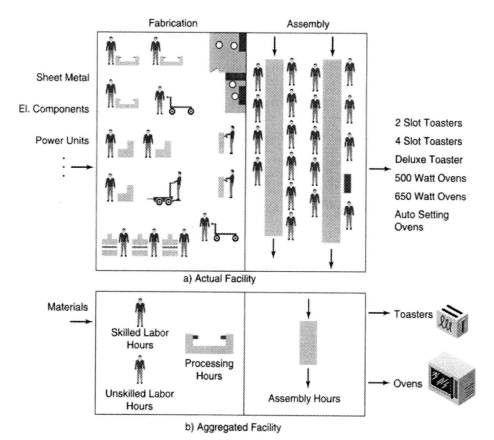

Figure 5.1 Example of Product and Resource Aggregation

materials such as primary metals, building supplies, or other raw materials. Dollars would be meaningful to a financially oriented management team but suffers from an inconsistent basis over time because inflation and competition change product prices. Nevertheless, some manufacturers find that dollars are a useful measure over a medium term horizon such as a year or 18 months.

The appropriate horizon length is determined by several factors. These include: 1) the lead time required to negotiate contracts with suppliers, and then to begin receiving parts and converting them into final products; 2) the time to plan and implement changes in capacity, either through the hiring and training of workers or by the acquisition and implementation of equipment; and 3) the length of the season or cycle for demand. The planning horizon should be long enough to accommodate each of these changes.

In the next section, we discuss the policy options or tactics usually considered in aggregate production planning. This is followed by a series of increasingly complex models. We begin with simple static, linear models and progress to dynamic, nonlinear models. We conclude with a discussion of decomposing the aggregate model output for production of families by month into production quantities for individual products by week. This disaggregation forms the master production schedule (MPS) that drives the operational planning of the factory. In some cases, the MPS is further broken down into a final assembly schedule (FAS) detailing the sequence of products to be assembled in the near future (a length of time ranging from one shift to several days).

5.1 PLANNING TRADE-OFFS: INVENTORY, WORKFORCE CHANGES, OVERTIME, SHORTAGES

Aggregate planning represents an attempt to determine the most economical choice of alternatives for meeting demand of the customers served by the production facility or facilities over the midterm. Inputs to the problem include demands for each product type over the horizon, production and distribution costs, the cost to modify capacity, and policy options. Policy options to be considered usually include letting inventory levels vary, varying production levels, allowing temporary shortages, and adding equipment or changing the workforce level to adjust capacity. Consider the situation shown in Figure 5.2. Cumulative demand is plotted over the next twelve months. Suppose current capacity was just adequate to meet total demand if we produce at capacity for the entire year. If we choose this route, as shown in the Figure, we will build up inventory levels early in the year and experience some backorders later in the year. The level of inventory surplus or shortage is given by the difference between the cumulative demand and production lines (plan 1) in the Figure. Another option would be to start off producing to demand, and then add capacity (workers and/or equipment) later in the year to increase the production rate (plan 2). Both choices incur costs. These costs are both tangible (salaries and interest payments) and intangible (customer satisfaction and employee morale). The optimal policy may be either of these or some combination, depending on the cost structure. The objective of aggregate planning is to find the best policy.

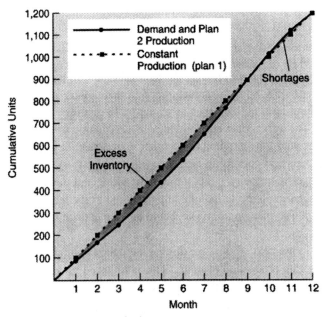

Figure 5.2 Production Options

Table 5.1 Monthly Demand for Example 5.1

5-Month	1	2	3	4	5	6	7	8	9	10	11	12
Demand	90	80	80	90	100	100	110	120	120	120	110	80

EXAMPLE 5.1

Consider Figure 5.2, with capacity of 100 units per month and demands as shown in Table 5.1. Inventory costs are $2 per unit per month, charged on end of month inventory. Shortage costs are $5 per unit per month. Capacity can be increased; however, each unit of capacity costs $3 per month, whether or not it is used. Once increased, capacity exists for each subsequent month. Compare the costs of the two solutions: 1) produce at a constant rate of 100 units per month; and 2) produce to demand, increasing capacity as needed.

SOLUTION

The two production plans and associated costs are shown in Tables 5.2 and 5.3. Maintaining constant production and letting inventory levels vary, as shown in plan 1, costs $4,470. Adding capacity as necessary and producing to demand, as shown in plan 2, costs $3,570. We can save $900 by selecting plan 2. Whether a third plan exists with less cost than plan 2, remains questionable.

Table 5.2 Constant Production Rate Plan for Example 5.1

Plan 1 month	1	2	3	4	5	6	7	8	9	10	11	12
Capacity	100	100	100	100	100	100	100	100	100	100	100	100
Production	100	100	100	100	100	100	100	100	100	100	100	100
Net Inventory	10	30	50	60	60	60	50	30	10	−10	−20	0
Inventory Cost	$20	$60	$100	$120	$120	$120	$100	$60	$20			
Shortage Cost										$50	$100	
Capacity Cost	$300	$300	$300	$300	$300	$300	$300	$300	$300	$300	$300	$300
Total Cost	$320	$360	$400	$420	$420	$420	$400	$360	$320	$350	$400	$300

Table 5.3 Increasing Capacity Plan for Example 5.1

Plan 2 month	1	2	3	4	5	6	7	8	9	10	11	12
Capacity	100	100	100	100	100	100	110	120	120	120	120	120
Production	90	80	80	90	100	100	110	120	120	120	110	80
Net Inventory	0	0	0	0	0	0	0	0	0	0	0	0
Inventory Cost												
Shortage Cost												
Capacity Cost	$300	$300	$300	$300	$300	$300	$330	$360	$360	$360	$360	$360
Total Cost	$300	$300	$300	$300	$300	$300	$330	$360	$360	$360	$360	$360

5.2 BASIC NETWORK MODELS

Suppose we know expected demand for each period of the planning horizon, and we have several options regarding how to meet this demand, each with associated fixed capacities in every period. For instance, we may be able to produce on regular time or overtime or by subcontracting some product to other facilities. If the corresponding production and inventory costs are known for each option, a network flow model can be used to find the optimal production levels for each period of the planning horizon. Inventory levels and backorders will be obtained as a by-product of our model. We use the following notation:

Model Parameters and Productivity Factors:

D_t = demand in period t, $t = 1, \ldots, T$;

c_{jt} = variable cost to produce a unit of product by method j, in period t, $j = 1, \ldots, m$;

h_t = holding cost per unit of product in period t;

K_{jt} = production capacity of method j, in period t;

π_t = shortage cost per unit back-ordered in period t.

Holding and shortage costs are presumed to be charged on the corresponding on-hand and backorders or lost sales each period. We can represent this formulation as a minimum cost network flow problem, as shown in Figure 5.3. Other than the source and sink nodes, we have one set of nodes to represent the sources of production and a second set to represent the demands in each period. Over the planning horizon (assuming no accumulation or depletion of inventory is expected) we must produce $\sum_{t=1}^{T} D_t$ units. The arcs show how the sources can be used to meet demands. Flow on the arcs represents allocation of capacity at the production sources to meet demand. We choose the most economical set of production options. From the Figure, we note that if we decide for each unit of demand which production option will be used, then the cost is easily computed. Thus, we can define our decision variables as $X_{jt_1t_2}$, the number of units produced by method j, in period t_1, to be used to meet demand in period t_2. The corresponding cost per unit is:

$$c_{jt_1t_2} = \begin{cases} c_{jt_1} + h_{t_1} + h_{t_1+1} + \ldots + h_{t_2-1}, \text{ if } t_1 \le t_2 \\ c_{jt_1} + \pi_{t_2} + \pi_{t_2+1} + \ldots + \pi_{t_1-1}, \text{ if } t_1 > t_2 \end{cases}.$$

If $t_1 < t_2$, we produce the item early and incur holding costs for $t_2 - t_1$ periods. If $t_1 > t_2$, we produce the item after it is demanded and, therefore, incur shortage costs for $t_1 - t_2$ periods. This model fits the format of the transportation problem of linear programming.

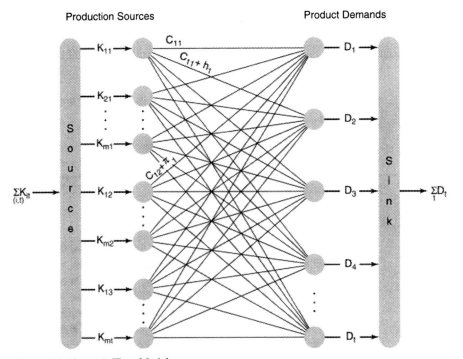

Figure 5.3 Network Flow Model

Table 5.4   Generic Transportation Tableau for Production Planning Problem

Production sources		Demand by period					
Method	Period	1	2	3	...	T	Capacity
1	1	c_{11}	$c_{11} + h$	$c_{11} + 2h$	...	$c_{11} + (T-1)h$	K_{11}
...	1	c_{i1}	$c_{i1} + h$	$c_{i1} + 2h$	...	$c_{i1} + (T-1)h$	K_{i1}
n	1	c_{n1}	$c_{n1} + h$	$c_{n1} + 2h$	...	$c_{n1} + (T-1)h$	K_{n1}
1	2	$c_{12} + \pi$	c_{12}	$c_{12} + h$	...	$c_{12} + (T-2)h$	K_{12}
...	2	$c_{i2} + \pi$	c_{i2}	$c_{i2} + h$	...	$c_{i2} + (T-2)h$	K_{i2}
n	2	$c_{n2} + \pi$	c_{n2}	$c_{n2} + h$	...	$c_{n2} + (T-2)h$	K_{n2}
1	3	$c_{13} + 2\pi$	$c_{13} + \pi$	c_{13}	...	$c_{13} + (T-3)h$	K_{13}
...	...	...	...	...	...	...	...
n	T	$c_{nT} + (T-1)\pi$	$c_{nT} + (T-2)\pi$	$c_{nT} + (T-3)\pi$	...	c_{nT}	K_{nT}
Demand		D_1	D_2	D_3	...	D_T	

We essentially ship (use) resources from the sources of supply (production options in each period) to the destination (demand periods). We can represent the model with a transportation tableau, as shown in Table 5.4. Each row is a production source. Each column is a demand period. The $c_{jt_1t_2}$ values in the upper right corner of each box indicate the cost per unit to meet the demand for the period indicated by the column heading, with the production source indicated by the row. The solution involves filling in the $X_{jt_1t_2}$ values in each box whereby a feasible solution is found with minimum cost. A solution is feasible if the sum of values in each row does not exceed the supply, and the sum of values in each column equals its demand. The cost is given by summing the product of the $C_{it_1t_2}$ and $X_{jt_1t_2}$ in each cell of the Table.

The optimal solution to the transportation model can be found relatively efficiently. We will show how to obtain a good feasible solution. (The interested reader may consult any introductory operations research text for details on how to improve on this solution.) A good feasible solution can be found by using the lowest cost cells. We will use a greedy procedure of first ranking cells based on their $c_{jt_1t_2}$ values. At each step we will try to fill as much of the demand as possible with the currently available lowest cost. Thus, we set the $X_{jt_1t_2}$ for the cell equal to the minimum of the column demand and row capacity. After each assignment, we update the remaining row capacity and unmet column demand, and then proceed to the next cheapest cell.

EXAMPLE 5.2

A firm is planning production for the next four months. Demand (in thousands of units) is 150, 150, 220, and 160, respectively. Production cost is $50 per unit on regular time, or $65 per unit on overtime. Inventory holding cost is $4 per unit per month and shortage cost is $6 per unit per month. Capacity is 180 per month regular time and 25 per month on overtime. Find a good production plan and the total cost for that plan.

SOLUTION

The transportation tableau for this problem is given in Table 5.5. As an example, consider the cell for overtime in period 2 (row 4) to meet demand in period 3 (column 3). The cost of this cell per unit is $69, which includes $65 for production, plus $4 for holding a unit one period. The lowest cost cells are the $50 values corresponding to regular time production that is used for demand in the period of production. Consider period 1: Because demand is 150 and capacity is 180 for this

Table 5.5 Transportation Tableau for Example 5.2

Production		Demand period				
Method	Period	1	2	3	4	Capacity
Regular	1	50 **150**	54	58	62	180
Overtime	1	65	69	73	77	25
Regular	2	56	50 **150**	54 **30**	58	180
Overtime	2	71	65	69	73	25
Regular	3	62	56	50 **180**	54	180
Overtime	3	77	71	65	69	25
Regular	4	68	62	56 **10**	50 **160**	180
Overtime	4	83	77	71	65	25
	Demand	150	150	220	160	

resource, we assign $X_{111} = 150$ the smaller of the two quantities. We then update the remaining capacity for this row to $180 - 150 = 30$. Next, we set the $X_{122}, X_{133}, X_{144}$ variables similarly to 150, 180, and 160, respectively, and update the availabilities. The only remaining demand is for 40 units in period 3. The lowest cost cell in the period 3 column with remaining capacity, is regular time production in period 2 at \$54 per unit. We would like to meet 40 units of demand; however, this source has only 30 units left. Thus, we set $X_{123} = 30$. The remaining demand in period 3 is 10, and capacity for regular time period 2 production is set to 0. The next cheapest alternative in column 3 is the cell for X_{143}. Capacity is 20; however, we need only 10 units. These units are allocated and we have our initial solution. Summing production quantities across rows, we see that all production is during regular time. We produce 150 units in period 1, 180 in periods 2 and 3, and 170 in period 4. The cost of this solution is: $Cost = (150)(50) + (150)(50) + (30)(54) + (180)(50) + (160)(50) + (10)(56) = 34,180$.

In the previous model, we assumed the objective was to produce precisely in accord with expected demand. In practice, other considerations may exist. We may have initial inventory. This can be easily accommodated by allocating this inventory to demand. Demand for the first period is then reduced by the amount of initial inventory (I_0). On the other hand, we may want to carry safety stock (SS), because in actuality we are using demand forecasts. In this case, any desired difference between current on-hand inventory, and desired safety stock, can be added to the first period's demand. Thus, we correct the demand in period 1 to D_1, where $D_1 = D_1 + SS - I_0$. In this way, the model will automatically produce the appropriate correction to bring inventory to the desired level. Thereafter, if we produce to expected demand in each period, we would expect the inventory level to remain at the safety stock level. Another approach, would be to define a service level for meeting demand, as we did in our earlier inventory models. Instead of using expected demand, we can use a specific point on the upper end of the cumulative demand distribution. Suppose, for instance, we wanted to make sure we do not have more than a 1% chance of shortages in any period. Let $F_t(D)$ be the cumulative distribution of demand in the first t periods. We could then set the demand level for the model at $D_1 = F_1^{-1}$ (0.99) $- I_0$, the 99th percentile point of demand corrected for any initial inventory. We then set: $D_2 = F_2^{-1}(0.99) - F_1^{-1}(0.99)$. This expression says to set the incremental demand in period 2 to the 99th percentile of cumulative demand for periods 1 and 2, minus the units considered for period 1.

We could also add multiple product types to the model. Each column in the table represents a demand. If we can represent the supply resources in common units, then we need only add additional columns to represent demand for the new items in each period. Thus, it is acceptable if demand and resource utilization for all products is measured in labor hours or pounds. However, if multiple resources or different units are needed, we will need to develop more general models such as those discussed in the next section.

EXAMPLE 5.3

Two production facilities are available to meet demand for two product families. Each plant usually services demand in its own region, however, shipments among plants can be made if necessary. Demands are measured in labor hours. Facility 1 has 1200 labor hours available per period and facility 2 has 1500 hours per period. Production costs per hour are the same at both plants. Shortages are not allowed, and holding costs are $1 per labor hour per period for either product. Shipment costs among plants are $1.50 per unit. Each unit of family 1 requires 2 hours of labor, and each unit of family 2 requires 3 hours of labor. Unit demands for the next two periods are given below. Form the transportation tableau for deciding how to meet demand.

	Demands (Units)			
Facility	Period 1, Family 1	Period 1, Family 2	Period 2, Family 1	Period 2, Family 2
1	300	250	400	200
2	200	250	350	200

SOLUTION

To build the model, we must define the sources and destinations. Sources correspond to production capacity by plant and period. Destinations are demands by region and product. Demands must be measured in the common units of labor hours. The corresponding tableau of transportation costs is shown in Table 5.6. A row (source) exists for each production facility in each pe-

Table 5.6   Transportation Problem Tableau for Example 5.3

Labor sources		Demand by period, product, region								
Facility	Period	1,1,1	1,1,2	1,2,1	1,2,2	2,1,1	2,1,2	2,2,1	2,2,2	Labor hours
1	1	0	0.75	0	0.50	1.00	1.75	1.00	1.50	1200
1	2	∞	∞	∞	∞	0	0.75	0	0.50	1200
2	1	0.75	0	0.50	0	1.75	1.00	1.50	1.00	1500
2	2	∞	∞	∞	∞	0.75	0	0.50	0	1500
Labor	Demand	600	400	750	750	800	700	600	600	

riod. A column (demand destination) exists for each product, in each region, in each period. Because shortages are not allowed, costs are infinite to produce in period 2 for use in period 1. Production costs are fixed and, therefore, not included. Costs in the Table refer to transportation and inventory carrying. Shipments occur when the production facility differs from the sales region. Each labor hour of product 1 results in a $0.75 shipment cost ($1.50 per unit divided by 2 production hours per unit), and each labor hour spent on product 2 would cost $0.50 to ship. Cells in the Table corresponding to production in period 1, for use in period 2, include a $1 per labor hour inventory holding cost, plus any relevant shipment cost. Consider, for instance, the cell for production of product 1 in period 1 at plant 2, for use at plant 1 in period 2. The relevant cost includes shipment among plants, and storage for one period. Thus, $c = \$0.75 + \$1.00 = \$1.75$. The far right column indicates available labor hours for the row source. Recalling the labor production stated in the problem, demand in labor **hours,** which is indicated along the bottom of the Table, is set to twice the demand for product 1, and three times the demand for product 2.

Finding the minimum cost method to accommodate flow in a network is a relatively easy problem to solve. The production planning model described above fits into the class of network problems. However, this is a very special problem form. When this structure does not hold, we must solve the problem as a general linear program. This is the subject of the next section.

5.3 LINEAR PROGRAMMING MODELS

Linear programming provides a powerful tool for modeling the aggregate planning problem. The network models of the previous section were efficient but not very robust with respect to the number of situations that could be modeled. Linear programming models (see Appendix I for more details) can describe any problem in which the decision variables may take on any value in a continuous interval, and the objective and constraints can be represented as linear functions (equalities or inequalities) of the decision variables. Our decision variables will typically be production and workforce levels. The objective is to minimize cost, and the constraints will come from the physical realities of conservation of mass, the limited resources available, and the policies of the organization. To describe the generic linear program, let $\underline{c}$ be an $n \times 1$ vector of cost coefficients, let $\underline{x}$ be an $n \times 1$ vector of decision variables, let A be an $m \times n$ matrix of technological coefficients defining resource usage per unit of the decision variables, and let $\underline{b}$ be an $m \times 1$ vector defining the amount of each resource available. The general linear programming model is:

$$Min \ \underline{c}^t \underline{x}; \tag{5.1}$$

subject to:

$$A\underline{x} = \underline{b} \tag{5.2}$$

$$\underline{x} \geq 0 \tag{5.3}$$

With a little ingenuity, many real-world situations can be modeled with this general structure. Even nonlinear cost or resource usage functions can often be modeled by using a linear approximation over the likely range of the decision variables. The most common factor that cannot be easily handled by linear models is the presence of setup costs or setup times. However, we will see that with clever formulation of the problem, even setup costs and times can be accommodated under certain conditions. To convey the power of linear programming models, we present a series of increasingly complex models. These models are typical of the factors often encountered in practice; however, as always, the reader should realize that every problem has its own local idiosyncracies. Each model should be built to answer the questions of local interest while incorporating the factors of local importance.

Note the presence of five basic components in each model. They are:

1. An objective function. In each model, we specify the relationship to minimize or maximize. In most cases, this will be an economic objective involving revenues and costs, and we will minimize costs or maximize profit.

2. Resource availability constraints. There are resource availabilities that limit our options. Maximum demand limits how much we can sell. The available machine and labor hours limit how much we can produce, as does the amount of raw materials we can acquire. In some models, financial investment or storage space may be a critical resource. In practice, it is sufficient and simpler to include in the model only those resources that cannot be easily expanded over the time horizon. Thus, if we can always buy more steel, then we can treat this as a variable cost component in the objective function, and not limit production to the use of steel on hand.

3. Conservation constraints. Physical stocks (be it inventory or workers) follow the basic law of conservation, which states that the amount available at the end of a period is given by the number available at the start of the period, plus what we add (produce or hire), minus what we use (sell or layoff). The exception to this rule is in labor hours. Labor hours are usually a "renewable" resource in the sense that we can use 40 new hours per week from each worker, and we do not need to rehire continuing workers each week.

4. Decision variables. Decision variables are the factors we want to determine such as production, workforce, and inventory levels. Worker levels and on-hand inventories cannot be negative. Accordingly, the corresponding variables must be constrained to be greater than or equal to zero. Setups may or may not occur in a period and hence are generally modeled using 0-1 variables. For each variable, we must define its type (non-negative, 0-1, general integer, etc.). The type definition may also include bounds on the range of the variable such as $0 \leq x \leq 1$.

5. Technological coefficients. We must know the resource availabilities, the amount of each resource required for the unit produced, and the variable production costs. These technological coefficients are used to convert the values of the decision variables into total costs and resource consumption. These values are usually available from standards held by the industrial engineering department or through accounting systems that track costs for financial purposes.

5.3.1 Dynamic, Single-Stage Models

In our initial model, we assume a single product type is being produced; however, several alternative methods exist to produce the product. These methods may be different machine groups, different work shifts, or different subcontracting options. We also include the presence of several important resource types that may limit production. Limiting resource types might include machine time, material availability, and/or worker hours. We begin by defining the decision variables for the model. This is followed by a list of the technological parameters. These parameters indicate the specific data that must be collected and maintained to operate the model.

Model Decision Variables:

X_{jt} = the number of units of product produced by method j in period t; and

I_t = net inventory at the end of period t.

The basic unit of inventory must be defined. This could be a pallet load, a specific number of units such as 12 dozen, an hour's worth of production, or a specific weight or volume of product such as 100 pounds. The unit, however, must be understood and consistently used throughout the modeling process. Net inventory at the end of an arbitrary period t is given by $I_t = I_t^+ - I_t^-$, the difference between the on-hand inventory and the backorders at time t. The standard approach is to charge inventory holding cost on the number of units left in inventory at the end of each period. This convention is used not only for simplicity, but also represents fairly the variable holding cost in most cases. If inventory is the same on average throughout the period, then this provides an unbiased estimate of actual cost. Suppose, on the other hand, that production occurs throughout the period, with shipments coming at the end of the period. Actual-time average inventory levels may be higher than the end-of-period level. The difference, however, is attributable to the flow time of each unit in the system. Because total production is determined by demand, the inventory cost not accounted for in the model (approximately $0.5h\overline{D}$ each period) is a constant. Adding or subtracting a constant value to the objective function will not change the optimal decision variables; therefore, our model is still satisfactory.

Model Parameters: We add the following notation for technological parameters used by the model:

π_t = shortage cost per unit per period charged based on end-of-period inventory level;

a_{jk} = the amount of resource k required to make one unit of product by method j;

c_{jt} = unit variable production cost for each unit of product made by method j in period t;

h_t = unit variable holding cost per period charged to each unit of product on hand at the end of period t;

R_{kt} = the number of units of resource k available for use in period t.

We can formulate the planning problem as:

$$Min \sum_{t=1}^{T} \sum_{j=1}^{m} c_{jt} X_{jt} + \sum_{t=1}^{T} h_t I_t^+ + \sum_{t=1}^{T} \pi_t I_t^- \tag{5.4}$$

subject to:

$$(I_t^+ - I_t^-) = (I_{t-1}^+ - I_{t-1}^-) + \sum_{j=1}^{m} X_{jt} - D_t \cdot \forall t \tag{5.5}$$

$$\sum_{j=1}^{m} a_{jk} X_{jt} \le R_{kt}, \forall k,t \tag{5.6}$$

$$0 \le X_{jt}, \forall j,t \tag{5.7}$$

The objective function (Equation 5.4) minimizes the sum of production costs, holding costs, and shortage costs. The first set of constraints [Equation (5.5)] tracks the inventory levels through balance equations. The constraints view inventory as a storage point, with the change in level from one period to the next being equal to the difference between the production that was added to inventory and the sales that were taken away. Next, we add the capacity constraints. For each key resource $k = 1, ..., K$, we must ensure that the amount of that resource used in period t does not exceed the amount available. The left hand side of Equation set (5.6) describes the amount of resource k needed as a function of the amount of product produced. The right hand side of this Equation set lists the amount of that resource available for use. Resources could be types of labor or machines. If the supply of a raw material was limited, it could also be modeled as a resource. For instance, if sheet metal rolls of a particular steel alloy were temporarily unavailable because of backorders with the supplier, we could model the limited production of products using this material in the near-term until delivery problems could be solved. The final set of constraints [Equation (5.7)] defines the variable types and limits.

We are assuming that resource limits are known a priori for each period. This is typically true for renewable resources such as labor. In some cases, a limit will exist on total resource consumption over the horizon. For instance, we may have only a limited supply of a material available for use in the first τ periods. This situation is easily handled by accumulating total resource usage as follows:

$$\sum_{t=1}^{\tau} \sum_{j=1}^{m} a_{jk} X_{jt} \le R_k.$$

The development of a model often goes through several stages. The first step is to define the set of questions the model should answer. Are we interested in exploring the effect of capacity changes or in just determining next week's production plan? Do we want production quantities for the 1,000 main products, the five aggregate product families, or all 100,000 possible end-item configurations, with all options specified? Will the model be used for determining financial needs in the next quarter or only for issuing purchase orders for bulk raw materials? This should determine the set of decision variables. It is usually helpful to draw a schematic diagram representing the flow of material. Nodes or icons can be used for the production facilities. Each facility to be included in the model should be shown in the diagram. Typically, this involves all production sources, particularly if the source has limited capacity. We will use triangles to represent inventory stores. If the number of products is small, we will include a triangle for each product type at each location. Otherwise, we will let triangles represent all part types stored at a specific location. Both raw material and finished goods inventory will be identified. Arcs represent material flows. These flows indicate the possible paths for materials to be shipped through the production-distribution system. A diagram of the example above is shown in Figure 5.4. We strongly urge that the modeler construct a schematic such as that in Figure 5.4 when constructing an aggregate model of any production-distribution system. The diagram may change as the modeler changes the level of aggregation or scope of decisions being considered, but it will always provide a quick visual representation of the model, and it will greatly facilitate the construction of the balance and resource availability constraints in the model.

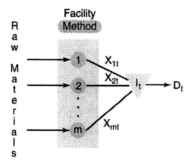

Figure 5.4 Schematic for Single Facility, Multisource, Dynamic System

The outputs of the model will either be values for the decision variables or functions of the variables that can be combined with other knowledge to compute the values. For instance, we may want production quantities for all 1,000 items. However, if each item is a known percentage of its family production, we need only directly model the five aggregate family production quantities. It is quite likely we will want inventory information for financial planning and checking space feasibility. However, we can obtain this information from a simpler formulation. Note that inventory at any point is the difference between cumulative production and cumulative demand. Thus,

$$I_\tau^+ - I_\tau^- = \sum_{t=1}^{\tau} \left(\sum_{j=1}^{m} X_{jt} - D_t \right) + I_{i0}. \tag{5.8}$$

Using this result in Equation (5.5), we can eliminate the inventory variables from the formulation. This gives us a smaller and consequently more easily solved linear program. Inventory values can then be recovered after the solution is found for the X_{jt} values by using Equation (5.8).

Linear programs are relatively easy to solve. However, as we progress through this chapter, models become larger and larger. In practice, we may have hundreds of thousands of continuous variables and perhaps even thousands of integer variables. In these cases, transforming the model to make it easier to solve becomes necessary because of the computational burden imposed by many continuous, or even a few integer variables. A full discussion of efficient formulations that provide the fewest variables and tightest constraints is beyond the scope of this text. The reader facing this problem should consult the references at the end of the chapter and consult their software provider.

The basic linear program for aggregate planning described in Equations (5.4) through (5.7) would allow determination of the optimal fluctuation in inventory level and production rate over the planning horizon. If capacity is determined by equipment, our options may be limited to the best way to accumulate excess inventory in slow periods and to endure shortages in heavy demand periods. We might be able to subcontract some production or increase production rate of machines at the expense of greater machine wear and accompanying higher variable production costs. If capacity is labor-oriented, we will probably have the option of changing the workforce level and/or scheduling overtime and undertime. (Undertime represents less than full use of capacity. We may pay workers for a 40-hour week despite planning only 30 hours of productive usage. The other hours could be used for routine maintenance and training. If workers can be dismissed early without pay, there may be no loss because of undertime.)

Next, we present a model that aggregates production methods into a single method but allows production capacity to be changed through the use of overtime or by hiring or firing workers. As noted previously, we also include the option of allowing inventory levels to vary over time. The solution procedure will automatically determine the optimal combination of strategies to be used to meet demand over time. Finally, we permit several product types to be modeled separately. These will be indexed by $i = 1, ..., n$. X_{it} is therefore the quantity of product i produced in period t. In addition to the previous notation, we add the following decision variables:

F_t = the number of workers fired, effective the beginning of period t;

H_t = the number of workers hired to begin work in period t;

O_t = the number of overtime hours planned for the workforce in period t;

U_t = the number of undertime or unused but paid hours of labor in period t;

W_t = the number of workers employed and available during regular time in period t.

The added parameters are:

θ = the retention factor for employees ($1-$ the natural turnover rate);

r_t = the number of regular time work hours scheduled for each worker in period t;

o_t = the cost of labor per hour of overtime in period t;

u_t = the cost per hour of unused labor (this is negative if hours are not guaranteed to labor and unneeded workers can be dismissed early without pay);

w_t = the loaded labor rate (salary plus benefits) for period t;

w_t^+ = the hiring cost per worker in period t (includes searching and training);

w_t^- = the cost per worker laid off in period t (severance pay, unemployment tax, local goodwill).

The formulation is then:

$$\text{Minimize} \sum_{t=1}^{T} \left[\sum_{i=1}^{n} (c_{it} X_{it} + h_{it} I_{it}^+ + \pi_{it} I_{it}^-) + w_t W_t + o_t O_t + u_t U_t + w_t^+ H_t + w_t^- F_t \right] \quad (5.9)$$

subject to:

$$I_{it}^+ - I_{it}^- = I_{i,t-1}^+ - I_{i,t-1}^- + X_{it} - D_{it}, \, i = 1, ..., n; \, t = 1, ..., T \quad (5.10)$$

$$W_t = \theta W_{t-1} + H_t - F_t, \, t = 1, ..., T \quad (5.11)$$

$$\sum_{i=1}^{n} a_i X_{it} \le r_t W_t + O_t - U_t, \, t = 1, ..., T \quad (5.12)$$

$$X_{it}, I_{it}^+, I_{it}^-, H_t, F_t, O_t, U_t \ge 0 \quad (5.13)$$

The objective function [Equation (5.9)] includes the previous costs of production and inventory, but we have also added the costs of regular time labor, overtime, undertime, and hiring and firing. Note that we have removed the labor cost from the variable cost component c_{it}. The variable cost still includes the cost of raw material and other nonlabor processing costs. The constraints are similar to those noted earlier in the chapter. Equation (5.10) is the inventory balance constraint. We have added a workforce balance constraint [Equation (5.11)] to account for the changes in regular time workforce level in each period. The workforce available in period t is given by the θW_{t-1} workers remaining from last period, plus those we hire, minus those we fire. Constraints [Equation (5.12)] treat labor as the primary resource. On the left, the amount of labor needed to make the planned production level is given. On the right, the actual hours provided by the model's decision variables are given, namely, regular time hours, plus overtime hours, minus unused hours. The r_t parameters account for period to period schedule differences due to holidays, changeovers, and other factors.

EXAMPLE 5.4

A manufacturing plant produces small kitchen appliances—toasters and microwave ovens. The plant operates a single shift with 200 line workers. Each worker is scheduled for 7.5 hours per day. The next three months have 22, 18, and 20 working days, respectively. Each toaster requires 0.6 labor hours, and each microwave requires 2.3 labor hours. Average worker cost is $28 per hour. Although some presses and injection molding machines are used in the process, adequate capacity exists and production is regulated by labor capacity. Workers are guaranteed 7.5 hours per day but can be asked to work up to 2 hours of overtime per day. Overtime wages are 50% more. Workers can be hired at a cost of $1250, but they are only 50% effective the first month. The severance pay package for laying off a worker is $2500. Holding costs are estimated at $0.50 per month for toasters and

$1 per month for ovens. Using the following initial inventory and demand data, formulate the aggregate planning problem.

Product	Initial Inventory	Month 1 Demand	Month 2 Demand	Month 3 Demand
Toasters	500	25,000	26,000	26,000
Microwave	1,250	5,000	5,500	5,500

SOLUTION

The key modeling decisions concern the level of aggregation of actual products into n product families for the model and the choice of limiting resources. From the problem statement, we can surmise that it is meaningful to let toasters and ovens be the two product families modeled. This decision may not be so straightforward in practice if different oven or toaster types varied somewhat in demand trend or labor usage. Labor appears to be the only resource that needs to be modeled explicitly. Let X_{it} and I_{it} be the production and end-of-period inventory levels, and let W_t, H_t, and F_t be the workforce, hiring, and firing decision in month t, measured in workers. O_t will represent overtime hours. We can formulate the problem as:

$$Minimize\ 28(7.5)(22)W_1 + 28(18)(7.5)W_2 + 28(7.5)(20)W_3$$

$$+ \sum_{t=1}^{3} (1250H_t + 2500F_t + 42O_t + 0.5I_{1t} + I_{2t}) \tag{5.14}$$

subject to:

$$I_{11} = 500 + X_{11} - 25000$$
$$I_{21} = 1250 + X_{21} - 5000$$
$$I_{12} = I_{11} + X_{12} - 26000$$
$$I_{22} = I_{21} + X_{22} - 5500 \tag{5.15}$$
$$I_{13} = I_{12} + X_{13} - 26000$$
$$I_{23} = I_{22} + X_{23} - 5500$$

$$W_1 = 200 + H_1 - F_1$$
$$W_2 = W_1 + H_2 - F_2 \tag{5.16}$$
$$W_3 = W_2 + H_3 - F_3$$

$$0.6X_{11} + 2.3X_{21} \le 7.5(22)W_1 + O_1 - 3.75(22)H_1$$
$$0.6X_{12} + 2.3X_{22} \le 7.5(18)W_2 + O_2 - 3.75(18)H_2 \tag{5.17}$$
$$0.6X_{13} + 2.3X_{23} \le 7.5(20)W_3 + O_3 - 3.75(20)H_3$$

$$O_1 \le 2(22)W_1;\ O_2 \le 2(18)W_2;\ O_3 \le 2(20)W_3 \tag{5.18}$$
$$X_{it}, I_{it}, W_t, H_t, F_t \ge 0\ \forall it$$

The objective function [Equation (5.14)] contains regular-time labor, hiring, firing, overtime, and inventory costs. The regular-time labor cost coefficients give cost per worker per month and are thus obtained by multiplying ($/worker-hr) × (hr/day) × (days/mth). The other costs are taken directly from the problem statement. Note that the units on each term in the objective is dollars. The first six constraints are the inventory balance constraints, and they account for the inventory of each product at the end of each period. Ending inventory is starting inventory, plus production, minus demand shipments. The next three constraints [Equation (5.16)] regulate the workforce level balance in each period. The three constraints [Equation (5.17)] ensure that sufficient labor hours are scheduled in each period. Hours required are given by 0.6 times the number of toasters produced plus 2.3 times the number of ovens produced in the period. The right side of the Equation lists available hours, consisting of regular-time paid labor plus overtime corrected for the loss of worker productivity in their first period after being hired. The inequality prevents the model from using more hours than are available in each period. Finally, we limit overtime in each period to 2 hours per worker day and force all variables to be non-negative.

EXAMPLE 5.5

Solve the problem described in Example 5.4 and determine the shadow prices of scarce resources.

SOLUTION

The formulation was solved using a commercial software package. The values of the decision variables in the optimal solution are given in Table 5.7. We see that the company currently has more workers than it needs. The optimal solution is to fire 22.6 workers in the first period to bring the

Table 5.7   Optimal Solution for Example 5.5.

	Variable	Value	Cost	Red. cost	Status
		Example 5.5 Optimal solution—Detailed report			
1	X11	24500.0000	0.0000	0.0000	Basic
2	X12	26000.0000	0.0000	0.0000	Basic
3	X13	26000.0000	0.0000	0.0000	Basic
4	X21	6334.4200	0.0000	0.0000	Basic
5	X22	3629.3480	0.0000	0.0000	Basic
6	X23	4786.2320	0.0000	0.0000	Basic
7	W1	177.3889	4620.0000	0.0000	Basic
8	W2	177.3889	3780.0000	0.0000	Basic
9	W3	177.3889	4200.0000	0.0000	Basic
10	H1	0.0000	1250.0000	5566.9930	Lower bound
11	H2	0.0000	1250.0000	4279.9640	Lower bound
12	H3	0.0000	1250.0000	3732.9710	Lower bound
13	F1	22.6111	2500.0000	0.0000	Basic
14	F2	0.0000	2500.0000	986.0145	Lower bound
15	F3	0.0000	2500.0000	1734.0580	Lower bound
16	O1	0.0000	42.0000	19.9758	Lower bound
17	O2	0.0000	42.0000	19.5411	Lower bound
18	O3	0.0000	42.0000	19.1063	Lower bound
19	I11	0.0000	0.5000	0.2391	Lower bound
20	I12	0.0000	0.5000	0.2391	Lower bound
21	I13	0.0000	0.5000	14.2362	Lower bound
22	I21	2584.4200	1.0000	0.0000	Basic
23	I22	713.7681	1.0000	0.0000	Basic
24	I23	0.0000	1.0000	53.6556	Lower bound
Slack Variables					
34	WF CAP T = 1	0.0000	0.0000	22.0242	Lower bound
35	WF CAP T = 2	0.0000	0.0000	22.4589	Lower bound
36	WF CAP T = 3	0.0000	0.0000	22.8937	Lower bound
37	OT CAP T = 1	7805.1110	0.0000	0.0000	Basic
38	OT CAP T = 2	6386.0000	0.0000	0.0000	Basic
39	OT CAP T = 3	7095.5560	0.0000	0.0000	Basic

Objective Function Value = 2294926

(*continued*)

Table 5.7   (*continued*)

<table>
<tr><th colspan="6">Example 5.5
Optimal solution—Detailed report</th></tr>
<tr><th></th><th>Constraint</th><th>Type</th><th>RHS</th><th>Slack</th><th>Shadow price</th></tr>
<tr><td>1</td><td>INV BAL 11</td><td>=</td><td>24500.0000</td><td>0.0000</td><td>13.2145</td></tr>
<tr><td>2</td><td>INV BAL 21</td><td>=</td><td>3750.0000</td><td>0.0000</td><td>50.6556</td></tr>
<tr><td>3</td><td>INV BAL 12</td><td>=</td><td>26000.0000</td><td>0.0000</td><td>13.4754</td></tr>
<tr><td>4</td><td>INV BAL 22</td><td>=</td><td>5500.0000</td><td>0.0000</td><td>51.6556</td></tr>
<tr><td>5</td><td>INV BAL 13</td><td>=</td><td>26000.0000</td><td>0.0000</td><td>13.7362</td></tr>
<tr><td>6</td><td>INV BAL 23</td><td>=</td><td>5500.0000</td><td>0.0000</td><td>52.6556</td></tr>
<tr><td>7</td><td>WF BAL T = 1</td><td>=</td><td>200.0000</td><td>0.0000</td><td>2500.0000</td></tr>
<tr><td>8</td><td>WF BAL T = 2</td><td>=</td><td>0.0000</td><td>0.0000</td><td>1513.9860</td></tr>
<tr><td>9</td><td>WF BAL T = 3</td><td>=</td><td>0.0000</td><td>0.0000</td><td>765.9420</td></tr>
<tr><td>10</td><td>WF CAP T = 1</td><td><=</td><td>0.0000</td><td>0.0000</td><td>−22.0242</td></tr>
<tr><td>11</td><td>WF CAP T = 2</td><td><=</td><td>0.0000</td><td>0.0000</td><td>−22.4589</td></tr>
<tr><td>12</td><td>WF CAP T = 3</td><td><=</td><td>0.0000</td><td>0.0000</td><td>−22.8937</td></tr>
<tr><td>13</td><td>OT CAP T = 1</td><td><=</td><td>0.0000</td><td>7805.1110</td><td>0.0000</td></tr>
<tr><td>14</td><td>OT CAP T = 2</td><td><=</td><td>0.0000</td><td>6386.0000</td><td>0.0000</td></tr>
<tr><td>15</td><td>OT CAP T = 3</td><td><=</td><td>0.0000</td><td>7095.5560</td><td>0.0000</td></tr>
<tr><td colspan="6">Objective Function Value = 2294926</td></tr>
</table>

workforce level down and then to maintain that level over the horizon. Toasters are produced to demand in each period. All remaining labor is used to produce ovens. The first month has extra work days, and this capacity is used to build up an inventory of 2584 ovens at the end of the period. This inventory is consumed in periods 2 and 3 in which demand exceeds capacity. Note that inventory was stored as ovens because the cost of inventory per labor hour is only half that of toasters, i.e., $1/2.3 hours instead of $0.5/0.6 hours. Note also that the model solution caters precisely to the data provided. We are left with zero inventory and a workforce sized to meet demand over these three periods. If future demand is higher, we will be forced to rehire workers or use overtime. In practice, of course, the decision maker may decide not to lay off as many workers. Undertime may be used, or some inventory build-up may be planned.

The shadow prices for the constraints provide useful information. The price of $13.2 for the first inventory balance constraint tells us how much more it would cost to meet demand for one additional toaster in period 1. The price of $2,500 on the first workforce balance constraint reconfirms that if we started with an extra employee, we would just want to lay him off at a severance cost of $2,500. A free volunteer hour of labor in period 1 would save us $22; however, there is no value in increasing the limit on overtime (as we would expect because overtime is never used). In a similar manner, the reduced costs indicate how the total cost would change in the optimal solution if we started to increase the corresponding variable.

5.3.2   Dynamic Multistage Models

Most products go through multiple stages during production. Raw material may pass through several facilities as it is transformed into a component, then subassembled, and then turned into a finished product. Within each facility, the materials or components may pass through several departments for processing. In each department, several workstations may be used. Even on a single machine, such as a progressive die, the part may undergo several processing steps. When the total path spans multiple periods and product may be stored as inventory between stages, we need to model these stages explicitly. This is often the case in multifacility networks in which material is transferred between several facilities as it progresses from raw material to delivered, finished product. In this environment, we need a model that helps coordinate the flow of products between facilities over time.

5.3.2.1   Multistage Production Facilities

Imagine a manufacturing plant where materials pass through several departments during production. Batches of material flow directly between work centers within each department. This material, however, enters a controlled storage facility between departments. For instance, the part begins with basic rough shape forming, then goes through machining operations. After machining, the part is sent to a satellite facility for painting. Later, painted parts are withdrawn from storage and used in final assembly. If products flow through the production process in a continuous manner, we may use the models discussed in the previous section. However, when product is produced at different rates at different stages and some product types stored for long periods at various stages of completion, we need to explicit include these production stages in our model. As input data for the model, we require resource availability, demand for each part family, storage locations, and resource usage per unit of production for each family. Current inventory and costs are also needed, along with any organizational constraints. The model then consists of minimizing all relevant costs of production, inventory, and shortages subject to inventory balance and resource capacity constraints. We can illustrate this with an example.

EXAMPLE 5.6

A plant produces three product families using the same facilities. There are two key departments—plating and finishing. All products go through plating first, and then finishing. Each of these departments has 300 hours of available time per month. Historical shop data indicate that the following time is required per unit of each product family:

Product Family	Plating Hours/Unit	Finishing Hours/Unit
A	0.1	0.5
B	0.3	0.2
C	0.25	0.1

Demand is normally distributed with the means and variances (μ, σ^2)

Product	Month 1	Month 2
A	(250, 100)	(300,100)
B	(500, 225)	(650,225)
C	(350, 100)	(300,100)

Company policy is to meet at least 95% of demand in any month. Current inventory levels are 0, but inventories of each item may be kept in either plated or finished form. Carrying costs per unit per month are:

Product	After Plating	After Finishing
A	$1.50	$2.25
B	$1.00	$2.00
C	$2.00	$3.75

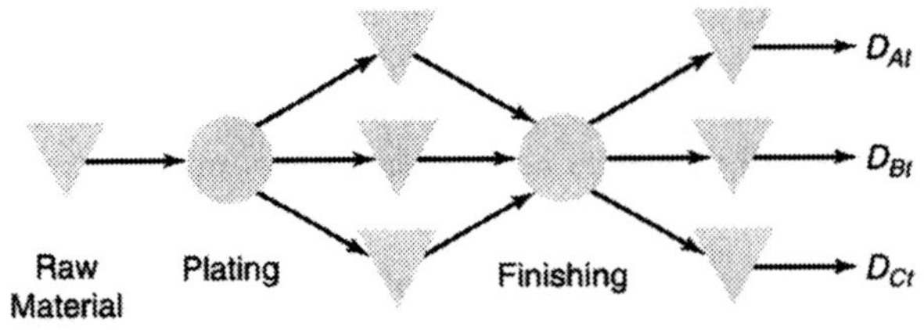

Figure 5.5 Schematic Diagram for Example 5.6

In addition, only 800 units of a certain raw material are available in the first month. Part B uses 1 pound of this material per unit, and each unit of product C requires 1.5 pounds. Formulate this planning problem as a linear program.

SOLUTION

We begin by representing the system in a schematic diagram showing production stages, inventory locations, and material flows (Figure 5.5). Next, we define the decision variables. Let X_{ijt} be the amount of product i ($i = A,B,C$), sent through department j ($j = 1,2$), in period t ($t = 1,2$). Accordingly, I_{ijt} will be the amount of inventory of product i that has completed process j and is on hand at the end of period t. Because of the desire to satisfy at least 95% of demand, our approach is that no shortages will be planned, and only non-negative inventory variables are needed. Instead we will adjust the demand values. The formulation becomes:

$$Minimize \sum_{t=1}^{2} (1.5I_{A1t} + 2.25I_{A2t} + 1I_{B1t} + 2I_{B2t} + 2I_{C1t} + 3.75I_{C2t}) \qquad (5.19)$$

subject to:

$$0.1X_{A1t} + 0.3X_{B1t} + 0.25X_{C1t} \leq 300; \; t = 1,2 \qquad (5.20)$$

$$0.5X_{A2t} + 0.2X_{B2t} + 0.1X_{C2t} \leq 300; \; t = 1,2 \qquad (5.21)$$

$$I_{i1t} = I_{i1,t-1} + X_{i1t} - X_{i2t}; \; t = 1,2; \; i = A,B,C \qquad (5.22)$$

$$I_{i2t} = I_{i2,t-1} + X_{i2t} - D_{it}; \; t = 1,2; \; i = A,B,C \qquad (5.23)$$

$$X_{B11} + 1.5X_{C11} \leq 800 \qquad (5.24)$$

$$X_{ijt} \geq 0, I_{ijt} \geq 0 \qquad (5.25)$$

The objective function [Equation (5.19)] accumulates inventory costs. The first set of constraints limits production in the plating department based on the 300 hours available per period. The left hand side of the constraint shows the hours required for the selected production quantities for products A, B, and C. Constraints (5.20) limit plating time and (5.21) limit finishing time. The next pair of constraints account for the inventory of each product type at the end of plating and finishing, respectively. The constraints have the standard balance equation form that ending level = beginning level + additions − subtractions. The amount of product i, plated in period t, is added to inventory I_{i1t} in period t. Subtractions for plating in each period are given by finishing production. This model assumes product can flow through both stages in a single period. Subtractions for finishing are given by demand. Demand must be set to ensure the service level. We set demand in each period such that cumulative demand for that item is equal to the 95TH percentile. For a normal distribution, the 95TH percentile falls at $+1.645\sigma$. Thus, $D_{A1} = 250 + 1.645(10) = 266.45 \approx 267$. $D_{A2} = (250 + 300) + 1.645(100 + 100)^{0.5} - 267 = 306.26 \approx 307$. In both cases, we rounded up to ensure satisfying the service constraint. Demand for products B and C would be set in a similar manner. Finally, we account for the limited raw material in period 1 and add the common sense restriction that production levels, as well as inventories in this case, cannot be negative.

In aggregate planning, our intent is to determine the final production of major product types. It may not be necessary to model all lower-level production stages. This would be the case when the flow process of material through the system always follows the same schedule. The product release schedule or completion schedule will then provide complete planning information. We will address multistage item coordination and inventory planning in this environment in later chapters. However, if inventory is stored at different levels and production quantities can vary from stage to stage, then it may be advisable to model each stage as we have done in this section. The same is true when the length of time spent in a stage is equal to one or more planning periods.

5.3.2.2 Multifacility Production-Distribution Networks

When operations occur at different facilities with parts being transported between facilities and into and out of controlled storage areas at each facility, these stages should be separated in the model. Capacity constraints should represent the key bottleneck resource(s). We demonstrate multifacility network models with two examples. In the first example, products go through fabrication and assembly at different plants, and assembly requires several types of components in different quantities to make a final product. We must decide when to make each component type and also when to transport them to the assembly facility. In the second example, we have multiple sources of production for the first two stages; thus, we must decide how to assign workload to plants. In this case, transportation costs become relevant.

EXAMPLE 5.7

An assembly plant makes two product types. Product 1 combines one part A and two parts B to make each finished unit. Product 2 requires one part A and one part B per finished unit. Each of the component types is manufactured at its own component facility and shipped to the assembly plant. Each plant operates 200 hours per period. Plant A produces 100 parts As per hour. Plant B produces 150 parts Bs per hour. It takes 0.02 hours to assemble a unit of product 1 and 0.01 hours to assemble each unit of type 2 product. Inventories can be kept at the production plants at a cost of $1 per unit per period for an A or a B, $4 per unit of product 1 and $3 per unit of product 2. Given the final product demands indicated below, find the optimal production plan for the next three periods. No shortages are allowed. No initial inventories exist, and the length of a planning period is large relative to throughput time.

Product	Demand in Period		
	1	2	3
1	10,000	11,000	12,000
2	8,000	11,000	10,000

SOLUTION

The key decision variable is the amount of parts manufactured at plants A and B and the amount of each product assembled in each period. For simplicity, we will also model the inventory variables explicitly. Thus:

X_{it} = the amount of item i produced in period t, i = A, B, 1, 2 (A, B are component parts, 1 and 2 are finished products);

I_{it} = the amount of inventory of item i on hand at the end of period t.

A network diagram of the problem is given in Figure 5.6, which shows the flow of parts between facilities and then to customers. The model becomes:

$$Min \sum_{t=1}^{3} (I_{At} + I_{Bt} + 4I_{1t} + 3I_{2t}) \qquad (5.26)$$

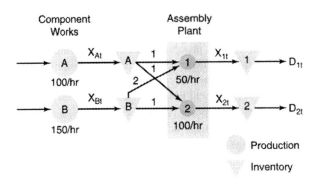

Figure 5.6 Schematic Diagram of Example 5.7

subject to:

$$X_{At} \leq 20{,}000; \ t = 1{,}2{,}3 \tag{5.27}$$

$$X_{Bt} \leq 30{,}000; \ t = 1{,}2{,}3 \tag{5.28}$$

$$0.02X_{1t} + 0.01X_{2t} \leq 200; \ t = 1{,}2{,}3 \tag{5.29}$$

$$I_{At} = I_{A,t-1} + X_{At} - X_{1t} - X_{2t}; \ t = 1{,}2{,}3 \tag{5.30}$$

$$I_{Bt} = I_{B,t-1} + X_{Bt} - 2X_{1t} - X_{2t}; \ t = 1{,}2{,}3 \tag{5.31}$$

$$I_{1t} = I_{1,t-1} + X_{1t} - D_{1t}; \ t = 1{,}2{,}3 \tag{5.32}$$

$$I_{2t} = I_{2,t-1} + X_{2t} - D_{2t}; \ t = 1{,}2{,}3 \tag{5.33}$$

$$X_{it}, I_{it} \geq 0$$

The objective function [Equation (5.26)] accumulates the cost of inventory in each period. Because no mention was made of production costs or workforce level and cost, we will assume labor is fixed, as are operating costs. Likewise, because total transportation amounts and paths are known, transportation cost will be fixed regardless of our plan and therefore does not affect the decisions. The first two sets of constraints limit production of component parts at each plant in each period to the product of the hourly production rate and hours available. The third constraint set ties together production of both finished products at the assembly plant. We have 200 hours available, and the left side of the equation computes hours consumed as a function of the assembly schedule. This is followed by an inventory balance constraint for each inventory point in each period. The equations represent the simple conservation equation that inventory level is equal to the starting level, plus the amount produced, minus the amount consumed. Both finished products consume component inventories. Equation (5.31) shows that two component Bs are used in each unit of product 1 as well as one B in each product 2. Consumption for the component parts inventories is dictated by the final assembly schedules. Demand sets the consumption in each period for the finished products. The demand values are provided in the data table.

When multiple plants produce the same item, the cost of transporting goods throughout the production chain becomes relevant in medium term production planning. Our final example models a three-stage process. Components are manufactured in stage 1 and shipped to assembly plants. After assembly, products are shipped to warehouses. Regional demand is met from warehouse stocks.

EXAMPLE 5.8

A manufacturing company produces two lines of consumer products. Each product line contains two major component types. Product 1 is made from one component type A and one type B. Each unit of product 2 requires one component B and two type C components. There are two components plants, and each of these plants can produce any combination of components. Two other assembly plants combine the subassemblies into products that are then shipped to the three regional warehouses to meet demand. Raw materials for the component works plants are commodity items that can be obtained at the same market price by both plants. Production costs are determined primarily by the cost of these raw materials and labor. The supply and cost of labor are fixed at each plant over the planning horizon. The assembly plants use similar technology and have identical production costs. Storage space exists at the component works but not at the final assembly plants. Demand is estimated in cases; where each case contains 100 products. Using the data below, formulate the production and distribution plan for the next three periods. Throughput times are short, thus, subassemblies can be used in assembly during the same period in which they are produced. Production capacity is the same at both assembly plants. Each plant can produce up to 375 cases, in any combination of the products, in each period. The component plants have 40,000 and 30,000 labor hours available per period, respectively. Plant 1 requires one labor hour per component (any type). Component plant 2 requires 1.3 labor hours per component owing to its lower-level of automation. Workforce level and labor costs are fixed over the planning horizon. Changeover costs and times are negligible both at the component and the assembly plants. Assume that initial inventories are zero. Unit holding costs per period are $0.1, $0.2, and $0.2, respectively, for components A, B, and C, and $1.0 and $1.3 for products 1 and 2.

Shipment Cost per Subassembly From Component Plant to Assembly Plant

From Component/To Assembly	1	2
1	$1.00	$2.00
2	$3.00	$0.50

Shipment Cost per Product From Assembly Plant to Warehouse (Product 1, Product 2)

From Assembly Plant/To WHSE	α	β	γ
1	($5.00, $6.50)	($2.00, $3.50)	($7.00, $9.00)
2	($1.50, $2.50)	($1.00, $1.50)	($2.00, $2.00)

Final Product Demand in Cases (Product 1, Product 2)

Whse. Region	Period 1	Period 2	Period 3
α	100, 150	110, 90	120, 125
β	50, 200	50, 220	60, 180
γ	75, 95	75, 100	80, 110

SOLUTION

It is helpful to begin by sketching the multiproduct production-distribution system (Figure 5.7). Each node corresponds to a production facility, and material flows are represented by the arcs. Triangles represent inventory stocks. The system has three stages: component production, assembly, and warehousing. Note that we may store subassemblies at the component plants but that finished products are shipped directly to the warehouses. The schematic can guide our model development. We will use $i = A,B,C$ to index the component types, $j = 1,2$ for the component plants, $k = 1,2$ for the assembly plants, $l = 1, \ldots, 3$ for the warehouses, and $m = 1,2$ for the final products. We then define decision variables corresponding to production at each node, shipment along each arc, and inventory level at each storage point. These variables are separated by product, location, and period. Thus, let X_{1ijt} be the production of component i at plant j in period t, and let X_{2mkt} be the production of completed units of m at plant k in period t. Demand at the warehouses will be indicated by D_{mlt}. Component and warehouse inventories will be denoted I_{1ijt} and I_{3mlt}, respectively. The first subscript indicates the stage. The remaining subscripts indicate component type (i), plant (j), and period (t). Finally, shipments of components to assembly plants will be denoted S_{1ijkt}, and shipments of products to the warehouses will be indicated by S_{2mklt}.

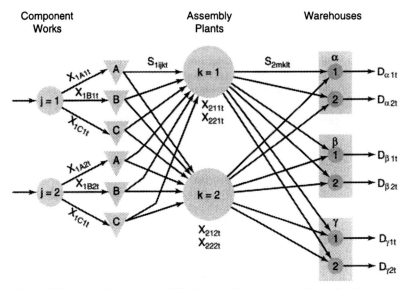

Component Works   Assembly Plants   Warehouses

Figure 5.7 Schematic Diagram of Production Operations for Example 5.8

The model formulation requires a balance constraint for each inventory point, including the zero accumulation points at the assembly plants, and a capacity constraint for each production facility. According to Figure 5.7, we need a balance constraint for each node-triangle pair in each period and a capacity constraint for each box in each period. The generic formulation is given next. We leave it as an exercise for the reader to fill in the exact numbers from the Tables in the problem statement and to find the optimal solution. For now, let shipment costs from component plant to assembly plant be denoted c_{1ijkt}, and let shipment costs from assembly plant to a warehouse be denoted c_{3mklt}.

$$Minimize \sum_{t=1}^{3} \left[\sum_{i=A}^{C} \sum_{j=1}^{2} \sum_{k=1}^{2} (c_{1ijkt} S_{1ijkt}) + \sum_{m=1}^{2} \sum_{k=1}^{2} \sum_{l=1}^{3} (c_{2mklt} S_{2mklt}) \right.$$
$$\left. + \sum_{i=A}^{C} h_{1i} \cdot \sum_{j=1}^{2} I_{1ijt} + \sum_{m=1}^{2} h_{mt} \cdot \sum_{l=1}^{3} I_{3mlt} \right] \tag{5.34}$$

subject to:

$$X_{1A1t} + X_{1B1t} + X_{1C1t} \le 40{,}000, \ t = 1{,}2{,}3 \tag{5.35}$$

$$1.3X_{1A2t} + 1.3X_{1B2t} + 1.3X_{1C2t} \le 30{,}000, \ t = 1{,}2{,}3 \tag{5.36}$$

$$X_{21jt} + X_{22jt} \le 375, \ j = 1{,}2; \ t = 1{,}2{,}3 \tag{5.37}$$

$$I_{1ijt} - I_{1ij,t-1} + X_{1ijt} - S_{1ij1t} - S_{1ij2t}, \ i = A{,}B{,}C; \ j = 1{,}2; \ t = 1{,}2{,}3 \tag{5.38}$$

$$X_{2mkt} = S_{2mk1t} + S_{2mk2t} + S_{2mk3t} + S_{2mk3t}, \ m = 1{,}2; \ k = 1{,}2; t = 1{,}2{,}3 \tag{5.39}$$

$$S_{1A1kt} + S_{1A2kt} = 100X_{21kt}, \ k = 1{,}2; \ t = 1{,}2{,}3 \tag{5.40}$$

$$S_{1B1kt} + S_{1B2kt} = 100X_{21kt} + 100X_{22kt}, \ k = 1{,}2; \ t = 1{,}2{,}3 \tag{5.41}$$

$$S_{1C1kt} + S_{C2kt} = 200 \cdot X_{22kt}, \ k = 1{,}2; \ t = 1{,}2{,}3 \tag{5.42}$$

$$I_{3mlt} = I_{3ml,t-1} + S_{2m1t} + S_{2m2t} - D_{mlt}, \ l = 1{,}2{,}3; \ m = 1{,}2; \ t = 1{,}2{,}3 \tag{5.43}$$

All variables ≥ 0.

The objective function [Equation (5.34)] accumulates shipping costs and holding costs. Shipping occurs from component plants to assembly plants and then from assembly plants to warehouses. Inventory is kept at the component works and warehouses. Constraints [Equations (5.35) (5.36)] ensure that adequate capacity exists at component works 1 and 2, respectively. Constraints [Equation

(5.37)] handle capacity at the assembly plants. As before, the left side of the capacity constraints details resources (hours) used as a function of the production decisions, and the right side states the amount of resources available. Constraints [Equation (5.38)] provide inventory balance for each stage 1 component, at each component plant, in each period. The standard balance constraint format of ending inventory equals starting, plus production, minus withdrawals is used. Balance constraints at stage 2 are slightly different because no accumulation of inventory occurs here. Constraints [Equation (5.39)] ensure that all assembly plant production is shipped immediately to one of the warehouses. Constraints [Equations (5.40) through (5.42)] ensure that all units of each component shipped to each assembly plant in each period are used for production in that period. Note that the production variables at stage 1 represent units and at stage 2 and 3 represent cases. This is why the factor of 100, the number of items per case, appears as coefficient on the production variables. The 200 used for X_{22kt} in constraint [Equation (5.42)] reflects the need for two components of type C for each type 2 finished product. Finally, Equation (5.43) is a standard inventory balance for each product, at each warehouse, in each period. Overall, we have 78 constraints plus the non-negativity restrictions.

The objective function in Example 5.8 included shipping and holding costs but not variable production costs. Because labor was fixed, we assumed direct production costs were also fixed over the planning horizon at each plant. Fixed or sunk costs need not be considered in deciding how to best use the production resources to meet demand. Of course, when making long-range strategic plans, these costs help determine which facilities should be included in the production-distribution network. If production costs had varied across plants, then it would have been necessary to account for these in the objective. There are many reasons why costs might vary. Plants could have different material, labor, tax, or utility rates depending on regional location. Efficiencies may depend on the age of the plant, the processing technology used, or even the management and information systems. If the actual cash flow varies with the choice of where to produce the parts, these costs should be included in the model. We can simply add cost parameters such as c_{1ijt} for unit variable production costs of producing a unit of i at plant j in period t. The objective function then includes additional terms such as $c_{1ijt} \cdot X_{1ijt}$.

Accounting for Interstage Lead Time   The zero lead time assumption given in the product statement (Example 5.8) may seem like a limiting assumption of this model. However, in actuality we need only assume a fixed lead time. If, for instance, there is a delay of τ periods between stages, we can solve the model as above. However, we must recognize that actual time is offset by τ periods between production stages. What is called period 1 in the model is chronologically period 1 for the subassembly stage but is period $\tau + 1$ at the assembly stage and period $2\tau + 1$ for warehouse demand. For instance, suppose τ is one month. Period 1 in the model may represent January for component production, February for the assembly plants, and March for the warehouses. Demand values in the model must be modified accordingly. To implement this redefinition of time, we must increase the length of our forecast time horizon and fix certain decisions. The assembly plan for January would have been determined back in December. The initial inventory used in the model for the warehouse would actually be the projected end of February (beginning of March) inventory level, based on planned production and forecasted demand for January and February.

An alternative approach is to include the time delays explicitly in the model. We could define the X_{1ijt} as the quantities of part i, started into production at component plant j, in period t. If it takes τ_1 periods for units to be produced at the component plant, then the inventory balance constraint [Equation (5.38)] would be changed to:

$$I_{1ijt} = I_{1ij,t-1} + X_{1ij,t-\tau_1} - S_{1ij1t} - S_{1ij2t}, \; i = A,B,C; \, j = 1,2; \, t = 1,2,3 \qquad (5.44)$$

to reflect the delay between when units are started and when they can be added to inventory. Units currently in process on the shop floor would be represented by values of

X_{1ijt} for $t < 0$. Delays in shipping and production at stage 2 could be incorporated in a similar manner by carefully adjusting the subscripts for time. Resource usage constraints may also be affected because we could model the amount of resource consumed in each of the τ_1 periods the product is in production.

EXAMPLE 5.9

Suppose in Example 5.8, component production required two periods and there was a one-period delay for shipments to reach the assembly plant from the component works. Show the required modifications to the formulation.

SOLUTION

Using the explicit approach, we modify Equation (5.38) to become:

$$I_{1ijt} = I_{1ij,t-1} + X_{1ij,t-1} - S_{1ij1t} - S_{1ij2t}, \forall i, j, t \qquad (5.45)$$

Note that we used $X_{1ij,t-1}$ and not $X_{1.ij,t-2}$ in Equation (5.45) although the delay time is two periods. This is because we defined the X_{1ijt} as units started at the beginning of a period and the I_{1ijt} as end-of-period inventory. Thus, the units added to inventory at the end of week 2 are those started at the beginning of week 1. This presents an additional complication, however. We must ensure that we do not ship more units than those on hand at the start of the week. Thus, we add constraints of the form:

$$S_{1ij1t} + S_{1ij2t} \leq I_{1ij,t-1}, \forall i, j, t \qquad (5.46)$$

stating that total shipments to the assembly plants for any component type from any component stock in a period is less than or equal to the starting inventory of that component in that period. Finally, we need to incorporate the one-week shipping delay. This is done by changing the S_{1ijkt} terms on the left side of Equations (5.40) through (5.42) to $S_{1ijk,t-1}$, reflecting the fact that assembly production in period t uses components shipped in period $t - 1$.

Accounting for Setup Time   Aggregate planning becomes more complex if we must explicitly account for setup time. This would be the case if the plant must shut down temporarily to change over from one product line to another. If the changeover schedule is known in advance, then we can simply correct the available production time by subtracting the time required for setups. If setup time is significant but we do not know the sequence of changeovers in advance, then we must include setup loss in the model. The simplest approach is to inflate average processing times by a percentage equal to the normal percent of time spent in setup. Thus, if we usually produce 1,000 units per setup and each setup takes 20 hours of resource, we could simply add 20/1000 hours to the standard for variable processing time per unit of product. Although not exact because the number of setups may vary, this will usually provide an adequate approximation for the aggregate model. The second approach explicitly includes setup in the model by introducing integer variables. This is shown in the following Example.

EXAMPLE 5.10

Example 5.9 implicitly assumed that plants could change over between products with only minimal effort. Suppose component works incurred 100 hours of downtime when changing over between products. Model this extension.

SOLUTION

Let us assume that multiple products are made each period and production time is a possible bottleneck constraint. (If ample time exists and the labor cost is fixed, then we need not be concerned with the time spent in changeovers). With this assumption, the main change is that we must define new variables to indicate whether a component is being produced at a plant in a given period. Let:

$$Z_{ijt} = \begin{cases} 1, & \text{if product } i \text{ is produced at plant } j \text{ in period } t. \\ 0, & \text{otherwise} \end{cases}$$

We must add constraints to ensure that the Z_{ijt} variables are valued properly and that we have adequate capacity at each plant in each period. Thus, we add the constraints:

$$X_{ijt} \le M \cdot Z_{ijt}, \forall i, j, t \tag{5.47}$$

where M is a large number on the order of total demand. The constraints [Equation (5.47)] will prevent the component production levels X_{ijt} from being positive unless we plan for the setup. We must also add the term:

$$100 \sum_{i=1}^{n} Z_{ijt}$$

to the left side of the resource capacity constraints [Equation (5.35)]. This adds the setup time to the time used at each plant in each period. Note that this model formulation would also allow us to include a charge for setups in the objective function. We would add the terms $S_{ijt} \cdot Z_{ijt}$, where S_{ijt} gives the cost of a setup.

In general, inclusion of setup time and setup cost in dynamic multiproduct planning models poses significant computational problems. Solving large linear programs is relatively easy. Indeed, the most difficult task is often data collection and management. However, considering setups usually requires 0-1 variables. Either we set up to produce product i in period t, or we do not. The binary variables allow us to model this either-or situation. Unfortunately, it is generally difficult to solve problems with many 0-1 variables. In the next section, we describe one modeling approach that will allow us to avoid this problem in some cases.

5.4   SCHEDULE GENERATION WITH LOT-SIZING

Manne [1959] proposed an approach for including setup considerations. The approach exploits the property that for multi-period, deterministic problems with no capacity constraints, the optimal solution satisfies the property that $I_{i,t-1} \cdot X_{it} = 0$ for any product i. The property states that we should not carry inventory into a period and also produce in that period. Consequently, when we produce, our production level should be equal to the demand over an integer number of periods. Although we will consider capacity constraints, if the number of products is large compared with the number of capacitated resources and time periods, this model can provide good solutions. The modeling strategy used involves defining a set of possible production schedules for each product and then letting the decision variables indicate whether we select a particular production schedule for a product. For each product, we will select exactly one schedule. Schedules are defined by the periods in which we produce the product. For a T period horizon, a schedule is represented by a vector of size T, where the tth element represents the amount to be produced in period t. Consider, for instance, a three-period horizon. Assuming we must produce the item in period 1, we have four possible schedules for an item as

$$\begin{bmatrix} D_1 \\ D_2 \\ D_3 \end{bmatrix}, \quad \begin{bmatrix} D_1 \\ D_2 + D_3 \\ 0 \end{bmatrix}, \quad \begin{bmatrix} D_1 + D_2 \\ 0 \\ D_3 \end{bmatrix}, \quad \begin{bmatrix} D_1 + D_2 + D_3 \\ 0 \\ 0 \end{bmatrix}$$

Assuming demand in period 1 exceeds current inventory, and, therefore, we must produce in period 1, we have 2^{T-1} possible production schedules for each product. It will help to define additional parameters:

X_{ijt} = the amount of product planned for item i in period t according to schedule j;

$$\delta(X_{ijt}) = \begin{cases} 1, & \text{if } X_{ijt} > 0 \\ 0, & \text{otherwise} \end{cases},$$

S_i = the setup cost for product i;

c_{ij} = the unit variable production cost for producing i in period t;

C_{ij} = the total cost to produce item i by schedule j;

a_i = the amount of resources used per unit of product i;

R_t = the amount of resource (possibly time) available in period t;

r_{ijt} = the amount of resource used by product i in period t if schedule j is selected;

s_i = the amount of resource used to set up product i.

Given a schedule for an item, we can compute the cost to produce this schedule (on regular time) and the resource requirement because we know the periods in which setups will occur and the production quantities. We obtain these quantities off-line:

$$C_{ij} = \sum_{t=1}^{T} (S_i \delta(X_{ijt}) + c_{it} X_{ijt} + h_i I_{it}) \text{ and}$$

$$r_{ijt} = s_i \cdot \delta(X_{ijt}) + a_i \cdot X_{ijt}.$$

We then can let $\theta_{ij} = \begin{cases} 1, & \text{if item } i \text{ is produced by schedule } j \\ 0, & \text{otherwise} \end{cases}$

and solve the formulation

$$\text{Minimize Cost} = \sum_{i=1}^{n} \sum_{j=1}^{J} C_{ij} \cdot \theta_{ij} \tag{5.48}$$

subject to:

$$\sum_{j=1}^{J} \theta_{ij} = 1, \ \forall i = 1, ..., n \tag{5.49}$$

$$\sum_{i=1}^{n} \sum_{j=1}^{J} r_{ijt} \cdot \theta_{ij} \leq R_t; \ \forall t = 1, ..., T \tag{5.50}$$

$$\theta_{ij} \in \{0,1\} \tag{5.51}$$

The objective [Equation (5.48)] selects out the total costs corresponding to the schedule for item i. Constraints [Equation (5.49)] ensure that exactly one schedule will be selected for each item. Constraints [Equation (5.50)] ensure that capacity is not exceeded by the set of schedules selected. If multiple resources exist, we need to replicate the constraint set [Equation (5.50)] for each key constraining resource.

At this point, you are probably questioning the advantage of this formulation. After all, we now have $n \cdot 2^{T-1}$ binary variables. However, suppose we relax the binary restrictions [Equation (5.51)] and solve a linear program with the θ_{ij} restricted only to the range [0,1]. We only have $n + T$ constraints, and at least one θ_{ij} must be selected for each of the n items i to satisfy Equation (5.49). This means that at most T items can have more than one positive θ_{ij} in any basic solution for the linear program. If n is large compared with T, then our solution is almost feasible to the model with integer restrictions [Equation (5.51)]. For the few products with unacceptable schedule combinations, we can either select the schedule with the largest θ_{ij} or perform a branch and bound investigation on these variables. The key to success for this formulation is a large number of products relative to capacitated resources and time periods.

In practice, we would most likely be operating the system on a rolling schedule. Each period we solve for the planned aggregated production schedule over the next T periods, but only implement the period 1 decision. Although we only implement the decisions for period 1, it is important to look into the future to ensure that we batch demands for products where this is economically advantageous or required because of capacity restrictions and setup time losses. There may be some products for which the initial inventory exceeds period 1 demand. For example, this would be expected if we produced several periods worth of demand for this product family in the previous period. For these products, we only consider schedules with $X_{ij1} = 0$. This reduces the number of schedules from which to choose for this item. However, we must include these items in the formulation to ensure that we account for their required resource usage in future periods.

EXAMPLE 5.11

Two products are to be planned for the next four months. There are 4,500 labor hours available each month, and it takes one hour of labor to make a unit of either product. Demands are forecast at 1,000, 1,500, 1,800, and 2,200 for product 1 and 2,000 per period for product 2. Currently, we have 1,200 units of product 1 in inventory, although the company does not wish to carry any excess inventory if it can be avoided. Setup takes 500 labor hours for either product. Because of tool wear and machine drift, a setup only lasts one period, and we must resetup the item to produce it in a second consecutive period. Each setup costs $1,000 in addition to the fixed labor cost. Holding costs are $0.50 per item per month for product 1 and $0.70 for product 2. Plan production for the next four months.

SOLUTION

First, note that product 1 does not need to be produced in period 1 because of the on-hand inventory, and we can reduce period 2 demand to 1,300. Thus, we have four possible schedules for product 1:

$$\theta_{11} = \begin{bmatrix} 0 \\ 1300 \\ 1800 \\ 2200 \end{bmatrix}, \theta_{12} = \begin{bmatrix} 0 \\ 1300 \\ 4000 \\ 0 \end{bmatrix}, \theta_{13} = \begin{bmatrix} 0 \\ 3100 \\ 0 \\ 2200 \end{bmatrix}, \theta_{14} = \begin{bmatrix} 0 \\ 5300 \\ 0 \\ 0 \end{bmatrix}.$$

Eight possible schedules exist for product 2, corresponding to the independent decisions of whether we produce in periods 2, 3, and 4. Because at most, 4,500 units can be produced in any period, we can eliminate variable θ_{14} and the product 2 schedules that call for producing for three or four periods at a time. This leaves us with the product 2 choices of:

$$\theta_{21} = \begin{bmatrix} 2000 \\ 2000 \\ 2000 \\ 2000 \end{bmatrix}, \theta_{22} = \begin{bmatrix} 2000 \\ 2000 \\ 4000 \\ 0 \end{bmatrix}, \theta_{23} = \begin{bmatrix} 2000 \\ 4000 \\ 0 \\ 2000 \end{bmatrix}, \theta_{24} = \begin{bmatrix} 4000 \\ 0 \\ 2000 \\ 2000 \end{bmatrix}, \theta_{25} = \begin{bmatrix} 4000 \\ 0 \\ 4000 \\ 0 \end{bmatrix}.$$

We must compute the cost and resource usage for each schedule. Consider schedule θ_{12}. Setup cost is $2,000 and inventory cost is $1,100 for the 2,200 units carried over at the end of period 3. Resource usage is 1,300 labor hours in period 2 and 4,000 in period 3, plus setup. Computing other values in the same manner, the model becomes:

Minimize $3000\theta_{11} + 3100\theta_{12} + 2900\theta_{13} + 4000\theta_{21} + 4400\theta_{22} + 4400\theta_{23} + 4400\theta_{24} + 4800\theta_{25}$

subject to:

$$\theta_{11} + \theta_{12} + \theta_{13} = 1$$
$$\theta_{21} + \theta_{22} + \theta_{23} + \theta_{24} + \theta_{25} = 1$$
$$2500\theta_{21} + 2500\theta_{22} + 2500\theta_{23} + 4500\theta_{24} + 4500\theta_{25} \leq 4500$$
$$1800\theta_{11} + 1800\theta_{12} + 3600\theta_{13} + 2500\theta_{21} + 2500\theta_{22} + 4500\theta_{23} \leq 4500$$
$$2300\theta_{11} + 4500\theta_{12} + 2500\theta_{21} + 2500\theta_{22} + 4500\theta_{22} + 2500\theta_{24} + 4500\theta_{25} \leq 4500$$
$$2700\theta_{11} + 2700\theta_{13} + 2500\theta_{21} + 2500\theta_{23} + 4500\theta_{22} + 2500\theta_{24} \leq 4500$$
$$\theta_{ij} \in \{0, 1\}$$

Solving this as a linear program, we find the solution:

$$\theta_{11} = 0.362, \; \theta_{12} = 0.259, \; \theta_{13} = 0.378, \; \theta_{21} = 0.808, \; \theta_{24} = 0.192 \text{ with cost } \$7,065.$$

Clearly, this solution is not feasible because we are attempting to use fractional schedules. (Recall that this model is best for problems with many more products than periods.) By adding the integer restrictions and resolving, we find the solution $\theta_{13} = 1$ and $\theta_{25} = 1$ for a cost of \$7,700. Thus, we produce product 1 in periods 2 and 4 and produce product 2 in periods 1 and 3.

The formulation can be extended to include workforce level and scheduling. Define a set of worker schedules. A schedule lists the number of workers in each period. The difference between periods indicates hiring and firing. The workforce level indicates direct labor cost. We then include another selection variable into the model to select a workforce schedule. (Note that with multiple labor grades, we could use multiple selection variables—one per grade.) A constraint is added to ensure that resources consumed in each period do not exceed available labor. Available labor can be found by summing over the schedules.

5.5 PARAMETRIC PRODUCTION PLANNING

In a dynamic environment, it is doubtful that plans for future production will ever be fulfilled or that forecasts for future demand will ever be precisely realized. Given demand forecasts, a knowledgeable manager could specify the ideal inventory and workforce levels. Unfortunately, this would seldom coincide with the actual current values. Parametric production planning attempts to continuously "close the gap" between current status and desired status. Let α be the proportion of the gap between current and ideal levels that we would like to close in a single period. I_t^* and W_t^* are the ideal inventory and workforce levels for period t. These values, along with the ideal production level, X_t^*, would typically be set based on forecasted demand. The workforce (W_t) and production levels (X_t) for upcoming period t would then be set to:

$$W_t = W_{t-1} + \alpha_1[W_t^* - W_{t-1}] \tag{5.52}$$

and

$$X_t = \frac{W_t}{K} + \alpha_2\left[\left(X_t^* - \frac{W_t}{K}\right) + \delta(I_t^* - I_{t-1})\right] \tag{5.53}$$

where K is the number of labor hours required per unit of product and δ $(0 \leq \delta \leq 1)$ is an additional weighting factor that may be used if desired.

The workforce level in Equation (5.52) is set to the current workforce plus a fixed proportion of the difference between the desired and current levels. Thus, starting at the current workforce level W_{t-1}, we add α_1 proportion of the difference between the current and ideal workforce levels. The first term in the production level equation sets production to fully utilize capacity. We then modify this value by α_2 proportion of the difference between the desired production level and full-time capacity. This is again modified to correct for any deviation between desired and actual current inventory. The optimal production level in the upcoming period, X_t^* is a smoothed average of future demand. Let F_t be the forecast for period t and T the length of the forecast horizon. As a weighed average,

$$X_t^* = \sum_{r=t}^{t+T-1} b_{r-t} \cdot F_t,$$

where

$$b_k = \frac{\beta^k}{\sum\limits_{i=0}^{T-1} \beta^i}$$

for some $0 < \beta \le 1$.

The b_k are weighting factors that geometrically decline with age (length of time into the future) and are normalized to sum to 1. The ideal production level is therefore set equal to the weighted estimate of average demand. The desired safety stock or inventory level is generally set as a multiple of the standard deviation of demand, σ_D. A typical value would be to let $I_t^* = 2\sigma_D$. The ideal workforce level is set to the level required to meet average demand plus an adjustment for desired inventory corrections. Thus,

$$W_t^* = K \cdot X_t^* + \delta \cdot K \cdot (I_t^* - I_{t-1}). \tag{5.54}$$

EXAMPLE 5.12

A manufacturing plant makes sheet metal frames for a variety of final assembly plants. Products range from filing cabinet panels to instrument stands to lamp bases. The operation is labor-intensive, and it is important to plan ahead to maintain an appropriate workforce and level of raw material (metal sheets and coils). Production is measured in pounds, and workforce is measured in hours. A worker produces an average of 20 lbs. of material per hour. The current workforce has 49 workers, and last week 41,600 lbs of material were produced. Workers are scheduled 40 hours per week. The forecast for the next four weeks is for 40,000, 38,000, 37,000, and 37,000 lbs of production. The current inventory level is 4,500 lbs. This is kept in the form of several high-demand products sold on a continuous basis to small manufacturers of industry standard items. The company would prefer to eliminate inventory and produce to order. Plan workforce and production levels for the next week assuming $\alpha_1 = \alpha_2 = \delta = 0.5$. Use equal weights over the next four periods to estimate average demand.

SOLUTION

From the problem statement, we have $I_t^* = 0$ and:

$$X_t^* = \frac{40,000 + 38,000 + 37,000 + 37,000}{4} = 38,000 \text{ lbs.}$$

Thus, because each worker produces (20 lbs/hr) · (40 hrs/week) = 800 lbs/week, and from Equation (5.54)

$$W_1^* = \frac{1}{800} \cdot [38,000 + (0.5)(-4,500)] = 44.69 \text{ workers.}$$

Using Equation (5.52), the workforce level is then set to $W_1 = 49 + 0.5 (44.69 - 49) = 46.85$, or 47 workers. Using Equation (5.53), we will plan production at $X_1 = 47(800) + 0.5[38,000 - 47(800) + 0.5(-4500)] = 36,675$ lbs.

5.6   OTHER APPROACHES

Several other approaches have been proposed for aggregate planning. Holt et al. (1955) suggested using an estimated quadratic cost function to model the costs of inventory (carrying costs if inventory is positive and shortage costs if inventory is negative), hiring, and changing production rates. The optimal production and workforce levels subject to inventory balance constraints can be found by standard optimization techniques. This is sometimes referred to as the "linear decision rule" model because the expressions for the workforce and production levels can be written as linear functions of demand, previous workforce, and inventory levels.

Multiple linear regression analysis may also be used to determine the underlying structure of historical decisions. Workforce and production levels are the dependent variables, with current workforce, current inventory, and future demand as the independent variables. A typical model for setting the new workforce level using last period's workforce, last period's demand, and forecast for this and the next period might take the form:

$$W_t = \beta_0 + \beta_1 \cdot K \cdot X_{t-1} + \beta_2 \cdot W_{t-1} + \beta_3 F_t + \beta_4 \cdot F_{t+1} + \varepsilon_t \qquad (5.55)$$

Production level decisions likewise could be modeled as a function of workforce and forecasts. Historical planning decisions are used as data to fit the model parameters (β_i). This approach can quickly rationalize past practice and help make decisions more consistent. However, the method simply explains past behavior. As such, it is only appropriate when past decisions are felt to be effective but not well-understood. This might be the case if an experienced employee has been making subjective decisions but is now ready for retirement or reassignment, and we need to retain the internalized knowledge.

5.7 DISAGGREGATION

The aggregate production plan sets the level of workforce and production by product family for each major period over a medium to long-term horizon. This plan must be disaggregated into operational production plans for each product family, detailing production quantities for each item in the family over shorter scheduling periods. The aggregate plan may assert that we will produce 2,000 tables and 4,000 chairs next month. The disaggregated plan breaks down these totals into plans for each model in each week of the month. It may set production to 500 Queen Anne tables in week 1, 600 oak contemporary in week 2, 400 oak contemporary and 100 cherry formal in week 3, and 250 mahogany chenile and 150 mahogany country in week 4. On the other hand, it may state that we should produce one-quarter of this quantity of each table type each week.

The disaggregated plan is constrained to use the assigned resources to meet the cumulative production quota each month for each product family. Adhering to these constraints, the disaggregated plan should minimize the cost of inventory, setup, and shortages. If the aggregate plan assigns resources to each family separately, we can then disaggregate for each product family independently. The output of disaggregation is the quantity of each finished product to be completed in each short-term period, such as a day or week, over a short-term planning horizon. This is referred to as the **master production schedule,** or MPS. The MPS serves as the input to subsequent detailed scheduling and order release planning for the shop. The MPS may also be used for ordering raw materials because we now know when those materials will be needed.

Disaggregation answers the question of how much to make of each product model in each period. A related issue concerns the production frequency or batch size to use for each item. This issue is covered in greater detail in the next chapter, however, we add a few relevant comments in this section as well.

5.7.1 Single-Cycle Run-Out Time Model

A standard practice for disaggregation divides production across items in the family to maximize the time until the first item drops to its desired safety stock. This allows us to delay machine setups while avoiding stockouts. Let the items in the product family be indexed by i; $i = 1, ..., n$. $I_{i,t-1}$ is the inventory position of item i at the start of period t (the end of period $t - 1$). SS_i is the desired safety stock for item i. Safety stock values are typically set equal to expected demand over some fixed time such as a week. The length of time selected depends on the level of variability in demand. The intent is to achieve some desired service level based on the distribution of demand over the replenishment lead time. (Detailed procedures for determining this safety quantity can be found in the next chapter.) Let $\overline{D}_i$ be the average demand rate for item i. If we produce Q_{it} units of item i this period, the anticipated **run-out time** at which the inventory position will drop to the safety stock level will be:

$$r_i = \frac{Q_{it} + I_{i,t-1} - SS_i}{\overline{D}_i} \tag{5.56}$$

For convenience, we measure inventory and production levels in units of resource usage. Thus, if the actual inventory was 300 chairs, and each chair requires 2.5 labor hours, then our inventory would represent $300 \cdot 2.5 = 750$ hours of resource. The total resource to be allocated to this family in period t is given by the aggregate production plan. We label this quantity P_t units of resource. (We deviate from the previous notation X_t to emphasize that this is now a parameter and not a decision variable.) If P_t is total time available, then it should be reduced by any setup time required to produce the product family. Our objective is to equalize the run-out times for all items in the wood chair family. We do this by solving:

Maximize R

subject to:

$$R \le (Q_{it} + I_{i,t-1} - SS_i)/(\overline{D}_i),\ i = 1, \ldots, n$$
$$Q_{1t} + Q_{2t} + \ldots + Q_{nt} = P_t \tag{5.57}$$
$$Q_{it} \ge 0$$

This model sets the production levels to consume all assigned resource while delaying as long as possible the need to produce this family of items again. R represents the earliest run-out time for any item in the family. Because of the special structure of this linear program, the optimal solution is easy to obtain. The solution is to set the run-out time the same, and as large as possible, for all products that need to be produced.[1] Thus, we set $Q_{it}^* = \max(0, Q_{it}^0)$, where

$$Q_{it}^0 = \left[P_t + \sum_{j=1}^{n}(I_{j,t-1} - SS_j)\right] \cdot \frac{\overline{D}_i}{\sum_{j=1}^{n}\overline{D}_j} + SS_i - I_{i,t-1} \tag{5.58}$$

The term inside the brackets contains the total effective production resource for the family attributable to both starting inventory and new production. This is first allocated to items based on their proportional demand rates. We correct these quantities to take into account initial inventories and desired safety stocks. Items are ordered for production whereby the items with the earliest run-out times before production go first.

EXAMPLE 5.13

The aggregate plan calls for 1,000 hours of production of tables next week. Safety stock is set to 25% of average weekly demand for each item. Each table requires 2 hours of production time.

---

[1] We assume that all items will be produced. If some items are already overstocked, remove them from set $i = 1, \ldots, n$.

Setup time is 3 hours for each table model. Using the data in the following Table, determine run sizes and expected run-out time for each item.

Table Model	Initial Inventory	Average Weekly Demand
1	0	10
2	30	40
3	80	60
4	25	90

SOLUTION

First note that we have $1000 - 3(4) = 988$ effective hours of production after setup time is subtracted. This allows for a total production of $988/2 = 494$ tables. The desired safety stocks are 2.5, 10, 15, and 22.5, respectively, for the four models. We then find:

$$P_t + \sum_{j=1}^{4}(I_{j,t-1} - SS_j) = 494 + (-2.5) + (30 - 10) + (80 - 15) + (25 - 22.5) = 579.$$

Batch sizes are then:

$$Q_{11}^0 = 579 \cdot \frac{10}{200} + 2.5 = 31.$$

$$Q_{21}^0 = 579 \cdot \frac{40}{200} + 10 - 30 = 96.$$

$$Q_{31}^0 = 579 \cdot \frac{60}{200} + 15 - 80 = 109.$$

$$Q_{41}^0 = 579 \cdot \frac{90}{200} + 22.5 - 25 = 258.$$

Expected run-out time can be found from any of the items. Using item 1,

$$r = \frac{Q_{it} + I_{i,t-1} - SS_t}{\overline{D}_1} = \frac{31 + 0 - 2.5}{10} = 2.85 \text{ weeks.}$$

5.7.2 Multiple Cycle Disaggregation

The previous model computed a production quantity Q_{it}^* for each item. You most likely assumed we would produce one batch of size Q_{it}^*. The single lot size may not be appropriate in a multiple product production system in which we produce a batch of item 1s and then a batch of item 2s, and so on, until a batch of all n items is completed. In this case, we should be careful about the choice of cycle times. It may be better to produce each item multiple times during the period to reduce cycle inventory. Recall that in Chapter 2, we defined the Economic Order Quantity $Q_i^* = \sqrt{\dfrac{2AD}{h}}$. We can likewise solve for the natural production cycle length T. Because the time between two orders for an item is $T = Q/D$, we can define the natural cycle as:

$$T_i^* = \frac{Q_i^*}{D_i} = \sqrt{\frac{2A_i}{h_i D_i}} \tag{5.59}$$

If $T_i^* < 1$ when measured in aggregate periods, we can repeat the cycle $\dfrac{1}{T_i^*}$ times.

Essentially, we divide the aggregate period into $\dfrac{1}{T_i^*}$ subperiods. In each subperiod, we produce each item in the family once. Thus, the actual batch sizes are $Q_{it}^* \cdot T_i^*$. We would usually compromise on a common cycle length, T, for all n products. If family setup times are sequence-dependent, then we may sequence items in production to minimize total setup time. Otherwise, product families should be ordered whereby the family with a product that will run out first (minimum $\dfrac{I_{it}}{D_i}$ value) should be run first, and so on. Within each family, individual items should also be sequenced, based on earliest run-out time, given existing inventory. It is also possible to have items with different natural cycles in the same family. Let T_f be the compromise natural cycle time of product family f and n_{if} be the relative frequency of item i within family f, i.e., $n_{if} \cdot T_f \approx T_i^*$. We can then choose to produce item if once every n_{if} times that we set up family f. When we produce i, we produce $Q_{it}^* \cdot n_{if} \cdot T_f$ units. Generally, at the start of each cycle for the family, at which these cycles begin T_f time units apart, we produce each item in the family that has a run-out time less than T_f. If item i fits this criteria, we produce approximately $Q_{it}^* \cdot n_{if} \cdot T_f$ units, making a possible modification based on current inventory levels to equate run-out times.

In the ideal case, with negligible setup time and cost, we produce in batches of size 1. Essentially, we produce to replace direct sales or in response to customer orders. This system adapts very quickly to changes in demand and carries a minimum of inventory. This situation has similarities to the mixed-model assembly line sequencing problem addressed in Chapter 12.

5.7.3 Run-Out Time with Dynamic Demand

The basic concept of run-out time can be extended to more complex disaggregation problems. For instance, if relative demand for individual items varies from period to period, we can replace the $\overline{D}_i$ values in the expression above with actual item-demand forecasts, D_{it}. In computing run-out time, we find the first point in time at which the current inventory, in excess of desired safety stock, will reach zero. In other words, as measured from the current time 0, the run-out time for item i satisfies:

$$I_{i0} - SS_i - \int_{t=0}^{r_i} D_{it}\, dt = 0. \qquad (5.60)$$

5.8 SUMMARY

Aggregate planning involves specifying anticipated workforce, production, and inventory levels for production facilities over a medium horizon, generally six to eighteen months. Production levels are set for each family of related products. A family consists of products that use the same resources (materials, labor, and machines) and have similar demand patterns. In some cases, products in a family may share a common machine setup as well. In addition to setting production goals for the factory, the aggregate plan outlines financial, human resource, and purchasing plans.

Various techniques have been proposed for aggregate planning. Linear programming and the transportation model are powerful tools. They have the capability to solve relatively large problems and to provide an efficient mechanism for investigat-

ing multiple scenarios, as well as to obtain economic information on the value of scarce resources. Linear programming models can be used to determine the minimal cost method to satisfy demand in the presence of resource constraints. In addition, key planning factors such as inventory levels, hiring/firing decisions, and production quantities are easily obtained from the models. Linear programs can include a variety of considerations such as multiple products; multiple production stages or facilities; capacity limitations caused by workforce, material, or machine availability; capacity change costs; transportation costs; inventory accumulation and storage; and planned back orders. The models include an objective function, resource balance constraints, resource capacity constraints, organizational policies and timing considerations. Whereas the linear programming models are flexible, certain considerations such as setup times are difficult to include and require binary integer variables. One approach in this case is to define a set of possible production schedules and then to select a single schedule for each product family. Other approaches to aggregate planning include the use of quadratic objective functions, parametric production planning, and multiple linear regression models.

Aggregate plans must be disaggregated to become operational. In disaggregation, demand rates, inventory levels, safety stocks, and setup costs and times drive the breakdown of product family aggregate plans into production schedules and batch sizes for individual items. The near-term plan for completion of finished products forms the master production schedule that will drive subsequent short-term production schedules. A general rule sets individual item batch sizes to equalize the expected run-out time of all items in a product family. Available time is devoted to setup to minimize the inventory cost over the planning horizon.

5.9   REFERENCES

BITRAN, G.R., & TIRUPATI, D. Hierarchical production planning. In: Graves, S., RINNOOY KAN, A.H.G., & ZIPKIN, P.H., eds. *Logistics of Production and Inventory,* New York: North-Holland, 1993: 523–568.

BOUCHER, T. & ELSAYED, E.A. *Analysis of Production Control Systems,* Englewood Cliffs, NJ: Prentice-Hall, 1994.

BOWMAN, E.H. Production scheduling by the transportation method of linear programming, *Operations Research,* 1956, 4, 100–103.

CRUICKSHANKS, A.B., DRESCHER, R.D., & GRAVES, S.C. A study of production smoothing in a job shop environment. *Management Science,* 1984, 30(3), 368–380.

HAX, A.C., & MEAL, H.C. Hierarchical integration of production planning and scheduling. In: Geisler, M., ed. *TIMS Studies in Management Science, Vol. 1 Logistics,* New York: Elsevier, 1975.

HOLT, C.C., MODIGLIANI, F., & SIMON, H.A. A linear decision rule for employment and production scheduling. *Management Science,* 1955, 2, 1–30.

JOHNSON, L.A., & MONTGOMERY, D.C. *Operations research in production planning, scheduling and inventory control.* New York: John Wiley & Sons, Inc., 1974.

MANNE, A.S., Programming of Economic Lot Size. *Management Science,* 1958, 4(2), 597–607.

5.10   PROBLEMS

5.1.   Discuss the advantages of aggregating multiple items into a single aggregate product for determining mid- to long-range production plans.

5.2.   List the standard outputs of an aggregate production plan. How does this differ from the output of the disaggregated plan?

5.3.   List the major tradeoffs examined in aggregate production planning models.

5.4.   A plant produces 1,500 different products using twelve types of workers and 180 machines. Management would like to build an aggregate production plan showing weekly production and levels for each product and weekly employment levels for each worker type over the next eighteen months with each month comprising four weeks. If each item and worker class are modeled, how many decision variables would the model need? Suppose you could aggregate the items into four types, treat all workers as simply labor hours, and use weekly time periods for the first six months and monthly time periods thereafter. How many decision variables would then be needed?

5.5.   Propose an alternate production schedule for Example 5.1 and evaluate the cost.

5.6.   Apply the least-cost first heuristic described in the text to solve the transportation problem given in Example 5.3 (see Table 5.6).

5.7.   Describe the types of cost factors and decisions that can and cannot be included in a transportation model for aggregate

planning. For instance, is it possible to allow back orders in the model? Is it possible to allow for hiring and firing of workers to change capacity?

5.8.   A firm has forecasted demand as 230, 250, 250, 300, and 280 units, respectively, for the next five months. Production cost is $45 per unit on regular time and $60 per unit on overtime. Inventory holding cost is $3 per unit per month. Capacity is 250 units per month on regular time and 50 units per month on overtime.

(a)   Set up the transportation cost table and suggest a production schedule that avoids shortages.

(b)   Suppose shortages were permitted at a cost of $5 per unit per month. Update the cost table and resolve the problem.

5.9.   A plant produces five families of product. Because of large setup times, not all families are produced each week. Capacity for the facility is measured in labor hours. The facility currently has 1,000 labor hours available per week, and each family setup requires 100 labor hours. Demands are also stated in labor hours (not including setup time). Using the data below, formulate the planning problem for the next three periods as a mathematical program using 0-1 variables for each product family in each period to indicate whether that family is produced. Clearly define all variables, constraints, and the objective function. Shortages are to be avoided whenever possible.

Family	Initial Inventory	Week 1 Demand	Week 2 Demand	Week 3 Demand
1	10	200	150	100
2	200	100	150	200
3	150	140	150	160
4	50	160	160	160
5	100	100	150	200

5.10. A plant produces two product families. You want to plan production for the next three months. Completed items can be stored at the plant or shipped to regional warehouses. The plant has unlimited storage capacity, but the warehouses can only hold 3,000 items each. Storage costs at the plant are $0.80 per item per month but only $0.60 per item per month at the warehouses. Family 1 requires 5 labor hours per unit to produce. Family 2 requires 6 labor hours per unit. Each worker can work 120 hours regular time per month and up to 20 hours overtime. Regular time wages are $20 per hour and overtime wages are $30 per hour. The initial workforce has 275 workers. Workers can be hired at a cost of $400 but are only 50% as productive in their first month as the experienced workers. One percent of the workforce quits each month. Workers can be fired at a cost of $500 each. Hiring and firing occurs at the start of the month. No workers were hired last month. Shortages are to be avoided whenever possible. There are no initial inventories at either the plant or warehouses, but the company would like to keep a safety stock of at least 500 units for each family at each warehouse. Demand occurs at the warehouses and is given in the Table below. Define the decision variables and write out a linear programming formulation for this problem. Give the complete formulation explicitly stating all constraints and parameter values.

Warehouse	Product Family	Month 1	Month 2	Month 3
1	1	1,100	1,500	2,000
1	2	900	1,500	2,000
2	1	1,200	2,000	2,000
2	2	1,300	1,000	2,000

5.11. Find the optimal solution for the first two periods in Problem 5.10, assuming that workers may not be hired or fired. How much would you be willing to pay on a monthly basis to rent more space at the warehouses? Do you wish you had more or less workers?

5.12. Suppose that in Problem 5.10, a setup time and cost existed for each product family. Describe the necessary modifications to the formulation and how you would solve the problem.

5.13. A production facility makes six different product families. The plant operates five days per week. An aggregate production plan has selected two product families to be run this week. These families are usually run every other week. The Table below shows average demand rate, processing time, and initial inventories for each item in these families. The aggregate plan scheduled 1,200 labor hours for this period. It takes 10 labor hours to set up to run a family. Once the family is set up, any item in the family can be run with an additional setup requiring 1 labor hour. Desired safety stocks are set at 20% of weekly demand for each item. Determine the appropriate disaggregated production policy for this week. Include daily production quantities for each item. The company tries to guarantee shipment of all items within one day after receiving an order.

Family	Item	Initial Inventory	Average Weekly Demand	Unit Processing Time (Labor Hrs)
A	1	100	150	0.1
A	2	250	300	0.2
B	1	50	500	0.2
B	2	0	500	0.2
B	3	245	1,000	0.3

5.14. Find the appropriate demand values for use in the model described in Example 5.6. Solve the model.

5.15. Write out the complete objective function, filling in all numerical values, for the problem described in Example 5.8. State the number of variables and constraints in the model (not counting non-negativity restrictions). Solve the linear program.

5.16. Formulate Problem 5.9 as a schedule generation model, as described in Section 5.4. List the possible production schedules for each product.

5.17. A company has 100 workers. Each worker produces four units of product per week. Demands for the next four weeks are 350, 380, 390, and 380. Forecast error variance is such that the company tries to keep a safety stock of 100 units of product. Current inventory is 30 units. Use parametric production planning to set production and workforce levels for the next week assuming $\alpha_1 = \alpha_2 = 0.5$ and $\delta = 1$.

5.18. Repeat the previous problem assuming $\alpha_1 = \alpha_2 = 0.25$ and $\delta = 0.5$. How would you decide in practice which set of parameter values was better?

5.19. You have been asked to develop an automated planning model for setting production quantities. You have collected the following historical data (at top of following page) relating the state of the system to decisions made. Build a model that will allow you to automatically set production levels in the future

	Historical Production Decisions				
Period (*t*)	Production Decision (Units)	Starting Inventory	Number of Workers	Forecast for Period *t* + 1	Forecast for Period *t* + 2
1	670	−10	112	650	680
2	540	23	106	680	525
3	590	100	106	525	610
4	630	62	107	610	630
5	575	15	103	630	580
6	615	31	104	580	625
7	680	−20	110	625	695
8	595	25	105	695	500
9	625	37	106	500	645
10	635	39	106	645	645

given values for the independent variables current inventory level, workforce, and forecasts.

5.20. The aggregate plan specifies 1,300 hours of labor for the next week.

(a) Using the data below, find the disaggregated production plan. Each item requires 1 hour of labor per unit produced. In which order would you suggest producing the items? Why?

Model	Initial Inventory	Average Weekly Demand	Safety Stock
1	110	500	50
2	4	600	50
3	73	300	25

(b) Repeat part (a) assuming models 1 and 2 require 1 hour of production per unit, but item 3 requires 1.3 hours of labor per unit.

5.21. Repeat Problem 5.20, assuming each item requires 45 hours of setup time.

5.22. A company operates three fabrication plants, two assembly plants, and four regional warehouses. Production is being planned for the next three months. The finished product is assembled from three subassemblies: SA, SB, and SC. Each unit of finished product requires one SA, one SB, and two SCs. Two of the fabrication facilities can make any combination of SA and/or SB subassemblies. All SC units are made at the third fabrication facility. It takes 3 machine hours to make an SA, 4.5 machine hours to make an SB, and 2 hours to make an SC. Capacity at the three facilities is 15,000, 42,000, and 25,000 machine hours per period, respectively.

The assembly plants have identical production costs of $50 per unit. It takes 2 labor hours to assemble the final product,

and capacities per period at the assembly plants are 6,000 and 8,000 labor hours, respectively.

Inventory can be stored at the facility where it is produced to cushion the effect of demand changes over time. The unit holding costs per month are $1.20, $0.95, and $1.50, respectively, for the subassemblies and $7.40 for finished products. Transportation costs and demand are as follows. The unit transportation costs from fabrication plants 1 and 2 are the same for both SA and SB subassemblies.

(a) Draw a schematic diagram of the production-distribution system of facilities.

	Demand by Region			
Period	Region 1	Region 2	Region 3	Region 4
1	1,000	1,500	800	2,000
2	1,200	1,500	800	2,500
3	1,350	1,750	1,000	3,000

Subassembly Transportation Costs/Unit

From Subassembly Plant	Assembly Plant 1	Assembly Plant 2
1	$1.80	$2.20
2	$2.90	$0.60
3	$1.50	$2.25

Finished Product Transportation Costs/Unit

From Assembly Plant	To Region 1	Region 2	Region 3	Region 4
1	$1.10	$2.15	$5.40	$6.10
2	$7.15	$6.70	$4.50	$2.15

					(Warehouse, Period)							
Product	(1,1)	(2,1)	(3,1)	(1,2)	(2,2)	(3,2)	(1,3)	(2,3)	(3,3)	(1,4)	(2,4)	(3,4)
A	1000	1400	1000	1000	1500	1000	1000	1600	1100	1000	1700	1200
B	500	800	600	500	800	700	500	800	700	400	800	800

Demand by Product, Warehouse and Period for Problem 5.27

(b) Draw the product structure.

(c) Formulate the production scheduling and distribution problem for this company.

(d) Solve the production scheduling and distribution problem.

(e) Reformulate the problem for the case in which there is a large setup cost and time to produce any subassembly type at a fabrication plant in a period.

5.23. Consider the disaggregation model shown in Equation (5.57) and solved in Equation (5.58). Suppose that each item in the family has its own resource requirement p_i per unit produced. Modify these equations to take this into account. Note that you must still consume a total of P_t units of resource.

5.24. A family of five products has current inventory levels of 112, 34, 67, 98, and 43 units, respectively. All safety stock levels are 30 units. Product 1 is the most popular item and accounts for 50% of sales. Demand for the other four products is equal. The aggregate production plan calls for producing 600 units of product this week. Find the appropriate production quantities for each item to equalize run-out times.

5.25. Suppose that in the previous problem, products 1 and 2 require 0.5 hours per unit to produce, and the other products require one unit of resource per unit of product. If 600 hours are available, find the production quantities that will equalize run-out times.

5.26. Three product families are produced in the same facility. The cost to set up any family is $1,000. Once the family is set up, there is no cost to change over from one item in the family to another. Current inventory levels and demand rates for individual items are shown in the Table. Holding cost is $1 per unit per month for all items. The aggregate plan calls for producing 25,000 units of product this month. For each family, determine the number of times the product should be produced

during the month and the total production quantity for each item in the family.

5.27. A manufacturing plant assembles two product families: A and B. Customer demand occurs at three regional warehouses. The assembled product can be stored temporarily at the plant or shipped directly to the warehouses. The product can be produced and shipped in a single week, but it is not available to meet demand at the warehouse until the week after it is shipped. Thus, if the product is made the first week of the month, it cannot be used to satisfy demand until the second week of the month. Demand is shown in the table at the top of the page.

To make a unit of product, family A requires one part type α and two parts type β. α and β parts are fabricated in the plant. There is a 2-week time span between starting α and β parts and when they actually become available for use at assembly. Completed α and β parts can be stored between the fabrication and assembly departments, if desired. Holding cost per week is $0.10 for a part of type α or β. Currently in process are 1,200 part type α and 3,000 part type β, which were started one week ago and are now half-completed. There are also 1,050 completed α parts and 2,230 completed β parts ready for use in assembly.

The raw material for product family B is purchased from vendors. It is available in ample supply and can be obtained on short notice.

Initial inventories at the warehouses are 525 of family A and 650 of family B at warehouse 1, 1,200 of family A and 465 of family B at warehouse 2, and 750 of family A and 345 of family B at warehouse 3. The current inventories of finished products at the plant are 900 As and 1,575 Bs. Holding cost per week is $0.65 for a unit of product A and $0.70 for a unit of B. This cost is the same for the plant and all warehouses.

The plant has 3,800 labor hours available per week for assembly. Each unit of A requires 0.5 hour of labor to assemble, and each unit of B requires 1 hour of labor to assemble. Parts fabrication for α and β is primarily automated. The plant can handle production starts of up to 11,000 units per week of α and β, in any combination, and still keep its 2-week delivery schedule.

Currently, there are 95 assembly workers. You may hire extra assembly workers at a cost of $1,500 each. Each worker is available 40 hours per week and each labor hour costs $12.50. You may also fire workers, but each worker fired costs $2,500. Formulate the plant's aggregate planning problem as a linear program.

Family	Item	Current Inventory	Monthly Demand
A	a1	10	1,250
A	a2	133	4,500
A	a3	16	2,500
B	b1	−40 (backlog)	2,500
B	b2	129	5,000
C	c1	325	5,000
C	c2	173	2,500

The 3F Manufacturing Company produces a family of appliances ranging from air conditioners to refrigerators. The company's production facilities are in the eastern United States, and marketing targets the entire United States east of the Mississippi River. The company operates three fabrication plants, two assembly plants, and four regional warehouses. Demand is somewhat seasonal but has been growing at the rate of 5 to 10% annually for the past five years. This trend is expected to continue. Jim Cool, the operations manager, would like to develop a system for setting medium range goals for the next three months. This plan will be updated monthly, but Jim feels it is important to set the plan with an eye toward the future.

Finished products are assembled from three subassemblies: SA, SB, and SC. Each unit of finished product requires one SA, one SB, and two SCs. Based on the current equipment at the facilities, two of the fabrication facilities can make any combination of SA and/or SB subassemblies. All SC units are made at the third fabrication facility. It takes 3 machine hours to make an SA, 4.5 machine hours to make an SB, and 2.0 machine hours to make an SC. Using the current schedule of one full shift and one light (half) shift at each site, capacity at the three facilities is 15,000, 42,000, and 25,000 machine hours per period, respectively.

The assembly plants have identical production costs of $50 per unit. It takes 2 labor hours to assemble the final product, and capacities per period at the assembly plants are 5,000 and 6,000 labor hours, respectively. Overtime can be used, but this increases production costs by 25%. Inventory can be stored at the facility where it is produced to cushion the effect of demand changes over time. The unit holding costs per month are $1.20, $0.95, and $1.50, respectively, for the subassemblies, and $7.40 for finished products. Transportation costs and demand are as follows. The unit transportation costs from fabrication plants 1 and 2 are the same for both SA and SB subassemblies.

Devise a production-distribution system planning model that Jim can use on a monthly basis. You should begin by drawing a schematic diagram of the production-distribution system

Demand by Region				
Period	Region 1	Region 2	Region 3	Region 4
1	1,000	1,500	800	2,000
2	1,200	1,500	800	2,500
3	1,350	1,750	1,000	3,000

Subassembly Transportation Costs/Unit

From Subassembly Plant	Assembly Plant 1	Assembly Plant 2
1	$1.80	$2.20
2	$2.90	$0.60
3	$1.50	$2.25

Finished Product Transportation Costs/Unit

From Assembly Plant	Region 1	Region 2	Region 3	Region 4
1	$1.10	$2.15	$5.40	$6.10
2	$7.15	$6.70	$4.50	$2.15

showing all facilities and the potential flows among them. You will need to formulate a general model and then test it on the data for the next three months. You should also consider whether Jim's plan is appropriate. Is three months an adequate horizon length? Should other factors, such as workforce planning, be integrated into the model? Currently, demand, cost, and productivity data are only available in the form provided. However, the facilities do experience setup time in switching between product families and subassemblies. Historic averages of batch sizes are used to apportion change over time over individual items. How should the model be maintained over time? How should the model be linked with strategic planning above and operational planning decisions below?

Chapter 6

Single-Stage Inventory Control

In Chapter 2, we provided a basic introduction to the types of inventory, the associated costs, and the basic trade-offs. In this chapter, we present a more technical explanation of these trade-offs, along with decision models for determining optimal policies. We begin with the basic economic order quantity (EOQ) concept and then extend that approach to a number of models available for guiding inventory planning decisions in various production and distribution environments. As assumptions are relaxed and changed, the models become increasingly complex. Two factors remain constant throughout the chapter, however. We consider only **single-stage** production systems. If multiple stages exist, then the coordinated strategies addressed in later chapters are preferable. Second, we consider only **independent-demand** items—each item's demand is unaffected by other products. These models have the advantage of not requiring sophisticated information systems and require only a limited amount of data. Although each model's assumptions indicate its intended use, in reality, many of these models are applied in various environments due to their relative simplicity and historical notoriety. Finally, the models in this chapter are intended primarily for situations in which items are ordered or produced intermittently and then **held in stock** until used.

6.1 REORDER POINT INVENTORY MODELS FOR STATIC, DETERMINISTIC DEMAND

The most basic inventory models assume demand occurs continuously over time and at a constant rate. We begin by analyzing these simple cases because they are the foundation for more general situations and illustrate key principles. We then progress to more sophisticated models.

6.1.1 Economic Order Quantity (EOQ) Model

Basic inventory control models that determine how many parts to order or produce at a time represent one of the earliest applications of mathematical modeling to scientific management. The widely used EOQ model is the basis for more advanced models. The EOQ specifies a (Q, r) inventory policy for determining how much and when to order. When the inventory position drops to r units, we initiate an order for Q units. To build the cost model, we need to visualize how the inventory level will vary over time. The scenario assumed for the classic EOQ model is as follows. When the ordered product arrives, it is placed in inventory and used to satisfy demand. Demand is assumed to occur at a constant, continuous rate, D. All replenishment orders are for exactly Q units, and all Q units are delivered simultaneously. Inventory cost is proportional to on-hand inventory. A plot

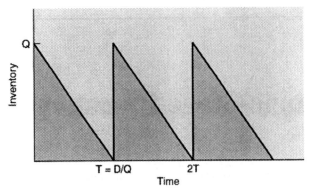

Figure 6.1 Cycle Inventory Level over Time

of the inventory level over time appears in Figure 6.1. This is commonly referred to as the "sawtooth function" and indicates instant replenishment (the entire lot is placed into inventory at once on receipt) and continuous demand (constant withdrawal rate from the inventory stock). To summarize, we make the following assumptions:

Assumptions:

1. Known, static, continuous, deterministic demand
2. Constant, fixed cost to place an order
3. Inventory carrying cost based on value of average on-hand inventory
4. Instantaneous batch delivery following a deterministic lead time
5. All replenishment orders are for the same quantity, and that quantity is not limited
6. No shortages are allowed

The total cost of the inventory per time is composed of ordering cost, product purchase cost, and inventory holding cost. The assumptions of fixed demand and lead time allow us to avoid all shortages. Holding cost will be charged as a fraction of item cost per time based on average on-hand inventory. The period covered by each order is called a **cycle.** It is apparent from the Figure that the inventory level decreases constantly from Q to 0 each cycle and, thus, averages $Q/2$. If this geometric argument is not convincing, then note that the process repeats each time Q units are sold (i.e., every $T = Q/D$ time units) and integrate over this cycle to find the average inventory, $\bar{I}$. In this case, we have:

$$\bar{I} = \frac{1}{Q/D} \int_0^{Q/D} (Q - tD)\, dt = \frac{D}{Q}\left(Qt - \frac{Dt^2}{2}\ \bigg|_0^{Q/D}\right) = \frac{D}{Q}\left(\frac{Q^2}{D} - \frac{Q^2}{2D}\right) = \frac{Q}{2}. \quad (6.1)$$

These ordering, purchase, and holding costs are represented by:

$$Cost/Time = \frac{AD}{Q} + CD + \frac{hQ}{2}; \quad (6.2)$$

where:

A = fixed cost to place an order

D = demand rate in units per time

C = unit purchase cost of product

h = inventory holding cost per unit-time = iC

τ = lead-time for order delivery

To find the optimal order quantity, we differentiate Equation (6.2) with respect to Q and set the result to 0. This yields:

$$\frac{dCost/Time}{dQ} = \frac{-AD}{Q^2} + \frac{h}{2} = 0 \tag{6.3}$$

To check whether the cost function is convex, we must check the second derivative:

$$\frac{d^2 Cost/Time}{dQ^2} = \frac{2AD}{Q^3} > 0 \text{ for } Q > 0.$$

Because the second derivative is positive for $Q > 0$, the function is convex, and we can find the global minimum by setting Equation (6.3) equal to 0. By rearranging terms, we find the optimal order quantity, Q^*, to be:

$$Q^* = \sqrt{\frac{2AD}{h}} \tag{6.4}$$

This is the well-known economic order quantity (EOQ). We reorder to receive the items just as on-hand inventory reaches 0. This translates into placing an order when our inventory position drops to the lead-time demand of (τ time periods) · (D units/time). Thus,

$$r = \tau D \tag{6.5}$$

Once the order quantity Q and reorder point r are known, we can evaluate Equation (6.2) to determine related costs.

EXAMPLE 6.1

A manufacturing facility uses lumber to construct shipping crates for products. The facility uses 250 8-foot boards per week. The supervisor estimates a fixed cost of $80 to place an order for paperwork preparation. Lumber costs $2.25 per board, and the company uses a carrying cost rate of 50% per year. The plant operates fifty weeks per year. It takes two weeks to receive a shipment. Find the optimal order quantity and reorder point.

SOLUTION

We use boards as the product unit and years as the time period. Using Equation (6.4),

$$Q^* = \sqrt{\frac{2AD}{h}} = \sqrt{\frac{2(\$80)(250\ boards/week)(50\ weeks/yr)}{(\$2.25/board) \cdot (0.50/yr)}} = 1333.\ boards$$

The reorder point is $\tau D = (2\ weeks) \cdot (250\ boards/week) = 500\ boards$. The cost per year is:

$$Cost/yr = \frac{AD}{Q} + CD + \frac{hQ}{2} = \frac{(80)(250)(50)}{1333} + 2.25(250)(50) + \frac{(1.125)(1333)}{2} = \$29,625.$$

Equation (6.4) provides clarification of the economic approach to inventory control. To begin, note the impact of the square root relationship. Demand must quadruple to justify doubling the order quantity. Likewise, setup cost would have to be reduced by 75% to cut the order size in half. If we do reduce A by 75%, the total setup cost incurred per time is reduced by only 50%. This follows because if we divide A by 4, the resultant Q^* from Equation (6.4) is only divided by 2. However, if the order size is cut in half, then we must order twice as often. If we make twice as many orders now as before, and each costs one-fourth as much, then total cost is half the original cost. Another interesting aspect of the

EOQ is that at the optimal Q^*, ordering cost and holding cost are equal. To understand this, compare these costs at Q^*. Ordering cost per period is:

$$\frac{A \cdot D}{Q^*} = \frac{A \cdot D}{\sqrt{\dfrac{2AD}{h}}} = \sqrt{\frac{A \cdot D \cdot h}{2}}.$$

Inventory holding cost per period is:

$$\frac{h \cdot Q^*}{2} = \frac{h \cdot \sqrt{\dfrac{2AD}{h}}}{2} = \sqrt{\frac{A \cdot D \cdot h}{2}}.$$

The purchase cost of parts does not really affect our inventory policy because we have the fixed cost of $C \cdot D$ no matter what policy we select. Total relevant inventory cost per time is, therefore, given by the sum of ordering and holding costs. This sum is simply:

$$2\sqrt{\frac{ADh}{2}} = \sqrt{2 \cdot A \cdot D \cdot h}.$$

Thus, if the cost to place order A is cut in half, total ordering cost, holding cost, and total inventory cost are all $0.5^{1/2}$, or 70.7% of their original cost. This translates to a reduction of almost 30%.

The question that arises is: How important is it to order exactly Q^* units each time? It may be more likely to order in multiples of 10 or 144 or some other prepackaged amount. Also, because our parameters h, A, and even D, are only estimates of the true parameters, we would like to know the sensitivity of the cost to the optimal choice of Q and the sensitivity of the optimal Q to the parameter estimates. We examine the first question here and leave the second question until the end-of-chapter problems. We can measure the sensitivity of the cost by comparing the *total relevant costs* (TRC) at Q^* and any other value for Q. The ratio is given by:

$$\frac{TRC(Q)}{TRC(Q^*)} = \frac{\dfrac{AD}{Q} + \dfrac{hQ}{2}}{\sqrt{2 \cdot A \cdot D \cdot h}} = \frac{\sqrt{AD}}{Q\sqrt{2h}} + \frac{Q\sqrt{h}}{2\sqrt{2AD}} \tag{6.6}$$

$$= \left[\frac{Q^*}{2Q} + \frac{Q}{2Q^*} \right] = \frac{1}{2}\left[\frac{Q^*}{Q} + \frac{Q}{Q^*} \right]$$

Suppose we overestimate Q^* and use a value that is 50% too large. Using this ratio:

$$\frac{TRC(Q = 1.5Q^*)}{TRC(Q^*)} = \frac{1}{2}\left[\frac{1}{1.5} + \frac{1.5}{1} \right] = 1.083.$$

Being off by even 50% in our choice of Q only increases our costs by 8.3%. In general, costs are relatively robust to choice of Q. Although it is important to know the approximate size of Q^*, we have some freedom to adapt our order quantity to the natural characteristics of our process.

6.1.2 Economic Manufactured Quantity Model

A similar approach can be used in the case of products manufactured in-house. In the EOQ, items were ordered and immediately became available for use in meeting demand when they were received. If we produce the items instead, then we must first obtain the raw materials and process these through the facility before the products can be used to satisfy demand. Thus, we should also account for the work-in-process inventory cost incurred between release of the material to the shop and completion of the batch. This leads to a modified model for determining the batch size.

Define s as the time required to set up the process to make the product and p as the variable processing time required per unit. The levels of inventory in process and finished products are shown in Figure 6.2. The Figure illustrates the case in which the entire batch is available during setup and production of all items (in the end-of-chapter exercises, we ask the reader to build a slightly different model). After production, the batch is added to finished goods. As in the EOQ, the finished goods level decreases at the constant rate, D.

To account for the WIP cost, we let M be the cost of the raw material (per unit of product) and v be the value added by the process. Thus, the cost of a completed item is $C = M + v$. Because the product increases in value from M to $M + v$ during production, we approximate the value of an in-process unit by $M + v/2$. We base WIP inventory cost on the realization that each unit of product demanded goes through the production process. Actual flow time through the shop can often be approximated by a multiple of batch-processing time. For example, it is commonly reported that throughput time averages twenty times batch processing time for a job shop. Assembly operations and balanced flow lines, on the other hand, may have a throughput time only two or three times the actual processing time. This ratio can be estimated from historical data or by noting Little's Law, which states that for any stable process:

$$\text{Average Throughput Time} = \frac{\text{Average Number of Batches in Process}}{\text{Arrival Rate of Batches}}$$

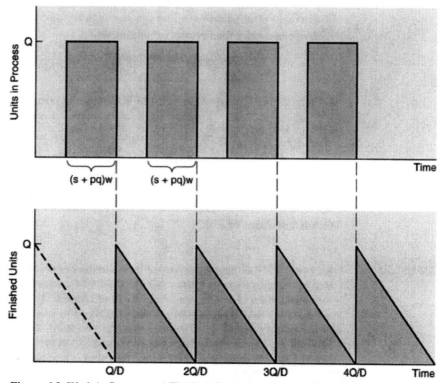

Figure 6.2 Work-in-Process and Finished Goods Inventory for Manufactured Product

where the arrival rate is measured as the number of batches either released to the shop floor per time or completed per time. We let w be the ratio of throughput time to processing time for the facility.

EXAMPLE 6.2

A facility produces 100 batches per 40-hour work week. Actual processing time per batch averages four hours. Records show an average of 30 batches are in process at any time. Estimate throughput time and w.

SOLUTION

Using Little's Law:

$$\text{Average Throughput Time} = \frac{30 \text{ batches}}{100 \text{ batches/week}} = 0.3 \text{ weeks} = 12 \text{ hrs.}$$

Thus,

$$w = \frac{12 \text{ hrs throughput}}{4 \text{ hrs production}} = 3.$$

The inventory cost model then follows as:

$$\frac{\text{Total Cost}}{\text{Time}} = \frac{A \cdot D}{Q} + M \cdot D + i \cdot \left(M + \frac{v}{2}\right) \cdot (s + pQ) \cdot D \cdot w + \frac{i(M + v) \cdot Q}{2} \quad (6.7)$$

where total cost in Equation (6.7) is composed of setup, material purchase, WIP inventory, and finished goods inventory cost terms, respectively. The five factors in the WIP cost include: 1) i, the carrying cost rate $\left(\frac{\$}{\$/unit/time}\right)$; 2) $(M + v/2)$, the value of an average in-process part (\$); 3) $(s + pQ)$, the throughput time for a batch (*time*); 4) D, the number of units sent through the production process (*units*); and 5) w, the throughput time factor. Differentiating this total cost per time function:

$$\frac{dTC/time}{dQ} = \frac{-A \cdot D}{Q^2} + i\left(M + \frac{v}{2}\right)pDw + \frac{i(M + v)}{2} = 0. \quad (6.8)$$

By rearranging terms, we derive from Equation (6.8) that:

$$Q^* = \sqrt{\frac{2 \cdot A \cdot D}{i \cdot \left[(M + v) + 2 \cdot p \cdot D \cdot w \cdot \left(M + \frac{v}{2}\right)\right]}} \quad (6.9)$$

Equation (6.9) closely resembles Equation (6.4) with the addition of the final term in the denominator for WIP cost. We refer to the rule in Equation (6.9) as the economic manufactured quantity, or EMQ.

EXAMPLE 6.3

Each year, 1,500 castings are purchased from a foundry as needed for \$18 each. They undergo finishing machining operations before being stocked for final assembly. The machine takes fifty minutes to set up, and each piece is machined for fifteen minutes. The machining work center is charged out at \$30 per hour, including labor. A study has shown that jobs spend about three-fourths of their time in the shop waiting and in-transport (the other one-fourth is spent in setup and processing time). In addition to machine setup, each batch incurs an \$80 charge for ordering castings. The carrying cost rate for the company is 40% per year. The plant is scheduled to operate 2,500 hours per year. Find the optimal production lot size.

SOLUTION

The fixed cost per lot includes $80 for ordering plus $30/$hour$ \cdot \left(\dfrac{50 \text{ min}}{60 \text{ min}/hr} \right)$ for machine setup, yielding a cost of $A = \$105$. We approximate item cost as:

$$C = M + v = \$18 + (0.25 \ hours) \cdot \$30/hour = \$25.5.^{[1]}$$

Applying Equation (6.9):

$$Q^* = \sqrt{\frac{2AD}{i\left[C + 2pDw\left(M + \dfrac{v}{2}\right)\right]}} = \sqrt{\frac{2(105)(1500)}{(0.4)\left[25.5 + 2\left(\dfrac{0.25}{2500}\right)(1500)(4)\left(18 + \dfrac{7.5}{2}\right)\right]}} = 123.5$$

Although it is important to optimize decisions, it is probably more important to take advantage of improvements in system design. Consider, for example, the impact of using a nonoptimal batch size. Suppose that in Example 6.3, we were off by 20% and produced in batches of 100. At $Q = 123.5$, using Equation (6.8), the total cost per year is:

$$\frac{Cost}{Year} = \frac{(105)(1500)}{123.5} + (18)(1500) + 0.4\left(18 + \frac{7.50}{2}\right)$$

$$\left(\frac{0.833}{2500} + \frac{0.25}{2500}(123.5)\right)(1500)4 + \frac{0.4(18 + 7.50)(123.5)}{2} = \$29,567.22$$

At $Q = 100$, the cost per year is:

$$\frac{Cost}{Year} = \frac{(105)(1500)}{100} + (18)(1500) + 0.4\left(18 + \frac{7.50}{2}\right)$$

$$\left(\frac{0.833}{2500} + \frac{0.25}{2500}(100)\right)(1500)4 + \frac{0.4(18 + 7.50)(100)}{2} = \$29,624.39,$$

an increase of just $52.17. Now, suppose we could redesign the setup procedures to reduce setup cost by 20% to $84. (This corresponds to an eight-minute setup time plus the $80 ordering cost.) The optimal batch size is now:

$$Q^* = \sqrt{\frac{2AD}{i\left[C + 2pDw\left(M + \dfrac{v}{2}\right)\right]}} = \sqrt{\frac{2(84)(1500)}{(0.4)\left[25.5 + 2\left(\dfrac{0.25}{2500}\right)(1500)(4)\left(18 + \dfrac{7.5}{2}\right)\right]}} = 110.5$$

with an annual cost of:

$$\frac{Cost}{Year} = \frac{(84)(1500)}{110.5} + (18)(1500) + 0.4\left(18 + \frac{7.50}{2}\right)$$

$$\left(\frac{0.133}{2500} + \frac{0.25}{2500}(110.5)\right)(1500)4 + \frac{0.4(18 + 7.50)(110.5)}{2} = \$29,283.41,$$

an annual savings of $283.81. In addition, shorter setup times mean that we can produce more items on the same equipment and use shorter production runs; thus, we can promise customers quicker delivery.

---

[1]Assuming the lot size will be large, we ignore the contribution from setup and ordering cost when figuring the value added.

6.1.3 The Case of Fixed Setup Cost

Most machines are used to produce multiple part types and require some changeover effort to convert from production of one part type to another. The EMQ model included the AD/Q term assuming that operating cost varies with the number of setups. If most setup cost is labor, and labor is fixed over the short-term, then at the time we make a decision to produce a batch, the incremental cost to changeover to the new product type may be insignificant. Labor cost would be a relevant factor in a higher-level planning model but not when choosing how to use the machine today (assuming the number of labor hours is fixed). If the variable setup cost is $A = 0$, then the model would suggest making a single item at a time. This may be impractical because changeovers do have to be performed, and we would soon run out of machine time. In this environment, we should set the batch size to the smallest value that will still permit enough productive machine time to make all the required items. To specify the batch sizes precisely, we need information on all items produced in this work center: their demands, setup times, and processing times. This issue is addressed in section 6.3.4 on multiproduct constrained models. For now, divide the setup time for job i into two portions: s_{1i}, the internal setup time and s_{2i}, the external setup time. Internal setup consists of the activities that can be performed only while the machine is idle. This might include placing a die on a press or running a test part. External setup consists of preparatory setup activities that can be carried out while the machine is running the previous batch. Examples include retrieving the materials and tooling for the next job, prepositioning and preheating tooling, and examining processing documentation. Let i be the current job being run and $i + 1$ the next scheduled job. The unit processing time for job i is defined as p_i, the external setup time for the next job is $s_{2,i+1}$. We consider the case in which the machine attendant performs internal and external setup, but the machine can run without an operator. We could then set $Q_i = Q_i^1$, where Q_i^1 satisfies:

$$p_i \cdot (Q_i^1 - 1) < s_{2,i+1} \leq p_i \cdot Q_i^1. \qquad (6.10)$$

This rule selects the smallest lot size that will keep the machine busy while the external setup for the next job is completed. This rule avoids unnecessary idle machine time. We should also consider the additional constraint that we not exceed the available time in the period. Suppose the rule noted above leads to a solution whereby we set up each item K times per period. Then, total machine time required for setup and production of all n products at this work center this period will be $\sum_{i=1}^{n} (K \cdot s_{1i} + p_i \cdot D_i)$. To be feasible, with respect to the available time T in a period, we must have $\sum_{i=1}^{n} (K \cdot s_{1i} + p_i \cdot D_i) \leq T$, or on rearranging terms:

$$K \leq \frac{T - \sum\limits_{i=1}^{n} p_i D_i}{\sum\limits_{i=1}^{n} s_{1i}} \qquad (6.11)$$

For K production batches of each product during the period, we must produce in batch sizes $Q_i^{min} = \dfrac{D_i}{K}$. If the period is of length T, then we have the decision rule that the optimal batch size is Q_j^{opt}, where:

$$Q_j^{opt} = max(Q_i^1, Q_i^{min}) \qquad (6.12)$$

The rule [Equation (6.12)] will ensure that unnecessary idle machine time that could limit production capacity and that overloading the machine are both avoided.

EXAMPLE 6.4

A work center produces three products each period. A period has 40 hours. Setup times, processing times, and period demands are given in the Table. Labor is fixed. Find the appropriate batch sizes for the upcoming period to minimize inventory cost.

Item	Internal setup (hrs.)	External setup (hrs.)	Unit processing (hrs.)	Demand
1	1.0	2.0	0.1	100
2	2.0	2.0	0.2	50
3	2.0	2.0	0.2	50

SOLUTION

Let's assume that items are produced in the order 1, 2, and 3. Using Equation (6.12), the batch size for product 1 should satisfy $p_1 \cdot (Q_1^1 - 1) < s_{2,2} \le p_1 \cdot Q_1^1$. Because $\frac{s_{2,2}}{p_1} = \frac{2.0}{0.1} = 20$, we would set $Q_1^1 = 20$. Likewise, we find $Q_2^1 = \frac{s_{2,3}}{p_2} = 10$ and $Q_{3,1} = \frac{s_{2,1}}{p_3} = 10$. This policy would require $\frac{D_i}{Q_i} = 5$ setups of each product during the period and require a total of $5(1 + 2 + 2) = 25$ machine hours for internal setup plus $100(0.1) + 50(0.2) + 50(0.2) = 30$ hours for production. From Equation (6.11), we find that we are limited to:

$$K \le \frac{40 - (10 + 10 + 10)}{(1 + 2 + 2)} = 2 \text{ setups/item.}$$

Time is thus the constraining factor, and we must use batch sizes of ($Q_1^* = 50$, $Q_2^* = 25$, $Q_3^* = 25$).

6.1.4   Price Breaks

Manufacturers recognize the value of major customers and the economies of scale associated with producing large quantities. Price breaks, therefore, are often offered to customers ordering large quantities of a product. In this section, we discuss the case in which the unit price is based on the quantity ordered, with this price applying to each unit purchased. (It is also possible to use an incremental scheme whereby the first q items are charged one price and additional items may be purchased at a lower price.)

We will let J be the number of levels (unit prices) in the pricing scheme. The cost per item is then:

$$C = \begin{cases} C_1 & \text{, if } Q < q_1 \\ C_j, \text{ if } q_{j-1} \le Q < q_j \text{ for } j = 1, ..., J - 1 \\ C_J & \text{, if } q_J \le Q \end{cases}$$

The relevant total cost function now must include the unit cost and appears as:

$$\frac{Total\ Cost}{Time} = \frac{AD}{Q} + C_j D + \frac{iC_j Q}{2}.$$

The part and total cost curves are shown in Figure 6.3. The thick line indicates the relevant cost curve for each order quantity. Note that if $C_{j_1} < C_{j_2}$, then the total cost curve for price C_{j_1} lies completely below the curve for C_{j_2}. Also, as the price decreases, the optimal order quantity found from:

$$Q_j^* = \sqrt{\frac{2AD}{iC_j}} \tag{6.13}$$

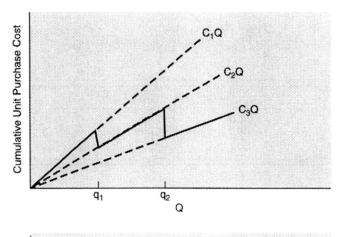

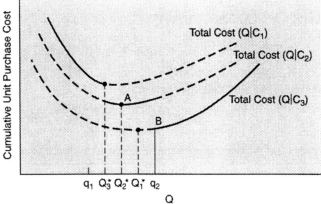

Figure 6.3 Unit Purchase and Total Cost Curves with Price Breaks

increases. This follows because the setup cost is unchanged; however, the holding cost is lower. The optimal feasible order size can be only one of a few values. It occurs either at the minimum of one of the cost curves such as point A in the Figure or at the minimum allowable quantity for some cost such as point B. Note that if we find a price for which the solution to Equation (6.13) is feasible, then no larger price can possibly be optimal. This follows because this point must lie below all points on the curves for higher prices. This leads to a relatively simple algorithm for solving the all-units discount problem. Start at the lowest price. Find the best allowed order size at that price and the associated total cost. Increase the price and repeat. Continue until we reach a curve where its optimal order size is also feasible. We then compare these alternatives and choose the best.

Solution Procedure:

1. Set $j = J$.
2. Find Q_j^* using Equation (6.13). If $Q_j^* \geq q_{j-1}$, compute $TC(Q_j^*)$ and go to 3. If $Q_j^* < q_{j-1}$, this price is not available for this quantity, and we must compute $TC(q_{j-1})$ for price C_j, set $j = j - 1$ and repeat step 2.
3. Compare the total cost values generated thus far and select the order size with minimum $TC(Q)$.

EXAMPLE 6.5

An assembly plant purchases 35,000 air filters per year. Fixed ordering cost is $100. The company uses an inventory holding cost rate of .4 per year. The vendor has offered the following price schedule. For orders of more than 5,000, unit cost is $2.50. For orders between 1,000 and 5,000, filters cost $2.65. Orders less than 1,000 have a cost of $2.85 per filter. Find the optimal order quantity.

SOLUTION

The first option is:

1. $C_3 = \$2.50$, and the optimal order quantity is $Q_3^* = \sqrt{\dfrac{2(100)(35,000)}{0.4(2.50)}} = 2,646$. Because this amount is not feasible, we check the cost of ordering the minimal amount at $2.50.

$$TC(5,000) = \frac{100(35,000)}{5,000} + 2.50(35,000) + \frac{0.4(2.50)(5,000)}{2} = \$91,400.$$

The next option is:

2. $C_2 = \$2.65$, and the optimal order quantity is $Q_2^* = \sqrt{\dfrac{2(100)(35,000)}{0.4(2.65)}} = 2,570$. This amount is feasible and has total cost of:

$$TC(2,570) = \frac{100(35,000)}{2,570} + 2.65(35,000) + \frac{0.4(2.65)(2,570)}{2} = \$95,474.$$

Because the optimal solution was in the feasible range for this price, we need not consider the higher-priced options. Comparing the two candidate solutions, we should purchase lots of 5,000 filters at $2.50 each.

6.1.5 Multiproduct Coordination Models

In some cases, replenishment decisions for multiple products are interdependent. The items may be ordered from the same vendor, produced on the same equipment, or share warehouse space. We discuss several common situations, but many others are possible. Hopefully, the reader will be able to apply the modeling and analysis techniques presented in this section to other related situations.

6.1.5.1 Organizational Constraints

Suppose we have n different stock items, and an inventory policy is needed for each. We could model the entire problem as:

$$\textit{Minimize Total Cost/Time} = \sum_{i=1}^{n}\left(\frac{A_i D_i}{Q_i} + C_i D_i + \frac{h_i Q_i}{2}\right). \tag{6.14}$$

If we differentiate Equation (6.14) with respect to Q_i, we obtain:

$$\frac{\partial Cost/Time}{\partial Q_i} = \frac{-A_i D_i}{Q_i^2} + \frac{h_i}{2}.$$

We have n equations of this form to set equal to 0, one for each product. The key fact here is that each of the n equations contains only the data and decision variable for that single product. Thus, we can use the original EOQ expression and solve for each item's inventory policy separately. However, when constraints exist linking these equations, the solution is not so straightforward.

Consider the situation in which the items are stored in a warehouse with F cubic feet available for the cycle stock of these products. Let f_i be the space occupied per unit of product i. If we must save space for the maximum possible inventory of each item, then we have the constraint:

$$\sum_{i=1}^{n} f_i Q_i \leq F. \tag{6.15}$$

Now, the choice of Q_1 limits the possible choices for the remaining products. The first step in solving the constrained problem involves solving the unconstrained problem and checking the constraints. That is, we compute the optimal unconstrained batch size for each product using Equation (6.4). We then check the constraint [Equation (6.15)]. If the constraint is satisfied, then we have our solution. Otherwise, we need to use a coordinated solution strategy. One approach is to minimize the lagrangian relaxation function:

$$\textit{Minimize } L(\lambda, Q_1, \ldots, Q_n) = \sum_{i=1}^{n} \left(\frac{A_i D_i}{Q_i} + C_i D_i + \frac{h_i Q_i}{2} \right) + \lambda \left(\sum_{i=1}^{n} f_i Q_i - F \right) \tag{6.16}$$

We can think of λ as a positive penalty factor for violating the constraint. The optimal solution to the original constrained problem can be shown to be equal to the optimal solution to the lagrangian problem [Equation (6.16)], where λ is selected such that the constraint is satisfied. The parameter λ also has a more direct economic interpretation. Essentially, λ is the value of an extra unit of floor space. If we could increase F by one unit, the allowed increase in the Q_i would reduce total cost by λ.

For any value of λ, we can solve Equation (6.16) by taking the partial derivatives with respect to the Q_i and setting them to zero. This gives the set of expressions:

$$\frac{\partial L}{\partial Q_i} = \frac{-A_i D_i}{Q_i} + \frac{h_i}{2} + \lambda f_i = 0$$

or, on rearranging terms:

$$Q_i = \sqrt{\frac{2 A_i D_i}{h_i + 2\lambda f_i}}. \tag{6.17}$$

Note that in Equation (6.17), as λ increases, Q_i and the space required for product i decrease. We then need to merely search over λ until the resulting Q_i, as computed in Equation (6.17), satisfies Equation (6.15).

EXAMPLE 6.6

A 10,000-square foot warehouse stocks four items. Product data are summarized as:

Item	Annual demand	Ordering cost	Holding cost/unit-yr	Sq. ft./unit
1	5,000	$ 75	$5.00	8.0
2	5,000	$ 75	$2.50	5.0
3	10,000	$100	$5.50	6.0
4	20,000	$ 75	$3.35	6.0

Assuming the warehouse must reserve space for each item's maximum inventory, find the appropriate order quantities.

SOLUTION

We begin by checking the unconstrained solution. For the first item:

$$Q^* = \sqrt{\frac{2(75)(5,000)}{5}} = 387.3,$$

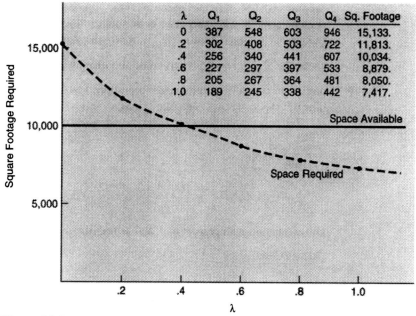

λ	Q_1	Q_2	Q_3	Q_4	Sq. Footage
0	387	548	603	946	15,133.
.2	302	408	503	722	11,813.
.4	256	340	441	607	10,034.
.6	227	297	397	533	8,879.
.8	205	267	364	481	8,050.
1.0	189	245	338	442	7,417.

Figure 6.4 Square Footage Required as a Function of λ

which requires $(387.3)*(8) = 3,098$ square feet for storage. Continuing for items 2, 3, and 4, we find a need for 15,133 square feet. This exceeds the 10,000 square feet available. Thus, we need to search over λ. If $λ = 1$ for the first item, we obtain:

$$Q* = \sqrt{\frac{2(75)(5,000)}{5 + 2(1)(8)}} = 189.0$$

which requires $(189.0)*(8) = 1511.9$ square feet. The order sizes for items 2 through 4 are 245, 338, and 442, respectively. These translate into 7,417 square feet of storage. Figure 6.4 illustrates a plot of total square footage required versus λ. We see that the smallest λ that yields a feasible solution is approximately 0.405. This yields order sizes of 256, 338, 439, and 604, respectively. These will fill 10,000 square feet of storage.

Other types of constraints occur. Inventory requires financial investment, which could also be limited. In manufacturing, setups take time and therefore reduce the productive capacity of equipment. This would lead to a constraint such as $\sum_{i=1}^{n} \frac{s_i D_i}{Q_i} \leq S$, where S is the setup time available per period. The lagrangian should be formed by adding a lagrangian multiplier, $λ_j$, for each violated constraint.

6.1.5.2   Rotation Cycle Policies

Many manufacturing facilities use the same equipment to produce multiple products. A press, for example, may form side, top, and bottom panels for a variety of models. Dies must be changed for each part type. A bottling line may fill bottles of several sizes with different mixes. Part feeders (bottles, caps, labels, boxes), mix feeders, and machine stops must be changed for each product. These changeover activities can consume a significant portion of available processing time. Thus, determining economic production quantities becomes vital for efficient operation. One approach is to use a rotation cycle. Each item is constrained to have the same cycle time. In a cycle of length T, the facility produces $Q_j = D_j \cdot T$ for each item $j = 1, \ldots, n$. To ensure that adequate capacity exists, we must check that $\sum_{j=1}^{n} p_j \cdot D_j \le 1$. This constraint indicates that the sum over all items of the proportion of time the system needs to produce that item is less than 1 (100% of available time). If this condition is satisfied, we can solve for the optimal rotation cycle, T.

Total cost is a function of T, and is composed of setup and inventory cost for all items:

$$\frac{Total\ Cost}{Time} =$$

$$\sum_{i=1}^{n} \left(\frac{A_i}{T} + C_i \cdot D + \frac{i \cdot (M_i + v_i) \cdot T \cdot D_i}{2} + i \cdot \left(M_i + \frac{v_i}{2} \right) \cdot D_i \cdot (s_i + p_i \cdot T \cdot D_i) \cdot w \right)$$

Differentiating with respect to T and setting the result to 0 yield:

$$\frac{\partial Total\ Cost/Time}{\partial T} =$$

$$-\frac{\sum_j A_j}{T^2} + \frac{\sum_j [i \cdot (M_j + v_j) \cdot D_j]}{2} + \sum_j \left[2 \cdot i \cdot \left(M_j + \frac{v_j}{2} \right) \cdot w \cdot D_j^2 \cdot p_j \right] = 0,$$

or:

$$T^* = \sqrt{\frac{2 \sum_{j=1}^{n} A_j}{i \cdot \sum_{j=1}^{n} D_j[(M_j + v_j) + 2 \cdot w \cdot D_j \cdot p_j \cdot (M_j + v_j/2)]}} \qquad (6.18)$$

Equation (6.18) provides the optimal coordinated production cycle, *if* sufficient time exists. We must make sure, however, that the machine has adequate capacity to produce $Q_j = D_j \cdot T$ of each item every T time units. We assumed at the start that $\sum_{j=1}^{n} p_j \cdot D_j \le 1$ to ensure sufficient production time, but setup time must also be included. This is the same concern modeled in Section 6.1.3. We ensure feasibility by adding the constraint:

$$\sum_{j=1}^{n} (s_j + p_j \cdot D_j \cdot T) \le T. \qquad (6.19)$$

The left side of the expression is total time per cycle spent on setup and production. The right side is the available time per cycle. The machine must produce for a fixed proportion of time. For the cycle to be feasible, the remaining time must be sufficient to allow each item to be set up once. We can rearrange Equation (6.19) to find a lower bound on T.

$$T^{min} = \frac{\sum_{j=1}^{n} s_j}{1 - \sum_{j=1}^{n} (p_j \cdot D_j)}. \qquad (6.20)$$

The optimal feasible cycle time, T^{opt}, is found from:

$$T^{opt} = \max(T^*, T^{min}). \qquad (6.21)$$

By reformulating this problem into a single-variable problem, namely, as a function of T, we have created a rather easily solved problem.

EXAMPLE 6.7

A bottle-filling machine fills 20-oz. plastic bottles with four different syrups. The machine usually operates 200 hours per month, and each changeover between syrups takes one hour and costs $100. The machine can bottle 3,000 bottles per hour. The production system operates as a flow line, thus bottles enter the line and are immediately filled, capped, packaged, and shipped to the warehouse. Material cost is $0.16 per bottle, and the value added from bottling is calculated at $0.04. The holding cost rate is 5% per month. Additional details are provided in the Table. Find the optimal rotation cycle time and batch sizes.

Item	Demand/mth
1	250,000
2	25,000
3	25,000
4	125,000

SOLUTION

For this continuous flow process, we set $w = 1$. Using Equation (6.18) and putting values into time units of months, we find the desired cycle time to be:

$$T^* = \sqrt{\frac{2(100 + 100 + 100 + 100)}{(0.05)\left\{(250,000)\left[(0.20) + \frac{2(250,000)(1)(0.18)}{(3000)(200)}\right] + \ldots + (125,000)\left[(0.20) + \frac{2(125,000)(1)(0.18)}{(3000)(200)}\right]\right\}}} = 0.347 mths$$

Converting setup and production time to months and using Equation (6.20), the minimum cycle time is:

$$T^{\min} = \frac{\sum_{j=1}^{4} s_j}{1 - \sum_{j=1}^{4} p_j D_j} = \frac{[(1 + 1 + 1 + 1)/200]}{1[(200)(3000)]^{-1}(250,000 + 25,000 + 25,000 + 125,000)} = 0.0686 mths$$

Therefore, because setup time is not a constraint, we select the preferred cycle time, $T^{opt} = 0.347 mths$. The corresponding batch sizes for the syrups are $T^{opt} \cdot D_j = (86,833, 8,683, 8,683, 43,417)$ bottles, respectively.

In our rotation cycle policy, we forced each item to be produced with the same frequency. If the natural cycle times[2] for items differ, it may be preferable to produce some items less often. This economic lot-scheduling problem has received considerable attention. (See Elmaghraby [1978] and Gallego [1990] references at the end of the chapter for more detail.) However, a simple heuristic is to first find the natural cycle $T_j^* = Q_j^*/D_j$ for each item, where Q_j^* is the independent EOQ value for item j. Then, look at the ratio $\dfrac{T_j^*}{\min_k T_k^*}$ for each item j. Round off this ratio to the nearest integer n_j. These n_j integers are the rel-

---

[2]Natural cycle time means the optimal time between replenishments if the item is considered independently of others, and order size is given by the EOQ or EMQ.

ative cycle times for items. We define a base period T^b and produce item j every n_j base periods. The base period must be at least:

$$T^{b.\min} = \frac{\sum_{j=1}^{n}(s_j/n_j)}{1 - \sum_{j=1}^{n} p_j D_j} \tag{6.22}$$

to allow enough slack time to handle setup time requirements. We would like to have a base period as brief as the natural cycle for each item, if possible. Therefore, we set the base period to:

$$T^b = \max\{T^{b,\min}, \min_j T_j\}. \tag{6.23}$$

Finally, we attempt to find a repeating cycle that produces item j, every n_j base periods. This may require some manual modification because some base periods may not be sufficient to handle all the products scheduled for that period. For instance, if $n_1 = 1$, $n_2 = 2$, and $n_3 = 3$, then we would produce item 1 in every base period, item 2 every second base period, and item 3 every third base period. The entire schedule would repeat every set of six base periods, with the list of items planned for production in each base period as follows:

Period	Items produced
1	1
2	1, 2
3	1, 3
4	1, 2
5	1
6	1, 2, 3

It may not be possible to produce the desired amount of all three items in every sixth base period. Thus, we may need to modify the schedule to start producing in period 5 and maybe even extend over to the next base period 1.

6.1.6 Lead-Time Minimization

In addition to cost, the responsiveness of the production system defines competitiveness. Thus, the lead-time from when an order is placed until the batch is completed may be an important criterion. Consider a work center that produces multiple part types in batches. Batches enter the work center's queue when the order is received. We find the batch size that minimizes the time in the work center. To model this problem, we let $\bar{s}$ be the average setup time for a part batch and D the total number of items produced in the work center per time period. The setup time average is taken across part types. We assume that part types have an average unit processing time, $\bar{p}$. If the setup and processing time structure is such that most production runs are short but a few part types have long run lengths, the performance of this work center can be modeled as a Poisson arrival, exponential service + time M/M/1 queuing system. Expected service time per batch is given by $\mu^{-1} = \bar{s} + \bar{p}Q$. The batch arrival rate is $\lambda = D/Q$, which leads to a utilization factor of

$\rho \equiv \dfrac{\lambda}{\mu} = \dfrac{D\bar{s}}{Q} + D\bar{p}$. For an M/M/1 queue, it can be shown that the throughput time (or

lead time) is $\tau = \dfrac{1}{\mu(1 - \rho)}$. Combining these terms, we find:

$$\tau = \frac{\bar{s} + \bar{p}Q}{\left(1 - \bar{p}D - \dfrac{D\bar{s}}{Q}\right)} \tag{6.24}$$

Differentiating Equation (6.24), we obtain:

$$\frac{d\tau}{dQ} = \frac{[Q(1 - pD) - D\bar{s}](\bar{s} + 2\bar{p}Q) - Q(\bar{s} + \bar{p}Q)(1 - \bar{p}D)}{[Q(1 - \bar{p}D) - D\bar{s}]^2}.$$

After equating to zero, rearranging terms, and simplifying, we eventually find:

$$Q^* = \frac{\bar{s}D\left(1 + \sqrt{\dfrac{1}{D\bar{p}}}\right)}{(1 - D\bar{p})} \tag{6.25}$$

In completing this derivation, we make use of the fact that we must have $\rho < 1$ to ensure

adequate time to set up. This places the lower bound on the batch size of $Q > \dfrac{D\bar{s}}{1 - \bar{p}D}$.

Note that $1 - \bar{p}D$ gives the proportion of time per period that is not needed for processing. This extra time can be used for setup. Equation (6.25) optimizes the tradeoff between utilization and batch size. Smaller batches can go through the work center faster when congestion is not a problem. However, smaller batches mean more batches, and this increases total setup time, which in turn increases utilization. At some point, utilization becomes large enough that additional decreases in processing time per batch are offset by increased waiting time at the work center. Figure 6.5 shows the impact of batch size on

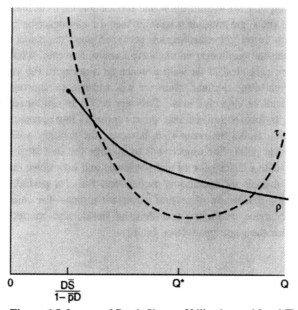

Figure 6.5 Impact of Batch Size on Utilization and Lead Time

machine utilization and lead time. In the end-of-chapter problems, we ask the student to explore this relationship further and to determine the effect of the assumption of exponential processing times. We also note the interesting result that the optimal utilization is equal to the square root of the minimum utilization.

EXAMPLE 6.8

A work center produces a family of parts that is produced to stock. Setups take 1 hour. Unit processing time at the work center varies by part type, but averages 0.01 hours. Demand is for 2,000 items per 40-hour work week. Customer orders for parts arrive randomly. Find the batch size that minimizes lead time.

SOLUTION

Demand per hour is $2,000/40 = 50$ units. The minimum feasible batch size is given by: $Q^{min} = \dfrac{D\bar{s}}{1 - \bar{p}D} = \dfrac{(50)(1)}{1 - (0.01)(50)} = 100$. With this batch size, we will spend half the time in setup for the $2,000/100 = 20$ batches per week and the other 20 hours producing the 2,000 items. The minimum lead-time quantity is:

$$Q^* = \frac{\bar{s}D(1 + \sqrt{(D\bar{p})^{-1}})}{1 - \bar{p}D} = \frac{1(50)(1 + \sqrt{[(50)\,(.001)]^{-1}})}{1 - (0.01)(50)} = 241.42$$

With this batch size, we have $2,000/241.42 = 8.28$ setups per week, the work center has a utilization of $28.28/40 = 0.707$, and the corresponding batch lead time is:

$$\tau = \frac{\bar{s} + \bar{p}Q}{1 - \bar{p}D - \dfrac{D\bar{s}}{Q}} = \frac{1 + (0.01)(241.42)}{1 - (0.01)(50) - \dfrac{50(1)}{241.42}} = 11.66 hrs.$$

6.2 REORDER POINT INVENTORY MODELS FOR STOCHASTIC DEMAND

It is unlikely that we would know the precise values of demand and lead time. It is more likely that these are random variables that we can describe by probability distributions. The models in the previous section provided good decisions for order sizes and reorder points when the parameter values were known; however, what about cases in which they are random variables? If we wait to place an order until the inventory position is equal to expected lead-time demand, then we will experience shortages whenever lead-time demand exceeds its expected value. Delivery delays or random demand increases will cause shortages. In this section, we take the randomness into account in setting the reorder point. We will also model the order size; however, the primary goal is to determine the appropriate reorder point. The objective is to manage the cost attributable to product shortages. We begin with a discussion of service levels and how these can be measured. This is followed by models for continuous review and then for periodic review systems. We conclude with a discussion of several important topics—the case of an item that can be ordered only once, modeling low-demand items, and modeling multiple products that compete for the same production facility.

6.2.1 Service Levels

For inventory systems, service relates to having products available for customer on demand. A shortage or stockout occurs when a customer requests an item that is not currently in stock. In general, when items are being consumed on a continuous basis, we can improve the service level by ordering early, before the inventory level drops to the level of expected lead-time demand. This will provide a safety stock of units usually still on hand when a new order arrives. In some cases, it may be possible to avoid this by expediting outstanding replenishment orders. If demand is customary, but the order is being delayed, we may be able to exert influence to keep the replenishment order on schedule. If the order is on schedule, but demand temporarily increases during the lead time, we may be able to negotiate for an earlier than usual delivery date. Such actions may be expensive or out of our control, however. We will therefore concentrate primarily on the case in which we can set the reorder point. However, the lead-time demand is an uncontrollable random variable.

Several measures are used when addressing service levels. The first measure is stockout probability in any cycle. This measure is the proportion of cycles in which a stockout occurs. For a periodic review system, this indicates the number of shortage periods per year. For a continuous review environment, we can multiply this shortage probability by the expected number of cycles per year to find the number of times per year we will be out of stock. In a continuous review system, the cycle stockout probability is just the probability that lead-time demand exceeds the reorder point.

The second measure of service level gives the proportion of demand that is met from stock. This is also referred to as the **fill rate.** To compute this measure, we will need the additional information on the expected number of shortages per cycle. If we define this as $\bar{S}(r)$ for the reorder point r, then the service level is $\dfrac{Q - \bar{S}(r)}{Q}$.

Customers may be interested in the expected time they will have to wait for an order to be filled. This time includes order entry, order processing, and delivery time, as well as delay caused by inventory stockout. Order entry, processing, and delivery time are determined by internal procedures. It may be advisable to invest time in reducing these factors instead of carrying extra inventory. Indeed, all customers will be affected by the time required for these activities, whereas only those who happen to order during a stockout period are affected by the inventory system's service level. Simple changes in procedures can make significant improvements in customer delivery times. For example, converting a sequential data processing system to a parallel system wherein the shipping department receives customer orders simultaneously with the financial order entry system can reduce the time to fill orders.

6.2.2   Continuous Review Systems

We begin with the situation in which inventory levels are known at all times. To manage inventory, we use a (Q, r) policy. In this policy, whenever the inventory position drops to r units, we place an order for Q units. If the lead time is short relative to the cycle time, this policy can be implemented in a simple manner using two bins: a primary bin and a reserve bin. Workers always remove parts from the primary bin first. The reserve bin holds r units. Whenever the primary bin becomes empty, an order for Q units is placed, and the workers then start removing from the reserve bin. When new units arrive, we fill up the reserve bin first and place the remaining items in the primary bin (Figure 6.6). Of course, this process also can be automated within a computerized inventory tracking system.

Consider, first, the situation in which shortages are backordered. Figure 6.7 shows a sample path of inventory versus time. We always place a new order when the inventory position drops to r units. For simplicity we assume cycle time exceeds lead time in the Figure. Usually, we have a few units left in stock when the order arrives; occasionally,

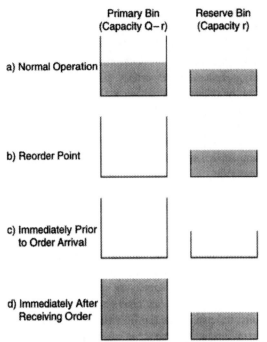

Figure 6.6 Two Bin System: Expected Inventory Levels

however, as shown in the third cycle in the Figure, shortages occur. We assume that back-orders are expensive. Shortages are assumed to cost $\$\pi$ each in our model. The model we derive is only appropriate when $\pi > \dfrac{hQ}{D}$. This condition implies that it costs more to incur a shortage of 1 unit in a cycle than to carry an extra item in inventory for an entire cycle. In this case, the reorder point will exceed the expected lead-time demand, and the

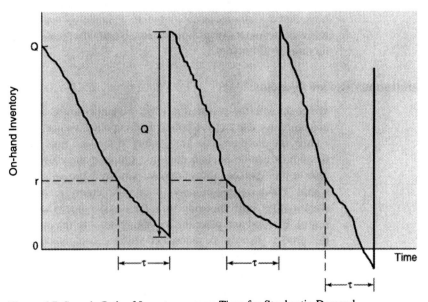

Figure 6.7 Sample Path of Inventory versus Time for Stochastic Demand

number of expected shortages per cycle will be small. The relevant costs to minimize include ordering, purchase, shortage, and cycle inventory costs. To guard against shortages, we will plan to have a positive expected on-hand inventory when an order arrives. This positive quantity is referred to as the **safety stock.** The expected demand during the lead time is $D \cdot \tau$, in which, as before, τ is the lead time but D is now the *expected* demand rate. Thus, the safety stock is equal to $r - D \cdot \tau$. Safety stock is an estimate of the minimum inventory on hand each cycle. When an order arrives, the inventory level increases by Q units. Thus, we expect inventory to vary uniformly between safety stock and the safety stock plus Q each period. Average inventory can be estimated by:

$$\bar{I} = \frac{Q}{2} + r - D \cdot \tau. \tag{6.26}$$

Shortages may occur if lead-time demand is random. This happens when lead-time demand exceeds the reorder point. Let x represent lead-time demand and $f(x)$ the probability density function of lead-time demand. The expected shortages per cycle are:

$$E(shortages/cycle) \equiv \bar{S}(r) = \int_{r}^{\infty} (x - r)\, f(x)dx \tag{6.27}$$

The cost model reflecting ordering, purchase, inventory, and shortage costs is:

$$Minimize\ E(total\ cost/cycle) = \frac{A \cdot D}{Q} + C \cdot D + h\left(\frac{Q}{2} + r - D \cdot \tau\right) + \frac{\pi \cdot D \cdot \bar{S}(r)}{Q}.$$

To find the optimal order quantity and reorder point, we differentiate and solve the equations:

$$\frac{\partial E(TC/cycle)}{\partial Q} = \frac{-A \cdot D}{Q^2} + \frac{h}{2} - \frac{\pi \cdot D \cdot \bar{S}(r)}{Q^2} = 0 \tag{6.28}$$

$$\frac{\partial E(TC/cycle)}{\partial r} = h + \frac{\pi \cdot D}{Q} \cdot \frac{\partial \bar{S}(r)}{\partial r} = 0. \tag{6.29}$$

Using the rule from basic calculus that if:

$$G(z) = \int_{a(z)}^{b(z)} g(y, z)\, dy,\ then\ \frac{dG(z)}{dz} = \int_{a(z)}^{b(z)} \frac{\partial g(y, z)}{\partial z}\, dy + g(b(z),z)\, \frac{db(z)}{dz} - g(a(z),z)\, \frac{da(z)}{dz}$$

and the definition of $\bar{S}(r)$, we realize by letting $x = y$ and $r = z$, that:

$$\frac{\partial \bar{S}(r)}{\partial r} = \int_{r}^{\infty} (-1)\, f(x)\, dx + 0 - 0 = -\int_{r}^{\infty} f(x)\, dx = -[1 - F(r)]$$

where $F(x)$ is the cumulative density function for x. Placing this result in Equation (6.29), we obtain the optimality conditions (we leave it to the reader to check the second-order derivatives to prove that the expected cost function is convex):

$$Q^* = \sqrt{\frac{2 \cdot D \cdot [A + \pi \cdot \bar{S}(r)]}{h}} \tag{6.30}$$

and

$$[1 - F(r^*)] = \frac{h \cdot Q}{\pi \cdot D} \tag{6.31}$$

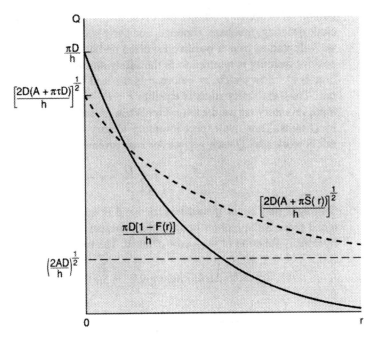

Figure 6.8 Illustration of Optimality Equations

Equation (6.30) gives the optimal value of Q for any r. Equation (6.31) provides the optimal r for any value of Q. A sample plot of Equations (6.30) and (6.31) is shown in Figure 6.8. The form of these curves facilitates finding a fixed point that satisfies both conditions by applying an iterative technique. We can start with the EOQ, $Q^0 = \sqrt{\dfrac{2AD}{h}}$, and solve for our first estimate of the reorder point, $r^{(1)}$, using Equation (6.31). With $r^{(1)}$, we first compute $\bar{S}(r^1)$, using Equation (6.27). We then estimate the order quantity by finding the value $Q^{(1)}$ that solves Equation (6.30). Next, return to Equation (6.31) and find $r^{(2)}$, using $Q^{(1)}$, to estimate Q. We continue this iterative process until the estimates stabilize. This usually requires a small number of iterations.

Before illustrating the solution procedure with an example, let us examine the expression for expected shortages per cycle. Thanks to the central limit theorem, lead-time demand can often be approximated by a normal distribution if the mean is large. (The Poisson distribution is often used when the mean is small.) For the normal case:

$$\bar{S}(r) = \int_{r}^{\infty} (x - r) \cdot \frac{1}{\sqrt{2\pi}\sigma} e^{-\frac{1}{2}\left(\frac{x-\mu}{\sigma}\right)^2} dx$$

If we make a change of variable from x to the standardized normal $z = \dfrac{x - \mu}{\sigma}$, then

$$\bar{S}(r) = \int_{\frac{r-\mu}{\sigma}}^{\infty} (\mu + z\sigma - r) \cdot \frac{1}{\sqrt{2\pi}\sigma} e^{-\frac{1}{2}z^2} (\sigma \cdot dz) = \int_{\frac{r-\mu}{\sigma}}^{\infty} [z\sigma - (r - \mu)]\phi(z)\, dz = \sigma \cdot \int_{\frac{r-\mu}{\sigma}}^{\infty} \left(z - \frac{r - \mu}{\sigma}\right)\phi(z)\, dz$$

This latter integral, $L(k) = \int_{k}^{\infty} (z - k) \cdot \phi(z) \cdot dz$, is known as the unit normal linear loss expression and is tabulated in Appendix I. Thus, for normally distributed lead-time demand, we can easily compute:

$$\overline{S}(r) = \sigma \cdot L\left(\frac{r - \mu}{\sigma}\right) \tag{6.32}$$

Table 6.1 contains computational expressions for the linear loss integral for the normal and several other distributions.

EXAMPLE 6.9

Demand is Normally distributed with a mean of 40 per week and a weekly variance of 8. Weekly demands are independent. Ordering cost is $50, lead-time is two weeks, and shortages cost an estimated $5 to expedite the order and appease the customer. Holding cost is $0.0225 per week. Find the optimal inventory policy.

SOLUTION

Because lead time is two weeks, mean lead-time demand is $\mu = 2(40) = 80$ units, with a standard deviation of $\sigma = \sqrt{2(8)} = 4$. We initially set $\overline{S}(r) = 0$ and solve Equation (6.30) for:

$$Q^{(1)} = \sqrt{\frac{2AD}{h}} = \sqrt{\frac{2(50)(40)}{0.0225}} = 421.6$$

Using Equation (6.31):

$$1 - F(r^{(1)}) = \frac{hQ}{\pi D} = \frac{(0.0225)(421.6)}{(5)(40)} = 0.0473$$

Thus, $F(r^{(1)}) = 0.9527$. From the standard normal distribution table, $\Phi(1.67) = 0.9527$. To convert from standard normal Z units to r, we use the relation $Z = \frac{r - \mu}{\sigma}$ and set $r^{(1)} = 80 + 1.67(4) = 86.68$. The expected number of shortages per cycle is:

$$\overline{S}(r) = \sigma \cdot L\left(\frac{r - \mu}{\sigma}\right) = 4 \cdot L\left(\frac{86.68 - 80}{4}\right) = 4 \cdot L(1.67) = 0.0788$$

We can now update the order size by:

$$Q^{(2)} = \sqrt{\frac{2D[A + \pi \cdot \overline{S}(r)]}{h}} = \sqrt{\frac{2(40)[50 + (5)(0.0788)]}{0.0225}} = 423.3$$

However, resolving Equation (6.31) with $Q = 423.3$, we find $r = 86.68$. Because this agrees with our previous estimate, we can stop. When the inventory position drops to 87, order 423 units.

Specifying Service Levels   Unfortunately, it is difficult to accurately estimate shortage cost parameters such as cost of a stockout (π). Instead, management may set target service levels based on its understanding of the market and relative costs of meeting service levels. Two common approaches are to constrain the number of times per year a stockout occurs and to set a goal on the percentage of orders that should be met from stock (fill rate). The logic of the manager's policy can be checked by computing the "imputed" shortage cost for which this service level is optimal. Essentially, we solve the inventory model to find the optimal Q and r for a variety of shortage cost parameter values. Each of these

Table 6.1   Linear Loss Expressions for Common Distributions

Distribution	Prob. density	Mean	Variance	$L = \int_k^\infty (x-k)f(x)\,dx$
Normal	$f_{\mu,\sigma}(x) = \dfrac{1}{\sqrt{2\pi}\,\sigma}\, e^{\frac{(x-\mu)^2}{2\sigma^2}},\ -\infty < x < \infty$	μ	σ^2	$\left(\dfrac{k-\mu}{\sigma}\right)\left[\Phi\left(\dfrac{k-\mu}{\sigma}\right)-1\right] + \phi\left(\dfrac{k-\mu}{\sigma}\right)$
Exponential	$f(x) = \lambda e^{-\lambda x},\ x > 0$	λ^{-1}	λ^{-2}	$e^{-\lambda k}(1-\lambda^{-1})$
Uniform	$f(x) = \dfrac{1}{b-a},\ a \le x \le b$	$\dfrac{a+b}{2}$	$\dfrac{(b-a)^2}{12}$	$\dfrac{(b-k)^2}{2(b-a)}$
Gamma	$f_{r\lambda}(x) = \dfrac{\lambda}{\Gamma(r)}(\lambda x)^{r-1} e^{-\lambda x},\ x>0,\ r>0,\ \lambda>0$	$\dfrac{r}{\lambda}$	$\dfrac{r}{\lambda^2}$	$\dfrac{r}{\lambda}[1 - F_{r+1,\lambda}(k)] - k[1 - F_{r,\lambda}(k)]$
Beta	$f_{a,b}(x) = \dfrac{\Gamma(a+b)}{\Gamma(a)\Gamma(b)}\, x^{a-1}(1-x)^{b-1},\ 0 \le x \le 1,\ a>0,\ b>0$	$\dfrac{a}{a+b}$	$\dfrac{ab}{(a+b+1)(a+b)^2}$	$\dfrac{a}{a+b}[1 - F_{a+1,b}(k)] - k[1 - F_{a,b}(k)]$
Triangular	$f(x) = \begin{cases} \dfrac{2(x-a)}{(c-a)(b-a)}, & a \le x \le b \\[2ex] \dfrac{2(c-x)}{(c-a)(c-b)} & b < x \le c \end{cases}$	$\dfrac{a+b+c}{3}$	$\dfrac{a(a-b)+c(c-a)+b(b-c)}{18}$	$\dfrac{a+b+c}{3} - k + \dfrac{k^3-ak^2-a^3+ka^2}{3(c-a)(b-a)};\ a \le k \le b$ $\dfrac{5k^3 - 9ck^2 + 3kc^2 + c^3}{3(c-a)(c-b)};\quad b < k \le c$

solutions will yield an associated service level. Management can then check the shortage cost parameter that leads to their desired service level to determine whether this parameter value is logical. For example, the manager may say, "I want a 99.9% fill rate." The model can then be exercised to find the associated shortage cost. Suppose this figure turns out to be $10,000 per item short. This fact is presented to the manager. If the imputed cost seems too high or too low, management should consider lowering or raising its service level goals accordingly.

To implement this approach, we replace the recorder point Equation (6.31) with the desired relationship. The number of stockouts per year is given by $\dfrac{D}{Q} \int\limits_{r}^{\infty} f(x)\, dx$. Therefore, if α_1 is the maximum allowable stockouts per year, we use the constraint:

$$1 - F(r^*) = \frac{Q \cdot \alpha_1}{D}.\tag{6.33}$$

On the other hand, $\dfrac{Q - \overline{S}(r)}{Q}$ gives the proportion of demand met from stock. Thus, if we want to meet the proportion α_2 of demand from stock, we can use the relation:

$$\overline{S}(r^*) = Q \cdot (1 - \alpha_2)\tag{6.34}$$

to find r^* for any Q.

Given a solution pair (Q, r) that satisfies the optimality condition [Equation (6.30)] and a service level constraint, we can find the implied shortage cost. For this pair to be optimal, we must satisfy Equation (6.31), and thus the imputed value of shortage cost per unit back ordered becomes:

$$\pi^* = \frac{h \cdot Q}{D \cdot [1 - F(r^*)]}.\tag{6.35}$$

Lost Sales Case   The previous analysis assumed shortages resulted in backorders that were eventually satisfied. If customers go elsewhere when inventory is unavailable, sales are lost. In this case, gross revenue is no longer a constant equal to demand times selling price. We must therefore take lost profit into account. One approach is to let the shortage cost now include the lost contribution to profit and overhead, i.e., the difference between selling price and variable production plus delivery cost. This is not the only change, however. Because the inventory position never becomes negative, the actual inventory position just before receipt of an order is:

$$E(I_{\min}) = \int\limits_{0}^{r} (r - x)f(x)\, dx + 0 \cdot \int\limits_{r}^{\infty} (x - r)f(x)\, dx = \int\limits_{0}^{\infty} (r - x)f(x)\, dx + \int\limits_{r}^{\infty} (x - r)f(x)\, dx$$

$$= r - \tau \cdot D + \overline{S}(r)$$

In other words, the inventory position is $\overline{S}(r)$ more than in the backorders case, both before and immediately after receipt of Q items. Also, because each cycle has Q sales and an expected $\overline{S}(r)$ lost sales, we will have an average of $\dfrac{D}{Q + \overline{S}(r)}$ cycles per year. We will again assume $Q >> \overline{S}(r)$ and use the approximate expected cost function:

$$E(cost/time) = \frac{A \cdot D}{Q} + C \cdot D + h\left(r - \tau \cdot D + \overline{S}(r) + \frac{Q}{2}\right) + \frac{\pi \cdot D \cdot \overline{S}(r)}{Q}\tag{6.36}$$

Differentiating Equation (6.36) with respect to Q and r and setting these expressions equal to 0, we obtain the optimality conditions:

$$Q^* = \sqrt{\frac{2D[A + \pi \cdot \overline{S}(r)]}{h}} \tag{6.37}$$

and:

$$1 - F(r^*) = \frac{h \cdot Q}{h \cdot Q + \pi \cdot D} \tag{6.38}$$

With regard to the backorders case, we will order more units at a time (π is larger in the lost sales case) and order earlier. We order earlier both because the shortage cost is larger and because the average inventory level is increased by the number of expected shortages per cycle. To solve for the optimal policy, we use the same procedure as in the backorders case except that we substitute Equation (6.37) for Equation (6.30), and Equation (6.38) for Equation (6.31).

6.2.3   Periodic Review Systems

Some environments inherently dictate the order frequency. A parts supplier may deliver every morning, or packing materials may arrive by train once per week. Instead of keeping continuous records of inventory, it would be sufficient to run a computer program once per period, updating all transaction records and providing information on the desired order quantity. Internally, production schedules may be set weekly or at other fixed intervals. If demand was constant and deterministic, a fixed order quantity would be used for each period. However, in a stochastic environment, a decision model is needed to guide choice of order quantity. It is important to note that once the order quantity is conveyed to the supplier, we cannot reorder for an entire period. Thus, we cannot affect our available inventory for another period, plus the lead time. This is in sharp contrast to the continuous review environment in which we only needed to guard against uncertainty during the lead time.

6.2.3.1   Order-Up-to R Policies

Suppose we are allowed to place an order every T time periods. The inventory level at the time of placing an order will vary from cycle to cycle. Instead of always ordering Q units, it will make more sense to order enough units to take the inventory position up to a given level R. R will be our decision variable. The consequence of not keeping track of inventory levels is seen in our exposure to demand variability. We were previously concerned only with variability over the lead time. Now, once we place an order, it must cover demand for the next $T + \tau$ time periods. The current order will arrive in time τ and increase our inventory level to $R - D \cdot \tau$, on average. T time units after, as we prepare to receive the next order, the inventory level will be $R - D \cdot \tau - D \cdot T$. More formally:

$$E(I_{\min}) = \int_0^\infty (R - x)f(x; T + \tau)\, dx = R - D \cdot (T + \tau)$$

for average demand rate D, in which $f(x; T + \tau)$ is the distribution of demand over a time interval, $T + \tau$ with corresponding cumulative distribution $F(x; T + \tau)$. The average inventory is thus:

$$\bar{I} = R - D \cdot \tau - \frac{D \cdot T}{2}.$$

The expected cost per time as a function of T and R is given by:

$$E(Cost/time) = \frac{A}{T} + C \cdot D + h \cdot \left[R - D \cdot \tau - \frac{D \cdot T}{2} \right] + \frac{\pi \cdot \overline{S}(R, T)}{T} \quad (6.39)$$

Once again, we find the optimal policy by equating the derivative of the expected cost function to zero. For the decision variable R, we have:

$$\frac{\partial E(Cost/Time)}{\partial R} = h + \frac{\pi}{T} \cdot \frac{\partial \overline{S}(R, T)}{\partial R} = h - \frac{\pi}{T} \cdot [1 - F(R; T + \tau)] = 0.$$

Thus, after rearranging terms we find that the optimal order-up-to quantity satisfies:

$$F(R^*; T + \tau) = \frac{\pi - h \cdot T}{\pi} \quad (6.40)$$

Equation (6.40) has an intuitive interpretation. The hT term is the cost to carry one unit of inventory for one cycle. If this cost exceeds the unit shortage cost, π, then the expression for Equation (6.40) is negative, implying that we should allow all units to be back ordered. As π becomes very large relative to hT, the ratio approaches one implying that we should set R large enough to prevent all shortages.

EXAMPLE 6.10

A special control board is used in one version of the product made on a production line. The board costs $122.50. The holding cost rate is 30% per year. Reorders are placed at the start of each week, and the supplier delivers these parts the following week. Stockouts of the board are expensive. It is necessary to maintain the order of products on the production line, and six workers become idle until a new board can be delivered by courier. This cost is estimated at $100 per board when shortages occur. Daily demand is distributed uniformly between 10 and 15 units. Find the optimal order-up-to quantity. The cost to check inventory level and place an order is $120.

SOLUTION

We will convert all time units to weeks. Holding cost per week is:

$$h = (\$122.50) \cdot \frac{0.30/yr}{52 weeks/yr} = \$0.7067/wk.$$

Thus, using Equation (6.40),

$$F(R^*) = \frac{100 - (0.7067) \cdot 1}{100} = 0.993$$

Clearly, we should avoid most shortages. Drawing on the central limit theorem, demand over the two-week period (one week for cycle time, plus one week for lead time) is normally distributed with a mean of $(10 \text{ days})(12.5/\text{day}) = 125$ and variance of $(10)(25/12) = 20.83$. From the standard normal Table $\Phi(2.46) = 0.993$, thus $R^* = 125 + (2.46) \cdot \sqrt{20.83} \approx 136$.

EXAMPLE 6.11

An employee has suggested letting parts continue on the production line when parts are short and then retrofitting the assemblies afterward. An engineer has estimated that this will reduce the shortage cost to $25. Determine the savings from instituting this new procedure in an optimal manner.

SOLUTION

With $\pi = 25$, the optimal solution is to set the order-up-to level to $F(R^*) = \frac{25 - 0.7067}{25} = 0.972$ which translates to $R^* = 125 + (1.91)(4.56) = 134$.

Using $\pi = 100$ and R $= 136$, expected cost per time is:

$$E(Cost/time) = \frac{A}{T} + C \cdot D + h \cdot \left[R - D \cdot \tau - \frac{D \cdot T}{2} \right] + \frac{\pi \cdot \bar{S}(R, T)}{T}$$

$$= \frac{120}{1} + 122.50(62.5) + (0.7067) \cdot \left[136 - (62.5)(1) - \frac{(62.5)(1)}{2} \right]$$

$$+ \frac{100(\sqrt{20.83})L(2.46)}{1}$$

$$= \$7,807.16$$

Using $\pi = 25$, and $R = 134$, expected cost per time is \$7,805.93, a savings of \$1.23 per week. You may wonder why the savings are so small. In both cases, shortages are expensive relative to carrying inventory, and thus we set reorder points high enough to eliminate most shortages.

We have assumed that a natural cycle, such as one week, is implicitly known. If this is not the case, we could treat T as a decision variable. If T was unconstrained, we could differentiate the expected cost function with respect to T and solve for the optimal cycle length. More than likely, however, T is constrained to be an integer whereby we could order every week or every two weeks, etc. We can try to find the best value of R using Equation (6.40) for each feasible value of T and then select the best (R, T) combination by evaluating the expected cost function [Equation (6.39)] for each of these values.

EXAMPLE 6.12   Using the data in Example 6.11, find the appropriate ordering frequency. Orders are always placed at the start of a week.

SOLUTION   Instead of searching over T, we quickly estimate T using the EOQ expression. For our cost parameters, the EOQ gives $Q = \sqrt{\frac{2AD}{h}} = \sqrt{\frac{2(120)(12.5)}{0.7067}} = 65$, which corresponds to just over five weeks of demand. Then, using $T = 5$ and noting that demand over six weeks has a mean of 30 (12.5) = 375 units and a variance of 30(25/12) = 62.5, we have:

$$F(R^*) = \frac{\pi - hT}{\pi} = \frac{100 - (0.7067)(5)}{100} = 0.965.$$

Because $\Phi(1.81) = 0.965$, $R = (30)(12.5) + 1.81(62.5)^{0.5} = 389$. This gives a cost per week of:

$$E(cost/week) = \frac{A}{T} + CD + h \cdot \left[R - D\tau - \frac{DT}{2} \right] + \frac{\pi \cdot \bar{S}(R, T)}{T}$$

$$= \frac{120}{5} + 122.50(62.5) + 0.7067 \left[389 - (62.5)(1) - \frac{(62.5)(5)}{2} \right]$$

$$+ \frac{(100)\sqrt{62.5}L(1.81)}{5}$$

$$= \$7,802.76$$

6.2.3.2   Single-Period Model

A special case of periodic review models occurs when we can only place a single order for a product over its life. This single-period, stochastic demand inventory problem is typically referred to as the **news vendor problem** because it describes the situation of deciding how many papers to pick up each morning for selling that day on a street corner. Each day, the vendor has a different product to sell, and demand for that paper disappears after the day of its issue.[3] Despite its rather specific name, the news vendor problem occurs in various production environments. A manufacturer may be planning a one-time production run of a product for a special promotion. A retailer may be purchasing a seasonal item, such as holiday greeting cards, clothing, or snow blowers. The model we develop may even be used for selecting the level of plant capacity to acquire for the next several years. In each case, the decision-maker must decide how much to order before demand is known. Extra items must be disposed of at a loss or left idle. Shortages result in a loss of potential profit, both for this item and possibly for future items, if the customer decides to go elsewhere. Our decision is therefore a tradeoff between the cost of disposing extra inventory and the loss in profit from shortages.

If this were truly a one-time decision, the relevant decision criterion would probably be minimization of the probability of some catastrophic loss or maximization of the probability of reaching some aspiration level of profit. However, it is more likely that this decision is repeated every period (such as daily for the news vendor) or even many times per period (as in the case of selecting the amount of seasonal stock for each retail item). Thus, it may be logical to maximize expected profit. Expectation provides the best guide when we are interested in how the outcomes aggregate over many random experiments. We need to modify some aspects of our notation for this model; accordingly, we use the following notation:

C = purchase or production cost of each item

p = price, i.e., revenue per item sold

s = salvage value of each unsold item ($s < 0$ if we must pay to dispose of remaining items)

D = demand, a random variable with probability density function $f(x)$

π = shortage cost for each unmet item demand

Q = number of units ordered (the decision variable)

Profit is a function of the quantity ordered before the period starts (Q) and the quantity demanded (D). We can write profit as:

$$E(Profit) = \begin{cases} (p - c)Q - \pi(D - Q) \text{ , if } Q \le D \\ pD - cQ + s(Q - D) \text{ , if } Q > D \end{cases}. \tag{6.41}$$

Using Equation (6.41), expected profit as a function of the order size is set by:

$$\textit{Maximize } E(Profit) = p \cdot \int_0^Q x \cdot f(x)\, dx + s \cdot \int_0^Q (Q - x) \cdot f(x)\, dx + p \cdot \int_Q^\infty Q \cdot f(x)\, dx$$

$$- \pi \cdot \int_Q^\infty (x - Q) \cdot f(x)\, dx - cQ \tag{6.42}$$

The first two terms in Equation (6.42) give the revenue and salvage value when demand does not exceed the supply of items. The next two terms define the revenue and shortage cost when demand exceeds the supply of items. In either case, the cost of goods purchased

---

[3]This problem has also been called the Christmas tree problem because a vendor may receive only one shipment of Christmas trees each Fall.

is cQ, and this is shown by the last term. The only unknown in this expression is the order quantity Q. To find the optimal Q, we differentiate Equation (6.42) and set the result to 0. Differentiating, we obtain:

$$\frac{dE(Profit)}{dQ} = (0 + p \cdot Q \cdot f(Q) - 0) + \left(s \cdot \int_0^Q f(x)\,dx + 0 - 0\right)$$

$$+ \left(p \cdot \int_Q^\infty f(x)\,dx + 0 - p \cdot Q \cdot f(Q)\right) - \left(\pi \cdot \int_Q^\infty (-1) \cdot f(x)\,dx + 0 - 0\right) - c$$

$$= s \cdot F(x) + (p + \pi)[1 - F(x)] - c \tag{6.43}$$

Setting the derivative in Equation (6.43) to 0 and rearranging terms, we obtain:

$$F(Q^*) = \frac{\pi + p - c}{\pi + p - s} \tag{6.44}$$

The optimal order quantity represents balancing between the costs of stockouts and extra inventory. The numerator of Equation (6.44), $\pi + p - c$, contains the lost net revenue, and the shortage penalty. The denominator contains this same cost, plus the loss from each unit unsold, namely $c - s$. Thus, the ratio is just the proportion of the sum of these two costs that is attributed to shortages. If we define the shortage cost per unit as $c_s = \pi + p - c$ and the overstock cost per unit as $c_o = c - s$, we could then write the optimality condition as:

$$F(Q^*) = \frac{c_s}{c_s + c_o}. \tag{6.45}$$

Note that if $s > c$, Equation (6.45) specifies an optimal order size Q^* for which the cumulative distribution of demand is greater than 1. To account for this inconsistency, we should specify $s \leq c$ as a model assumption. The reader should note the intuitive explanation of this assumption. If we can always dispose of extra units for at least as much as they cost, then we should order an infinite supply to start the period. Why not, we can't lose!

To confirm that Equation (6.44) yields the optimal solution, we must show that the expected profit function is concave. We can establish this by showing that the second derivative is negative. By differentiating Equation (6.43), we obtain:

$$\frac{d^2 E[Profit]}{dQ^2} = s \cdot f(Q) - p \cdot f(Q) - \pi \cdot f(Q)$$

Because $f(Q) > 0$, we require only that $\pi + p > s$. This result holds for any well-formulated problem in which salvage value does not exceed selling price. Optimality of the Q^* defined in Equation (6.45) is thus established.

EXAMPLE 6.13

A stationery company is planning the production run for a holiday greeting card. Because of production and delivery lead times, seasonal cards can be produced only once. Demand is estimated to be normally distributed with a mean of 5,000 boxes and a standard deviation of 250. The variable production cost for a box of cards is $0.95, and a box is sold for $2.25. Management believes that selling leftover boxes to discount houses cuts primary sales and thus prefers to scrap extra cards at no cost. Shortages hurt the company image and name recognition. In addition to lost profit on the current potential sale, marketing surveys place this cost at approximately $1.50 in future lost profit per one-box shortage. How many boxes should be produced?

SOLUTION

We are given the parameter values $c = 0.95$, $p = 2.25$, $s = 0$, $\pi = 1.50$, and $f(x) \approx N(5,000, 250^2)$. Note that the \$1.50 loss per one-box shortage exceeds the loss per leftover box, \$0.95. Thus, we expect to purchase more than the average demand. From Equation (6.44), we have:

$$F(Q^*) = \frac{\pi + p - c}{\pi + p - s} = \frac{1.50 + 2.25 - 0.95}{1.50 + 2.25 - 0} = 0.75$$

Let $\Phi(z)$ be the cumulative distribution of the standard normal $(0,1)$ random variable. From the standard normal distribution table in Appendix I, $\Phi^{-1}(0.75) = 0.675$, thus $Q^* = 5,000 + 0.675(250) = 5,169$.

In addition to order quantity, we might also like to know expected profit. This is found by evaluating Equation (6.42) for the selected Q. To facilitate this calculation, Table 6.1 contains expressions for the "partial means" of the normal and several other distributions. The "lower" partial mean is defined as $\mu_l(x, k) = \int_{-\infty}^{k} x \cdot f(x) \cdot dx$. The "upper" partial mean, $\mu_u(x, k) = \int_{k}^{\infty} x \cdot f(x) \cdot dx$, can be readily found as well by observing that for a random variable X, $E(X) = \mu_l(x, k) + \mu_u(x, k)$.

EXAMPLE 6.14

Find the maximum expected profit from Example 6.13.

SOLUTION

Using $Q^* = 5,169$ and Equation (6.42), we have:

$$Expected\ Profit = (p - s) \cdot [\mu_1(x, Q) - \mu_1(x, 0)] + s \cdot Q \cdot F(Q)$$
$$+ (p + \pi) \cdot Q \cdot [1 - F(Q)] - \pi \cdot [E(X) - \mu_1(x, Q)] - c \cdot Q$$

From Table 6.1 and Appendix I, we have:

$$\mu_1(x, Q^*) = \mu \cdot \Phi\left(\frac{Q^* - \mu}{\sigma}\right) - \sigma \cdot \phi\left(\frac{Q^* - \mu}{\sigma}\right)$$
$$= 5,000 \cdot (0.75) - 250 \cdot \phi(0.675)$$
$$= 3,670.58.$$

Using this result, we obtain:

$$Expected\ Profit = 2.25(3,670.58) + 0 + 3.75(5,169)(0.25) - 1.50(5,000 - 3,670.58)$$
$$- 0.95(5,169) = \$6,200.04$$

6.2.4   Discrete and Slow Demand

We have been treating the lot size as a continuous variable. This is a reasonable approximation when lot sizes are large, and we would most likely be rounding off in practice anyway. Suppose, however, that the demand rate is small relative to holding costs, and we will be ordering batches of ten or fewer units. Although the cost models we have discussed are unchanged, the solution method must restrict Q to be an integer. Let $E[TCT(Q)]$ be the expected total cost per time for a lot size of Q. Instead of differentiating the cost function as in Equation (6.43), we take "1st differences" of the cost function and rely on the optimality condition:

$$E[TCT(Q^*)] \le E[TCT(Q^* - 1)] \text{ and } E[TCT(Q^*)] \le E[TCT(Q^* + 1)].$$

We define the "1st difference" $\Delta TCT(Q) = E[TCT(Q)] - E[TCT(Q - 1)]$ and select the largest Q for which $\Delta TCT(Q) \le 0$.

The cost model assumes a constant expected demand rate over time. Any valid probability mass function could be used for demand. However, for many situations, the Poisson distribution serves as a reasonable description of the demand process. The Poisson corresponds to an exponentially distributed time between receipt of orders. It occurs naturally when we have a large population of potential customers each of whom acts independently with a constant probability of initiating a demand for a unit of product during any non-overlapping, fixed interval, length of time.

EXAMPLE 6.15

A supply depot receives requests for an average of two spare engine fuel pumps per week. Orders are placed to the factory once per week and arrive one week later. Fuel pumps cost $12, and the depot has a holding cost rate of 0.5% per week. Shortages result in a special courier delivery costing an additional $35 per item. Find the optimal order quantity.

SOLUTION

We assume that interarrival times for demands are independent, thus, demand is Poisson-distributed with a mean of $\lambda = 4$ over the two-week period for $T + \tau$. For discrete demand, the 1st difference expected total cost per time function is:

$$\Delta TCT(R) = E[Cost/time\ (R)] - E[Cost/time\ (R - 1)]$$

$$= \frac{A}{T} + CD + h\left[R - D\tau - \frac{DT}{2}\right] + \frac{\pi \cdot \bar{S}(R, T)}{T}$$

$$-\left(\frac{A}{T} + CD + h\left[R - 1 - D\tau - \frac{DT}{2}\right] + \frac{\pi \cdot \bar{S}(R - 1, T)}{T}\right)$$

$$= h + \frac{\pi}{T} \cdot \left(\bar{S}(R, T) - \bar{S}(R - 1, T)\right)$$

$$= h + \frac{\pi}{T}\left(\sum_{x=R+1}^{\infty} (x - R)p(x) - \sum_{x=R}^{\infty} (x - R + 1)p(x)\right) = h - \frac{\pi}{T} \sum_{x=R}^{\infty} p(x),$$

where $p(x)$ is the probability mass function of demand over $T + \tau$. Thus, we select the largest R for which $h - \frac{\pi}{T} \sum_{x=R}^{\infty} p(x) \le 0$, or $\sum_{x=R}^{\infty} p(x) \ge \frac{hT}{\pi} = \frac{(0.005)(12) \cdot 1}{35} = 0.00171$. This condition is equivalent to finding the largest R for which $1 - \sum_{x=0}^{R-1} p(x) \ge 0.00171$ or $\sum_{x=0}^{R-1} p(x) < 1 - 0.00171 = 0.99829$. Noting that for a Poisson with a mean of 4:

$$p(x) = \frac{4^x e^{-4}}{x!}, x = 0,1, ..., \text{ we find } \sum_{x=0}^{10} p(x) = 0.997 \text{ and } \sum_{x=0}^{11} p(x) = 0.999$$

Thus, we set $R = 10$.

6.2.5   Multiproduct Systems

In the case of deterministic demand and multiple products being produced at the same facility, we developed a rotation cycle policy in Section 6.1.4.2. If demand is stochastic, we may want to let our batch size decisions vary from cycle to cycle based on actual inventory. Finding the optimal solution to this problem is not easy. However, we can implement relatively simple yet effective policies. First, we need to select a basic production period. The base period here represents the usual time spent producing a single item and not the entire rotation cycle. Typical values might be one hour or a half shift. At the end of each period, we decide to continue producing the same product, switch to a new product, or let the system remain idle. We will assume that once we decide to produce a product in a period, we set up (if necessary) and produce that product for the entire period. Let I_i be the on-hand inventory for item i and let N be the number of product types produced. At the start of each period, we decide whether to produce based on the total stock of products on hand. The stock of inventory measured as average number of periods worth of supply is: $\bar{I} = N^{-1} \sum_{i=1}^{N} \frac{I_i}{D_i}$. If this value exceeds some threshold, we let the facility remain idle for the upcoming period. Otherwise, we produce the item with the largest backlog or smallest on-hand stock measured in periods of average demand. Thus, if we produce, we produce the item for which $\frac{I_i}{D_i}$ is minimum. This item is expected to run out first. We determine $\bar{I}$ by treating the facility as a single-product entity and finding the desired safety stock for a periodic review system.

EXAMPLE 6.16

A facility produces three items. Demand is random. Inventory is measured in hours of production. Every hour, a decision is made as to what item to produce for the next hour. The organizational rule is to produce unless more than eight production hours of inventory are on hand. Current inventory levels are 4, 1, and 2 hours of production, respectively. Demand rates are for 0.4, 0.2, and 0.3 per hour. What should be produced during the next hour?

SOLUTION

Because total supply is $4 + 1 + 2 = 7 < 8$, we will produce. Ranking items based on time until expected runout, we have $4/0.4 = 10$, $1/0.2 = 5$, and $2/0.3 = 6.7$ hours of supply, respectively. Thus, we select item 2 to be produced in this period.

The above-noted rule implicitly assumes that shortage costs are similar for all products. If this is not the case, we can define a minimum safety stock, SS_i, for each product i and produce the item with the smallest ratio of $\frac{I_i - SS_i}{D_i}$.

6.3   DYNAMIC MODELS

At this point in the chapter, we have assumed that demand is stationary, i.e., that mean demand rate is constant over time. We now examine the situation in which our forecasting system has provided us with an estimate of demand or our aggregate planning system has set production levels that vary from period to period over a foreseeable **planning horizon.** Our objective is to select an inventory policy over this horizon. We begin with models that assume we can order (or make) as many as we want in any period. Afterward, we will consider the effect of capacity constraints.

Choosing an appropriate length for the planning horizon can be very important. Inventory orders are planned, period by period, for the entire horizon. If the horizon is too short, we may not use enough information and may leave ourselves in a bad situation for the long term. We may, for example, stockpile too little or too much inventory by ignoring future demand changes. If the horizon is too long, we are forced to perform unnecessary computations to select our policy and may even make our decisions worse by using unreliable long-term forecasts. It is best to choose a planning horizon equal to an integral number of natural ordering cycles and one that exceeds the lead time by at least one natural cycle. We briefly discuss conditions for ensuring an adequate length for the planning horizon later in this chapter. In most cases, we plan for the entire horizon, but only implement or "freeze" our decisions for the beginning of the cycle. The length of time for which we actually fix the inventory-ordering decision will be called the **firm-order horizon.** In effect, we use a **rolling schedule.** Each period, we add an additional period onto the end or the horizon and resolve for the best inventory policy. Decisions over the firm-order horizon are permanently fixed. The remaining decisions are tentative and used only as a guide to future plans. Although we produce a plan for the entire horizon, we only implement the plan for period 1.

The firm-order horizon must be at least as long as the lead time to allow us to receive the first order on time. If we want to coordinate replenishment decisions, the firm-order horizon should be at least as long as the cumulative lead time between when we begin the process of reordering parts and when we receive finished products. To illustrate, suppose we must order raw materials three weeks before use and release an order for manufactured parts two weeks before they are available for final assembly. If the firm-order horizon for final assembly is at least five weeks, we can order materials and plan part production to match final assembly. If the firm-order horizon is only two weeks, we must have preordered the materials and must use a standard EOQ-type model to control the material inventory.

6.3.1 Infinite Capacity, Dynamic Lot-Sizing Models

We begin by examining the case in which adequate capacity exits. We develop lot-sizing models to balance setup and inventory costs while avoiding shortages. We then extend the models to consider the case of limited production rate capacity. In practice, we may also want to smooth (equalize over time) the workload induced by orders for manufactured items. This makes the production facility easier to operate.

6.3.1.1 Continuous Review: A Simple Runout Time Model

The main objective in dynamic demand inventory models centers on using demand information to avoid shortages during high demand periods and excess inventory during periods of low demand. A simple rule, similar in spirit to the (Q, R) policy with static demand, places an order whenever the inventory position drops to the level of safety stock, plus expected lead-time demand. The order quantity would be set, based on the current or average demand rate. The point in time at which inventory will drop down to the safety stock level is again called the **runout time.**

EXAMPLE 6.17

A product has the following forecasted demand for the next six weeks (100, 90, 80, 70, 80, 100). Holding cost is $0.50 per week. Ordering cost is $50. Lead time is one week, and the current inventory level is 150 units. Safety stock is nominally kept at two days of demand (each week represents five days).

SOLUTION

With a lead time of one week and safety stock of two days, we order whenever the expected runout time of current on-hand plus on-order inventory drops to expected demand for the next seven (5 +

2) days. Expected demand for the next seven days is $100 + (0.4) 90 = 136$. Thus, it is almost time to order. Later today, when we have used 14 units and our inventory level drops to 136 from the present 150, it will be time to order. Using the basic EOQ and estimated average demand of 87 units per week (the average of the next six weeks), the order quantity is:

$$Q = \sqrt{\frac{2AD}{h}} = \sqrt{\frac{2(50)(87)}{0.50}} = 132.$$

6.3.1.2 Periodic Review: The Wagner-Whitin Algorithm

In many cases, the production planning system is designed to specifiy a production quantity for each period over the planning horizon. This corresponds to a periodic review model wherein our task is to determine the amount to make in each period. We assume that demand has been estimated for each of these planning periods. Let the amount of product to be made in period t be denoted by X_t, and let $\delta(t) = \begin{cases} 0, \text{ if } X_t = 0 \\ 1, \text{ if } X_t > 0 \end{cases}$ be an indicator of whether we place an order in period t. Period length should reflect the minimum time between reorders for an item and any natural frequency in the system, such as weekly planning and order release. I_t is the expected on-hand inventory at the end of period t. The mathematical model would appear as:

$$Minimize\ Cost = \sum_{t=1}^{T} (A \cdot \delta(t) + CX_t + hI_t) \qquad (6.46)$$

subject to:

$$I_t = I_{t-1} + X_t - D_t \qquad (6.47)$$

$$X_t \geq 0, \forall t = 1, ..., T. \qquad (6.48)$$

The objective function [Equation (6.46)] includes the cost of setups, variable production (or purchase) costs, and carrying inventory. We assume the variable production cost is fixed over the horizon at $\$C$ per unit. Constraints [Equation (6.47)] document the changes in inventory from period to period. The on-hand inventory at the end of a period is given by the starting inventory, plus the units produced, minus those demanded by customers.

The presence of the fixed setup charge when we produce complicates finding the optimal production plan. Essentially, we must consider every possible plan. Fortunately, we can show an important property about the optimal solution. The solution to the problem of minimizing a concave cost function subject to linear constraints occurs at an **extreme point** of the region formed by the constraint set. Figure 6.9 provides a geometric illustration of this property. This result is vitally important because it tells us that in the absence of capacity constraints on production in a period, there is a solution that satisfies the relationship:

$$X_t \cdot I_{t-1} = 0 \text{ for all } t_1, ..., T. \qquad (6.49)$$

The property [Equation (6.49)] states that we never produce in a period if we start that period with inventory. Thus, when we do produce, we must produce just enough to cover demand for an integral number of periods. We could produce exactly D_1 or exactly $D_1 + D_2$; however, it would never be optimal to produce $D_1 + D_2/2$. It is easy to understand why. Compare the two production schedules $(X_1, X_2) = (D_1, D_2)$ with $(X_1, X_2) = (D_1 + 0.5D_2, 0.5D_2)$. In both cases, we produce in period one and two so that setup cost is the

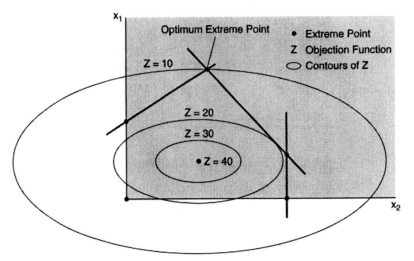

Figure 6.9 Minimum of Convex Function in a Convex Set

same. Purchase or variable production cost is also the same. However, the second option incurs an additional inventory cost of $hD_2/2$ at the end of period 1. We could save inventory cost and incur no additional cost by waiting until period 2 to produce all the demand for period 2. In the end-of-chapter exercises, we ask the reader to show the correspondence to solutions satisfying equation (6.49) and extreme points of the feasible region.

We can make use of the property [Equation (6.49)] to solve this problem using dynamic programming. Visually, we must find the **shortest path** to move from time 0 to T (Figure 6.10). Travel is along arcs, and each arc represents a single order. Arcs connect regeneration points. A **regeneration point** is a point in time (the end of one period and the start of the next) at which inventory is zero. The regeneration point, therefore, marks the time at which we must produce again or incur shortages. Arc costs in the figure include setup for the period at which the arc begins and holding costs until the next regeneration point indicated by the head of the arc. To begin, we find the best solution if the planning horizon is one period. We then solve the two-period problem and then the three-period problem, and so forth, until the T-period problem is solved. Each period becomes a **stage** in the solution procedure. We solve this sequence of problems in a manner that allows us to use the earlier calculations to solve the next stage.

We start at time (end of period) 0, a regeneration point. If j is a regeneration point, then we must produce in period $j + 1$ to prevent shortages. We can compute the cost to

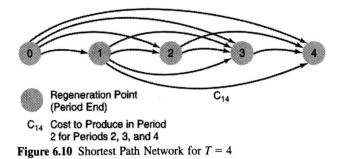

 Regeneration Point
(Period End)

C_{14} C_{14} Cost to Produce in Period
2 for Periods 2, 3, and 4

Figure 6.10 Shortest Path Network for $T = 4$

operate between successive regeneration points. This is the cost to produce in period $j + 1$ to cover demand in periods $j + 1$ up to k and to carry inventory over this horizon. We denote this cost as G_{jk}, where, for the case of nonstationary costs

$$G_{jk} = A_{j+1} + C_{j+1} \cdot \sum_{t=j+1}^{k} D_t + \sum_{t=j+1}^{k-1} h_t \left(\sum_{l=t+1}^{k} D_l \right)$$ (6.50)

If setup cost is A, holding cost is h, and unit production cost is C for all periods, Equation (6.50) becomes:

$$G_{jk} = A + C \cdot \sum_{t=j+1}^{k} D_t + h \sum_{t=j+1}^{k} (t - j - 1)D_t.$$

As an example, suppose demands for the first three periods are 100, 80, and 90, respectively. We have $G_{03} = A + C(100 + 80 + 90) + h(1 \cdot 80 + 2 \cdot 90) = A + 270 \cdot C + 260 \cdot h$. We interpret this result to mean that if we produce in period one to cover demand in periods one, two, and three, then we incur the period one setup cost, production cost to produce the 270 units demanded over these three periods, plus the holding cost for the 170 units carried over at the end of period one, plus the holding cost to carry over 90 units at the end of period two.

Let F_k be the cost of the optimal solution for the first k periods. The dynamic programming recursion begins with $F_o = 0$ and then iteratively finds $F_t = \min_{0 < j < t-1}\{F_j + G_{jt}\}$ for $t = 1, ..., T$. The cost of the optimal solution is given by F_T. To find the optimal production quantities, we trace backward from period T to period 1, finding the regeneration points.

Before demonstrating the solution approach, note that under certain conditions, we can shorten the computations required to solve this problem. An important area of research on dynamic production planning models has been the development of planning horizon results. These results identify sufficient conditions to guarantee that the solution for the beginning periods will not change as we increase the length of the planning horizon. Such results allow us to stop the solution procedure early, knowing that we have the correct decisions for the beginning periods covering the firm-order horizon. Let $r^*(t)$ be the optimal last regeneration point before t, in a t-period problem. Two important planning horizon results follow: [see Wagner and Whitin, (1959)].

Planning Horizon Theorems

Assume that production cost consists of setup cost, variable cost, plus holding cost of the form:

$$C_t(X_t) = A + cX_t + hI_t, \text{ with } A, c, \text{ and } h \geq 0,[4] \text{ then}$$

1. If we increase the length of the planning horizon, the optimal period of the last production stays the same or increases, i.e. if $t_1 > t_2$, then $r^*(t_1) \geq r^*(t_2)$; and

2. If the optimal t period solution says to produce in period t, then the optimal solution for any longer horizon will also schedule production in period t, i.e., if $r^*(t) = t - 1$, then $X_t^* > 0$ for $T > t$.

The first result simply states that as we consider more periods, the period in which we last produce cannot decrease. Intuitively, suppose in the two-period problem, it was

---

[4]Actually, it is sufficient to assume the cost parameters are non-increasing instead of constant over time.

optimal to produce D_2 in period 2. Then, when we solve the three-period problem, we must produce at least once in periods 2 and 3. To understand why, note that if the three-period problem solution was to produce $D_1 + D_2 + D_3$ in period 1, then:

$$A_1 + C_1(D_1 + D_2 + D_3) + h_1(D_2 + D_3) + h_2 D_3 \leq$$
$$A_1 + C_1(D_1) + A_2 + C_2(D_2 + D_3) + h_2 D_3 \text{ or} \tag{6.51}$$
$$C_1(D_2 + D_3) + h_1(D_2 + D_3) \leq A_2 + C_2(D_2 + D_3).$$

Suppose this first result was incorrect, and the solution to the two-period problem stated that we should produce D_1 in period one and D_2 in period 2. Then, from the two-period solution, we know $A_2 + C_2 D_2 < C_1 D_2 + h_1 D_2$. Combining this with Equation (6.51), then it must be true that $C_1 D_3 + h_1 D_3 < C_2 D_3$, which violates the assumption of positive holding cost and nonincreasing production costs. This contradiction establishes the first result.

The second result states that if in the T_0 period problem, we produce in period T_0, then for all horizons greater than T_0, we will also produce in period T_0. Thus, if the solution to the three-period problem states that we should produce in period 3, then the solution to the four-, five-, six-period problems, and so forth, will also set the end of period 2 as a regeneration point and schedule production in period 3. Thus, we need not consider options such as $F_1 + G_{14}$ for the four-period problem because this option corresponds to producing in period 2 for periods 2, 3, and 4.

EXAMPLE 6.18

A product family has estimated demands of 120, 50, 95, 110, and 75 over the next five weeks. Setup cost is \$500, and holding cost is \$1.50 per unit per period. Unit production cost is \$5 initially; however, it will increase to \$6 in week 2 because of a new labor contract and then is expected to drop to \$4 in week 4 when a new production process comes on line. Find the optimal production plan.

SOLUTION

Initially, $F_0 = 0$. For the one-period problem:

$$F_1 = A + C_1 \cdot D_1 = 500 + 5(120) = 1100.$$

For the two-period problem:

$$F_2 = \min \begin{cases} F_0 + G_{02} = 0 + A + C_1 \cdot (D_1 + D_2) + h \cdot D_2 = 500 + 5(170) + 1.5(50) = 1{,}425^* \\ F_1 + G_{12} = 1{,}100 + A + C_2 \cdot D_2 = 1{,}100 + 500 + 6(50) = 1{,}900 \end{cases}$$

For the three-period problem:

$$F_3 = \min \begin{cases} F_0 + G_{03} = 0 + A + C_1 \cdot (D_1 + D_2 + D_3) + h \cdot (D_2 + 2D_3) = 2{,}185^* \\ F_1 + G_{13} = 1{,}100 + A + C_2 \cdot (D_2 + D_3) + h \cdot D_3 = 2{,}612.50 \\ F_2 + G_{23} = 1{,}425 + A + C_3 \cdot D_3 = 2{,}495. \end{cases}$$

For the four-period problem:

$$F_4 = \min \begin{cases} F_0 + G_{04} = 0 + A + C_1 \cdot (D_1 + \ldots + D_4) + h \cdot (D_2 + 2D_3 + 3D_4) = 2{,}970^* \\ F_1 + G_{14} = 1{,}100 + A + C_2 \cdot (D_2 + D_3 + D_4) + h \cdot (D_3 + 2D_4) = 3{,}602.50 \\ F_2 + G_{24} = 1{,}425 + A + C_3 \cdot (D_3 + D_4) + h \cdot D_4 = 3{,}320. \\ F_3 + G_{34} = 2{,}185 + A + C_4 \cdot D_4 = 3{,}125. \end{cases}$$

By now, you may have noticed several patterns. We may save on computations by noting that $G_{jk} = G_{j,k-1} + C_{j+1} \cdot D_k + h \cdot (k - j - 1) \cdot D_k$. In other words, as we add an extra period to this cycle, we must produce the extra period's demand and hold it until the end of the new cycle. For the five-period problem, then:

$$F_5 = \min \begin{cases} F_0 + G_{05} = F_0 + G_{04} + C_1 \cdot D_5 + 4h \cdot D_5 = 3{,}795. \\ F_1 + G_{15} = F_1 + G_{14} + C_2 \cdot D_5 + 3h \cdot D_5 = 4{,}390. \\ F_2 + G_{25} = F_2 + G_{24} + C_3 \cdot D_5 + 2h \cdot D_5 = 3{,}995. \\ F_3 + G_{35} = F_3 + G_{34} + C_4 \cdot D_5 + h \cdot D_5 = 3{,}537.50^* \\ F_4 + G_{45} = F_4 + A + C_5 \cdot D_5 = 3{,}770. \end{cases}$$

We work backwards to identify the optimal solution. We begin with the regeneration point at period 5. F_5 then identifies the end of period 3 as the previous regeneration point. F_3 indicates time 0 as the regeneration point before that. Thus, we produce in periods one and four.

We produce 265 units in period one to cover the first three periods and 185 units in period 4 to cover periods four and five. The total cost is \$3,537.50. It is interesting to note that if we had used a four-period planning horizon the decision would have been to produce all 375 units in period one. Extending the horizon to include a fifth period caused us to produce a smaller quantity in period one. We should also notice that if we had nonincreasing costs, the planning horizon theorem would have begun at period five. If we added a sixth period to the horizon, then for F_6, we would produce in period four or later. Thus, we would only have to evaluate:

$$F_6 = \min \begin{cases} F_3 + G_{36} \\ F_4 + G_{46} \\ F_5 + G_{56} \end{cases}$$

Although our dynamic programming solution does provide an optimal solution to this planning model, the model suffers from several shortcomings. When used to plan for a single product or part family, the Wagner–Whitin model tends to increase the variability in workloads from period to period. This can put a strain on the manager, who is trying to smooth the requirement for renewable resources, such as labor and machines over time.

A second concern is the sensitivity of the solution to the length of the planning horizon and forecasts used for demand far into the future. In practice, we are most concerned with the solution we need to implement in period one, and we are most sure of our cost and demand estimates for the early periods in the planning horizon rather than for those at the end of the horizon. We would be uncomfortable if our decision to make 100 or 500 units this week depended on whether we were planning for the next 100 or 101 weeks or whether forecasted demand in week 100 was 99 or 100 units. We would like a method that provides the same solution for period 1 for similar values of future demand and costs. However, this is not always the case with the Wagner–Whitin model. The planning horizon theorems provide some cushion against this problem; however, we are not certain that the assumptions and conditions of the theorem will apply in practice. Consider, for example, the problem with $h = \$1$, $A = \$150$, $c = \$1$, $D_1 = D_2 = 110$, and $D_t = 100$ for $t > 2$. In general, we see that we would like to produce every two periods because the setup cost exceeds the cost to carry period demand for one period, but it is less than the cost of carrying inventory for two periods. If the planning horizon is three periods, the dynamic programming algorithm tells us to produce 110, 210, and 0, respectively, in periods 1, 2, and 3. If the planning horizon is four periods, the algorithm says to produce 220, 0, 200, 0. This is a much different result for the first two periods, and these are the periods of concern because we must act now to schedule this production and order the raw materials. So which planning horizon is best? Consider what happens if we extend the horizon. For any horizon with an odd number of periods, we will produce 110 in period one and 210 in period 2. For any horizon with an even number of periods, we will produce 220 units in period one. It is disconcerting to have such a change in period one plans as we change the length of the planning horizon from approximately 1,000 to 1,001 weeks. Because we do not know exactly when the life of this product will end, we cannot determine the best answer.

Nevertheless, the Wagner–Whitin model is useful as a subproblem when we have many products or families produced on the same resources. In this case, we can alternate between items from period to period so that we are always producing something in our

Table 6.2   Heuristic Lot-Sizing Methods

Procedure	Criterion for including periods into order
Economic Order Period	Order every Q/D periods
Least Unit Cost	Minimize (Order Cost + Holding Cost)$/Q$
Part Period Balancing	Equate Total Holding and Order Cost
Silver Meal	Minimize Ave. Cost per Period
Bounded Holding Cost	Maximum Periods with Total Holding Cost $< A$

shop. However, we do not need to set up for every item, every period. In the next section, we discuss an alternative heuristic solution method that is less sensitive to end-of-horizon forecasts.

6.3.1.3   Rolling Schedules and the Silver-Meal Heuristic

In practice, we do not expect the company to go out of business at the end of the planning horizon. The Wagner–Whitin procedure can be sensitive to these end-of-horizon effects. A number of heuristic procedures have been proposed that depend more heavily on the relatively accurate forecasts of demand in the near-term periods and are robust, assigning less importance to forecasted values for periods later in the planning horizon. We look at one of these procedures briefly. Other procedures are listed in Table 6.2. Total relevant order cost (TROC) in this Table includes the fixed order cost plus holding cost for the periods covered by the current order being planned. All the procedures take a "greedy" approach of starting with the first period and adding the demand for the next period, provided some rule is satisfied. Once the rule is violated, the order size is set, and we move to the next period with a new order. In each case, we produce the demand for an integral number of periods, relying on the optimality principle:

$$I_{t-1} \cdot X_t = 0.$$

This approach works well, in some cases even better than the Wagner–Whitin model, because of its potential sensitivity to end-of-horizon effects and the fact that the solution is implemented on a rolling horizon basis. Bitran et al. [1984] examined the worst-case performance of several heuristic procedures. However, most can be arbitrarily bad.[5] The part-period balancing approach will never cost more than three times the optimal, and the bounded holding cost approach is never more than twice the optimal.

The Silver-Meal heuristic is a popular alternative to the Wagner–Whitin model. The Silver-Meal approach uses the decision rule that we should continue to add another period's demand to the current order as long as the average cost per period decreases. If we produce for the next T periods, the average relevant cost per period is:

$$Ave.\ Cost/Period = \frac{A + h \sum_{t=1}^{T} (t-1)D_t}{T} \tag{6.52}$$

---

[5]Arbitrarily bad implies that pathological data sets exist for which the ratio of the cost of the heuristic solution to the cost of the optimal solution approaches infinity.

We select T as the smallest value, for which:

$$\frac{A + h \sum_{t=1}^{T+1} (t-1)D_t}{T+1} > \frac{A + h \sum_{t=1}^{T} (t-1)D_t}{T}$$

or equivalently, the smallest T that satisfies:

$$T^2 \cdot D_{T+1} > \frac{A}{h} + \sum_{t=1}^{T} (t-1)D_t \qquad (6.53)$$

EXAMPLE 6.19   Find the production schedule for the data from Example 6.18, using the Silver-Meal model.

SOLUTION   For $T = 1$, $T^2 \cdot D_{T+1} = 50 < \dfrac{A}{h} = 333.3$.

For $T = 2$, $T^2 \cdot D_{T+1} = (4)(95) = 380 < \dfrac{A}{h} + D_2 = 383.3$.

For $T = 3$, $T^2 \cdot D_{T+1} = (9)(110) = 990 > \dfrac{A}{h} + D_2 + 2D_3 = 573.3$

thus, we stop and set $X_1 = D_1 + D_2 + D_3 = 265$. We then roll forward to period four and start over, assuming the problem horizon starts there, and the old period four becomes the new period one. For this example, we eventually obtain the same solution as with the Wagner–Whitin model. However, this will not always be the case.

6.3.2 Limited Capacity Models

The dynamic models in the previous section are most useful when applied to purchased items or when producing a large number of manufactured parts so that no single item consumes a significant proportion of capacity. For a single item, the model increases the variability in production levels from period to period by combining demands from individual periods together into batches for production. This can cause a problem for items produced internally, when trying to balance workloads over time. Figure 6.11 illustrates the "lumpiness" in production resources that may result from combining demands into production batches. With a large number of manufactured items, we can hopefully schedule whereby each period produces several items, and the sum of the required setup and processing times is relatively equal. To help ensure this, we can solve a multiproduct constrained production planning problem. Letting X_{it} be our decision variables indicating the amount of product i to produce in period t, the model becomes:

$$Minimize \sum_{i=1}^{n} \sum_{t=1}^{T} (A_{it} \cdot \delta(X_{it}) + C_{it}(X_{it}) + h_{it} \cdot I_{it}) \qquad (6.54)$$

subject to:

$$I_{it} = I_{i,t-1} + X_{it} - D_{it}, \forall it \qquad (6.55)$$

$$\sum_{i=1}^{n} (s_{it} \cdot \delta(X_{it}) + p_{it} \cdot X_{it}) \leq K_t, \forall t \qquad (6.56)$$

$$I_{it} \geq 0, X_{it} \geq 0 \qquad (6.57)$$

The multiproduct models add a capacity constraint to the model shown in Equations (6.46) through (6.48). The objective function [Equation (6.54)] minimizes the sum of setup cost,

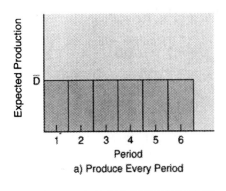

a) Produce Every Period

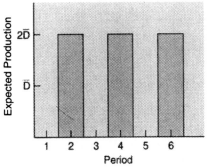

b) Produce Every Two Periods

Figure 6.11 Lumpiness from Demand Batching for a Single Item

material cost, and holding cost. Inventory balance is kept by Equation (6.55). For resource availability K_t in period t, Equation (6.56) restricts our decision space to the production plans that satisfy the capacity limitations. The left side of the Equation sums the total resource required for setup and production as a function of our planned production levels. This model may have a large number of binary variables if the number of periods or products is large. With the addition of the capacity constraints, it is no longer true that we must have $X_{it} \cdot I_{i,t-1} = 0$. As a result, the efficient dynamic programming scheme used for the single-product unconstrained case does not apply. Instead, we present a heuristic method for minimizing inventory costs for multiple items subject to a capacity constraint. We assume that the problem is feasible in the sense that a production schedule exists that will meet all demand without shortages. If setup time is negligible and the production efficiencies are constant ($P_{it} = p_i \ \forall t$), the problem will be feasible if:

$$\sum_{t=1}^{t_0} K_t \ge \sum_{i=1}^{n} p_i \left(\sum_{t=1}^{t_0} D_{it} - I_{i0} \right), \ \forall t_0 = 1, \ldots, T. \tag{6.58}$$

The condition [Equation (6.58)] states that the sum of available production hours for the first t_0 periods is at least as large as net cumulative production time demand for those periods. If this is not true for some period in the planning horizon, we will have a shortage in that period.

Our heuristic solution procedure begins at period one and moves ahead one period at a time. At each period, we first make sure that all current demand is met and then attempt to increase current batch sizes to eliminate future setups if economical and feasible. If capacity is insufficient to meet demand for the current period, we increase production levels in an earlier period to eliminate shortages. A more complete description follows.

For period $t = 1$ to T:

1. *Satisfy Current Demand.* Examine the set of products that must be produced this period and produce at least enough to avoid shortages. Accordingly, if the starting inventory in period t is insufficient to meet demand for that period, namely if $I_{i, t-1} < D_{it}$, then initially set:

$$X_{it} = D_{it} - I_{i,t-1}.$$

2. *Reduce Future Setups When Economical and Feasible.* If excess capacity remains because $K_t > \sum_{i=1}^{n} p_{it} X_{it}$, then we consider potential savings from increasing one or more batch sizes in this period. For each item being produced, i.e. $X_{it} > 0$, we consider increasing the batch by one additional period's worth of demand. While capacity still exists in period t, we add an extra period's demand for the item with the largest savings per unit of capacity required. Let r_{it} be the reward (savings) if we pull demand for one more period into X_{it}. If we add demand for period $t + l_i$ to X_{it}, then the savings r_{il_i} represent the saved setup cost in period $t + l_i$, minus any holding cost incurred by the early production of these units, i.e., $r_{il_i} = A_i - h_i \cdot D_{i,t+l_i-1} \cdot (l_i - 1)$. We consider adding production only for item-period combinations with a positive reward. Let k_{il_i} be the increase in resource usage from this decision, typically $k_{il_i} = p_{i,t+l_i} \cdot D_{i,t+l_i}$. For those items with $r_{il_i} > 0$ and k_{il_i} not exceeding the remaining slack capacity, add $D_{i*,t+l_{i*}}$ to X_{i*t} where $i*$ maximizes$_i \left(\dfrac{s_{il_i}}{k_{il_i}} \right)$. We repeat this process while there remains at least one item with positive potential savings and adequate capacity in period t. We then move forward one period and repeat the process for period $t + 1$.

3. *Remove Infeasibilities.* If at any period t we have an infeasible solution because the capacity required to produce $\sum_{i=1}^{n} (D_{it} - I_{i,t-1})$ exceeds the available capacity in period t, then we enter a backward phase. To avoid the problem of insufficient capacity in this period, we must transfer production of $\sum_{t=1}^{n} p_{it}(D_{it} - I_{i,t-1}) - K_t$ hours of demand backward in time. For each item with insufficient initial inventory in this period, i.e., items for which $I_{i,t-1} < D_{it}$, we find the cheapest method for moving demand backward to an earlier period. Several options exist for each product type. We could move units back to the period in which this item was last produced and for which slack capacity exists, or we could just move it back to the closest period with excess capacity. In the first case, we incur holding cost $h_i \cdot (D_{it} - I_{i,t-1})$ for each period that we move this production backward. In the second case, we must also include the cost of a new setup in that previous period. We choose to move production of the item with the minimal cost increment per unit of reduction of production resource infeasibility in period t. The amount of production moved cannot exceed the excess capacity available in that earlier period. We will usually move the minimum {infeasible demand in t, excess capacity in earlier period}. However, if the slack capacity in the previous period to which we are adding demand is sufficient to add $D_{it} - I_{i,t-1}$, we should also consider moving the entire $D_{it} - I_{i,t-1}$ backward to save a setup. In summary, for each product that needs to be produced in period t, we find the least-expensive alternative for pushing some of its production backward. Looking across all products, we choose the item with the smallest increase in cost per unit of resource requirement reduced in period t. This process repeats until a feasible solution is obtained for period t. At that point, we begin to move forward again at period $t + 1$. The process is demonstrated best with an example.

EXAMPLE 6.20

Production is to be planned for three items over the next three periods. Demands, setup costs, and holding costs (per unit per period) are shown in the Table. Assume setup time is minimal. Demand is measured in factory hours, and we have 150 hours available per period. Find a good schedule.

	Demand (Hrs.)				
Product	Period 1	Period 2	Period 3	Setup cost	Holding cost
A	30	30	50	100	1
B	50	50	60	100	2
C	40	30	80	100	3

SOLUTION

We begin with a check of feasibility. If we add demand for all three products, cumulative demands are 120, 230 and 420 hours respectively for periods one, two, and three. This is less than the 150 hours of capacity available each period over the horizon. Thus, a feasible solution exists. We can begin by scheduling period-one production to equal period-one demand. After loading these $30 + 50 + 40 = 120$ hours, we have 30 hours of capacity remaining in period one. If we add the period-2 demand for product A into the first period, we save the \$100 setup cost in period 2 but incur a holding cost of \$30. This yields a net reward of $r_{12} = \$70$, and the return per unit of resource consumed is $\frac{r_{12}}{k_{12}} = \frac{70}{30} = 2.3$. Because we can only produce 3/5 of D_{22}, we cannot save a setup, and there are no savings. If we add D_{32} to the period one production plan, we save $r_{32} = \$100 - 3(30) = \$10$ for a savings per resource ratio of $\frac{r_{32}}{k_{32}} = \frac{10}{30} = 0.3$. The best option is to produce the extra 30 units of product one. Thus, we set period one production to (60, 50, and 40) for products (A, B, and C), respectively. Because no capacity remains in period one, we move to period 2.

In period 2, we are forced to produce at least (0, 50, 30) units of the three products to meet demand. This leaves a slack capacity of $150 - 50 - 30 = 70$ hours. We do not choose to produce any of product 1 because this would incur an extra setup cost of \$100 with no corresponding savings. If, at period 2, we move D_{23} into the period-2 production plan, our savings are $s_{22} = 100 - 2(60) = -\$20$—not an advisable choice. Product 3 is excluded because we would require 80 hours of additional production time in period 2, and only 70 hours are available. Partial production will not save any setup cost and, thus, would not be desirable. We then move to period 3.

We first attempt to produce the remaining demand, (50, 60, and 80) units, respectively. This requires 190 units of capacity and is infeasible. We must move $190 - 150 = 40$ hours of demand back to an earlier period. We attempt to find a solution that will not incur any additional setup. If we move product 1 back, we must schedule it in period 1. However, period 1 did not have any slack capacity. If we produce an extra 40 units of product 1 in period 2, then we would incur an extra cost of \$100 for setup, plus \$40 for holding. If we move all 50 units of D_{13} back to period 2, then we save the setup in period 3; however, we still incur a \$50 holding cost. Adding a setup in period 2 and removing a setup in period 3 leaves the setup cost unchanged. If we have to move product 1, the best choice is then to move all of D_{13} back to period 2. Likewise, we can move 40 or 60 hours of D_{23} back to period 2. Either is feasible because we have 70 hours of extra capacity in period 2. If we move 40 hours back, this does not affect setups because we would still be setting up product 2 in each of the three periods. We would, however, incur an additional holding cost of $40(2) = \$80$. If we move all 60 units back, we save a setup cost but incur \$120 in holding costs, a net cost increase of \$20. Alternatively, we can move 40 hours of D_{33} back at a cost of $40(3) = \$120$. We cannot move all 80 units of period-3 demand backward because we do not have sufficient slack capacity in period 2. Our best choice is then to move $D_{23} = 60$ units of product-2 production back to period 2. We end up with the production schedule:

	Production		
Product	Period 1	Period 2	Period 3
A	60	0	50
B	50	110	0
C	40	30	80

Total cost is $700 for setups plus $180 for holding ($60 for product 1 in period 1 plus $120 for product 2 in period 2).

So far in the discussion, we have assumed deterministic demand. If demand is stochastic, we can specify a service level, α, for the minimum percentage of demand to be met from stock in any period. We then set the D_{it} whereby $\sum_{t_0=1}^{t} D_{it_0} = F_t^{-1}(\alpha)$ for $t = 1, \ldots, T$, where $F_t^{-1}(\alpha)$ is the upper α percentage point of the distribution of cumulative demand for the first t periods. Once we determine these modified demand values, we can solve the problem using the same procedure employed for the deterministic case.

6.4   MODEL IMPLEMENTATION

The models presented in this chapter cover a variety of situations and require various amounts of data to administer. It is difficult and expensive to impose the discipline to maintain accurate inventory records. Cost estimates are often very approximate. Indeed, as presented in Chapter 1, the data definitions used in the accounting system may preclude determining the proper costs with a reasonable amount of effort. In this section, we address the issues of how to apply these inventory models to a large set of items and judge the performance of an actual system when precise data are unavailable or prohibitively expensive.

6.4.1   ABC Analysis

Suppose you were to list your personal expenses. If you grouped these expenses into categories, you would most likely find that the vast majority of your expenses fall into a small number of categories such as housing, food, and tuition. This phenomenon of a "vital few and trivial many" occurs repeatedly in nature and is known as Pareto's Law[6]. The same is true for the volume and cost of inventory items. It is common to find that 80 to 90% of the dollar volume of business is associated with no more than 10% of the items. This result is shown in Figure 6.12.

It is common to divide the set of inventory items into A, B, and C categories based on their relative importance. Importance is usually based on dollar volume of activity

---

[6]Vilfredo Pareto was an Italian economist of the early twentieth century. He became famous for his observations and theories regarding how consumers choose preferences and the variability of importance of objects.

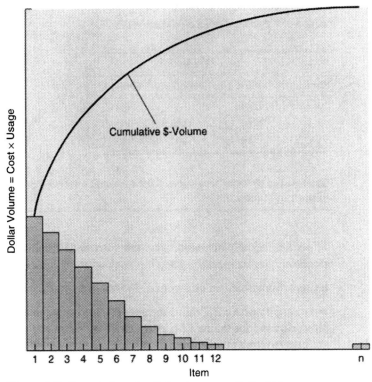

Figure 6.12 Pareto's Law Applied to Inventory Dollar Volume

(cost × usage) but may include other factors such as cost of a shortage. A items are the top 10% of items that often account for 80 to 90% of activity and cost. For these items, it makes economic sense to install monitoring systems, track status, identify cost parameters, and use sophisticated inventory control models. For the next group of B items, some control is warranted but probably with less frequent checking and analysis. Periodic review and joint replenishment strategies are used for this group. The majority of items fall into class C. These items are relatively inexpensive. Assuming it is even justified to carry them in inventory because of their long lead times or steady usage, they should be ordered in bulk and checked only occasionally. It is usually not necessary to keep track of the exact level of inventory in stock, and a heuristic, visually based, two-bin system is adequate. We may decide to extract parameter estimates from historical data and use the EOQ or EMQ; however, we would probably not invest too much in finding the optimal policy.

6.4.2 Exchange Curves

We have described a variety of models for determining optimal inventory policies. These models depend on knowledge of parameters such as ordering and shortage costs. In practice, these parameters are difficult to estimate. It can be useful to examine the trade-off between competing factors such as the number of shortages and investment in safety stock. Exchange curves display these trade-offs in the form of a curve, containing the optimal value of one inventory system characteristic relative to another. Figure 6.13 shows a trade-off curve for safety stock versus stockouts. Another useful curve shows cycle inventory

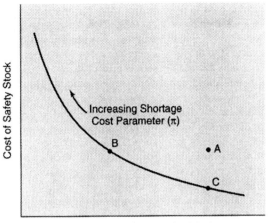

Figure 6.13 Safety Stock versus Stockouts Exchange Curve

versus setup cost (or time or $\frac{A}{i}$ ratio). After creating the curve, a company can locate its current performance on the axes. Consider point A in Figure 6.13. We see that the company is operating with a nonoptimal policy. The optimal point on the curve is debatable without precise knowledge of costs. However, the company can see that by improving its policies, they can: 1) reduce shortages without increasing cost of safety stock (point B); 2) decrease safety stock and supply the same service level (point C); or 3) improve both cost and service by selecting a point between B and C on the curve.

Exchange curves can be used in multiproduct systems in which it is impractical to determine the appropriate parameter value for each item. For instance, a single-unit shortage cost or setup cost parameter value can be assigned for all the items in a class. The curve then plots cumulative results for the entire class of items that would result from selecting different values of this parameter. Management then selects the parameter value that yields the preferred cost and service level.

6.4.3 Extensions

The literature on inventory models is quite extensive. Many different environments have been modeled in addition to those described in this chapter. For example, the case of perishable inventory items with a limited shelf-life is addressed in Nahmias [1982]. An extensive list of models for both discrete and continuous demand with a variety of service-level measures is given by Peterson and Silver [1979]. A more recent comprehensive description appears in Zipkin [2000].

6.5 SUMMARY

Inventory will always exist. It must exist because of processing time and transit time among suppliers, processing stations, and customers. The unpredictable nature of demand and machine availability make it advisable to carry safety stock. In addition, there is an optimal level of production cycle inventory attributable to the economics of setup and ordering costs. Economic decision models trade off the costs of ordering, shortages, and carrying inventory while incorporating organizational constraints such as space limits and available machine time to determine the optimal inventory policy. Customer service level or lead-time minimization objectives may also be used to determine an inventory policy. A policy indicates **when** an order is placed and **how much** is ordered.

A wide variety of models have been developed to fit specific environments. Environments vary based on data availability, item importance, ordering policies, capacity limitations, demand characteristics, and cost structures. Several cases of a single item facing continuous demand are modeled in the chapter. These include the economic order quantity for orders from external vendors and the economic manufactured quantity for internal production. Policies are derived for deterministic demand and stochastic demand both under continuous review conditions, for which orders could be placed at any time, and periodic review, for which orders can be placed only at discrete time intervals. A general result holds that the optimal order quantity grows in proportion to the square root of both the demand rate and the ratio of setup to holding cost. Setup time and costs are thus key factors in determining the optimal inventory policy. Reducing setup time and cost is often the most effective action for improving an inventory system. This results in increased capacity, shorter throughput times, lower inventory levels, and opportunity for product customization. As an alternative to minimizing cost, we could choose to minimize cycle time or inventory subject to meeting demand with the allocated resources.

Coordinated policies can be established for multiple items that share a common resource such as processor time, budget capacity, or warehouse space. Price breaks should be considered when available. Randomness in replenishment lead time or customer demand results in the need to carry some safety stock. For continuous review situations, the safety stock should protect against variability during the lead time; for periodic review models, demand variability over the lead time plus review cycle affects the optimal safety stock. Dynamic demand requires that we look ahead and forecast the future. For continuous review, a simple procedure was proposed that keeps track of expected run-out time for each item. With periodic review, models can be solved by noting that when adequate capacity exists, either the initial inventory or the production level will be zero in each period. Dynamic programming can be used to find the optimal set of production quantities over the planning horizon; however, simple decision rules such as the Silver-Meal heuristic work well in a rolling horizon and are less sensitive to end-of-horizon effects and forecast variability. If capacity is limited, we may need to adjust plans by producing early to make use of slack capacity in slow periods.

Finally, items should be prioritized based on their total usage cost. The degree of monitoring and sophistication used in inventory control should be based on the importance of the item. An organization can compare its performance and examine opportunities for improvement by using exchange curves.

6.6 REFERENCES

BITRAN, G.R., MAGNANTI, T.L., & YANASSE, H.H. Approximation methods for the uncapacitated dynamic lot size problem. *Management Science*, 1984, 30(9), 1121–1140.

ELMAGHRABY, S. The economic lot scheduling problem: review and extensions. *Management Science*, 1978, 24(6), 587–598.

GALLEGO, G. Scheduling the production of several items with random demands in a single facility, *Management Science*, 1990, 36(12), 1579–1592.

GRAVES, S.C., The multiproduct production cycling problem. *IIE Transactions*, 1980, 12(3), 233–240.

HADLEY, G.J., & WHITIN, T.M. *Analysis of inventory systems*. Englewood Cliffs, N.J.: Prentice-Hall, 1963.

JACKSON, P.L., MAXWELL, W.L., & MUCKSTADT, J.A. The joint replenishment problem with a powers of two restriction, *IIE Transactions*, 1985, 17, 25–32.

KARMARKAR, U.S. Lot sizes, lead times and in-process inventories. *Management Science*, 1987, 33(3), 409–423.

NAHMIAS, S. Perishable inventory theory: a review. *Operations Research*, 1982, 30, 680–708.

ROUNDY, R. 98% Effective integer ratio lot-sizing for one warehouse multi-retailer systems. *Management Science*, 1985, 31, 1416–1430.

SILVER, E.A., & PETERSON, R. *Decision systems for management and production planning*. New York: John Wiley & Sons, Inc., 1985.

WAGNER, H.M., & WHITIN, T.M. Dynamic version of the economic lot scheduling model. *Management Science*, 1959, 5(1), 89–96.

ZIPKIN, P.H., *Foundations of inventory management*. New York: McGraw Hill Inc., 2000.

6.7 PROBLEMS

6.1. Describe the difference in assumptions between the EOQ and EMQ models. How does this affect the development of the cost function in the two models?

6.2. An automotive repair shop stocks many sizes of tires. One particular size and model is purchased for $15 and sold for $30. The manager estimates the cost to order at $75, including the delivery charge and paperwork. Using the cost of rent, interest, and utilities, the manager estimates the cost of carrying inventory at approximately 50% per year based on average inventory value. The shop sells approximately 2,000 of these tires per year. Determine the optimal reorder quantity and reorder point. Orders are received two weeks after placement.

6.3. An office furniture warehouse sells 5,500 chairs of a popular style per year. Chairs are purchased from the factory at a cost of $45. The comptroller has estimated the inventory carrying cost rate at 0.4 per year. Labor, mail, and receiving costs per order are estimated at $125. Determine the optimal order quantity for the warehouse.

6.4. Suppose that in the previous problem, the warehouse and factory were jointly owned and the objective was to minimize combined cost. For each order, the factory sets up the machines and produces the requested amount. A setup costs $235 and takes two hours. Each unit requires one hour of processing time. The $45 cost comprises $20 of purchased material and $25 of value added. Throughput time in the plant is approximately five times total processing time. Determine the optimal order quantity and total system cost per year. Assume the plant operates 2,500 hours per year.

6.5. A bank branch must keep a stock of cash on hand for customer withdrawals. The net daily withdrawal rate at the branch is $5,000. The opportunity cost of cash is 10% per year. Replenishment of the cash stock from the central bank requires careful accounting and contracting to an armored delivery service. This costs $250 per replenishment and requires two days of notice to the carrier. Determine the optimal policy for obtaining cash.

6.6. Suppose the actual outflow of cash per day were random in the previous problem, with daily demand being normally distributed with a mean of $5,000 and a standard deviation of $500. The bank does not want to run out of cash more than once per year (assume 250 days per year). Find the optimal inventory policy and the imputed cost of a shortage.

6.7. Consider the bank's situation in the previous problem. What factors should be included in computing its cost of a shortage?

6.8. A warehouse sells 10,000 boxes of copy paper per year. It costs $150 to reorder from the factory. Each box costs $5, and the holding cost rate is 0.35 per year. Find the optimal order quantity, the average inventory cost, the number of orders placed per year, and the inventory cost per year.

6.9. A pharmacist stocks 250-milligram amoxicillin pills. Daily demand is uniformly distributed between 30 and 330 tablets. The pharmacist spends approximately 1 hour to place an order with the distributor and check it on arrival. Her salary plus benefits cost $35 per hour. Each pill costs $0.13. Because of the careful inventory control required for drugs and the limited shelf-life, the pharmacist estimates a 50% holding cost rate for inventory per year. Find the optimal ordering policy if the pharmacist wants to avoid all shortages and delivery lead time is one week.

6.10. Suppose that in the previous problem, shortages are handled by obtaining pills from a nearby pharmacy at an additional cost of $1 each. Find the optimal inventory policy.

6.11. A plant purchases components that are then combined into subassemblies. The subassemblies are produced in batches and stored until customers place orders. The subassemblies are then quickly combined into finished, customized products. For one particular subassembly, parts cost $50, and $12 is added during subassembly. Holding cost for the facility is 40% per year, and annual demand for subassemblies is 14,500. Each setup costs $125. Processing time is 1 hour per subassembly, and batches spend approximately two thirds of total time waiting at machines or for material handling. Find the optimal batch size for making the subassemblies.

6.12. A machine center produces 50 batches per 8-hour work day. Average batch processing time is 5 minutes. Records indicate that batches spend an average of a half-day in the work center. Estimate the average number of batches in process and the ratio of throughput time to processing time.

6.13. One item produced by a work center has a setup cost of $250 and an annual demand of 20,000. Demand is deterministic and occurs at a constant rate. Raw material costs $1.50, and the completed item is valued at $2.25. The holding cost rate is 0.5 per year. The work center could process 100,000 of these items per year if it did not make anything else. Historically, batches at the work center spend approximately 80% of the time waiting and 20% of the throughput time in process. Find the batch size that minimizes cost, and compute the inventory cost per year. What reorder point should be used to avoid shortages?

6.14. A work center with fixed labor cost is scheduled to operate 800 hours per month. The work center could produce up to 35,000 units in 800 hours; however, demand averages 24,000 parts per month. Each setup takes 15 minutes, and each month has twenty 8-hour working days.

a. Find the batch size for minimizing inventory cost.

b. Find the batch size that minimizes production lead time.

6.15. A machine produces three products per 40-hour period. Setup times, processing times, and demands are shown below. If setup cost consists only of labor, and total labor cost is fixed, find the appropriate batch sizes to minimize inventory cost. All times are in hours.

Item	Internal Setup	External Setup	Unit Proc. Time	Demand
1	1.0	1.5	0.1	100
2	1.5	1.0	0.1	80
3	2.0	2.0	0.2	60

6.16. Consider setup and inventory holding cost terms of the basic deterministic EOQ model (ignore WIP costs). Examine the sensitivity of the cost function to deviations from the optimal Q^* as derived in the text. Plot the increase in total relevant costs as a function of Q/Q^*.

6.17. In the basic EOQ model, the cycle time between consecutive orders is $T = Q/D$. Thus, $T^* = \sqrt{\dfrac{2A}{hD}}$. Show that for any other cycle time, T,

$$\frac{\text{Annual Cost for Cycle Time } T}{\text{Annual Cost for Cycle Time } T^*} = \frac{1}{2}\left[\frac{T}{T^*} + \frac{T^*}{T}\right] \quad (6.59)$$

6.18. Suppose we decide to use a cycle time of $T = 2^m \cdot T*$ for some integer (positive or negative) m. Using the result from Equation (6.59), show that there will always exist an m for which

$$\frac{\text{Annual Cost of } T}{\text{Annual Cost of } T*} \leq \frac{1}{2}\left[\sqrt{2} + \frac{1}{\sqrt{2}}\right] = 1.06.$$

6.19. Actual costs are sometimes difficult to determine exactly. Suppose we have an error in our estimate of holding cost and that the actual cost is $a \cdot h$ for some constant a. Using the basic deterministic EOQ model, derive an expression for the ratio of the true optimal batch size to that found by the model when using our estimate of h. Likewise, derive an expression for the ratio of the sum of the true holding and setup costs to the model's estimate under these conditions. Plot these relations as a varies from 0 to 2.

6.20. Repeat the previous problem for the case that the error is in the estimation of the setup cost A.

6.21. A plant purchases 150,000 units of an electrical component per year. The fixed ordering cost is $60. The unit cost is $0.25 if ordered in quantities less than 1000, $0.24 if order size is between 1,000 and 10,000, and $0.23 for orders of more than 10,000. The holding cost rate is 0.4 per year. Find the optimal order quantity and annual cost.

6.22. A warehouse of 6,000 square feet stocks three items. Using the item data below, find the appropriate order quantities. Random access storage is used so that space planning can be based on average inventory levels. How much would you pay for an extra square foot of storage space?

Item	Annual Demand	Cost/ Order	Holding Cost/Unit-Yr	Sq. Ft/ Unit
1	12,500	$150	$2.40	5.0
2	15,000	$ 80	$3.50	4.0
3	15,000	$ 80	$3.00	4.0

6.23. A work center produces the four items described below. Find the optimal rotation cycle inventory policy. The facility operates 4,000 hours per year. The inventory holding cost rate is 50% per year. The work center is charged at the rate of $20 per hour for setup or production.

Item	Material Cost	Annual Demand	Production Time (Hrs./Unit)	Setup Time (Hours)
A	$125.00	2,500	0.30	8.0
B	$ 76.50	5,000	0.15	4.5
C	$ 55.34	7,500	0.10	2.0
D	$ 49.68	12,000	0.10	2.0

6.24. A work center produces three product types. Changeovers cost $75 and take 1 hour. Holding cost is $3 per item per month. Monthly demands for items are 250, 500, and 600, respectively. The facility is scheduled 200 hours per month, and each item takes 0.15 hours to produce. Find the appropriate production schedule assuming constant, deterministic demand.

6.25. Apply the results from queuing theory to determine the effect of batch sizes on throughput time. Consider a single-stage production system. Compute the change in flow time at the production facility under the following conditions if batch size is doubled. Note that doubling the batch size will cut both the batch arrival rate and the service rate in half. Assume that batches arrive according to a Poisson process.

a. Processing time is exponential (note: for an M/M/1 queue with exponential interarrival and processing times, flow time is $[\mu(1 - \rho)]^{-1}$, where μ is the batch processing rate and ρ is the proportion of time the facility is busy).

b. Processing time is deterministic with mean time equal to μ^{-1} for each batch (note: for an M/D/1 queue, flow time is $\mu^{-1} + \dfrac{\lambda}{2\mu^2(1 - \rho)}$, where λ is the arrival rate of batches).

6.26. A machining center makes several types of parts in each 8-hour work day. It takes an average of 1 hour to change over the machine from one part type to another. Unit processing times average 5 minutes. Total demand for all part types averages fifty parts per day. Find the minimum feasible batch size whereby the work center will be able to meet demand. Plot the utilization of the work center as a function of the batch size.

6.27. A bottling line requires 30 minutes to change over between different mixes or product sizes. The line can operate at sixty bottles per minute, and it must bottle 20,000 bottles per 8-hour shift. Filled bottles are used to resupply finished goods that are ordered by customers. We can select the batch size; however, bottling orders will arrive randomly.

a. Find the optimal batch size to minimize replenishment lead time.

b. Plot lead time as a function of batch size.

c. Discuss the assumptions made in your answers to parts a and b.

6.28. Verify the result of Section 6.1.6 that for an M/M/1 queuing model, the batch size $Q = \dfrac{\bar{s}D \cdot \left(1 + \sqrt{\dfrac{1}{D\bar{p}}}\right)}{(1 - D\bar{p})}$ minimizes average time in the system (waiting time plus batch service time).

6.29. Using the result in Equation (6.25), show that the optimal workstation utilization is equal to the square root of the minimum utilization, i.e., $\rho* = \sqrt{D\bar{p}}$.

6.30. Suppose that in Section 6.1.6, batch processing times had a general distribution instead of exponential distribution.

Modify Equation (6.24) for the case of an M/G/1 queuing system. Note that for an M/G/1 queue, $E(\tau) = E(P) + \dfrac{\lambda \cdot E(P^2)}{2(1 - \rho)}$, where P is batch processing time including setup.

6.31. Below in the text, we learn that it is important to be able to predict lead time accurately. Suppose the objective is to find the batch size that minimizes the variance of lead time. Follow the approach used in Section 6.1.6 to find the appropriate batch size for this case. Note that for exponential batch arrivals and general service times:

$$V(\tau) = V(P) + \frac{\lambda \cdot E(P^3)}{3(1 - \rho)} + \frac{\lambda^2 [E(P^2)]^2}{4(1 - \rho)^2},$$

where P is the batch processing time including setup.

6.32. A business firm has limited cash. It is proposing to introduce a new product that is manufactured for $12 and sold for $25. To test the market, a single batch up to 2,000 units will be produced. It is estimated that items not sold at the regular price can eventually be sold at cost. Demand is estimated to be normally distributed with a mean of 1,000 and a standard deviation of 100. The capital reserve for the company is such that it wants to minimize its probability of losing more than $2,000. Unmet demand causes a loss of customers. Each lost customer means an average of $100 in future lost profit.

a. Plot the probability of a loss of at least $2,000 as a function of the batch size Q. What batch size would you recommend, and what is the expected profit for this Q?

b. Find the batch size that maximizes expected profit. For this Q, what is the probability of losing at least $2,000?

6.33. Demand for an item is normally distributed with a mean of 500 per week and standard deviation of 50. Each order costs $200. Holding cost is $0.4 per unit per week. If each shortage is estimated to cost $5, find the optimal order quantity and reorder point.

6.34. Continuous review inventory control is being applied to purchased motors for an electric fan manufacturer. Demand is uniformly distributed between 500 and 600 motors per week. Each order costs $250 to prepare, place, and receive. Motors cost $2.75 and the holding cost rate is 1% per week. Management proposes using the EOQ order quantity and setting reorder points to ensure a 97% fill rate. Find the imputed cost of a shortage and the expected number of shortages per year.

6.35. Consider the periodic review model in section 6.2.3. Suppose that shortages resulted in lost sales. Redefine the pertinent parameters, and derive expressions for the expected cost per time and optimal order-up-to point.

6.36. A company is planning to distribute a souvenir with the company's logo, free of charge, to customers during a special promotion. Demand during this promotional period is estimated to be normally distributed with a mean of 10,000 and a variance of 2,500. The supplier has agreed to make a single batch of the souvenirs and to sell them to the company for $.75 each.

They have no salvage value. The giveaway will be advertised, and the cost of a disappointed customer, should the company run out of souvenirs during the period, is estimated at $3 each. How many items should be ordered?

6.37. Suppose that in the news vendor problem, a limit existed on the number of units that could be disposed at value s. Additional units are scrapped with no revenue or additional cost. Let S be the maximum number of units that can be sold for s. Write out the new expected profit expression, and find an expression for the optimal order quantity.

6.38. A strategic decision is being made about the level of capacity to acquire for a new facility. Production will match demand in each period to avoid shortages and inventory accumulation. The objective is to make most product on regular time, but overtime may be used in a period if demand is above average. Demand per period is expected to be normally distributed with mean of 100,000 pounds and standard deviation of 5,000 pounds. Workers will be guaranteed 40 hours per week leading to a wasted cost of about $200 for every pound of regular-time capacity not used in a month. Overtime production will cost $100 per pound more than regular-time production. Assuming that capacity is continuous (many workers and machines), determine the optimal amount of capacity for the plant.

6.39. Below is a list of item demands for the past 24 weeks. Determine an appropriate probability model for weekly demand.

Week	Demand	Week	Demand	Week	Demand
1	867	9	611	17	783
2	754	10	780	18	817
3	632	11	694	19	746
4	697	12	813	20	711
5	715	13	723	21	648
6	1432	14	745	22	912
7	823	15	1937	23	825
8	536	16	653	24	687

6.40. A single facility produces five items. Changeovers take a half-hour, and the facility will only consider changing over between products once per day. Given the following demand rates and inventory levels, select the item to be produced today.

Item	Current Inventory	Demand/Day
1	100	50
2	100	20
3	45	30
4	74	20
5	38	12

6.41. Demand for an item is forecasted at 100, 150, 75, 75, 50, and 60 for the next 6 weeks. Each setup costs $80, and inventory holding cost is $1.25 per week.

a. Find the optimal order quantities for each week.

b. Find the order quantities selected by the Silver-Meal heuristic.

6.42. Demand for a product is estimated at 1,250, 1,000, 1,000, 900, 900, and 900 for the next six periods. Each setup costs $500, and inventory carrying cost is $3 per unit per period. Find the optimal production policy.

6.43. Resolve the problem above using the Silver-Meal heuristic.

6.44. Make up an example problem for which the Wagner–Whitin and Silver-Meal procedures result in different production policies. Explain why the results are different.

6.45. Explain why the Wagner–Whitin procedure may not work well in practice. (Indicate the effect of the explicit and implicit assumptions made by the method.)

6.46. Develop the expression for an exchange curve that shows the trade-off between the number of setups (or orders) per year versus the cycle inventory holding cost. Sketch the curve.

6.47. A work center produces three part types. Demand is stochastic with a mean of 1,200 units per month. Fifty percent of demand is for part type 1; the other two items each account for 25% of demand. Part changeovers take 1 hour, and cost $75. Each item takes 0.15 hours to produce, and the facility is available 200 hours per month. Current inventory levels are 125, 13, and 78, respectively. Shortages are to be avoided if possible. Develop a scheduling rule for this facility.

6.48. Consider the planning horizon theorem for the Wagner–Whitin problem. Show that the results hold if the variable production costs are nonincreasing, i.e., $C_t \geq C_{t+1}$.

6.49. In developing the minimum feasible base period for a multiproduct rotation cycle policy, we assumed that the resource was available the entire T time units in a period. Suppose the resource is unavailable α percent of the time because of maintenance and breakdowns. Derive the result for the minimum feasible period length.

6.50. Describe two shortcomings of the Wagner–Whitin model and its optimal solution.

6.51. Consider the constrained optimization problem defined by Equation (6.15). Suppose production of items was phased and space could be shared in such a way that it was sufficient to plan for average inventory plus F_i square feet for item i. Find an expression for the optimal order size.

6.52. Consider the result from section 6.3.1.2 that $X_t \cdot I_{t-1} = 0 \; \forall t$. Show that this corresponds to an extreme point of the constraint set defined in (6.47).

Chapter 7

Decentralized Pull Systems

You just purchased a new personal computer (PC). When you open the box, you find the monitor, tower, keyboard, mouse, software diskettes, manuals, cables, and assorted packing materials. If you open the main tower, you find a power module, circuit boards, connecting straps, fan, and other assorted components. (You could find another list of parts by opening the monitor, keyboard, or mouse.) Each of the components had to be produced, passing through several production stages to transform basic raw materials into that component. For the screws holding the tower together, this is a fairly basic machining operation, starting with bar stock. Consider, however, a printed circuit board. Each of the integrated circuits on the board, such as the central processing unit (CPU) and memory chips, required potentially hundreds of operations. Likewise, the board had to be produced with each layer of circuitry added gradually and the holes had to be drilled. Finally, the components had to be mounted and secured, again requiring multiple steps of component preparation, placing, and soldering. This complex structure typifies product manufacture. Each part potentially requires a series of technologically challenging operations. The manufacturing system includes a network of machines and workers operating in parallel, series, and cyclical configurations. The complexity of properly executing all part operations and coordinating the work assignments to bring together the parts at the time for assembly suggests the need for a complex organizational planning and control structure. It's unlikely that you would have bought the PC if the salesperson had said, "There are thousands of parts and all that it's missing is the fan blade that prevents the unit from overheating."

In the next several chapters, we explore various strategies for coordinating the planning of this flow of materials through the fabrication and assembly stages. In this chapter, we cover systems that pull materials through from final assembly back to raw materials. Subsequent chapters cover push systems that initiate the flow at the beginning of the process and force materials through to completion. In those chapters, we learn that the essential difference between these approaches stems from the manner in which information is transmitted to workstations.

7.1 TECHNIQUES FOR PRODUCTION AUTHORIZATION

We begin to address multistage production control systems by describing the set of information and physical moves necessary to authorize and coordinate production. First, let's clarify the difference between throughput time and lead time. **Throughput time,** sometimes referred to as flow time, refers to the time from when a job is released to the shop-floor until it is completed. During this time, we consider the job to be part of work-in-process (WIP). **Lead time** is the time between recognizing the need to plan and execute a job, and the completion of that job. Lead time includes throughput time, job-planning time, and the time the job spends waiting to be released. The release of jobs to the shop-floor has a significant impact on total production and the productive use of key resources. A job may not be immediately released to the production shop if resources are already occupied. It is easier to set meaningful priorities for processors, to require less floor space for storage, and to lower the investment in inventory if we wait until the shop is ready before we release a job to the shop. By delaying release, the ordering of materials can be delayed. Even if material is held in stock, materials have less value than partially finished jobs. Thus, delaying production as long as possible minimizes inventory investment and frees production capacity for current high-priority operations. Excess WIP inventory on the shop-floor only serves to inflate flow times, increase material handling frequency (because we must move extra WIP out of the way), increase material handling distances (because we must provide greater separation between workstations to accommodate the extra WIP), delay identification of quality problems (because large queues at machines cause longer delay times between completion of one operation and use of that product at the next stage), and provide opportunity for product damage, loss, and obsolescence.

7.1.1 Information Requirements for Production Operations

The fundamental operating decisions for each work center are, "What to make" and "When to make it." The "what" involves the specific part type, including all options and the quantity. The "when" is most likely driven by an objective of having the parts **completed** at a specific time. However, the more relevant operational issue to the work center management is when to **start** the part batch by releasing it to the shop-floor. We describe a basic framework for comparing the various approaches to answering these questions in multistage production scheduling. Within the framework, the entire fabrication/assembly system is regarded as a set of production cells. We refer to each group of machines that are scheduled and set up together to perform one or more operations as a **cell** or **work center.** The key factor in our definition of a work center (cell) is when a batch of material is dispatched to the work center, it is authorized to continue through the cell until the scheduled tasks are complete. Each individual machine or operator area is called a **workstation.** A work center may have one or more workstations, and operations performed in workstations may be identical or different. After each cell, there is a controlled storage area for holding completed parts. Parts are held until they undergo production as raw material at the successor cell. We refer to the ordered sequence of cells required to produce the product as production stages. Coordination requires a protocol for authorizing production at a stage and transferring parts to successor stages for continued processing. Likewise, we may want to pass information between stages to ensure availability of the appropriate parts as needed.

 Production authorizations (PAs) give the workstations permission to make parts. A mechanism must exist in all systems to communicate this authorization to the work center. A PA instructs the workstation to produce a given quantity of some part either immediately or when certain other conditions are met. These conditions are usually in the form of a start time or reorder-point level. The PA also includes an explicit or implicit indication of the intent to remove completed parts either now or at some point in the future. Material **requisitions** control the physical movement of parts. Requisitions authorize the removal of parts from controlled storage. They are issued when the parts are actually needed to produce the successor product. Buzacott and Shanthikumar [1992] present a more complete description of a unifying framework for comparing production authorization systems on the basis of intents, PAs, and requisitions.

Imagine each work center as a set of processors followed by a storage area for completed parts. Figure 7.1 illustrates production authorization and information transmittal protocols for various production systems. Material flows are shown with a solid line and information flows with a dotted line. Figure 7.1a shows a decentralized kanban system (Section 7.2). Information regarding demand is passed upstream (from finished products toward raw materials) one stage at a time as product withdrawals occur. Figure 7.1b shows a constant work-in-process system (see Section 7.3). The plant is usually treated as one large work center, and only the first and last work centers are connected by information. As completed products leave the system, a signal is sent to authorize the release of new

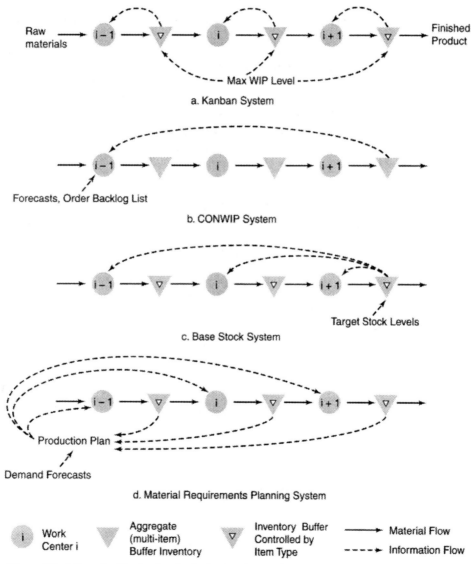

Figure 7.1 Alternative Production Authorization Approaches

part batches to the shop-floor. Once released, parts push their way automatically through their required workstations as those resources become available. Figure 7.1c presents a full feedback base-stock system wherein customer demand is broadcast immediately to all work centers. When a finished product is shipped, each work center related to this product is informed and begins work to replace the portion of the product that the work center produces. Base-stock systems are useful for large items with long lead times and low demand. Finally, Figure 7.1d describes a material requirements planning push system (Chapter 8). A centralized controller coordinates the production plans of all work centers and authorizes the work centers to make parts. In Figures 7.1a–c, actions are driven by removing (pulling) stock from finished goods inventory. In Figure 7.1d, authorizations push work through the system in anticipation of future need. The figure depicts a serial production system at which every work center has one predecessor and one successor. The information framework can be easily extended to a general system. Parallel requests can be sent to all predecessors indicating a need for raw materials. Figure 7.2 illustrates kanban and MRP systems with multiple feeder lines that produce parts used in a final assembly cell. This is a fairly common situation that describes the facility plan for many products such as furniture, appliances, computers, and automobiles. For large systems and modern supply chains, feeder cells may be housed in different plants. Nevertheless, in both systems, parts move downstream from part fabrication cells to the assembly cell. Both the MRP and kanban multicell systems of Figure 7.2 function similarly to their serial counterpart in Figure 7.1. Information flows move upstream from assembly to feeder cells in the kanban paradigm and spread out from a centralized controller to all cells in the MRP environment. In the next three chapters, we further describe the detailed opera-

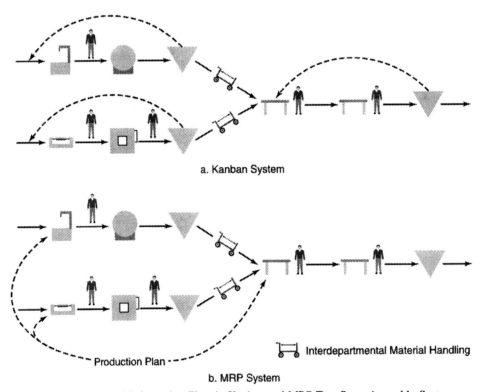

Figure 7.2 Product and Information Flow in Kanban and MRP Two-Stage Assembly Systems

tion of these production-scheduling systems. We compare the strengths and weaknesses of production scheduling systems in Chapter 10.

7.2 KANBAN SYSTEMS

Kanban production control systems are an elegant demonstration of the value of simplicity. With no advance notice or elaborate signaling system, production occurs as work centers react to their customers. Communication occurs by a successor workstation that issues a request for parts to the output buffer of a predecessor work center. When desired stock of completed parts is depleted, the work center reacts to replace the removed parts and to maintain a balanced, target level of finished parts. Coordination between the production levels occurs automatically by each work center as it strives to maintain a fixed output buffer.

The word "kanban" comes from the Japanese word for "card." Kanbans form the principal driver of the production control system perfected by Toyota [see Monden, (1981)]. The kanban production control system forms one of the four cornerstones of the widely discussed "just-in-time" approach to production systems. (The other cornerstones are addressed in Chapter 10.) A sample kanban is shown in Figure 7.3. The kanban includes information on the part type, the number of units authorized by the kanban, and the location of materials needed to produce this item. Modern kanban systems do not necessarily have physical kanbans. If a computerized production control system exists, kanbans can be electronic. In either case, each part type produced in a work center has its own set of kanbans, and each kanban authorizes a particular number of units of that part type. When work centers produce multiple part types, it can be helpful to associate colors with part types and their kanbans. Each part type has its own color code, and this color is used for its kanbans and perhaps even for the transport containers. Once the association is made, the likelihood of mixing different part types or producing the wrong part type is reduced.

In the remainder of this section, we describe various methods for organizing information and production authorization in kanban systems, as well as describe models for assisting in the selection of optimal parameter values for kanban systems.

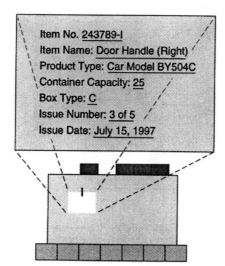

Figure 7.3 Production Ordering Kanban on a Container of Parts

7.2.1 The Kanban Authorization Concept

In a kanban system, each work center is charged with keeping a full container of parts for each kanban allocated to that work center. If a kanban is not accompanied by a full container, then the kanban authorizes production of those units. The maximum number of units in inventory at a work center for any part number is therefore given by the product of the number of its kanbans and the number of units authorized per kanban. When the decision is made to produce a product, the material handler proceeds to the predecessor to requisition the required materials. As these parts are removed and transported to the next workstation for processing, the predecessor automatically begins to replace this stock. The only announcement of intent is the material handler who appears at the output buffer to remove a container of parts. No other production scheduling information is supplied to the work center. By using this simple authorization procedure, kanban systems avoid the need to maintain a direct information link between the long-range production plan and the operational schedule. Several versions of kanban systems are used. We begin with a description of single-kanban systems and later describe dual-kanban systems.

7.2.2 Single-Kanban System

A single-kanban system is illustrated in Figure 7.4. The flow of materials is shown by a solid line, the movement of the kanbans by a dashed line. Consider stage i in Figure 7.4. The work center attempts to maintain one full container of parts in the output buffer for each kanban. When the successor workstation $i + 1$ requires parts, the operator or material handler transports a container of parts from the output buffer of stage i to stage $i + 1$. Discipline is required here because the operator *must* first remove the **production ordering kanban** (POKs) at stage i and place the kanban on the stage i collection box. Loose kanbans are arranged into a production sequence and moved to the **schedule board** immediately or at some fixed point in the near future. When the operator at i becomes available, the schedule board is checked for kanbans. The kanbans on the board are arranged into a production schedule. The details of forming a schedule are addressed in Section 7.2.4.

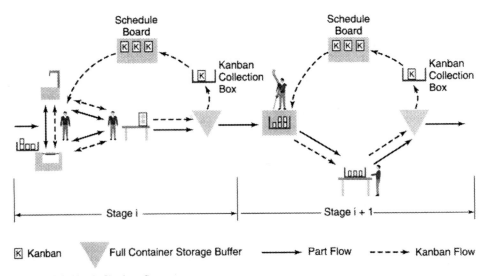

Figure 7.4 Single Kanban System

This type of single-kanban system is used when workstations are close together. Essentially, the output buffer at stage i is the input buffer at stage $i + 1$. The consecutive stages are linked by this intermediate buffer of parts. Because stage i is charged with maintaining a fixed quantity in the buffer for each part type, the system may be implemented by assigning a dedicated storage space for each part type. The size of the space corresponds to the space required to store the number of parts defined by the set of kanbans for that part type. The need for actual kanbans is eliminated because empty space is akin to kanbans in the collection box. This space-allocation concept gives rise to the term **kanban squares,** which is sometimes used to describe this single-kanban system.

Figure 7.4 implies that a single workstation exists at each stage. In actuality, each stage could be either a workstation or an entire department. Within the workstation or department, routing information must exist for each part type or product produced. Once authorized and released, parts begin production at the first workstation. The container of semifinished parts then moves from workstation to workstation based on its process plan until reaching the controlled output buffer. Movement between workstations within a stage is considered a push mode—on completion at a workstation, the parts are automatically authorized to proceed to the next workstation. Lead time in a kanban system extends from the time the kanban is removed from a container until that kanban re-enters the output buffer with a full container of parts. Several elements combine to form the lead time. First, there may be a delay in collecting the kanban and moving it to the schedule board. Once there, we must formulate a schedule, and the kanban must wait until its production is planned. This time will vary based on the work center's utilization and on whether a continuous or periodic discipline is used. Finally, the container must pass through each processing step in the cell. For each operation, there may be a wait for the processor, followed by processing, and a wait for a material handler, followed by a move to the next operation.

7.2.3   Dual-Kanban Systems

Dual-kanban systems are used when large distances between workstations dictate the need for input buffers at workstations to stage raw materials in addition to the output buffers used in the single-kanban system. System operation is illustrated in Figure 7.5. We continue to use the production-ordering kanban of the single system, but add **withdrawal ordering kanbans** (WOKs) for each part type. The system has two loops. The production-ordering kanbans follow the same loop as before—between the output buffer, the collection box, the schedule board, the workstations, and back to the output buffer. The withdrawal ordering kanbans, which are sometimes referred to as **transportation kanbans** or **conveyance kanbans,** serve as material requisitions. They loop between the input buffer at stage $i + 1$, their own collection box at stage $i + 1$, the output buffer at stage i and finally back to the input buffer at stage $i + 1$. This works in the following manner. The material handler periodically checks the withdrawal kanban collection box. If kanbans are present, the material handler moves from stage $i + 1$ to the output buffer at stage i to obtain the parts requisitioned by the withdrawal-ordering kanbans. At stage i, the handler removes the production-ordering kanbans and places them on the schedule board. A withdrawal-ordering kanban is added to each container and they are moved to the input buffer at stage $i + 1$. The withdrawal-ordering kanbans are removed from the container only when that container's first part is placed into production at stage $i + 1$. The two loops therefore control production and transport.

A minimum transfer batch, q_i', for example, may be defined for stage i materials. When this number of withdrawal kanbans for part type i is accumulated, the handler re-

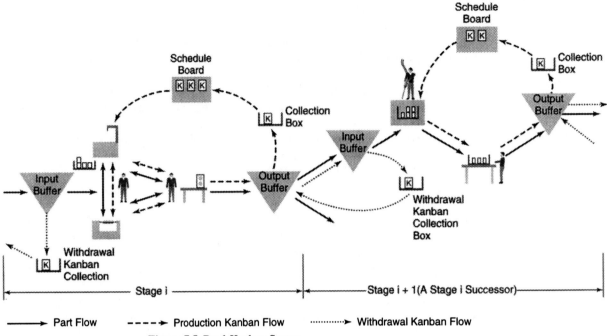

Figure 7.5 Dual Kanban System

turns to i to retrieve q_i' more containers of parts. Instead of a transfer batch, a periodic system may be implemented wherein at every fixed interval, such as twice a day, all the loose withdrawal kanbans are collected at stage $i + 1$, and the handler attempts to replenish the amounts of each raw material. Production-ordering kanbans remain in a single stage, recycling among the output buffer, the schedule board, and the processor(s). Withdrawal kanbans cycle among input buffers at stage $i + 1$, the stage $i + 1$ collection box, and the output buffer of stage i.

Single-kanban systems are used where the production facility is laid out such that material-handling distances are short, and frequent trips are feasible. Dual-kanban systems are useful when these distances are long, and it is not economical for the production worker to retrieve one container at a time. Whereas material transfer may be handled by line workers in a single-kanban system, dedicated material handlers would typically be used in a dual-kanban system to minimize lost production time while traveling.

We described the system as if it was serial, with stage i output going only to stage $i + 1$. In this case, it is often easy to place stages adjacent to one another and avoid separate input and output buffers. Dual-kanban systems are often used in more general systems. Stage $i + 1$ may have to collect raw material from several predecessors, and stage i may have its output used by multiple work centers. Successor work centers may not even be in the same facility. The number of production-ordering kanbans will depend on the replenishment lead time and total demand for a part. For a common component, this includes aggregate demand for the part among all its users. The number of withdrawal kanbans for a part will depend on the usage rate of that raw material at the work center and the material handling move time. Several work centers could have withdrawal kanbans for the same standard component.

7.2.4 Hybrid Single-Kanban Systems

In previous chapters, we saw how demand forecasts can filter through aggregate planning to create a master production schedule. In the next chapter, we describe material requirements planning, a technique for pushing the MPS into a plan for timed order releases for each part type. This may seem antithetic to our pull-oriented kanban approach. However, hybrid systems that incorporate both approaches exist. One common approach is a hybrid single-kanban system that uses push for order releases and pull for part conveyance. A production schedule is constructed, using MRP or other planning tools, to set the production goals for each department each day. The schedule lists the quantity of each item to be produced each day. The work center is charged with meeting this schedule over the course of the day but is given the flexibility to accomplish that task in the manner it chooses, provided its output buffer does not run out of any of its products. This allows the work center to choose to batch a full day's production for a particular part if that will save on setup. Production authorization is provided by the schedule, and production-ordering kanbans are eliminated. Withdrawal-ordering kanbans are still used to control the conveyance of parts between work centers. WOKs are used in the same manner as in the dual-kanban system to control input buffers, to smooth the flow of parts between work centers, and to inform workers where to find input parts and materials.

7.2.5 Rules for Scheduling Kanban Production

As kanbans are collected and moved to the schedule board, procedures are needed to convert these into a sequence of planned production jobs. Common rules for scheduling production include:

1. Single-Kanban Priority Rule. If setup time is brief, then production quantities can be set equal to the size of a container. The work center produces a batch size equal to the quantity specified on a single kanban. Any dispatching rule may be used for assigning priority to the kanbans pending production on the schedule board; however, **FCFS** (first-come, first-served) would be usually preferred to minimize both flow-time variability and maximum flow time. With FCFS, containers of parts are replaced in the same order in which the production-ordering kanbans are placed on the schedule board. The assumption here is that the time to process a container of each part type is similar, and thus a rule such as **SPT** (shortest processing time first) has little relevance. Likewise, we are assuming that the safety factor is similar for all part types and that container sizes are reasonably proportional to demand rates. Thus, due dates relate closely to time spent on the schedule board. If these conditions are not met, other rules may be used.

2. Family Rules. In some work areas, parts can be divided into families based on setup requirements. A family of parts may share the same fixtures or other tooling with quick changeovers being possible between parts in the same family. In this case, we fill all the kanban orders for parts in the same family at the same time. Any rule can be used to select the family that will run next if several families are waiting. Again, however, the most common approach is to select the family with the kanban that has been waiting the longest. Family rules are called "exhaustive" if we continue with the same family while any unfilled production kanbans exist. The rule is "non exhaustive" if we produce only for those kanbans available at the start of family production. The difference depends on whether we fill also those kanban orders that arrive while the family is being produced. Other setup-dependent rules are addressed below.

3. Minimum Economic Order Quantity. To justify a setup from an economic perspective, the operator may need to wait until a minimum number of kanbans of a particular type are on the schedule board. In this case, kanbans are first grouped. Then, one of the part types for which the number of kanbans exceeds the minimum production quantity is selected for processing. The selection may be based on order size, processing time, part priority, run-out time, or due date. With the run-out criterion, we select the part that is expected to run out soonest. Implementation requires selecting for production the part type with the minimum ratio of current stock of finished parts to average demand rate. Management must specify the minimal acceptable production quantity for each part type produced by the work center in accordance with standard lot-sizing decision rules and the amount of storage space available. The models in Chapter 6 are helpful here as well as the discussion in Section 7.3.2.

4. Signal Kanbans. Signal kanbans can be used when a minimum economic-order quantity is used. The minimum-order quantity would result from large setup cost or time. A signal kanban that authorizes production of the economic-order quantity is used for each part type. The reorder point is determined using standard inventory models as the lead-time demand plus safety-stock point. All units of a part type are stacked together in the output buffer (or tracked electronically). The signal kanban is placed at the reorder point level. When the inventory level drops to this level, the signal kanban is removed and used to initiate a replenishment order. When units arrive, the signal kanban is replaced at the reorder point. This system operates as a traditional "two-bin" system. The process is illustrated in Figure 7.6. One possible extension is to include a **material kanban** before

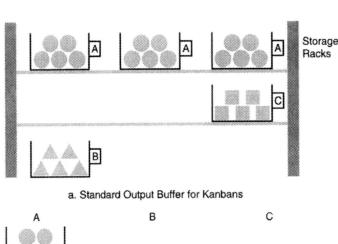

a. Standard Output Buffer for Kanbans

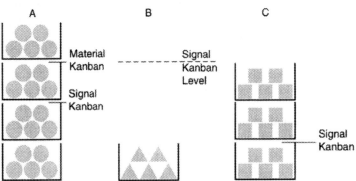

b. Signal Kanban Output Buffer

Figure 7.6 Output Buffer for Standard and Signal Kanban Systems

the signal kanban when a significant time is required to acquire raw materials. The material kanban allows the replenishment lead time used by the signal kanban to be shortened. When the inventory level drops to the level of the material kanban, the raw material used in making the item is ordered. However, actual production of the item will still wait until the signal kanban inventory level is reached. The number of units between the material and signal kanban should be sufficient to last during the lead time to acquire the raw material. Then, when the signal kanban is reached, production can begin immediately because the raw material has already been obtained. In this manner, if the production process has a high value added, then we can save the cost of carrying finished part safety stock due to variability in acquisition time for raw materials.

5. Cyclical Production–Continuous Time. If machine setups are dependent on the sequence of part types produced, part types may be produced in a fixed, repeating sequence, i.e., part A, then B, then C, and so on. The sequence is specified to minimize changeover times. Each time a part type completes processing, the kanbans on the schedule board for the next part type in the sequence are counted and production is set to this amount. On completion of each container, the parts are placed in the output buffer pending removal by the next stage. It is possible to set minimum and maximum production quantities for each part type to avoid excessive setups or tie ups of production capacity. If too few kanbans have been scheduled for a part, we wait until the next cycle. If too many kanbans have been scheduled, we may decide to hold some of these until the next cycle for replenishment.

6. Cyclical Production–Periodic Review. The periodic review uses the same sequence as the continuous review case but produces each part type once every fixed time interval. At the start of each period, the operator collects the kanbans on the schedule board. The quantity of each part type to be produced in the period depends on the number of kanbans collected and their container size. With periodic review operation, the workload is known at the start of the period. Thus, the rate of processing can be adjusted by adding workers or changing the processing speeds of the machines. This makes the system easy to administer, however, the system may be slower than the continuous review approach to adapt to changing production rates because we must wait a full period before producing the same part again. This will not be a problem, provided demand rates are relatively constant and periods are short.

EXAMPLE 7.1

A work center produces three part types. It just completed an order for part B. The set of kanbans that have been collected and are waiting for refill are shown in Table 7.1. If the work center operates on an FCFS basis, determine the next job to be started.

Table 7.1 Kanbans Available on Schedule Board

Part type	Quantity	Arrival time (day, hour, minute)
A	15	(121, 8:45)
B	15	(121, 9:10)
B	15	(121, 11:23)
C	10	(121, 13:23)
A	15	(121, 14:03)
C	10	(121, 14:15)
C	10	(121: 14:29)

SOLUTION

The oldest kanban on the schedule board is for 15 units of part type A. Thus, we will now produce this quantity.

EXAMPLE 7.2

Suppose that in Example 7.1, the workstation operates on a daily periodic review cycle. It is now day 122, and the kanbans on the schedule board are those that were available this morning. The cycle is to produce part type A, then B, and then C, each once per day. What is the next job?

SOLUTION

Because it was stated in the previous example that we were just finishing the run of part type B, the next job is to produce all type C parts on the schedule board. Thus, we start producing 30 type C parts.

7.2.6 Integration through the Supply Chain

Kanban systems are not restricted to internal suppliers. To operate effectively, members of the entire supply chain, from basic raw materials through finished products, must work together to ensure a continuous resupply of parts. This usually involves multiple plant locations and multiple companies. Typically, long-term contracts are established to ensure the supply of materials. The terms of the contract such as price, delivery frequency, and volume will be renegotiated periodically; however, the relationship endures. In this manner, both parties are interdependent, and it is in the best interest of both parties that a fair contract emerges whereby supplier and customer both profit and continue to exist. The long-term contract would usually guarantee the purchase of parts at a nominal daily rate ± some percentage such as 10%. This will not only allow the producer to adjust production to meet dynamic demand changes but will also minimize the schedule disruption to the supplier. Moreover, the supplier may store a target fixed quantity of parts at the customer's site. This material may still belong to the supplier and may not be billed until it is removed from the supplier's storage area located at the customer's plant. Frequent deliveries are made to replenish the consumed stock. In fact, if the supplier and customer are in close proximity, delivery trucks may be in constant motion between the sites for high-volume industries. An arriving driver will drop off the truck with parts and return to the supplier's location in a vehicle that has made a previous delivery. The vehicle that has just arrived will be emptied by receiving personnel at the consuming plant. When the driver returns to the home site, the returned vehicle is left to be restocked, and the driver immediately moves to a truck that has been restocked already and drives to a customer. Although a more complex cycle may be involved, the basic practice is shown in Figure 7.7. At any point, we may have one loaded vehicle in motion carrying parts to the customer, another vehicle in motion returning to the facility that produces the raw parts, yet another vehicle being filled at the production site, and a final vehicle being emptied at the customer location. Of course, the feasibility of this strategy depends on the total volume of

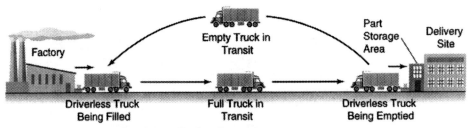

Figure 7.7 Integrating Supply into Kanban System

parts received from a supplier per time and the distance between supplier and customer. In Japan, where these strategies were first developed, dense industrial areas meant close proximity between supplier and customer. Although this is not always the case in the United States, production volumes tend to be larger, thus helping to make frequent deliveries feasible.

It has been argued that the kanban system simply pushes the inventory storage requirements back to the vendor. Note, however, that if the vendor is assured of a steady stream of demand, the kanban system can be cascaded all the way back to the original material source. The key is to keep material constantly flowing. The only reason to deviate from this system is the presence of high setup costs or unreliable processes at some stage.

7.2.7 Specifying Parameter Values

The kanban system operates without high-level coordination. The control parameters for the system are the container size and number of kanbans for each part type. In this section, we address methods for setting the values of these parameters.

7.2.7.1 Selection of Container Size

We begin our discussion of parameter specification by considering the number of units authorized by a kanban. We denote this as n_i for part type i. For many parts, the number authorized by a kanban has traditionally been set to the quantity that fits conveniently into a material-handling container. Because each kanban must be transported with its parts, this definition is easy to implement. Simply attach the kanban to the container indicating the number of parts authorized. One simple verification that the system is operating properly can be made from visually checking that each container has an accompanying kanban.

Inventory cost becomes a secondary concern when specifying the kanban quantity or container size. Inventory costs derive from the average inventory level, and inventory levels are usually small in kanban systems. The average buffer inventory level is determined by the maximum inventory level minus the expected number of units on order. The maximum inventory is $k_i \cdot n_i$ where k_i is the number of kanbans in the system for part type i. The units on order are those associated with kanbans currently in the replenishment loop. The replenishment lead time, and not the container size, determines the average number of units on order. As a result, the kanban authorization quantity can be based on material-handling considerations. The exception to this comes from the units that have already been removed from the output buffer but have not yet been put into production at the next stage. When the container is withdrawn, all the container's parts move to the next work center. This is not a problem if the entire container of parts is used at once to make a container of products at the next stage. However, if parts are entered into production at the successor work center at a slow, intermittent pace instead of by using a full container at a time, then, on average, there are $\dfrac{n_i - 1}{2}$ parts of type i pending use at the successor work center. If the container size is large, then we may have measurable inventory costs related to the container size. Nevertheless, in most cases, n_i is rather small, and the input buffer inventory resulting from the container size will not be substantial.

In many circumstances, several material-handling technologies are possible. For example, we might be able to ask the worker to carry one part at a time between operations, or to push a cart with a tub of parts between workstations, or to use a forklift to transport a pallet load. In each case, we have a maximum number of units that can be moved at a time using that method. For an annual demand rate of D_i and load size of n_i, we move

$\dfrac{D_i}{n_i}$ loads per year. Each material handling option has its own fixed cost per year, plus variable cost based on total distance traveled. Given the layout of the facility, we can measure the total distance traveled by a container from the beginning of the work center through processing and finally to the input point at the next stage. If we denote the technology options by j, then the cost model for selecting n_i becomes:

$$Minimize_j \left(c_{1ij} + c_{2ij} \cdot \frac{D_i}{n_i} \right) + \frac{h_i \cdot (n_i - 1)}{2},\qquad(7.1)$$

subject to:

$$n_i \leq N_j \qquad (7.2)$$

where c_{1ij} and c_{2ij} represent fixed cost per time and variable costs per move for part type i using technology j. N_j is the maximum load size for that material-handling technology. We have chosen to include in Equation (7.1) the holding cost of the partially used container of parts stored at the successor workstation. This is not significant in most cases; however, its presence in the equation provides a more complete model. In the problems at the end of the chapter, the reader is asked to show that if holding cost is not important, then Equation (7.1) is minimized by setting $n_i = N_j$ for some j. The part weight, size, shape, rigidity, and packaging all interact with the technology being used to determine N_j. Thus, we typically use the largest container size allowable by the chosen technology. With that observation, it becomes a simple matter of comparing the cost given in Equation (7.1) for each technology and selecting the cheapest option. A general solution can be found to Equations (7.1) and (7.2) for any technology j. By briefly ignoring the constraint, we can differentiate Equation (7.1) with respect to n_i to obtain:

$$\frac{\partial Cost}{\partial n_i} = \frac{-c_{2ij} \cdot D_i}{n_i^2} + \frac{h_i}{2},$$

and on setting this to zero (note that the second derivative is greater than zero),

$$n_{ij}^* = \sqrt{\frac{2c_{2ij}D_i}{h_i}}. \qquad (7.3)$$

Equation (7.3) should bring to mind the inventory expressions presented in Chapter 6. The problem of choosing between technologies resembles the price break decision. If the n_{ij}^* of Equation (7.3) is feasible for technology j, then it is a contender. If n_{ij}^* is too large, then our choice for technology j is to use N_j. Our choice, then, is from the best option with each technology. The final optimal container size is denoted by n_i^*.

EXAMPLE 7.3

The bottom of a printer body goes through three steps: molding, trimming, and detailing. Table 7.2 provides the annual fixed cost, cost per load, and maximum transport size for three handling options. Choose the container size and technology. The system will produce 200,000 printers per year. Annual inventory holding cost is $2 per unit of WIP.

Table 7.2   Material Handling Option Cost and Productivity Factors

Option	Annual cost	Cost per trip	Maximum load size
Manual Carry	$27,000	$0.15	2
Push Cart	$28,000	$0.16	20
Forklift	$50,000	$0.90	500

SOLUTION   The solution process involves finding the best container size for each technology, then comparing the costs of these options. Substituting data values into Equation (7.3) yields:

$$n_{ij}^* = \sqrt{\frac{2 \cdot c_{2ij} \cdot D_i}{h_i}} = \sqrt{\frac{2 \cdot c_{2ij} \cdot (200,000)}{2}} = 447.21\sqrt{c_{2ij}}.$$

Noting that batches move three times during their flow through the cell, $n_{i,manual}^* = (447.21) \cdot \sqrt{(3) \cdot (0.15)} = 300$, $n_{i,cart}^* = (447.21) \cdot \sqrt{(3) \cdot (0.16)} = 310$, and $n_{i,forklift}^* = 735$. In all three cases, we are constrained by the maximum load size. Thus, we compare the cost using these feasible load sizes as follows:

$$\text{Manual: } \$27,000 + \$0.15 \cdot \frac{(3)(200,000)}{2} = \$72,000$$

$$\text{Push Cart: } \$28,000 + \$0.16 \cdot \frac{(3)(200,000)}{20} = \$32,800$$

$$\text{Forklift: } \$50,000 + \$0.90 \cdot \frac{(3)(200,000)}{500} = \$51,080$$

The obvious choice is to use a push cart with containers of size $n_i^* = 20$.

7.2.7.2   Selecting the Number of Kanbans

The final control variable for each part type is the number of kanbans to keep active in the production system. We have noted that the maximum inventory for a part type i is given by the product of the number of kanbans, k_i, and the number of units authorized per kanban, n_i. When the first part is removed from a container in the workstation's output buffer, the kanban begins the replenishment process. Parts associated with that specific kanban will not be needed until $n_i \cdot k_i$ more units are requisitioned by the successor work centers. This follows because when the kanban goes into production, there are the remaining parts in the container plus $k_i - 1$ other containers of parts ahead of it in its efforts to rejoin the output buffer. This progression of kanbans and future customers is shown in Figure 7.8. Parts in storage and process are numbered to show how they will match

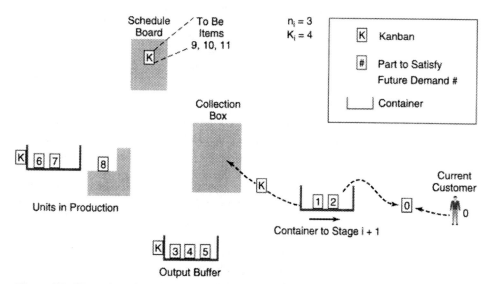

Figure 7.8 Illustration of Kanban Coverage: Matching Parts in the System to Future Demand for Four Kanbans with Three Units per Kanban

with future customer demand. To avoid shortages, this kanban must return to the output buffer with a full container of parts before $n_i \cdot k_i$ more units are demanded. However, parts are demanded at the rate D_i, and the expected lead time to replenish a container is τ_i. Shortages will be avoided on average then, if $n_i \cdot k_i \geq \tau_i \cdot D_i$. To handle randomness in demand and lead time, we add a safety factor of l. Combining these factors, we set the number of kanbans to the smallest integer, whereby:

$$k_i \geq \frac{\tau_i \cdot D_i(1 + l)}{n_i} \tag{7.4}$$

Essentially, we set the target inventory to equal the lead-time demand ($\tau_i \cdot D_i$) plus some safety factor.

EXAMPLE 7.4

A punch press forms a variety of sheet metal parts. A recent setup reduction team has succeeded in reducing setups to two minutes. It is now feasible to produce as few as six items at a time when the machine is set up to make a part. The six items can be stacked and moved as a single load. Demand for one item is contractually set at 75 units per day. With the small lot sizes, replenishment lead time is expected to be constant at two hours (0.2 days). Find the minimum number of kanbans needed to avoid shortages. How would your solution change if kanbans were collected once per day, and each part type was produced once per day in a quantity equal to the product of container size and the number of kanbans collected for that part type?

SOLUTION

To find a lower bound, we can assume that the variabilities of lead time and of demand are minimal and set $l = 0$. We then have $k = \left\lceil \dfrac{\tau \cdot D}{n} \right\rceil = \left\lceil \dfrac{(0.2) \cdot (75)}{6} \right\rceil = \lceil 2.5 \rceil = 3$.

Suppose that kanbans were collected and each part scheduled once per day. Lead-time demand now becomes equal to daily demand and $k = \left\lceil \dfrac{D}{n} \right\rceil = \left\lceil \dfrac{75}{6} \right\rceil = 13$. Clearly, the maximum possible inventory will be much larger in the second case. However, these extra kanbans would usually be on the schedule board waiting to be produced. Because they would not be associated with adding parts to WIP, why should we be concerned with their presence? The answer lies in the dynamics of how kanban systems adapt to change. Suppose that demand changes, and customers want other parts instead of the kanbans. With a kanban system, we will continue to refill the extra ten containers and store 13 full containers in the output buffer. While we are filling these containers, we are delayed from increasing the production rate of those items that are being demanded. Thus, backorders may accumulate as kanbans for the newly desired parts wait their turn on the schedule board.

We have presented the logic that the number of kanbans is set to ensure replenishment before lead-time demand is consumed. It seems natural, therefore, to measure lead-time demand. The simplest approach would be to track lead-time demand over time. This can be easily done by recording the timing of all withdrawals from the output buffer. When a kanban is removed from a container, the time is noted. When that kanban returns to the output buffer attached to a full container, we then count the number of units of that part type that have been removed from inventory during the lead time. These sample statistics can then be aggregated into a descriptive lead-time demand distribution. Although consecutive values may be correlated because of busy and slack periods for the processor, if the work center is producing a variety of parts, then the separation of time between consecutive production runs of any single part type should mitigate the effects of serial autocorrelation in the data values. If this is not the case, then we can sample only replenishments far enough apart in time to be practically uncorrelated.

EXAMPLE 7.5

Table 7.3 presents a sample record of withdrawals and replenishments for a part. Recorded times are days followed by a "/" and then the hour (measured in 1/100s of one hour). The workday is 10 hours. Each container holds six units of the part type. Estimate the lead-time demand distribution, then find the number of kanbans required.

Table 7.3 Record for Part Type 23A567

Withdrawal	Replenishment
14/8.45	15/1.15
15/0.85	15/9.80
15/8.95	16/2.20
16/7.10	16/9.30
17/3.60	17/6.70
18/5.75	18/9.10
19/1.35	19/4.15
14/9.37	15/3.60
15/6.20	16/0.50
16/2.35	16/8.10
17/0.10	17/3.90
17/8.55	18/0.25
18/7.20	19/1.45
—	

SOLUTION

We can compute lead times and estimate lead-time demand from the Table. The first lead time is 2.7 hours from hour 8.45 on day 14 until hour 1.15 on day 15. The 2.7 hours include the final $10.00 - 8.45 = 1.55$ hours on day 14 plus the first 1.15 hours on day 15. The second lead time includes 0.63 hours on day 14 plus 3.60 hours on day 15 for a total of 4.23 hours. Computing all 13 lead-time instances from the Table yields an average lead time of 3.88 hours with a standard deviation of 1.87 hours. The data appear to be approximately normally distributed. (We leave it to the reader to plot the lead-time demand values over time to check for correlation and to construct a histogram of lead-time demand values.)

From hour 8.45 on day 14 until hour 1.35 on day 19, 12 containers of six units each are consumed. At 10-hour days, this translates into 72 units consumed in 42.9 hours or 16.8 units per day. Converting both time and demand rate to hours, this yields an average lead-time demand of $\tau \cdot \overline{D} = (3.88) \cdot (1.68) = 6.5$ units. Assuming that the demand rate is fairly constant, the standard deviation of lead-time demand is $\tilde{\sigma}_{\tau D} = (1.87) \cdot (1.68) = 3.1$ units. Thus, for any desired level of confidence we can select the appropriate numerator in Equation (7.5). For example, to ensure a 95% chance that we will not incur a shortage, we must have $\tau \cdot D \cdot (1 + l) = 6.5 + 1.645 \cdot (3.1) = 11.6$ units. Because the number of kanbans must be an integer, this translates to $k = 2$ containers of six units each.

A key concern is how to determine the safety factor, l. Is 95% the correct distribution point to use as we did in Example 7.5? In a sense, we solved this problem earlier when addressing inventory models. A single-stage kanban system with continuous kanban collection and scheduling resembles a (Q, r) inventory policy for which the batch size

Q is set to the container size n. (The reader should also consider the analogy between a kanban system with periodic kanban collection and scheduling to a periodic review inventory model.) Recall that from Equation (6.31), for the continuous review inventory model with a fixed shortage cost for each unit short, the optimal reorder point satisfies:

$$F(r^*) = 1 - \frac{h \cdot n}{\pi \cdot D},$$

where r^* is the optimal reorder point, $f(x)$ is the lead-time demand distribution, h is the inventory holding cost per period, and π is the shortage cost per unit. In our kanban model, the reorder point for part type i, r_i corresponds to a drop in inventory position below the full buffer level of $n_i \cdot k_i$. As soon as a customer removes a container, we reorder. Thus, our policy is to maintain an inventory position of $r = n \cdot k$.

EXAMPLE 7.6

Using the lead-time demand distribution approximated in Example 7.5, find the optimal number of kanbans if each unit costs \$0.15 per day to store and shortages cost \$10 each.

SOLUTION

From the previous example, we have average daily demand of 16.8 units. Thus, $F(r^*) = 1 - \frac{(0.15) \cdot (6)}{(10) \cdot (16.8)} = 0.995$. For a lead-time demand distribution with a mean of 6.5 and a standard deviation of 3.1, the optimal reorder point is $r = 6.5 + \Phi^{-1}(0.995) \cdot (3.1) = 6.5 + (2.56) \cdot (3.1) = 14.4$ units, assuming a normal distribution for lead-time demand. (Your plot of the lead-time demands from Example 7.5 should look reasonably close to normal so that this modeling assumption is acceptable.) Because we must have $n \cdot k \geq r$, this would require three kanbans.

We have just applied a common engineering method to selecting the number of kanbans. We built an analytical model of the system and then optimized that model. In this case, the model was an economic model with decision parameters of lot size and reorder point. However, in doing so, it was easy to lose sight of the essential principle at work, specifically, **the amount of finished parts inventory required is driven by the variability of lead-time demand.** It is efficient to follow the path laid out above and to select the optimal parameter values based on the lead-time variability observed in practice. However, it would be more effective to reduce the variability. It is best to use the kanban system only in situations where total demand is relatively constant. However, coordinating the production schedule to achieve a smoothed production rate for each part type is also important. Suppose you are Toyota Motor Co. and you assemble two-door Corollas, four-door Corollas, Corolla hatchbacks, four-door Camrys, and Camry station wagons on the same production line. Some of the parts may be common to all the models, but others may differ. If we made two-door Corollas for a week and then switched to four-door Corollas for the next week, and so on, then the demand rate for the parts that are specific to an individual model would be very irregular. To avoid shortages during the period of high demand, we would be forced to carry large inventories (many containers and kanbans) the remainder of the time. In setting the number of kanbans, the average demand rate in this scenario is overshadowed by the large variability in the time between successive withdrawals. We could just accept this and use a large safety lead-time factor. However, it would be much better to smooth the production rate of individual models by rotating on a unit-by-unit basis, effectively making all models all the time. Thus, the order of production of individual vehicles may look like a two-door Corolla, followed by a Camry wagon, followed by a four-door Corolla, then a two-door Corolla, and then a four-door Camry, and so on. The production rate always matches customer demand, and the demand rate for parts varies only slightly over time as it evolves with the customer demand rate. The low variability is evidenced by a nearly constant interval between successive requests for a container of the same part type. In addition to reducing the variability of lead-time demand for individual parts and hence the number of kanbans needed, this staggering of production allows smoothing of the workload on the individual processors. Thus, if one workstation has a heavy workload for making wagon parts and is underutilized during the production of two-door Corollas, then we avoid periods of idleness followed by overwork. The smoothed production schedule is an effective way of improving workload profiles on workstations and inventory levels.

A variety of techniques for evaluating the performance of kanban systems along with models for establishing the number of kanbans have been published in recent years. Philipoom et al. [1987] identified total part demand, processing time variability, machine utilization, and autocorrelation in processing times as key factors for determining the number of kanbans required. Askin et al. [1993] provide guidelines for setting kanban levels in variable systems where shortage costs depend on the length of time parts are back ordered. Deleersnyder et al. [1989] and Mitra and Mitrani [1990] propose stochastic models for determining the number of kanbans required in a multistage serial system. Gunasekaran et al. [1990] review additional contributions from the literature on a variety of topics related to kanban production control.

7.2.7.3   Selection of Production Quantity

We have now selected the container size based on material-handling considerations and the number of kanbans based on the lead-time demand distribution. Pull systems usually invest in setup time reduction activities. However, in some cases, setup time will still be significant, and we may need to determine a minimum allowable production quantity. A signal kanban may be used in this case. In Chapter 6, we addressed the concept of internal and external setup. If capacity is limited at a work center, then we should produce a part long enough to cover the external setup of the next job. Let external setup time be denoted s_{2i}, and let unit processing time be t_i for part i. Assume part $i + 1$ will follow part i on the workstations. Then we require $Q_i \geq \dfrac{s_{2,i+1}}{t_i}$ for which Q_i is the lot size. This keeps the machine active while the next job is being prepared. Secondly, if internal setup time exists, then we may have to limit the number of setups per year to meet production goals. If s_{1i} is the internal setup time for item i, then we have the constraint:

$$\sum_i \left(s_{1i} \cdot \frac{D_i}{Q_i} + t_i \cdot D_i \right) \leq 1. \tag{7.5}$$

Let Q_i^0 be a smallest set of values that satisfies Equation (7.5). The production batch size is then set equal to the maximum of n_i^* (the container size found in the previous section), the minimum coverage of external setup, and Q_i^0, i.e.:

$$Q_i = Max\left\{ n_i^*, \left\lceil \frac{s_{2,i+1}}{t_i} \right\rceil, Q_i^0 \right\}. \tag{7.6}$$

Note that if the operator performs the setup and must attend to the machine while it is running, then we can consider all setup time to be internal. In a multioperation work center, we can evaluate Equation (7.6) for each workstation k individually and then select the largest of the n_i^* for any workstation in the cell.

Table 7.4   Operation Data for Example 7.7

Workstation	External setup (years)	Unit processing time (years)
Molding	0.0002	0.000003
Trimming	0.0003	0.000003
Detailing	0.0001	0.000003

EXAMPLE 7.7

Reviewing Example 7.3, suppose that several printer models are produced yielding a total of 200,000 per year. Internal setup time is 0.0001 years. For scheduling simplicity, the company would like to make the batch size the same as the container size and have it the same for all printer models. External setup and unit processing times for the three stages are given in Table 7.4. Find an appropriate container size.

SOLUTION

We must find the lower bound imposed by each of the three workstations based on material-handling technology, internal setup, and external setup. We recall from the previous example that $n_i^* = 20$, and this is true for any model. The external setup induced batch minimums are:

$$Q_i \geq \frac{0.0002}{0.000003} = 66.7, \; Q_i \geq \frac{0.0003}{0.000003} = 100, \; Q_i \geq \frac{0.0001}{0.000003} = 33.3,$$

respectively. Finally, if all batch sizes are the same, the internal setups require $(0.0001)\dfrac{200,000}{Q} +$ $(200,000) \cdot (0.000003) \leq 1$. The smallest feasible batch size is thus $Q = 50$ for setup time feasibility. The largest of these lower bounds is 100 as determined by the trimming operation. Thus, we set $Q = 100$. With a container size of 20, this means that whenever we set up to run a printer model, we must make 100 units—we wait until five containers or kanbans are available. Producing 100 at a time, we have $\dfrac{200,000}{100} = 2000$ setups per year, which will consume $(2000)(0.0001) = 0.2$ years.

Unit-processing time will consume $(200,000)(0.000003) = 0.6$ years. This yields a total utilization factor of 80% for the workstations.

7.2.8   Environmental Requirements for Kanban Systems

In kanban systems, parts are **pulled** through the facility because authorization to produce comes from downstream in the manufacturing process. The quantity of parts in the cell's output buffer dictates production priorities. As parts are pulled out of this stock by the successor work center, production is scheduled to replace these parts. No advance warning is usually provided until the parts are physically removed. If demand is fairly constant, then future demand for parts can be inferred, and this information does not need to be sent in advance. Instead, we set the number of kanbans whereby parts are replenished just-in-time (JIT) for the next timed requirement. Long-range production planning is still used, however, to set expectations for the rate of production. Actual production is then expected to fall within a range that is easily accommodated by the production facility, usually ±10%. Changes in inventory levels can handle small changes in demand, larger changes require adding or subtracting kanbans from the system.

Because of the limited information transfer, kanban systems are not designed for all production environments. Key environmental assumptions necessary for a kanban system to operate effectively include the following:

1. **Demand is approximately constant over a planning period.** Suppose that work center 1 makes two part types, A and B. Customers change and begin to purchase more

As and fewer Bs. The time between successive requests for part A becomes shorter than expected. The work center, not expecting this, has no available parts. The supply of A disappears, and the customer must wait. At this point, everyone rushes to expedite production of A. Meanwhile, full containers of part B sit in the output buffer. Hopefully, the system will correct itself quickly. We will stop producing part Bs if they are not requested. More capacity thus becomes available for As, and their cycle time is reduced. However, we still must wait the minimum cycle time to refill a container of part A, and if the demand rate over this minimum cycle exceeds the total number of units permitted, then perpetual backlogs occur. Eventually, the problem will be noticed, and more kanbans of part A are added, but in the meantime, supply problems will exist.

2. Small setup times. Although it is possible to set the minimum production quantity as some integer multiple of the container size, this requires batching of kanbans, which in turn increases lead time. Safety stock is proportional to lead-time demand variability, which generally increases with lead time. Cycle inventory increases linearly with batch size. Thus, large setup times can lead to large inventory levels. In addition to the standard disadvantages of large inventories, the lack of demand forecast information in most kanban systems accentuates the problem of mismatching production to demand. The potential exists for carrying large supplies of unneeded parts, whereas the hot items are back ordered and are waiting for available production capacity. Without advance warning of demand changes, kanban systems require small setup times to permit small production quantities and quick response to demand-rate changes.

3. Available, flexible capacity. Because inventory levels are kept to a minimum, and demand changes occur without warning, we must be able to replace parts and make minor adjustments to the relative production rates of the work center's part types rapidly. This requires that the processor be available on short notice. Machine breakdowns and quality problems become troublesome. Later in this section, we present an example of how the efficiency of a kanban system can be destroyed by unreliable processes. Two strategies are widely used here. First, extensive preventive maintenance is performed to reduce the variability of machine availability. Frequent short maintenance activities are scheduled to avoid long, unplanned breakdowns. Even if the total available time is not increased, the short down times improve performance of the kanban system by reducing replenishment lead-time variability. Second, workers are extensively cross-trained. If workstations, or even a work center, falls behind, additional help is assigned to speed up the workstation or work center's production rate temporarily. Overtime may also be used to extend the shift if a work center falls behind.

4. Disciplined workforce. The kanban system presumes that workers report withdrawals and wait for loose kanbans before producing. If a worker withdraws parts (or complete containers) from an output buffer without removing the kanban and requesting replenishment, then the replenishment process will be delayed, and eventually part shortages will result. This will cause workers to lose confidence in the system, and they will begin to produce without kanban authorization. Such a system soon leads to output buffer inventories that do not match needs. Excess levels of some parts are offset by shortages of others. Likewise, if pressure exists to keep utilization high, as is common in many systems with myopic productivity reporting and reward systems, workers will tend to overproduce unneeded parts. This creates excess inventory that cannot be used and delays production of the items that should be replenished.

A central theme in these environmental restrictions concerns the necessity for predictability and capability of production processes. Otherwise, kanban systems fail to operate smoothly and efficiently. The system must be able to respond in a timely manner to any request. This is achieved by limiting the range of production requests and ensuring consistency in production rate and yield. An example will help to emphasize the importance of process availability.

EXAMPLE 7.8

In a two-stage process, the first stage can produce 125 units per hour when operating. Unit processing time is fixed, but the workstation fails once every eight hours and is down for two hours. The second stage is reliable and can produce 80 units per hour. (Container size is five units.) Choose an appropriate number of kanbans for stage 1. Demand is relatively constant at 75 units per hour.

SOLUTION

Stage 1 is available eight of every 10 hours, or 80%. The long-run capacity of stage 1 is 125(0.8) = 100 units per hour. Capacity at stage 2 is 80 units per hour. Because both values exceed the demand rate, sufficient capacity exists to meet demand if an adequate buffer inventory is kept.

With a production rate that exceeds demand, stage 1 will tend to keep its output buffer full. However, the maximum lead time at stage 1 for a container of five units is two hours of down time plus $\dfrac{5 \text{ units}}{125 \dfrac{\text{units}}{\text{hour}}} = 0.04$ hours for production. Thus, the buffer must cover 2.04 hours worth of demand. With demand at 75 units per hour, this translates to 153 units, or 31 containers. The levels of production and inventory over time are shown in Figure 7.9b. Stage 1 will produce at the rate of 125 units per hour until all 31 kanbans are filled and in storage. At this point, the production rate will drop off to 75 units per hour to match the withdrawal rate. At eight hours, stage 1 goes down and stops producing for two hours. Stage 2 continues to withdraw units at the rate of 75 per hour. The buffer inventory drops to zero at 10 hours, and the process reinitializes.

The need for 31 kanbans sounds excessive, but suppose that we try to manage with a single container. For the first eight hours, a container is demanded every 5/75 hour or every four minutes. Stage 1 immediately sets about refilling the container and accomplishes this in 5/125 hours or 2.4 minutes. Thus, the container is idle for 1.6 minutes until a new withdrawal by stage 2. Conditions are stable during this period. Then, stage 1 goes down for cleaning. Stage 2 extracts its five units and after four minutes returns for a new container but finds the buffer empty. Stage 2 is then idle for two hours while waiting for stage 1 to return on line. The resulting inventory levels and production rates are shown in Figure 7.9c. During the downtime, demand backs up 150 units at the end of stage 2. Once stage 1 is available, the system produces at the rate of 80 units per hour for the next eight hours. Stage 1 could produce faster, but it is constrained by the rate of stage 2 and the limit of one container at most filled with parts in its output buffer. Over the next eight hours, the backlog is whittled down by $\left(5 \dfrac{\text{units}}{\text{hour}}\right) \cdot (8 \text{ hours}) = 40$ units. However, long-run system production is only 80(8) = 640 units every 10 hours. This average of 64 units per hour is insufficient to keep up with the external demand rate. The backorder level grows, eventually causing lost sales or the need for significant overtime.

EXAMPLE 7.9

Suppose that we could implement a preventive maintenance program whereby we would slow up the machine to produce at the rate of 110 units per hour and spend one half-hour for preventive maintenance after every two hours of operation. The preventive maintenance would prevent breakdowns.

SOLUTION

Initially, we appear to have lost capacity. The machine is still unavailable for two hours of every 10 hours while undergoing preventive maintenance. During the eight hours of operation, capacity has dropped from 125(8) = 1000 to 110(8) = 880 units. However, 880 still exceeds the demand of 75(10) = 750 units in a 10-hour period, and thus we should be able to meet demand. While operating, the workstation can outpace demand, and thus the output buffer will usually stay full. The maximum replenishment lead time is now one half-hour + 5/110 hour = 0.55 hours. Demand during this period is (0.55 hr)*(75 units/hr) = 41.3 units. At five units per container, we require only nine kanbans to cover maximum lead-time demand. This is a significant reduction over the 31 kanbans required in the original system.

a. Two Stage System

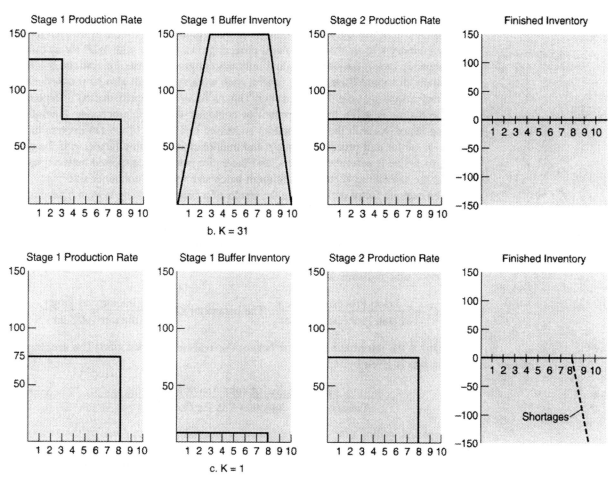

Figure 7.9 Effect of Unreliable Processes

These examples demonstrate the importance of processor availability for a kanban system. The lack of continuous availability of the processor caused a large increase in the number of kanbans required. At all times, we must carry enough kanbans to guard against the worst case of machine downtime.

7.2.9   Average Inventory Level

Because inventory costs are generally associated with average inventory levels, it is worthwhile to consider the average inventory in a kanban system. The inventory takes two forms: WIP in the form of parts passing through intermediate workstations in the cell and cell safety stock. The expected number of parts in the output buffer of the work center measures safety stock. If we had a genuine just-in-time system with no randomness, then we would schedule our container replenishments to complete processing and arrive at the output buffer just as the request for a container of parts arrived from the successor work center. In that case, the output buffer would always be empty. Nevertheless, WIP results from the flow of each unit through the system during processing. Although we can try to place bounds on this level of WIP, it is difficult to estimate precisely. For example, at a minimum, each unit remains at each workstation long enough to produce its container quantity. To the extent that queuing time exists at workstations, this time will be even longer. Queuing theory tells us that the time in process grows linearly with WIP for a constant utilization. In fact, if service times and intervals between requests for containers are exponentially distributed, the waiting time at each work center will also be proportional to the mean processing time for a container. Thus, a first-order approximation of replenishment lead time is given by the product of container size, unit processing time, and a queuing factor. As with the EMQ model addressed in Chapter 6, we can express this as $\tau = n \cdot p \cdot w$ for unit processing time p and multiplicative queuing factor, w. If the minimum batch size is several containers, and the entire batch is transported between operations at the same time, then, this minimum batch size replaces container size.

The queuing factor derives from the results of a general queuing system. Queuing approximations are often based on the first and second moments of arrival (a) and service (s) (processing time) distributions. This information is summarized in the coefficient of variation C, which is the ratio of standard deviation to mean for a process. Most models are then based on the square of this coefficient. With respect to processing of jobs, we use the service time measure:

$$C_s^2 = \frac{V(\text{Job Procesing Time})}{E(\text{Job Processing Time})^2} . \text{ The parameter } C_a^2 = \frac{V(\text{Interarrival Time})}{E(\text{Interarrival Time})^2}$$

summarizes the information on time between arrivals at a workstation. The average utilization rate is given by:

$$\rho = \frac{\text{Average No. of Jobs Arriving per Time}}{\text{Average No. of Jobs that Can Be Completed per Time}} .$$

The expected waiting time at a workstation for processing is then given by the general interarrival/general service time/single server queuing model, which is denoted by GI/G/1. An approximate estimate of expected waiting time in queue for this model is:

$$E(W_q) = E(\text{Job Processing Time}) \cdot \frac{\rho \cdot (1 + C_s^2) \cdot (C_a^2 + \rho^2 \cdot C_s^2)}{2 \cdot (1 - \rho) \cdot (1 + \rho^2 \cdot C_s^2)} . \tag{7.7}$$

We can make several useful observations from Equation (7.7). Note that the expected waiting time grows linearly with job-processing time at a fixed level of processor utilization and coefficients of variation for service and interarrival times. The service time distribution is determined by the set of parts produced and their operation requirements, processing speed, and container size. Accordingly, C_s^2 can be computed directly for each workstation from knowledge of its process technology and part container sizes. The interarrival time distribution is affected by preceding operations. When active, the service time at a workstation becomes the interarrival time at the successor workstation. When idle, the successor's arrivals are determined by the predecessor's remaining interarrival time after a container is completed plus the service time. In a serial system, the approximation:

$$C_{a_i}^2 = \rho_{i-1}^2 \cdot C_{s_{i-1}}^2 + (1 - \rho_{i-1}^2) \cdot C_{a_{i-1}}^2, \tag{7.8}$$

is often used to estimate the the squared coefficient of interarrival time at the ith workstation beginning with the second workstation. The exponential distribution, which has a squared coefficient of variation equal to 1, exhibits a relatively high level of variability. Thus, in practice, we would typically have C_a^2 and $C_s^2 \leq 1$. If both service and interarrival times are exponentially distributed, then Equation (7.7) reduces to $E(W_q) = E(\text{Processing Time}) \cdot \dfrac{\rho}{(1 - \rho)}$. We understand from Equation (7.7), that if C_a^2 and $C_s^2 < 1$, then we obtain a shorter waiting time. At the opposite end, consider a synchronous process in which the variance of processing time is zero, and dispatches to the cell are at fixed intervals, spread at least as far apart as the processing time. A paced assembly line would fit these conditions. In this case, both $C_s^2 = 0$ and $C_a^2 = 0$ and Equation (7.7) reduce to $E(W_q) = 0$.

To obtain total time at a workstation, we need to simply add the expected processing time to the expected waiting time. To estimate total WIP in a multistage cell, exclusive of the output buffer's safety stock, we can add expected waiting times from Equation (7.7) over all workstations. Safety stock exists in the form of containers pending extraction from the buffer. Average safety stock for a part type equals maximum inventory minus the expected number of units in the process of being replenished. Algebraically, this is $k_i \cdot n_i - \tau \cdot D_i$.

EXAMPLE 7.10

A work cell operating under kanban control produces an average of 50 containers of parts per eight-hour day. Each container holds 10 parts. On average, it takes six minutes to fill a container with parts, but part types vary, and container processing time has a variance of nine minutes². Containers are removed, seemingly at random, and an exponential distribution has been found to be a reasonable fit to the data for time between container withdrawals from the output buffer. Management prefers a 10% safety factor on inventory lead time. Find the utilization, lead time, and average inventory level of the work center. The time to collect kanbans and move parts within the work center is minimal.

SOLUTION

First, consider utilization. The work center is active for an average of (50 containers) (six min/container) or 300 minutes per day. In an eight-hour day, this yields a utilization of 0.625. The squared coefficient of variation for service is $C_s^2 = \dfrac{9}{6^2} = 0.25$, and the exponential interarrival time of kanbans makes $C_a^2 = 1$. Ignoring the minimal kanban collection and container transport times, lead time and flow time are both equal to waiting time plus processing time. From Equation (7.7),

$$\text{Flow Time} = E(\text{Proc. Time}) \cdot \left[1 + \frac{\rho \cdot (1 + C_s^2) \cdot (C_a^2 + \rho^2 \cdot C_s^2)}{2 \cdot (1 - \rho) \cdot (1 + \rho^2 \cdot C_s^2)} \right]$$

$$= 6 \cdot \left[1 + \frac{(0.625)(1.25) \cdot (1.098)}{2(0.375) \cdot (1.098)} \right] = 12.25 \text{ min.}$$

Finally, consider average inventory. The average number of parts in process at any minute is given by the total time spent in process by all parts divided by the number of minutes per day or

$E(\text{number of parts}) = (\text{parts produced per day})\, E(\text{lead time})$

$$= \frac{\left(500\, \dfrac{\text{parts}}{\text{day}} \right) \cdot (12.25 \text{ min})}{\left(480\, \dfrac{\text{min}}{\text{day}} \right)} = 12.76 \text{ parts (or 1.276 containers).}$$

If a 10% safety lead-time factor is added, then each part sits in the output buffer an average of 1.225 minutes for a total of 1.225 (500/480) part-days or 1.276 parts in the buffer on average.

Thus, total average inventory is 12.76 parts in process, plus 1.276 parts in the buffer, or 14 parts (1.4 containers).

EXAMPLE 7.11

A product passes through two workstations in a cell. The first workstation is the bottleneck. Requests to refill a container arrive 70 minutes apart on average with a standard deviation of 20 minutes. The second workstation is active only approximately 40% of the time and is always available to immediately work on the output of the first workstation. Processing time for a container is always 60 minutes at the bottleneck and 30 minutes at the second workstation. Finally, assume a 10% safety lead-time factor is used. Estimate average inventory for the cell.

SOLUTION

To compute total cell inventory, we must find the average number of containers at workstation 1, plus those at workstation 2, plus those in the output buffer. We begin with WIP at workstation 1. Workstation 1 has a utilization of $\rho_1 = \dfrac{60}{70}$. For arrivals,

$$C_{a_1}^2 = \frac{\sigma_{a_1}^2}{\mu_{a_1}^2} = \frac{20^2}{70^2} = 0.0816$$

where μ_a is the mean interarrival time. If processing time is always 60 minutes, $C_{s_1}^2 = 0$. Then, from Equation (7.7) at workstation 1,

$$E(W_q) = E(\text{Job Processing Time}) \cdot \frac{\rho \cdot (1 + C_s^2) \cdot (C_a^2 + \rho^2 \cdot C_s^2)}{2 \cdot (1 - \rho) \cdot (1 + \rho^2 \cdot C_s^2)}$$

$$= (60) \cdot \frac{(0.857) \cdot (0.082)}{2(1 - 0.857)} = 14.68 \text{ min.}$$

Total container time at workstation 1, then averages 74.68 minutes including service. Using Little's law, and noting that we produce one container every 70 minutes, the average number of containers at workstation 1 is $N_1 = \left(\dfrac{1}{70}\right) \cdot (74.68) = 1.07$ containers. For workstation 2, we need to estimate the interarrival time distribution. This is provided in Equation (7.8) as:

$$C_{a_2}^2 = \rho_1^2 \cdot C_{s_1}^2 + (1 - \rho_1^2) \cdot C_{a_1}^2 = (0.857)^2 \cdot 0 + (1 - 0.857^2) \cdot (0.0816) = 0.0217.$$

Applying Equation (7.7) to workstation 2,

$$E(W_q) = E(\text{Job Processing Time}) \cdot \frac{\rho \cdot (1 + C_s^2) \cdot (C_a^2 + \rho^2 \cdot C_s^2)}{2 \cdot (1 - \rho) \cdot (1 + \rho^2 \cdot C_s^2)}$$

$$= (30) \cdot \frac{(0.429) \cdot (0.0217)}{2(1 - 0.429)} = 0.245 \text{ min.}$$

Adding the 30 minutes of processing time per container and applying Little's Law,

$$N_2 = \left(\frac{1}{70}\right) \cdot (30.245) = 0.432 \text{ containers.}$$

Workstation 2 has the same production rate as workstation 1 attributable to conservation of matter. Each container moves through workstation 1 initially and then goes through workstation 2. Total lead time is $74.68 + 30.25 = 104.93$ minutes. Demand over this period is $\dfrac{104.93}{70} = 1.5$ containers.

With a 10% safety factor based on lead-time demand, this would yield an average buffer inventory of 0.15 containers. The total average inventory in the workstation is thus $1.07 + 0.43 + 0.15 = 1.65$ containers.

To summarize what we have learned in this section, note that the number of kanbans required is based on the expected lead-time demand plus the desired safety lead time. Average output buffer inventory depends on the desired safety time factor. The WIP, moving through the workstations in the work center, is based on container size, interarrival time of requisitions for containers of parts, workstation service time distributions, and workstation utilization.

7.2.10 Dynamic Management of WIP Levels

Kanban systems are designed for stable demand environments. However, the number of kanbans that are active in the system determines service and inventory levels. Thus, it is possible to use a dynamic kanban control strategy in production environments with seasonal demand or even make-to-order environments. Consider, first, the case of seasonal demand for an item. To facilitate a discussion of how the dynamic kanban strategy operates, consider the situation in which the work center makes a variety of products, and its total utilization is fairly constant over time. Perhaps we are assembling lawn mowers in the winter and spring (for spring and summer sales) followed by leaf blowers in the summer and snow blowers in the fall. We have noted in Section 7.2.6.2 that for a given lead time the number of kanbans needed is proportional to the item's demand rate. Thus, if demand forecasts are available, we can adjust the number of kanbans over time to satisfy the maximum inventory ≈ lead-time demand relation. When selecting the demand rate in a highly volatile situation, we should use the demand forecast for a period into the future equal to the expected lead time. Thus, at any point, we have a desired number of kanbans based on the demand forecast for the point in the future equal to current time plus lead time. At times, we may have extra kanbans and filled containers in the system. Once units are produced, we will not discard them. Instead, whenever a container of parts is removed and a kanban becomes free, we first check to see if that kanban should be kept in the system before placing it on the collection bin. If the number of kanbans currently in the system is greater than the desired number of kanbans, we remove that kanban instead of returning it to the schedule board. When the number of kanbans in the system for a particular part type drops below the desired number [Equation (7.4)], we simply enter more kanbans to the work center by adding the requisite number to the collection bin. Adding these kanbans will automatically schedule production as capacity becomes available. The resulting number of kanbans active in the system at any time is proportional to the current demand rate for the item. A similar strategy can be used if demand is dynamic, but no particular cyclical pattern exists. At any point, either with periodic or continuous review, we adjust the number of kanbans in the system to match the current demand rate. In implementing these policies, it makes sense to use a safety factor that guards against the adjustment time as well as randomness in demand. Thus, the safety factor, l, would usually be larger than for the constant demand case.

EXAMPLE 7.12

A company offers a constantly changing portfolio of consumer electronics. Audio equipment represents one line of products. A key circuit card that is used in speakers is produced in a work cell. Because of changing technology and various product price levels, three versions of the card are currently produced. Considering capacity and demand, the aggregate production plan for the three models over the next ten weeks is shown in Table 7.5. Containers hold five cards of the same type. Normal replenishment lead time is three days (half-week), and the company has found that a 20% safety factor is sufficient for its kanban operations. Determine a dynamic control strategy for the number of kanbans required of each type.

Table 7.5 Model Demand for the Next 10 Weeks

	Week									
Model	1	2	3	4	5	6	7	8	9	10
A	10	20	30	40	50	50	50	60	70	80
B	40	40	40	40	40	50	50	50	50	50
C	100	90	80	70	60	50	50	40	30	20

SOLUTION

First note that total planned production stays constant over the horizon. Thus, it seems reasonable to assume that lead times will be constant. Using Equation (7.5) and letting the subscript i represent the model and t denote the week:

$$k_{it} = \left[\frac{\tau_i D_{it}(1 + l)}{n_i} \right] = \left[\frac{(0.5) \cdot D_{it} \cdot (1.2)}{5} \right] = [0.12 \cdot D_{it}].$$

For model A in week 1, this translates to $k_{A1} = [0.12 \cdot (10)] = [1.2] = 2$. Similarly,

$$k_{B1} = [1.2(40)] = 5 \text{ and } k_{A2} = [1.2(20)] = 3.$$

Completing the entire Table in this manner, the schedule for the number of active kanbans per week becomes:

Number of Active Kanbans by Week

	Week									
Model	1	2	3	4	5	6	7	8	9	10
A	2	3	4	5	6	6	6	7	8	9
B	5	5	5	5	5	6	6	6	6	6
C	11	10	9	8	7	6	6	5	4	3

To implement the strategy, we can check the number of active versus desired kanbans for each product at the start of each week. If more kanbans are desired, we add the extra kanbans to the schedule board. Thus, for model A, we start with two kanbans and add one each week for four weeks. Another kanban is added at the start of weeks 8, 9, and 10. If fewer kanbans are desired, then we remove kanbans from the system as parts are withdrawn instead of returning them to the schedule board for replenishment. For part type C, we remove one kanban each week up through week 6.

7.2.11 Defining the System Control Points

The container size and number of kanbans regulate the inventory level at each stage of the production system. In some cases, the stages (work centers) are easily defined by a natural set of production activities or bill of materials. If component parts are fabricated in one area and then stored, that forms a stage. On the other hand, if part fabrication requires parts to visit several geographically dispersed machining centers with interspersed heat treating and inspections followed by painting, we might prefer to establish intermediate control points. We may also choose to define control points for strategic reasons. For example, suppose that we are manufacturing sliding closet doors for residential use. The door consists of a wood frame onto which we add oak, pine, or mahogany facing. To begin the process, the raw lumber is withdrawn from storage and flows through sawing, sanding, framing, and gluing operations. At the same time, thin sheets of facing material are sliced, interlaced, and glued together to form the front and back of the door. Holes are then punched (to eventually hold the doorknob). The facing is then added to the frame, the door is stained to the desired finish, and varnish is applied followed by accessorizing operations, which include inserting the doorknob and adding top and bottom runners. Suppose that doors are produced in lots of size 10. One option is to stock each of the three finished frame types as final products. When the output buffer drops, we schedule the replacement of 10 doors with that finish. Alternatively, we could divide this production process into two stages. The first stage produces raw door frames by sawing, sanding, and gluing. Door frames are stored in this form. The next stage withdraws door frames and proceeds to add the proper facing material, stain and accessorize as needed. A general depiction of one versus two work centers is shown in Figure 7.10a, b. The two-stage approach adds a second controlled storage point. In exchange for the extra WIP inventory point, we gain a reduction in cycle time to replace doors because semifinished doors are usually available, and fewer steps are needed for completion.

The choice of control points and stage definition depends on product variety, storage costs, and cycle time requirements. A common approach sets control points immediately before the bottleneck operation and also as far down the operation sequence as we can go before the product differentiates into many models. Thus, if the facing was the same for all models, and doors became different only after staining, then it might be appropriate to end a stage with a control point of faced but unstained doors. This allows storage of fewer in-process products but still allows for rapid customization to replenish demand for fin-

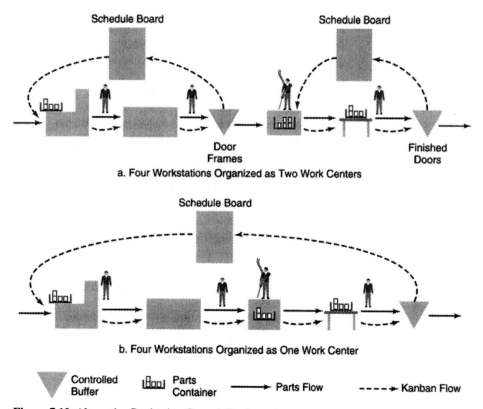

Figure 7.10 Alternative Production Control Configurations

ished products. Because WIP control points (or stages) serve to decouple and insulate the production process, and variability in lead time increases the number of kanbans and inventory required, it is also advisable to designate a highly variable process as a stage.[1] The variability could result from occasional breakdowns in the process or complexity of some models that require long processing times. In this manner, we can avoid depriving the next operation. Suppose, for example, that the staining tanks and spray nozzles have adequate capacity but need to be cleaned occasionally causing several hours of down-time. If a supply of stained doors is maintained (kanbans are defined and storage buffers are located at this point in the production system), then the final production can proceed in accord with demand. The safety factor in Equation (7.4) is set to allow replenishment before shortages occur even when it is affected by down-time.

7.3 CONWIP: A CONSTANT WORK-IN-PROCESS PULL ALTERNATIVE

The simplicity of kanban systems stems from their reliance on a fixed-target inventory level. If we consider the output buffer inventory as part of the WIP, we can view the kanban system as maintaining a constant level of total WIP for each part type. Suppose that we aggregate part types into a single item and aggregate all production areas into a single facility. Parts being produced, and those in the output buffers, count without regard to specific part type. We will control production and inventory levels by releasing jobs to the shop to maintain a level of N jobs in the system at all times. We call this **constant work-in-process system, CONWIP.** To operate the CONWIP system, we maintain a backlog list of parts that need to be produced, either because their inventory level is low, or we anticipate an increase in demand. As soon as one batch of parts is completed and leaves the system, the next batch on the backlog list for which all raw material is available is dispatched to the shop-floor. For communication of production authorization, we need to only have a mechanism for linking part usage with order release. When a job leaves the shop, a message is sent to release the next order. Note that it is relatively easy to keep track of the amount of WIP in the shop. In CONWIP, there is no need to worry about the internal distribution of work on the shop-floor. We simply keep a count of the number of jobs in process, adding one when the job is dispatched and decreasing the count by one when a job is removed from the output buffer.

With the kanban system, we maintained a set of kanbans for each part type. We defined a work center to include the resources needed to convert one or more basic materials or components into a higher level component or assembly. We envisioned a series or network of work centers that were linked through requisitions for parts but were separately controlled. Although we could impose a CONWIP limit on each of these same work centers, it is not necessary. By maintaining an overall backlog list, we can treat the entire production facility as a single work center when controlling WIP. Thus, in Figure 7.1, our depiction of CONWIP resembles a single work center and we control only the total inventory in the shop.

Spearman et al. [1990] describe the advantages of the CONWIP control strategy. CONWIP production control offers the advantages of simplicity of operation and simplicity of analysis. Kanban systems also control WIP, but the limited buffer capacities at each workstation make it more difficult to estimate production rate for the system. Kan-

---

[1] We should remember that the best choice is to reduce the causes of variability. However, this may not always be economically or technologically possible.

ban systems also require maintaining a set of cards and inventory for each part type produced by the system. This is neither simple nor efficient when individual part type demands are highly variable through time. This requirement is eliminated with CONWIP. Unlike kanban systems, the CONWIP system can immediately and automatically adapt to unforeseen changes in part mix. If parts are not needed in the near future, then the part type will be the backlog list, and batches will not be produced. Thus, CONWIP can be used in an environment of custom jobs or where part production is intermittent. Typical kanban systems would make and store unneeded items in these environments. CONWIP shares with kanban the ability to match production with actual demand. Both systems address the common tendency to want to see machines in use. That tendency leads to clutter and the premature production of unneeded products. By controlling WIP, machines with excess capacity will be forced to be idle part of the time.

On the downside, the CONWIP system requires a mechanism for maintaining the prioritized backlog list of jobs that are pending release. In addition, CONWIP requires ample storage space at each workstation. We must be able to accommodate all N jobs at any workstation since we do not control the location of WIP within the work center. In theory, therefore, each workstation must have storage space for all the jobs in process. In practice, the jobs will generally be distributed in a predictable manner with larger WIP levels at the bottleneck workstations. WIP will naturally tend to collect in front of the bottleneck resource(s).

The trick in implementing this CONWIP strategy, is to find the WIP level that produces the average demand rate as depicted in Figure 7.11. We maintain N jobs in the system. The objective is to find the value N, as shown in the Figure, that sets the average production rate equal to the average demand rate. Methods for estimating this relationship are covered in the next section. In Chapter 1 we explained that a primary reason for WIP was to ensure a continual input stream to the bottleneck (Figure 1.15). A simple approach to setting the control parameter N is to gradually increase N until deprivation is virtually eliminated at the bottleneck (the input queue is rarely empty). This is the efficient WIP level. Additional increases will be primarily translated to increased cycle time with minimal increase in throughput.

By maintaining a constant level of WIP, production rate per time and batch flow time remain relatively constant. Both the mean and variance of cycle time will usually be less with a CONWIP strategy than for a system with varying WIP levels that produces an

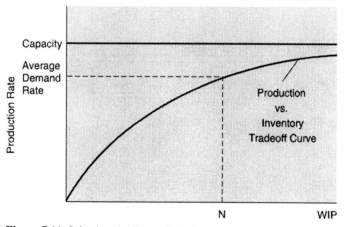

Figure 7.11 Selecting the Target WIP Level

equivalent long-run throughput. The lower variability in cycle time allows more accurate quotation of delivery dates to customers and purchasing of materials. The lower mean WIP level (tied by Little's Law to lower mean cycle time) means less space requirement and less congestion on the shop-floor. The constant pace of work also avoids employee burn-out and periods of underutilization followed by overtime.

Consider how a CONWIP system reacts to demand changes. If demand increases, as evidenced by rapid withdrawal of completed parts from the output buffer, then new jobs are dispatched to the shop faster, and the production rate increases. If demand drops, then completed jobs sit in the output buffer, and fewer new jobs are released to the shop-floor, slowing down the production rate. As with standard kanban, this is precisely the result we desire. We constantly produce at a rate equal to the current demand rate. There is a potential disadvantage to CONWIP operation. Over time, as demand changes, we can develop a core of unusable inventory in the output buffer. Orders were released with the expectation that demand would acquire the products, but the customers' preferences varied between our forecast and their consumption. These jobs will count toward N and thus impede the release of other jobs. Therefore, if we follow the CONWIP rules, we may decide that obsolete inventory prevents us from releasing important jobs to the shop. This can be handled by periodically assessing the value (and age) of inventory and discarding extraneous parts. In addition, we realize that our ability to meet demand will be only as good as the validity of our forecasted backlog priority list.

7.3.1 Prioritizing Order Releases

We have not yet addressed the construction of the backlog list. The backlog list controls the sequence of job releases to the shop. One option is to operate the shop as a large, single work center kanban system. Target output buffer levels for each part type are selected. The backlog list captures the difference between current on-hand and target levels. When a container of parts is consumed or a signal kanban level is reached, a replenishment order is added to the list for that part type.

EXAMPLE 7.13

A manufacturing system produces eight part types. The system is operated using CONWIP, with a target level based on total WIP on the shop-floor and in the output queue. Current inventory status and processing characteristics are shown in Table 7.6. The maximum target inventory levels shown in the Table are based on historical estimates of lead-time demand. The foreman has learned from experience that when the shop gets ahead of demand whereby 150 hours of total production is in finished inventory, it is time to reduce production. Using these data, form the backlog list.

Table 7.6 Current Shop Status

Part type	Batch size	Target inventory	Current units in inventory	Current units in process	Production hours/batch
1	50	100	95	50	3.5
2	50	200	240	0	8.4
3	100	500	378	100	9.3
4	50	50	23	50	6.6
5	100	300	268	0	8.1
6	25	75	47	25	1.5
7	25	50	30	0	11.5
8	50	100	84	50	2.2

SOLUTION   As a check on the target inventory levels and the foreman's observation, we realize that the targets call for two batches of part type 1, four batches of part type 2, five batches of part type 3, and so on up to two batches of part type 8. Multiplying these batches by their processing time yields target inventory = $2(3.5) + 4(8.4) + \ldots + 2(2.2) = 149.9$ hours. Current on-hand inventory in the output buffer corresponds to:

$$\text{Current Inventory} = \left(\frac{95}{50}\right)3.5 + \left(\frac{240}{50}\right)8.4 + \left(\frac{378}{100}\right)9.3 + \ldots + \left(\frac{84}{50}\right)2.2$$

$$= 127.2 \text{ production hours.}$$

Current WIP on the shop-floor corresponds to:

$$\text{Current WIP} = 3.5 + 0 + 9.3 + 6.6 + 0 + 1.5 + 0 + 2.2 = 23.1 \text{ hours or } 150.3 \text{ hours total.}$$

The backlog list contains batches for items with on-hand + in-process quantities that are less than target for that item. With 95 units on hand and 50 units in process, part type 1 does not qualify. Part type 2 does not qualify either because the 240 units on hand exceed the target of 200. As we go down the list, we find that items 3, 5, 6, and 7 qualify for the backlog list. As soon as demand removes parts from the output buffer that represent more than 0.3 hours of production, the excess of the 150.3 hours in WIP and the 150-hour target, it will be time to release a batch of one of these four part types.

Instead of maintaining fixed target inventory levels for each part, we can use demand forecasts and plan ahead. The backlog list is constructed from parts expected to be needed in the near future. Essentially, we are combining a **pull** order release rule with a **push** order prioritization scheme. Based on expected time of need, due dates are assigned to jobs. An earliest due date rule can be used for prioritizing the backlog whereby the next job dispatched to the shop is the job needed soonest. If flow times differ for jobs based on their routings, then target release dates can be set based on due date minus expected flow time. Jobs are then released based on an "earliest target release date goes first" basis. In the next section, we describe a method for estimating replenishment lead time. For now, assuming these estimates are known, we can set the target release date as:

$$\text{Release Date} = \text{Due Date} - E(\text{Flow Time}) \qquad (7.9)$$

EXAMPLE 7.14   A forecast of finished product demand combined with current inventory status records indicate the need to produce seven batches of parts in the forseeable future. Table 7.7 contains the part types and the work centers in the plant that each part type visits. Historical data indicate that throughput time averages two days for work center 2 and five days for work center 4. All other work centers have a one-day lead time. The facility operates under a CONWIP release rule. Form the backlog dispatch list for the cell.

Table 7.7   Anticipated Part Needs

Part number	Work centers visited	Due date
A2387	3, 4, 1	10
B3264	2, 5, 6, 3	6
A1119	3, 4, 1	8
D9871	1, 3, 5	7
R3481	1, 3, 7, 4, 5	12
M2220	1, 3, 5	4
F3450	1, 2, 3, 7, 4	7

SOLUTION

The first step involves finding the desired release date for each job. These dates will determine release priorities. Desired release date for job j is given by:

$$R_j = d_j - \sum_{k=1}^{K_j} C_{O(kj)}$$

where d_j is the due date for job j, $O(kj)$ is the work center used for the kth operation of job j; K_j is the number of operations for job j, and $C_{O(kj)}$ is the expected cycle time for the subscripted work center. Following this rule, we have $R_1 = d_1 - C_3 - C_4 - C_1 = 10 - 1 - 5 - 1 = 3$. Likewise, $R_2 = d_2 - C_2 - C_5 - C_6 - C_3 = 6 - 2 - 1 - 1 - 1 = 1$. Computing the other desired release dates yields $R_3 = 1$, $R_4 = 4$, $R_5 = 3$, $R_6 = 1$, and $R_7 = -3$. The seventh job, part type F3450, appears to be three days late already and should be released first. The second highest priority should go to jobs 2, 3, and 6. Jobs 1 and 5 are next. Job 4 has the lowest priority and is last on the backlog list. As batches of completed parts are removed from the system, jobs should be dispatched to the shop-floor in this order: 7, 2, 3, 6, 1, 5, and 4.

When final products are assembled from multiple parts, we have a choice in constructing the backlog list. We can ignore the dependent nature of part relationships and, as before, use either a target output buffer or demand forecast approach for each item separately. Demand forecasts can be obtained for component parts by looking at demand for the final products that use that part type (we explore this approach in more detail in the next chapter). We then prioritize and list orders for each part type on the backlog list separately and release them when their turn comes. Alternatively, if we know the set of component parts in each finished product, we would list finished items only on the backlog list. Planned releases of finished products link a coordinated release of each constituent component part. If different components have different flow times, we could make use of this information and release the component parts whereby they would arrive at final assembly approximately the same time.

7.3.2 Performance Evaluation of CONWIP Systems

In this section, we describe a technique for modeling the performance of CONWIP systems. The method is intended to provide a rough estimate of performance with only minimal data requirements. The model is based on basic results from queuing theory. Details on part routes and scheduling rules at work centers are not required. This makes the model easy to develop, but it limits the accuracy of the model and its ability to evaluate the effect of many policy changes. If more detail is available, then more accurate queuing models can be developed, or computer simulation can be used.

We begin by computing upper bounds on the throughput of a CONWIP system. For the CONWIP system, N jobs circulate in process at all times. M denotes the number of work centers. The average batch processing time at workstation j is t_j. If parts visit all M work centers, then a lower bound on the cycle time through the system is given by the cumulative processing time at all stations[2],

$$t_p = \sum_{j=1}^{M} t_j.$$

This lower bound would be achieved only in an ideal situation in which a job has the advantage of never having to wait for a machine at any stage. Balanced assembly lines

---

[2]This assumes that the job can be in process on at most one workstation at a time. If, simultaneous processing on several workstations is possible, then the t_j values need to be corrected to reflect this capability.

may come close to this ideal, but queuing is a fact of life for most systems. We know from Little's Law, that in general,

$$\text{Production Rate} = \frac{\text{Ave. No. of Jobs in System}}{\text{Ave. Time in System}}.$$

Thus, if we set the value of N, the lower bound on average time in the system leads to an upper bound on production rate, X_p.

This is given by $X_p \leq \dfrac{N}{t_p}$. This result applies to any system in steady-state with N jobs in process on average and average job throughput time of at least t_p.

The bound in the previous paragraph assumed that jobs never waited. If N exceeds the number of processors, then we know that this cannot be the case. Suppose that c_j is the number of jobs that can be served simultaneously at j. If work center j has a single workstation, and the workstation can only work on one unit at a time, then $c_j = 1$. In other cases, the work center may have multiple servers or be able to work on several jobs at a time. An oven, for example, may be able to bake multiple jobs simultaneously and thus c_j would be the oven capacity multiplied by the number of ovens. The maximum number of jobs being worked on at any time is thus $C = \sum_{j=1}^{M} c_j$. If $N > C$, then the maximum production rate is still limited by the fact that at any one time no more than C jobs are being worked on. The other jobs are idle. Adding this fact, we now have the improved bound:

$$X_p \leq \frac{\min(C, N)}{t_p}. \tag{7.10}$$

For the remainder of this section, we will assume work centers and workstations are synonymous, and there is at most one job only in process at a work center at a time. Thus, $C = M$.

The bound provided in Equation (7.10) is optimistic and may not be an accurate estimate of the production rate in many cases. However, we can use a similar logical argument to estimate the expected production rate. Think of yourself as one of the N jobs in the system. As you look around, you see $N - 1$ other jobs spread across the M workstations (and possibly the output buffer). If the system is balanced but random, then we expect to see $(N - 1)/M$ jobs at each workstation on average. Then, when you arrive at a workstation you see an average of $(N - 1)/M$ jobs ahead of you. If processing times are exponentially distributed, then, from the memoryless property of the exponential, the expected processing time remaining for the job currently being produced at workstation j is still t_j. In this case, the waiting time at workstation j should average $\left(\dfrac{N - 1}{M}\right) \cdot t_j$. Adding processing time, we find that the expected time spent for each visit to workstation j is $\left(1 + \dfrac{N - 1}{M}\right) \cdot t_j$. On the other hand, if processing times are constant and the same at each workstation, then you should arrive at workstation j just as the workstation completes a job and passes it on. In fact, this is happening at each workstation, thus there are M jobs, including yourself, in transit and only $N - M$ jobs left to be in queue at the workstations. This assumes that we have at least M jobs in the system $N \geq M$. Therefore, with a deterministic, balanced system, the waiting time would average $\left(\dfrac{N - M}{M}\right) \cdot t_j$. Waiting time plus processing time at the workstation would average $T_j = \left(1 + \dfrac{N - M}{M}\right) \cdot t_j$

$= \left(\dfrac{N}{M}\right) \cdot t_j$ per visit. In either case, a rough estimate of total flow time can then be found by adding the T_j values over all workstations. If we assume that a goal is to minimize inventory, and the safety stock kept in the output buffer is small, then we can approximate time in the system by flow time and, therefore,

$$T = \begin{cases} \left(1 + \dfrac{N - 1}{M}\right) \cdot t_p, & \text{exponential processing time} \\ \left(\dfrac{N}{M}\right) \cdot t_p, & \text{constant, synchronous processing with } N \geq M. \end{cases} \tag{7.11}$$

Accordingly, because $N = X_p \cdot T$, production rate would be:

$$X_p = \begin{cases} \dfrac{N \cdot M}{(M + N - 1) \cdot t_p}, & \text{exponential processing time} \\ \dfrac{M}{t_p}, & \text{constant, synchronous processing with } N \geq M. \end{cases} \tag{7.12}$$

EXAMPLE 7.15

Color printer cartridges are produced on a 50-step serial production line, two hundred cartridges are kept in process at all times. Processing times are highly random due to variation in material and machine condition. Average times also vary between stages but range from three to ten seconds with a total of 350 seconds. Find a rough estimate of cycle time and production rate for this line design.

SOLUTION

We are given $t_p = 350$ seconds. Using Equation (7.11) to estimate cycle time we obtain:

$T \approx \left(1 + \dfrac{N - 1}{M}\right) \cdot t_p = \left(1 + \dfrac{200 - 1}{50}\right) \cdot 350 = 1743$ seconds. Production rate is given by

(7.12) as: $X_p \approx \dfrac{N \cdot M}{(N + M - 1) \cdot t_p} = \dfrac{(200) \cdot (50)}{(200 + 50 - 1) \cdot 350} = 0.115$ cartridges per second. To verify that these estimates fit with Little's Law, note that $(1743)(0.115) = 200$ units or $T \cdot X_p = N$.

EXAMPLE 7.16

Suppose that in the previous example, processing times were fixed at seven seconds per workstation. What effect would this have?

SOLUTION

If we operate the line synchronously with no variability, we have from Equations (7.11) and (7.12) that $T \approx \dfrac{N}{M} \cdot t_p = \dfrac{200}{50} \cdot (350) = 1400$ seconds and $X_p \approx \dfrac{M}{t_p} = \dfrac{50}{350} = 0.14$ cartridges per second. Of course, if the process was truly ideal and without variation, we could reduce N to 50 and obtain $T = 350$ seconds without affecting the production rate.

CONWIP systems are controlled by the fixed WIP level. If we increase (or decrease) this WIP level N for a particular facility, we obtain some combination of increase (or decrease) in throughput time and production rate. Equations (7.11) and (7.12) provide an estimate of these effects. It is a relatively simple matter in practice to fine-tune this control value to obtain the desired output level. Spearman et al. [1989] studied the impact of knowing the optimal N. They discovered that CONWIP system performance is relatively insensitive to knowing the optimal value when jobs have different values and priorities. If N is set too small, the production rate will lag demand slightly. However, with job priorities, we should still be able to accomplish the most critical tasks on time. In practice, if we notice that low-priority jobs tend to be late, we could increase N until obtaining on-time completions. If N is too large, jobs finish early, and we have a little extra inventory cost. In practice, we may even find that we cannot maintain N at the higher level because the higher production rate puts us at a point where there are no jobs left to be released to the shop-floor. In this case, this exogenous constraint of job availability would cause the system to automatically control itself. There is one sobering observation that can be made from Equations (7.11) and (7.12). For the constant, synchronous processing case with $N >$ M, the production rate does not increase with N. Instead, adding more work to the shop-floor serves only to increase the throughput time. The machines are already constantly busy, and extra WIP does not serve any purpose. With exponential processing, there will be some idle time for machines because they will be starved for work occasionally. Increasing N can reduce starvation, but the increases in production diminish as N is increased.

EXAMPLE 7.17

A manufacturing system has 14 production stages. On average, jobs require two hours of processing at each stage, but processing times for individual operations are exponentially distributed. Estimate throughput time and production rate as a function of the number of jobs in process.

SOLUTION

Using Equation (7.11), we have:

$$T = \left(1 + \frac{N-1}{M}\right) \cdot t_p = \left(1 + \frac{N-1}{14}\right) \cdot (2) \cdot (14) = 28 + 2 \cdot (N-1) \text{ hours.}$$

Using Equation (7.12), we have:

$$X = \left(\frac{N \cdot M}{M + N - 1}\right)\left(\frac{1}{(2)(14)}\right) = \frac{1}{28}\left(\frac{14N}{13+N}\right) \text{ jobs per hour.}$$

A plot of the production rate is shown in Figure 7.12.

A general production system using CONWIP corresponds to a type of **closed queuing network (CQN)**. In a CQN, each workstation has its own processing time characteristics and may or may not have limited buffer space. The defining characteristic of a CQN relates to maintaining a fixed number of jobs in the system at all times, precisely as in CONWIP. The models developed in this section are rather simplistic, and we caution the reader from placing too much confidence in these estimates. We ignored the detailed distributions of processing times at each workstation and assumed the units in process were free to be distributed in any manner. In reality, limited storage space at workstations may result in blocking, causing workstations to become idle until space clears. Workstations with heavier loads will probably have larger input queues on average, and workstations with multiple servers may simultaneously process multiple jobs thus reducing waiting time as computed from the average number of jobs at the workstation. Nevertheless, the model above provides a rough estimate of cycle time and production rate. The queue lengths in a CQN have a tendency to be negatively correlated because the sum is constant. Thus, if the simple model overpredicts waiting at one station, it should underpredict waiting time elsewhere. Thus, when aggregated over the entire facility, the model estimates should be reasonably accurate. A substantial body of literature exists for obtaining more accurate estimates of throughput times and production rates in CQNs. Several sources for more extensive modeling assistance are provided in the references at the end of the chapter. As a

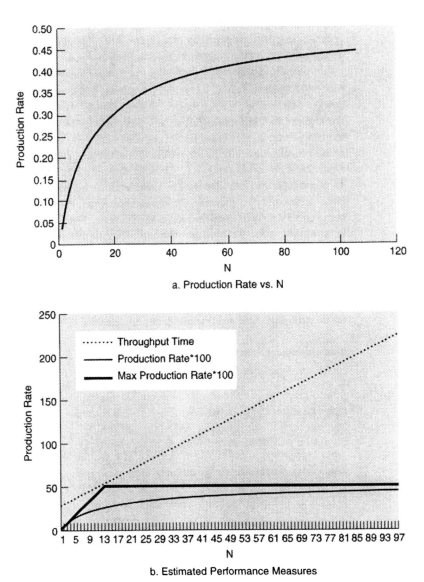

a. Production Rate vs. N

b. Estimated Performance Measures

Figure 7.12 Estimated Performance Measures for Example 7.17

brief introduction, we present a basic mean value analysis (MVA) model for exponential queues with infinite buffers (no blocking).

We assume P part types are being produced by M workstations using an FCFS service discipline. The routes for each part are known, and part type p visits workstation j V_{jp} times on its route each time requiring an average processing time of t_{jp}. We maintain a WIP level of N_p parts of type p in process at all times. Thus, total WIP is constant at $N = \sum_{p=1}^{P} N_p$. The unknown variables to be estimated are the production rates for each part type (X_p), the average times spent at each workstation per visit (T_{jp}), and the average number of parts at each workstation (L_{jp}). Fortunately, we have an equal number of equations that relate these unknown quantities to known quantities and to each other. By Little's Law:

$$L_{jp} = X_p \cdot T_{jp} \cdot V_{jp}, \qquad (7.13)$$

indicating that the average number of parts of each part at each workstation equals the product of part production rates, average visit time, and visits per unit produced. Average visit times are estimated by:

$$T_{jp} = t_{jp} + \left(\frac{N_p - 1}{N_p} \right) \cdot L_{jp} \cdot t_{jp} + \sum_{r \neq p} L_{jr} \cdot t_{jr} \qquad (7.14)$$

the sum of average processing time, plus time spent waiting in queue for parts of the same type to be produced, plus time spent waiting in queue for other part types to be produced. On arrival at a workstation, a type p part expects L_{jr} parts of type r to be ahead of it in queue. Also, the number of parts of type p ahead of the arriving part utilizes the correction factor $\frac{N_p - 1}{N_p}$ to denote that the L_{jp} term is based on N_p jobs in the system, whereas the arriving part views a system with only $N_p - 1$ other parts of type p. Finally, we can use Little's Law to estimate production rates by:

$$X_p = \frac{N_p}{\sum_j V_{jp} \cdot T_{jp}} \qquad (7.15)$$

Equation (7.15) reflects that production rate is the ratio of WIP levels to throughput time for a part type.

A set of values that simultaneously solves Equations (7.13) through (7.15) provides mean value estimates of production rates, queue levels, and throughput times. This is valuable information for a production planner attempting to determine whether a particular production schedule can be met and how to set kanban or CONWIP parameters accordingly. Numerous extensions are possible for factors such as multiple servers at a workstation, general service times, and part priorities. To solve the equations as presented, we can begin by estimating L_{jp} values, then solve for T_{jp} values using Equation (7.14) followed by estimating production rates with Equation (7.15). Initial values of L_{jp} must ensure the desired N_p values and can be found by equally distributing parts of each type. Setting:

$$L_{jp}^{(0)} = \frac{N_p}{M_p}, \qquad (7.16)$$

where M_p is the number of workstations visited by part type p will satisfy these requirements. We can then iterate among the L_{jp}, T_{jp}, and X_p equations until values converge and the right and left sides of the equations agree within an acceptable level of tolerance. Reasonable convergence usually occurs within a few iterations.

EXAMPLE 7.18

Fabrication, assembly, and test workstations manufacture two basic product types. Product type 1 averages 30 minutes per batch for fabrication, 15 minutes per batch for assembly, and 20 minutes per batch for testing. Product type 2 utilizes purchased parts and averages 20 minutes at assembly and 20 minutes at testing per batch. Most batches are small, but some require significantly longer processing times. A kanban system is used maintaining three batches of each product type in the system. Analyze system throughput.

SOLUTION

The skewed processing time distributions imply that the exponential assumption in MVA may be a reasonable approximation. To initialize, we assume the three jobs of type 1 are split evenly among the three workstations and the three jobs of type 2 are split between assembly and testing. Define fab, assembly, and test as workstations 1, 2, and 3, respectively. Then, $L_{11}^{(0)} = L_{21}^{(0)} = L_{31}^{(0)} = 1, L_{12}^{(0)} = 0$ and $L_{22}^{(0)} = L_{32}^{(0)} = 1.5$. From the product routes, the visit counts become $V_{11} = V_{21} = V_{31} = V_{22} = V_{32} = 1$ and $V_{12} = 0$. Using Equation (7.14), at the first iteration denoted by a superscript (1), we have:

$$T_{11}^{(1)} = 30 + \frac{2}{3} \cdot (1) \cdot (30) + 0 = 50$$

$$T_{21}^{(1)} = 15 + \frac{2}{3} \cdot (1) \cdot (15) + (1.5) \cdot (25) = 62.5$$

$$T_{31}^{(1)} = 20 + \frac{2}{3} \cdot (1) \cdot (20) + (1.5) \cdot (20) = 63.\overline{3}$$

$$T_{12} = 0$$

$$T_{22}^{(1)} = 25 + \frac{2}{3} \cdot (1.5) \cdot (25) + (1) \cdot (15) = 65.0$$

$$T_{32}^{(1)} = 20 + \frac{2}{3} \cdot (1.5) \cdot (20) + (1) \cdot (20) = 60.0.$$

All times are measured in minutes. To illustrate the equations, consider the expression for T_{21}, the time for a batch of product type 1 to go through the assembly station. This time comprises the 15-minute batch assembly time, plus time spent waiting for other product type 2 batches that are already at assembly when the new batch arrives (2/3 batches are expected), plus waiting time for the expected 1.5 batches of product type 2 that we will find at assembly on arrival. Now, we can estimate production rates by:

$$X_1^{(1)} = \frac{3}{50 + 62.5 + 63.\overline{3}} = 0.01706 \text{ and } X_2^{(1)} = \frac{3}{65 + 60} = 0.0240.$$

This indicates that we expect to complete a batch of product type 1 every $1/0.01706 = 58.6$ minutes and a batch of product 2 every $1/0.024 = 41.7$ minutes on average. Using Equation (7.13), the estimated production rates and visit times imply a WIP distribution in the shop of:

$$L_{11}^{(1)} = (0.01706) \cdot (50) = 0.8530; \quad L_{12} = 0$$
$$L_{21}^{(1)} = (0.01706) \cdot (62.5) = 1.066; \quad L_{22}^{(1)} = (0.024) \cdot (65) = 1.56$$
$$L_{31}^{(1)} = (0.01706) \cdot (63.\overline{3}) = 1.081; \quad L_{32}^{(1)} = (0.024) \cdot (60) = 1.44.$$

First, note that the distribution for each product retains the total WIP level of three batches for each product type. Second, note that the WIP levels have changed somewhat from the initial $L_{jp}^{(0)}$ values. Because the values have not stabilized, we should repeat the iterative process to improve the estimates.

7.3.3   CONWIP Variations and Extensions

A number of variants in CONWIP policies are possible. For example, measuring WIP by the number of jobs makes sense if all jobs contain approximately the same amount of work or if the ratio of jobs to machines is low. Otherwise, it may be preferable to fix the amount of work in the shop as measured by total production hours instead of number of jobs. Likewise, we could measure the load in the shop as the number of production hours needed at the bottleneck workstation. It is important to keep the bottleneck productively engaged when demand is close to production capacity. Idle time at the bottleneck work center resulting from the lack of available work translates into lost capacity for the entire facility. Measuring the WIP as the amount of work headed toward the bottleneck would help to ensure that the bottleneck will not be deprived of work. Thus, a new job is released whenever a job finishes at the bottleneck workstation. This policy can be very effective for a shop with a single-bottleneck resource. The CONWIP system described in detail previously in this chapter fits this description if demand is the bottleneck. This is true because we maintained N jobs in the system ahead of the bottleneck (demand). This will always be the case when each work center has adequate capacity to meet demand. If, however, the average demand rate exceeds the maximum production rate for one or more production stages, then one or more of these stages become bottlenecks. We could then release parts to ensure that a steady supply of WIP continually moves toward these bottleneck workstations.

Another option operates the system with a value of N that is proportioned across the part types whereby N_i batches of part type i are kept in process at all times. The backlog list contains the batches necessary to bring the WIP to level N_i for each part type. This approach would be appropriate if demand rates are constant for the part types produced by the cell. The N_i is set to match demand for each part type. This system then resembles the kanban approach. Controlling WIP only, allows us to smooth production over time.

CONWIP can be implemented in a decentralized manner. If parallel feeder lines merge into assembly, setting a CONWIP quota for each feeder line will automatically pace these lines at a compatible rate. The final assembly will pull parts from each of the feeders at the rate needed for demand, and this will automatically cause new jobs to be released in each feeder line. The trick, then, is to maintain a correct backlog list for each feeder based on the master assembly schedule. If a first-in-system-first-served (FISFS) dispatching rule is used within work centers whereby job priority at machines is based on the order in which they were released to the cell, jobs will finish in basically the order in which they were released to the work center. By releasing jobs in order of planned assembly time, this becomes equivalent to a due-date based internal scheduling system. If process routes or other job characteristics cause some jobs to have longer cycle times than others, then we need to have an estimate of cycle time by job. It is straightforward to estimate expected time through each workstation and to simply add these together for jobs with different routes. The backlog list is then compiled by giving each job a target release date based on planned assembly time minus expected cycle time minus some constant safety factor.

In the standard CONWIP system, the output buffer is considered part of the WIP. Thus, job release is controlled by **current** demand. If we want to smooth the production rate as we did in aggregate production planning, and we know the long-term demand rate, it would be possible to use the output buffer as a buffer against demand variability. Count only the jobs in process toward N, excluding those in the output buffer. Then, set the control variable N to produce the average demand rate. This will lead to an accumulation of finished parts when demand is slow. When demand increases, the inventory level drops. If inventory becomes negative, meaning shortages are experienced, then actions can be taken such as temporarily increasing N, adding overtime work hours to the schedule, running the processes at higher (and perhaps less economical) rates, and increasing the workforce temporarily. However, for the most part, this strategy will lead to smooth production and avoid underutilizing workers in some periods and overutilizing them in others.

7.4 SUMMARY

Pull inventory control provides a simple approach for coordinating production along a multistage production network. In a pull system, sequences of one or more operations are grouped into cells, and a controlled storage point is defined at the end of each cell. Removal of product from that controlled storage point authorizes production to replenish the supply of product.

In a kanban system, the storage point has a target inventory level for each product produced. When the on-hand level drops below the target, a signal is issued to begin replacing that product. Necessary raw materials are obtained, and production proceeds in a continuous manner through the group. When work centers are close together, the output buffer of one stage serves as the input buffer to the next stage. A single-kanban system may be used with production ordering kanbans. These kanbans stay within the work center. When a container of parts is removed for use at a successor workstation, the kanban is sent from the output buffer to the schedule board pending processing. After the new parts are completed, they are returned to the output buffer along with the kanban. When distances between groups are large, raw material inventories are also kept as input for each work center. Withdrawal-ordering kanbans are added to input storage containers. As these parts are consumed, the kanbans are collected and occasionally returned to the previous stage to authorize withdrawal of containers of parts to replenish the input buffer at the next stage. Signal kanbans, periodic cyclical scheduling, and minimum EOQ production quantities can be used when setup times prevent the production of single containers of parts.

In general, pull systems match order releases to current customer demand. By limiting WIP, the production rate and flow time remain relatively constant. Pull systems react automatically to minor changes in demand and minor breakdowns. However, without the foresight to plan ahead, pull systems rely on relatively constant demand rates and reliable processes.

Kanbans offer one common approach to production authorization in pull systems. Any kanban not attached to a full container of parts authorizes production. The key decision variables for system control are the number of units authorized per kanban and the number of kanbans kept in the system for each part. The total number of units of each part type authorized should be set equal to the replenishment lead-time demand plus some safety stock. The optimal safety factor can be determined by applying a probabilistic inventory control model or by company policy. In the latter case, the implied cost of a shortage can be identified and verified for logic. The utilization of the workstations, container processing time distribution, and inter-arrival time distribution for kanbans all determine the number of parts moving through a kanban system and its inventory levels. The variability of lead-time demand is the dominant factor in specifying the safety stock.

CONWIP attempts to fix the amount of WIP at an efficient level that will meet demand without excessive inventory requirements. Based on current inventory levels and expected demand, a backlog list of parts needing to be produced is maintained. As product exits the shop, orders recorded on the backlog list are dispatched to the shop-floor. CONWIP is easy to implement and can be used in environments in which the demand for part types changes over time. The WIP level is set whereby the system production rate matches the average demand rate. Safety stocks of finished goods are maintained to guard against variability in demand. Approximate throughput rate and time estimates can be easily computed for CONWIP systems. Production starts match demand. Flow times tend to have lower variability than in other systems. Detailed performance evaluations can be obtained by use of standard closed queuing network models. A number of variants of standard CONWIP such as measuring workload in hours instead of jobs, measuring only the bottleneck workload, and further smoothing the production rate by ignoring jobs in the output buffer when counting WIP are also possible.

7.5 REFERENCES

ASKIN, R.G., MITWASI, M.G., & GOLDBERG, J.B. Determining the number of kanbans in JIT systems. *IIE Transactions*, 1993, 25(1), 89–98.

ASKIN, R.G., & STANDRIDGE, C., *Modeling and analysis of manufacturing systems*, New York: John Wiley & Sons Inc., 1993.

BASKETT, F., CHANDY, K.M., MUNTZ, R.R., & Palacios, F.G. Open, closed, and mixed networks of queues with different classes of customers. *Journal of the ACM*, 1975, 22(2), 248–260.

BUZACOTT, J., & SHANTHIKUMAR, J.G. A general approach for coordinating production in multiple-cell manufacturing systems. *Production and Operations Management*, 1992, 1(1), 34–52.

DELEERSNYDER, J.-L., HODGSON, T.J., MULLER, H., & O'GRADY, P.J. Kanban controlled pull systems: An analytical approach. Management science, 1989, 35(9), 1079–1091.

GERSHWIN, S.B. *Manufacturing systems engineering*, Englewood Cliffs, NJ: Prentice-Hall Inc., 1994.

GUNASEKARAN, A., GOYAL, S.K., MARTIKAINEN, T., & YLI-OLLI, P. Modelling and analysis of JIT manufacturing systems. *International Journal of Production Economics*, 1993, 32, 23–37.

KIMURA, O., & TERADA, H. Design and analysis of pull system, A method of multi-stage production control. *International Journal of Production Research*, 1981, 19(3), 241–253.

MITRA, D., & MITRANI, I. Analysis of a kanban discipline for cell coordination in production lines. *Management Science*, 1990, 36(12), 1548–1566.

MONDEN, Y. *The Toyota production system*, Norcross, GA. Industrial Engineering and Management Press, 1983.

MUTH, E.J. The production rate of a series of work stations, *International Journal of Production Research*. 1973, 11(2), 155–169.

PHILIPOOM, P.R., REES, L.P., TAYLOR, B.W. III, & HUANG, P.Y. An investigation of the factors influencing the number of kanbans required in the implementation of the JIT technique with kanbans.

International Journal of Production Research, 1987, 25(3), 457–472.

REES, L.P., PHILIPOOM, P.R., TAYLOR, B.W., & HUANG, P.Y. Dynamically adjusting the number of kanbans in a JIT production system using estimated values of leadtime. *IIE Transactions,* 1987, 19(2), 199–207.

REISER, M. Mean value analysis of queuing networks: a new look at an old problem. In: A RATO, M., BUTRIMENKO, A., & GELENBE, E., eds. *Performance of computer systems.* New York: North-Holland, 1979: 63–77.

SINGH, N. *Systems approach to computer integrated design and manufacturing,* New York: John Wiley & Sons Inc., 1996.

SO, K.C., & PINAULT, S.C. Allocating buffer storages in a pull system. *International Journal of Production Research,* 1988, 26(12), 1959–1980.

SPEARMAN, M., & HOPP, W.J. *Factory physics: Foundations of manufacturing management,* Burr Ridge, IL: Irwin Professional Publishing, 1995.

SPEARMAN, M.L., WOODRUFF, D.L., & HOPP, W.J. CONWIP: A pull alternative to kanban. *International Journal of Production Research,* 1990, 28(5), 879–894.

WHITT, W. The queueing network analyzer. *Bell System Technical Journal,* 1983, 62(9), 2779–2815.

WHITT, W. Approximating the GI/G/m queue. *Production and Operations Management,* 1993, 2(2), 114–161.

7.6   PROBLEMS

7.1.   What are the disadvantages of carrying more WIP inventory than necessary?

7.2.   Describe the difference between a pull and push production system.

7.3.   Describe the difference between a single- and a dual-kanban system. Which system should a small manufacturer of toy cars use? Which method would work best for a large manufacturer that produces many versions of product with part fabrication, subassembly construction, and final assembly all in different buildings?

7.4.   How is the production of parts authorized in a kanban system?

7.5.   List the environmental requirements for a pull system to operate effectively. For each requirement, explain what will happen if the condition is not met.

7.6.   Show that if we can ignore holding cost in Equation (7.1), that the optimal container size must be the maximum allowable for the technology selected.

7.7.   Material handlers use push carts to move parts around a plant. A cart load is used as the minimum production quantity. Each cart holds 12 parts. Replenishment lead time for a part is a half-day and its demand rate is 100 units per day.

a.   Find the minimum number of kanbans needed for this part.

b.   Suppose that a 10% safety factor is desired. How many kanbans are needed?

7.8.   Demand for a particular model of dining room tables averages 35 tables per day. Each table requires four identical table legs. Each container holds eight legs. Kanbans are collected twice a day, and all empty containers are replenished. Find the minimum number of kanbans required for this process.

7.9.   A material-handling system is being selected to connect two work centers that are separated by 1000 feet. The manual system has a fixed cost of $35,000 per year plus a variable cost of $0.10 per load. The maximum load size is 12 units. Annual demand is estimated at 250,000 units, and the cost to hold one unit in inventory for a year is $0.20. The motorized system will cost $45,000 with a variable cost of $0.25 per load but can transport 100 units at a time. Compare the two options on the basis of material-handling cost and average inventory for the cases below. Which option would you choose? Each item has a value of $1.50, and the annual holding cost rate is 30%.

a.   The production batch size is equal to the material-handling load capacity.

b.   The production batch size is 1200 units as determined by an EMQ calculation.

7.10.   How would your answer to the previous question change if an option existed to build a conveyor between work centers at an annual cost of $50 per foot? The conveyor can handle up to 300,000 loads per year.

7.11.   Two material-handling options are being considered for connecting two work centers. A conveyor can be purchased and operated for an annual cost of $275,000. The conveyor can transport up to 500,000 containers per year with each container holding up to eight units. The alternative is a forklift/pallet system that would cost $85,000 per year plus $1.65 per move. The pallets can hold up to 16 units each. The facility expects to manufacture 1,300,000 parts per year. Select the best container size and material-handling option for this production system.

7.12.   A two-workstation cell uses a kanban system with a container size of 10 units. The workstation makes eight different versions of the same product. Total annual demand is 150,000 parts. The first workstation has a maximum production rate of 250,000 parts per year, and the second workstation could produce 325,000 parts per year working full time. Internal setup is

0.0004 years at the first workstation and 0.0003 years at the second. External setup time is 0.0002 years per batch at each workstation. What batch size would you use for this work center?

7.13. One workstation in a computer plant assembles and packs computer mice. Demand for the dual-button model for a specific brand runs approximately 60 mice per eight-hour shift. Average replenishment lead time is two hours. Each container holds five mice. If the company wants a 15% safety stock factor, determine the appropriate number of kanbans.

7.14. In the previous problem, suppose that lead time is exponentially distributed, ($f(\tau) = 0.5e^{-\frac{\tau}{2}}, \tau > 0$); the holding cost of a mouse is estimated at $0.05 per shift whereas the shortage cost is $150 per shift because all other production is halted. Estimate the optimal number of kanbans.

7.15. Table 7.8 below shows the record of withdrawals and replenishments for a part type over a three-day period. Each container holds 20 items. Estimate the lead-time demand distribution. How many kanbans are needed to ensure that at least 95% of withdrawal requests can be met from stock on hand in the output buffer? The plant operates 16 hours per day. Time is given by "day/hour/minute."

7.16. A department manufactures three different products using a kanban system. Data for the individual products are given in Table 7.9. Setup time is negligible.

a. Find the minimum number of kanbans required for each part type assuming replenishment lead time is always four hours.

Table 7.8 Withdrawal and Replenishment History for Problem 7.15

Withdrawal	Replenishment
1/2/34	1/10/41
1/4/12	1/11/39
1/8/38	1/16/03
1/12/09	2/3/27
2/2/14	2/6/50
2/4/56	2/11/30
2/11/23	2/15/50
2/14/45	3/4/48
2/15/51	3/6/55
3/1/36	3/10/10
3/4/06	3/10/10
3/10/21	3/14/48
3/12/45	In process
3/14/06	In process

Table 7.9   Part Data for Problem 7.16

Product	Demand/hour	Container size	Unit processing time (hours)
A	20	5	0.02
B	40	10	0.01
C	10	5	0.01

b. Suppose that lead time is exponentially distributed with a mean of four hours. How many kanbans are required for each part type to ensure that parts are available for at least 99% of the time that they are requested?

7.17. A kanban work cell produces 60 containers per 10-hour day. Each container holds eight parts. Processing time is fixed at one part per minute. The time between container withdrawals is exponentially distributed. Find the utilization, average cycle time, and average inventory of parts in the replenishment process for the work center.

7.18. Repeat the previous problem assuming processing time has a standard deviation of one minute but the interarrival time between container withdrawals is uniformly distributed between 5 and 15 minutes.

7.19. Using Equation (7.7), show that if $C_s^2 \leq 1$ and $C_a^2 \leq 1$, and at least one of these is a strict inequality, then expected waiting time in queue satisfies:

$$E(W_q) < E(\text{Processing Time}) \cdot \frac{\rho}{(1 - \rho)}.$$

7.20. You produce a line of electronic keyboards. All keyboards contain a platform onto which the white and black keys are attached. The platform is connected to a panel that senses when the keys are pressed. There are three platform sizes that vary only in the number of octaves (3, 5, or 7). The platform and sensor system are plugged into a second module that contains the memory and controls for setting sound features. This module must match the platform size, but it also comes in several price categories for each size. The price is affected by the level of sophistication of the control and power module. Keyboard covers match the memory/control module but they also come in several colors for each platform size and price class. These covers are added last. Consider the assembly process for keyboards. An engineer wants to create a kanban system. Decide what you think should be the definition of work centers for keyboard assembly, i.e. should you have one or more?

7.21. A kanban work cell produces two part types each with a container size of 15 units. Demand for part type A is currently 50 units per day and is growing by five units per day each week. Demand for part type B is 75 units per day but is declining by three units per week. Normal replenishment lead time is two days, and weeks are six days long. Using a 20% safety lead-time factor, find the number of kanbans required for each part type, each week, for the next six weeks.

7.22. A circuit card assembly area has 15 workstations configured in a mixed serial and parallel arrangement. The 15 workstations are split among four serial stages (three parallel workstations at stage 1, two at stage 2, six at stage 3, and four parallel workstations at stage 4). Each card must visit only one of the workstations at each stage. The current policy is to maintain 45 batches of circuit cards in process in the area. Processing times vary by card type, but the average processing time of a batch is 45 minutes at stage 1, 30 minutes at stage 2, 60 minutes at stage 3, and 45 minutes at stage 4. The system is managed by keeping 25 batches in process spread among the 15 workstations at all times. Estimate the flow time and production rate.

7.23. Several models of small stereo speakers are produced on a 40-step serial production line. At all times, 150 speakers are kept in process. Processing times are highly random. Average times vary between stages but range from 10 to 15 seconds, with a total of 500 seconds. Find a rough estimate of flow time and production rate for this line design.

7.24. A manufacturing system has twelve production stages. On average, jobs require one hour of processing at each stage, but processing times for individual operations are exponentially distributed. Construct graphs of throughput time and production rate as a function of the number of jobs in process for up to $N = 100$.

7.25. A three-stage production process requires 10 minutes of processing time at the first stage, 25 minutes at the second stage, and 20 minutes at the third stage. The first stage can process only one job at a time, but the other two stages can process up to four jobs simultaneously.

a. What is the maximum production rate for the system?

b. What is the maximum production rate if only six jobs are kept in-process at all times?

7.26. Section 7.3.4 addressed estimating average inventory in a single-multioperation cell. Consider, instead, a set of serial work centers in a kanban facility. Each work center uses as raw material the product(s) produced by the previous work center. The production process begins at the last work center when consumers purchase parts from finished goods inventory. As each work center requisitions raw materials, its predecessor work center begins to replenish those parts. Describe how the queuing model of section 7.3.4 can be used to estimate inventory for the entire facility.

7.27. A ten-stage serial production system operates under CONWIP control. The workload is balanced across workstations with each workstation requiring an average of one hour to perform its assigned tasks.

a. Construct a graph estimating the trade off between production rate and the assigned WIP level.

b. Construct a graph estimating the trade off between production rate and average throughput time.

7.28. A CONWIP system with five serial machines is currently managed by keeping 45 batches of work in the system. Processing times vary and operation is asynchronous. Total processing time in the system is 2.4 hours for a part.

a. Find an upper bound on the production rate for the system when it is operating.

b. Estimate flow time and production rate.

7.29. Find the ratio of flow time for the exponential case to that for the fixed processing, synchronous case in Equation (7.11). What is the limiting value of this ratio as N goes to infinity? What is the limiting value of this ratio as N approaches M?

7.30. Four machines are used to produce two part types. The first part type spends forty minutes at the first machine, twenty-five minutes at the second, and fifty-five minutes at the fourth machine. The second part type spends fifteen minutes at the second machine, thirty-five minutes at the third workstation, then returns to the second machine for another twenty minutes. Four parts of each type are kept in the system at all times. Perform one complete iteration of the mean value analysis algorithm to estimate the production rate of each part type, throughput times, and the average number of parts of each type at each workstation.

7.31. How do your estimates in the previous problem differ from results you would obtain using the simpler bounds in Equations (7.11) and (7.12)? Which estimates do you expect to be more precise? Does your answer depend on the exact distribution of processing times?

7.32. Two part types are produced on three machines. The first part type requires ten minutes at each workstation. The second part type requires ten minutes at workstation 1, fifteen minutes at workstation 2 and six minutes at workstation 3. The production manager releases jobs to maintain a constant supply of WIP for each part type.

a. Suppose that the demand for both part types is equal. Find an upper bound on the number of each part type that can be produced per hour.

b. Suppose that the demand is for two parts of each type per hour. Use the mean value analysis model to find the number of parts of each type that should be kept in the system to meet demand.

7.33. Consider the MVA algorithm in Section 7.3.2. We could incorporate parallel servers c_j at workstation j by dividing WIP levels by c_j in the second and third terms of Equation (7.14). Determine why this is merely an approximation. Will throughput times be overestimated or underestimated, and what will this do to the estimate of the production rate?

7.34. Modify the throughput time equation of the mean value analysis algorithm to allow for priority dispatching at workstations. In this case, each part type has a priority and parts in queue are able to proceed ahead of lower-priority part types. Assume that once the machine begins a job, it will complete the job even if a higher-priority job arrives, i.e. there is no preemption of the current job.

7.35. The MVA algorithm described in Section 7.3.2 assumes exponential service. Modify the model to allow for general service-time distributions. Hint: It can be shown that when a customer arrives to a queue, the distribution of the expected residual processing time for a job currently in process is given by: $R = \text{Pr (Server Busy)} \cdot E(\text{Remaining Service/Busy}) = \dfrac{\lambda E(S^2)}{2}$ in which λ is the arrival rate and S is the service time.

CASE STUDY 7.7   *Safe-T-Lock Company*

Safe-T-Lock Company is one of North America's premier manufacturers of residential door locks. Founded in California in 1898, the company specialized in small jewelry and furniture lock sets for most of its existence. Spurred by the construction boom and migration to the southwest, the company built a facility in Arivaca, AZ in 1970 to produce residential door locks. The plant was later expanded several times and reached its current size with 185,000 square feet of manufacturing and warehousing space in 1988. Business declined soon thereafter, and Safe-T-Lock operated at a deficit from 1988 through 1992. The company even considered closing the Arivaca facility. However, residential construction began to pick up again in the mid-1990s, and plans to close the facility were discontinued. The facility turned a net profit of $3.4 million in 1998 on sales of 3,900,000 lock sets grossing $21,000,000. This profit was realized despite absorbing a significant investment to switch to more environmentally friendly cleaning and plating processes.

A knob and door lock set comprises an internal tumbler frame, a lock/key mechanism, inner, side and outer faceplates, and handles (knobs). A single-door lock contains approximately 30 (mostly metal) parts. The company is known for its high-quality brass finishes and ornate face plate designs. The inner and outer faceplates are stamped in the facility whereas most of the other metal and plastic parts are purchased outside the facility. All metal parts are cleaned, polished and finished in-house. Tumbler frames connect the inner and outer handles and house the key mechanism. The frames and key mechanism subassemblies are produced in-house and final lock sets are also assembled and packaged in-house. Tumbler frame subassemblies are produced on a semi-automated line. Lock/key mechanisms are hand-cut to specific customer codes and then inserted into the tumbler frame before being transferred to final assembly. There are two standard frames and five grades of lock/key mechanisms varying from inexpensive designs to highly secure deadbolts. Tumbler frames come in a standard size, therefore, either frame can accommodate all lock/key mechanisms.

The company distributes a catalog with over 100 face plate and handle designs each with a choice of four finishes. Several new designs are added each season. The top six designs with the smooth brass finish usually account for approximately 40% of sales. The smooth brass finish with other catalog plate designs makes up an additional 18% of sales. The other three finishes each account for approximately 12% of demand with this being evenly divided among all 100 plate designs. The remaining business is in custom-design sets. These customers either supply their own design specifications or work with the company's designers to produce the engineering drawings and required tooling.

Currently, the high-demand items that account for 40% of sales are made-to-stock. Typically, an order quantity of 500 lock sets is used for these items. Any subset of a production batch can be supplied with a common master key if requested by the customer. The company tries to maintain a promise to ship orders for these items within twenty-four hours. Customer orders for other catalog items should be met within five days and custom items should be met within two weeks.

Final assembly and packaging involves sandwiching the tumbler frame and lock/key mechanism between the inner and outer face plates, locating that assembly into the bottom plastic wrapper along with the side plate and instructions, adding the top wrapper and sealing the package with a semi-automatic snap and seal process. To ensure a correct matching at assembly, the instruction set follows the lock/key mechanism throughout the process. A serial number is engraved on the lock/key mechanism and a matching bar code is removed during packaging and placed on the instruction sheet. All material handling in the plant is done with manual push carts. The carts are used to move bins of parts to work centers from the warehouse and to transport parts between work centers. The carts have three layers and each layer can hold a container.

A container can hold up to fifteen packaged lock sets. Summary data on the production processes appear in the Table below.

Production Capacity for Safe-T-Lock Company

Department	Capacity (Lock Sets/Shift/Workstation)	No. of Workstations	No. of Shifts/Day
Parts Cleaning	2500	8	1.5
Plating	8000	2	2
Finishing	7000	2	2
Tumbler Subassembly	600	15	2
Lock/Key Subassembly	145	35	3
Assemble and Pack	3000	3	2

The facility is laid out with a process organization. All workstations of the same type are located in the same area. A centralized shipping and receiving area uses a common warehouse for storing parts and completed lock sets. The two plating and finishing lines would be difficult to move, however, the other workstations are easily moved. Setup time is approximately a half hour to change over the finishing line but minimal for other processes.

Most of Safe-T-Locks sales are through national hardware and department store chains. Customers fall into two basic categories — developers and homeowners. The developers often order large sets of locks. A typical developer order through a retailer may be for one hundred homes with three exterior doors each. The three locks destined for a single home would match identically and all the locks for an order would typically have the same finish spread perhaps over several faceplate designs. The order may require a common master key for use by the developer during construction as well as individual key codes for each house.

In the past two years, Safe-T-Lock has been plagued with shortages. Despite increasing employment and inventory levels, the company is now losing business because delivery promises are not being met. In addition, although product quality is still high, shipments are frequently returned for not having the correct combination of products. Currently, the plant employs 220 workers spread over two shifts. The plant usually operates five days per week but extends to six days per week plus possible daily overtime during the busy season of January through March, which follows the publication of the new catalog and precedes the peak home building season.

Richard Bellin was hired from the Bensen Consulting Company one month ago to become the Production Control Manager for the Arivaca, AZ manufacturing site. Mr. Bellin had been with Bensen for five years after graduating with an undergraduate degree in Industrial Engineering. Bellin initially specialized in installing MRP software while with Bensen, however, in recent years he had gained expertise in just-in-time practices and lean manufacturing. He has spent the past month trying to understand the product flow and current production process at Safe-T-Lock. He has come to the conclusion that the facility layout is logical, but poor demand forecasting has resulted in large inventories of the wrong products and shortages of hot items. In addition, the order-priority system does not match customer demand or inventory and, thus, workers are continuing to prioritize the wrong products. Yesterday, when he realized that the entire week had been spent rushing through an order for a make-to-stock part while several large customer orders were cancelled because of late delivery, Richard Bellin recognized that he had to either change the system or start looking for his next job.

Activities and Discussion Questions:

1. Draw a sketch of the production flow process.

2. Does either kanban or CONWIP seem to be an appropriate production control system? How does the current situation compare to the environmental requirements for a kanban system?

3. Propose a production control system that you believe would work for this facility. What additional information would you want to collect before deciding to go ahead with this new system?

4. How would you define the work centers for this facility?

5. What kind of relationships would you try to develop with vendors, and how would you go about establishing those relationships?

6. How would you manage the balance between make-to-stock and make-to-order items?

7. If any items are to be operated using a kanban system, give the container size, and indicate the number of kanbans required.

8. What other actions need to be taken?

Chapter 8

Multistage Production Systems: Materials Requirements Planning for Dependent Demand

The kanban systems addressed in the previous chapter share much in common with reorder point based inventory control methods. In both cases, we look to the past to determine the average demand rate and then assume that the rate will continue on a relatively constant basis into the future. In both cases, the inventory policy for an item is affected by its expected lead-time demand plus some safety factor, and we place orders to try to bring inventory back to a specified level. Indeed, the kanban system resembles a traditional reorder point inventory system with small setups. CONWIP resembles a reorder point system where all parts are lumped together into a common pool. These systems are easy to operate. Once the operating parameters are specified, we require only current inventory status data to make ordering decisions. In this chapter, we use a fundamentally different underlying model. The underlying model is now one of **dependent demand** relationships between items. Orders for component parts are timed to coincide with the production schedule of the products in which they are used. We combine knowledge of where each part and subassembly appear in final products with the planned production of those final products and then attempt centralized control for coordinated production of all items. We push orders into work centers to coordinate the flow of related parts so that all parts come together in time and place as detailed by the product design. As a result, if everything goes according to plan, we do not need to carry any unnecessary safety stock or carry any cycle inventories during periods of low demand for a part. This approach to coordinated scheduling for dependent-demand items is known as materials requirements planning (MRP). Industrial surveys indicate that two of three manufacturing companies use MRP in their production planning.

The parts assembled into finished products in the typical modern production facility represent a combination of purchased, standard, and custom parts. Among the standard parts, annual usage may vary from high-volume parts consumed constantly during production to relatively low-volume parts used only in certain low-volume products. Consider, for instance, a company that manufactures household appliances. Many different product families are made including washing machines, dishwashers, dryers, refrigerators, and ovens. Each product family has a variety of models representing differences in size, shape, style, color, and features. Some models sell in high volume, and some sell at a low, steady pace, whereas other models may sell only once during special promotions. Some parts will be unique to each model; however, other parts may be used in multiple models of the same product family or even in different product families. For example, the same motor, insulation material, or door hinge may be used in different products. High-volume items may justify continuous production or be best controlled by a kanban system. However, because of setup or varying demand, it may be more efficient to produce other items in intermittent batches. Infrequent production runs or batch sizes that vary significantly from period to period based on the production schedule of the parent products that use that component may be the best strategy. This is the environment for which MRP was designed.

8.1 MATERIALS REQUIREMENTS PLANNING BASICS

Basic materials requirements planning consists of a data-processing system for specifying the quantity and timing of order releases for parts. MRP applies when demand for a part depends on the final assembly schedule laid out in the **master production schedule** (MPS). The foundational elements of MRP are: 1) the product configuration model that relates the usage of materials, parts, and subassemblies to the production of higher-level products; 2) the MPS for end-item assembly; and 3) the algebraic relationship between available inventory levels, part withdrawals to meet demand and shop orders to replenish stocks. In this section, we first address data needs and organization and then proceed to the central computational model of MRP.

8.1.1 Data and Support-System Requirements

MRP offers the promise of efficient production planning and inventory control. To accomplish this, a variety of data elements are required along with support from other business and engineering management systems. Data requirements include:

1. Item master data for each part. A unique part number is assigned to each separately purchased or produced item. The item master file can be indexed by the part number. For each part, the file contains standard data such as annual demand, nominal order quantity, cost, and scrap rate. If the item is produced internally, pointers to engineering drawings and process plans are included along with production lead time. A pointer to data on standard resource requirements and tooling may also be included. If the item is purchased, then vendor data are included along with ordering information and lead time.

2. Knowledge of the **lead time** to replenish each part number. Lead time measures the time from when an order to replenish a part is initiated until the part becomes available for use. Although lead time is a random variable, MRP systems are designed to accept a fixed value in the planning process. This is one of the weaknesses of MRP. It is best to maintain values for expected lead time, a reasonable upper bound on the likely lead time, and the minimum possible expedited lead time. Each of these measures has its use in developing and updating production schedules.

3. Inventory status records must be maintained for each item and made available to the MRP logic processing module. Whereas the item master file contains static data, the inventory status file is a dynamic list of currently available on-hand inventory for each item. Some on-hand inventory in storage may already be designated for specific shop orders that are in process or planned for the near future. This must be noted because these units are not free to be allocated to new orders.

4. A **bill of materials** (BOM) listing all the materials and component parts needed to make each item produced by the factory. This includes manufactured parts, subassemblies, and final products. The BOM is so crucial to the MRP process that we address it in greater detail in the next section.

5. A time-phased **MPS** showing planned production quantities for each end item in each period of the planning horizon. By *end item,* we mean the final form of a product that is controlled by the MRP system. In a make-to-stock system, the end items will correspond to final products. High-volume catalog items such as small motors, furniture, and household appliances fit into this category. Lower-level components that are produced for sale to other manufacturers or as spare parts would also be treated as end items. If products contain many optional configurations resulting from the customer's choice of options, then an assemble-to-order strategy is used. The major modules that make up the product are stored and then quickly assembled into customized products on customer order. Each stockpiled module would constitute an end item in this case. The MPS may list only total units produced with associated multiplicative factors for options. For example, 20% of the models may be bronze-plated, and 80% may be painted. Perhaps 40% of the models contain an oversized storage tank, and 60% contain the regular tank. In this case, MRP planning can be performed using the major modules as the end items. Computers and automobiles as well as many other products have customer options. If the customer has 10 options such as color, engine, detailing, brakes, and so on, each with three alternatives, that creates $3^{10} = 59,049$ end products. Instead, the MPS contains total products produced in a period. By multiplying this total by the proportional factors for each option, we determine the use of each module. These values then set the gross requirements for the modules. A final assembly build schedule then dictates the unit-by-unit use of the modules. Instead of planning for 59,049 end items, we have 31, the basic product plus the three module options for each of the 10 modules. The build schedule shows the specific sequence of final products to be produced, listing the choice of modules for each product unit. MRP can also be used for make-to-order business. When orders accepted, engineering produces the product structure, and the planned delivery schedule information is loaded into the MPS.

The length of the planning horizon for MRP must exceed the cumulative lead time to replenish a final product through all production stages controlled by MRP. This may extend from basic raw materials through final products or begin at some later stage of production. In the latter case, the initial stage(s) of production, beginning with the ordering of raw materials, is controlled by a kanban system or traditional reorder point inventory models. The cumulative lead time is demonstrated in Figure 8.1 for the case of a control box built around a circuit card. The components can be acquired while the board is being made in-house. Components are then added to the board, and then the final product is assembled. It takes three weeks to acquire parts and make bare boards in the shop. Circuit card assembly takes two weeks, and final product assembly requires another week. Therefore, it takes six weeks to produce a control box. We must know the demand this far into the future if we are to coordinate the production of these items to come together in time to meet demand. In addition to the MPS assembly schedule, forecasts of additional demand for each part number are required. We may, for example, sell spare boards to customers as replacements when components wear out.

Demand in the MPS must be broken into the same time periods used in MRP planning. A period is often referred to as a "time bucket" because of the tabular format used for displaying MRP data. The choice of time bucket has considerable effect on system performance. In MRP, we plan to have all parts available at the start of the period (time

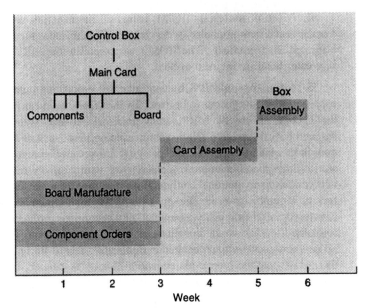

Figure 8.1 Cumulative Lead Time for Controller

bucket) in which they are to be used. Thus, both the computational effort required for MRP processing and the average inventory level are proportional to the length of the time bucket. This is a basic trade-off to be considered when designing the MRP system. Other factors must also be considered. In the dynamic environment of the typical plant, with processing times varying and customers calling to change orders, it can be difficult to plan the future day by day. If we choose a time bucket that is too short, random variation in events may make it difficult to maintain a schedule. When combining these competing factors, a common choice is to view a week as the time period.

8.1.2 Bill of Materials and Requirements Matrix

The bill of materials (BOM) serves as a road map for the MRP processing system. The BOM lists the components required for each item produced by the shop. The listing includes the item number and quantity. In most cases, the BOM is stored in a hierarchical manner whereby only the parts and materials used directly in making an item are listed. If the product has multiple module options, then we can either store each module as an end item with its own BOM or, preferably, we can list modules in the BOM and assign them proportions. The BOM may indicate then that the automobile needs four tires but needs only 0.4 speed control units because only 40% of customers order this option.

Consider the hand-held, solar-powered calculator shown in Figure 8.2. A finished calculator requires a case, top, base, keypad, display, solar cell, connecting screws, and logic board. The case is composed of 200 cm^2 of vinyl, two 5×8 cm pieces of cardboard, and top and bottom clear plastic straps. The logic board consists of a substrate on which various chips, electrical components, and connectors are added. These data are summarized by the BOM (Figure 8.3). The hierarchical BOM is illustrated in Figure 8.4. To determine the entire set of parts and materials needed to make a calculator, trace through the levels of the hierarchy from top to bottom. During assembly, the logic board is placed in the base, the solar cell and display are connected via straps on the board, and the top is added.

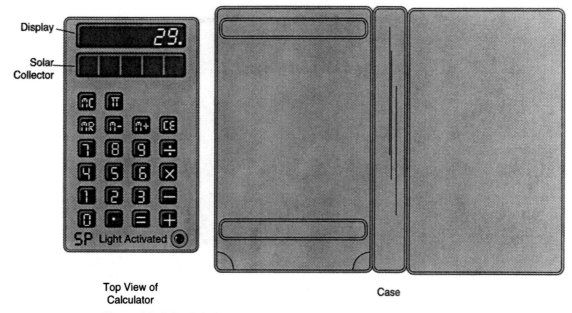

Display

Solar
Collector

Top View of
Calculator

Case

Figure 8.2 Solar Calculator

The top is connected to the base by attaching four small screws through the back of the base into outlets formed into the top. When addressing the relationship among items in the BOM structure, we refer to parent and child items. For any item, its parents are those items to which it is directly linked at a higher level. The children of an item are all those parts to which it is a parent. Thus, in Figure 8.3, the logic chip is a child of the logic board. The calculator is a parent of the circuit card, case, and other parts on level 1.

The BOM seems like a simple and straightforward concept. However, consider what happens in practice. Engineering designs the product and produces the BOM from the part drawing. The engineering BOM comprises a list of parts needed to make the final product, each part being designed separately. The engineering BOM may indicate how engineering expects the product to be assembled, but final assembly plans are usually deter-

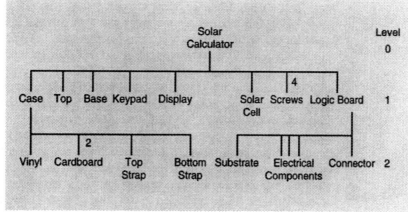

Figure 8.3 Schematic Bill of Material for Solar Calculator

```
Solar Calculator
    Case
        Vinyl (200cm²)
        5 × 8 Cardboard Plates - 2
        Top Strap
        Bottom Strap
    Top
    Base
    Keypad
    LCD
    Solar Cell
    1/8" Metal Screws - 4
    Logic Board
            Substrate
            Chip No. #####
            ...
            Radial Component No. #####
            ...
            Connector No. #####
```

Figure 8.4 Tabular Display of Hierarchical Bill of Material for Solar Calculator

mined by manufacturing planning and may differ because of ease of assembly and process capabilities. Overall, engineering is interested in the list of parts that needs to be designed to fit together and perform the required functions. Manufacturing is primarily interested in how the product will be made. Manufacturing may decide to modify processes or add features to facilitate assembly or eliminate impossible tolerances. Marketing may work with manufacturing to change assembly plans and subassembly design to facilitate the production of customized versions of final products. The hierarchical structure of how parts are combined into subassemblies and subassemblies into products becomes important. Also, each time an engineering change order (ECO) occurs, corresponding changes must be made in the BOM. Many engineering groups issue ECOs constantly to reduce costs, correct quality problems, take advantage of new materials or processes, and react to the competition. There is usually a lag between when the ECO is issued and when it becomes effective because of existing inventory supplies, vendor and test plan qualification, acquisition of tooling, and other factors. Thus, there is pressure to maintain multiple versions of the BOM across time and function. Next week's process plan for a part may differ slightly from this week's. Coordinating with engineering to issue the appropriate instructions with each shop order and with purchasing to acquire the correct raw materials and purchased parts becomes a complex task. Substitute items or material grades may be necessary at times because of material availability or cost issues. Factors such as tolerances on purchased parts become critical. Down on the shop floor, manufacturing problems may have led to adjustments in machining or assembly instructions that in turn imply changes in the BOM. Perhaps the torque required to securely fashion the top and base with screws results in warping of the plate or damage to board components. Thus, a different board thickness, cushioning material, or connection technology is used. If these changes are not communicated to purchasing and design, then we may have manufacturing trying to produce a product that is different from that which engineering is testing and analyzing and different from the product for which purchasing is acquiring parts. Finally, accounting is primarily interested in the value of parts in inventory and may not care to distinguish parts that have subtle differences but are used in different product models or positions. If one central BOM is maintained for the entire facility, then the problem of ownership also occurs. Who has the right to change the BOM and how does one stop manufacturing from making changes if they cannot produce the product that engineering designs? Moreover, each function requires certain data that are not of interest to the other functions, thus the full BOM model by itself, may be unnecessarily bulky for each group. This discussion is meant to illustrate how even fairly simple concepts become complex when placed in a dynamic, multifunctional environment.

We have determined that different functional areas require different information in the BOM. The form of the BOM may also differ. One approach is to list for each part or product only those lower-level components that are used directly. As an alternative, we can list total contents. Consider the calculator, for example. Do we list the circuit board as one component of the calculator, or do we list all the chips, resistors, and other components that make up the board? Either approach is possible because we can always reconstruct the complete list by moving progressively through the levels; however, a standard convention must be selected for the firm. The second approach of listing the fully exploded set of parts yields significant duplication across similar final products but keeps information more readily available to the user. We will use an indented format in which each parent item has its direct component parts listed (Figure 8.4).

An imploded version of the BOM also serves an important function. The normal exploded BOM, addressed above, lists the component or child items for each parent item. We "explode" or take apart the parent item to reveal the parts of which it is made. In implosion, the BOM lists for each item the parent items that use it. This information is very handy at times and thus the imploded BOM may also be stored. Suppose that we receive a call from a supplier who states that because of a production problem, an outstanding order for parts will be delivered late. Unless we know where those parts are to be used, we cannot adjust our internal production schedule. The imploded BOM allows us to trace up the hierarchy to see where those delinquent parts were to be used and, therefore, which internal shop orders need to be rescheduled. At times, we may even decide to disassemble subassemblies or halt production orders in process to cannibalize those parts for higher-priority final product production.

There is one convention that is relatively standard (but sometimes confusing). The part hierarchy is numbered from the top level down with the top level being level 0. Thus, end items are at level 0. Their main assemblies are at level 1 and so on. This was the convention used in Figure 8.3.

8.2   REQUIREMENTS EXPLOSION AND ORDER RELEASE PLANNING

Planning order releases constitutes the fundamental operation of MRP. The process involves moving down through the product hierarchy one level at a time, allocating existing inventory and determining remaining part requirements for each item at each level as we proceed. This process is often referred to as "requirements explosion." The key terms in performing the explosion for an item are **gross requirements, scheduled receipts, net requirements,** and **planned order releases.** Gross requirements indicate the number of units of item i that must be available at the start of period t to enable planned production to proceed. Gross requirements are determined by the number of units needed for the planned production starts of the parent item(s) in that period. For example, if a particular circuit card is used in two models of calculators, and we plan to assemble 100 units of one model and 250 units of another, then we have a gross requirement of $100 + 250 = 350$ circuit cards this week. Scheduled receipts account for units in process or on order. These are the units belonging to orders that have been placed with vendors that have not yet been received. We use the convention that the arrival period listed for a scheduled receipt is the period when we expect those units to be available for use. The current time will always be the end of period 0. Suppose that it takes two weeks to replenish circuit cards, and a batch of 400 was started last week. Those cards belonging to that open shop order should be completed by the end of this week and be available one week from now to satisfy gross requirements. These units would be listed as scheduled receipts for week 2. Current on-hand inventory represents units available for use at the start of the planning horizon and effectively represent receipts scheduled at time 0. When discussing on-hand inventory in future periods, we refer to the end-of-period inventory level. This will be the inventory after scheduled receipts have arrived and gross requirements consumed for that period. Net requirements are what is left of gross requirements after we take into account starting on-hand inventory and scheduled receipts. We then plan order releases in time to obtain parts to satisfy the net requirements.

The explosion process involves calculation of the gross and net requirements for each item i in each time period t. It is based on the inventory level identity:

$$\text{ending inventory} = \text{starting inventory} + \text{units received} - \text{units demanded}.$$

"Units received" are given by scheduled receipts in any period. "Units demanded" are the gross requirements. Our inventory balance can thus be written:

$$I_{it} = I_{i,t-1} + SR_{it} - GR_{it} \tag{8.1}$$

where GR_{it} is the level of gross requirements for part type i in period t, I_{it} is the on-hand inventory of item i at the end of period t, and SR_{it} is the number of units scheduled to be received of part type i in period t.

Net requirements for any period are then given by the excess of gross requirements over starting inventory (that inventory in stock at the end of the previous period) and scheduled receipts. The basic equation for MRP explosion in the first period of the planning horizon is thus:

$$NR_{i1} = GR_{i1} - I_{i0} - SR_{i1} \tag{8.2}$$

where NR_t is the net requirements for part type i in period t measured in number of parts. If initial inventory and scheduled receipts are sufficient to cover gross requirements, the net requirements in Equation (8.2) will be negative indicating that we end the period with inventory on hand. Thus, there is not really a requirement for part type i. By convention, we use only the term *net requirements* to refer to a positive need for parts. Thus, we revise Equation (8.2) to:

$$NR_{i1}^+ = \max\{0, GR_{i1} - I_{i0} - SR_{i1}\} \tag{8.3}$$

It is understood hereafter that net requirements refer to any actual positive number of requirements for part type i in period t. Likewise, I_{it}^+ refers only to the number of units on hand at the end of period t and will therefore be non-negative. If we are out of inventory, then $I_{it}^+ = 0$, even if back-orders exist. We can generalize these relationships beyond the first period. For any period t, the explosion process uses the relation:

$$NR_{it}^+ = \max(0, GR_{it} - SR_{it} - I_{i,t-1}^+) \tag{8.4}$$

and

$$I_{it}^+ = \max(0, I_{i,t-1} + SR_{it} - GR_{it}) \tag{8.5}$$

Once the net requirements are known, we can plan order releases to the shop to cover requirements. Net requirements indicate the total number of shortages we would have for that item if we do not take any corrective action. Thus, we must place orders to receive a quantity of units at least as large as the net requirements in that period (or earlier). If positive lead times exist, order releases must be scheduled before the period in which the net requirements occur. The lead time for order replenishment is typically set to an integral number of periods in MRP. Letting the lead time for part type i be τ_i, for now we place orders for NR_{it}^+ units in period $t - \tau_i$. For lead time τ_i, planned order releases are set to:

$$POR_{it} = NR_{i,t+\tau_i}^+ \tag{8.6}$$

to avoid shortages. Expression [Equation (8.6)] is used for items in a make-to-order production environment. It can also be used in a make-to-stock environment; however, in the sections below, we address modifications to this relation for make-to-stock items that allows us to trade off the cost of setups and carrying inventory.

Throughout the discussion, we have assumed that gross requirements are known. Gross requirements are determined by planned production of parent items plus independent demand of the item for spare parts and direct sales. Let $S(i)$ represent the set of parent items that require part type i. Now, it may take one unit of i per unit of $j \in S(i)$. On the other hand, if part i is a table leg and j is the table, each j requires four parts of type i. In general, we let a_{ij} be the number of units of i needed per unit of j for $j \in S(i)$. We also let d_{it} be the estimated independent demand for part type i in period t. We can then obtain gross requirements from:

$$GR_{it} = d_{it} + \sum_{j \in S(i)} a_{ij} \cdot POR_{jt} \tag{8.7}$$

EXAMPLE 8.1

Consider a common stapler, the BOM for which is listed in Figure 8.5. The stapler is assembled from three simple subassemblies. The base assembly consists of a jaw support plate that is inserted through slits in the metal base plate. A thin tension-modulating metal plate is then placed between the upright walls of the jaw support, and two rivets are staked through the assembly to secure the parts. (The plate helps absorb the hitting force when the stapler top is hit during stapling. One end of the modulating plate is bent upward to absorb the force.) The base plate is then inserted into the plastic base molding. Another major component of the stapler is the spring assembly. The spring assembly applies force to keep the supply of staples forward in the staple magazine. For the spring assembly, a copper spring wraps around a metal pin at the top, and the bottom of the spring wraps

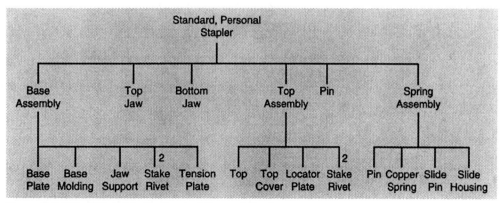

Figure 8.5 Bill of Materials for Standard Personal Stapler

around the end of the plastic slide housing. A slide pin is also inserted into the housing. The top assembly of the stapler consists of a locator plate that is staked onto the top plate and a plastic top molding cover. Assembly of the stapler is performed at a single workstation. The jaw top is positioned in the top assembly, and then the jaw bottom is positioned. These pieces are then slid over the jaw support, and a pin is inserted to hold the stapler together. During the final step, the spring assembly is inserted into the jaw assembly.

The MPS calls for producing 1000 staplers a week for the next three weeks. Production will then be halted for two weeks, followed by a release of 1500 staplers in week 6. Find the necessary planned order releases for the stapler, top assembly, spring assembly, locator plate, and slide pin. (In practice, we would want the order releases for all items; however, we restrict our attention to this subset of parts for demonstrating the principle of requirements explosion.) It is the start of week 1. Current on-hand inventory levels are 1212 top assemblies, 1150 spring assemblies, 146 locator plates, and 558 slide pins. It takes two weeks to acquire locator plates from a vendor; the lead time for all other parts and assemblies is one week. Currently, an order for 750 locator plates is outstanding and should arrive early in week 1. A batch of 650 slide pins is just about complete and will be received shortly as well.

SOLUTION

We begin with the planned order releases for the end-item stapler. Current time is the end of period 0. As described in the problem data, the stapler PORs are given in the MPS as:

Product: personal stapler no. 12468		Week					
	1	2	3	4	5	6	
Planned order releases	1000	1000	1000	0	0	1500	

The PORs for the stapler become the gross requirements for the base assembly, top assembly, spring assembly, jaw top, jaw bottom, and jaw pin. For the two assemblies of interest, we can now fill out their gross requirements, initial on-hand inventory, and scheduled receipts. Gross requirements match planned stapler order releases (Figure 8.6a). Next, applying Equation (8.4) we can compute net requirements for level 1 items including the top assembly and spring assembly (Figure 8.6a). Values in period 0 represent current on-hand inventory quantities. We plan for period 1 and thereafter. For the spring assembly, net requirements in period 1 are determined by subtracting the 1150 units on

	Week						
	0	1	2	3	4	5	6
Top Assembly: Lead time = 1							
Gross Requirements		1000	1000	1000			1500
Scheduled Receipts							
(Projected) On-Hand	1212	212					
Net Requirements			788	1000			1500
Planned Order Releases		788	1000			1500	
Spring Assembly: Lead time = 1							
Gross Requirements		1000	1000	1000			1500
Scheduled Receipts							
(Projected) On-Hand	1150	150					
Net Requirements			850	1000			1500
Planned Order Releases		850	1000			1500	

Figure 8.6a Level One Order Planning for Example 8.1

				Week			
	0	1	2	3	4	5	6
Locator Plate: Lead time = 2							
Gross Requirements		788	1000			1500	
Scheduled Receipts		750					
(Projected) On-Hand	146	108					
Net Requirements			892			1500	
Planned Order Releases		892*		1500			
Slide Pin: Lead time = 1							
Gross Requirements		850	1000			1500	
Scheduled Receipts		650					
(Projected) On-Hand	558	358					
Net Requirements			642			1500	
Planned Order Releases		642			1500		

Figure 8.6b Level Two Order Planning for Example 8.1

hand from the 1000 unit gross requirement. This leaves us with 150 free units expressed as projected on-hand inventory at the end of period 1. We do not have a net requirement in period 1 for spring assemblies. In period 2, there are no scheduled receipts but an additional gross requirement of 1000 units. Subtracting this gross requirement leaves us with a net requirement of 850 units and no units on hand. This is shown in the week 2 column and in the net requirements row in Figure 8.6a. The remaining net requirements for spring assemblies and top assemblies are calculated in a similar manner, one period at a time (Figure 8.6b). Planned order releases are obtained by offsetting any positive net requirements by the item lead time as given in the Table. By releasing orders for period t net requirements in period $t - \tau_i$, the units should arrive in time for use. The stated lead time for both top assemblies and spring assemblies is one week. Thus, whenever a positive net requirement exists for one of these items, we plan an order release one week in advance for the same number of units.

Once all level 1 items are planned, we can proceed to level 2. For our example, this includes the 13 items shown in Figure 8.5 that are used to make the base assembly, top assembly, and spring assembly. Consider two of these items—the locator plate and slide pin. The gross requirements for locator plates are set equal to the planned order releases of top assemblies. Gross requirements for slide pins are determined by spring assembly order releases. As before, we proceed one period at a time, subtracting initial on-hand units and scheduled receipts from gross requirements to obtain net requirements or on-hand inventory in each period. The gross requirement of 850 slide pins in week 1 is matched by the 558 units currently in stock plus the 650 about to be received leaving 850 − 558 − 650 = 358 slide pins in stock at the end of week one. These 358 units can be used against the 1000 additional units needed in week 2 leaving a net requirement of 1000 − 358 = 642 units needed. As before, to achieve positive net requirements, we plan orders for release, τ_i time units before part type i is needed. Our plan is feasible for slide pins. However, the locator plate has a two-week lead time. This presents a problem. Even with the on-hand and on-order units, we do not have enough to cover demand in week 2. We will be short 892 units next week as noted by the asterisk in Figure 8.6b. We can go ahead and place the order now, but if ordered in week 1, these units will not be available until week 3. Consequently, manual replanning is necessary. Perhaps we can expedite the order and reduce the lead time.

A few general comments are necessary before concluding our discussion of the explosion and requirements netting process. First, the example assumed that the MPS was given in terms of order releases for end items. In some cases, the MPS lists desired product completions. In this case, we would have to treat these values as net requirements and offset the planned order releases based on the lead time. It is less likely but possible that the MPS will be given in terms of gross requirements. In this case, we would have to adjust these values by allocating current inventory and outstanding orders.

Second, suppose that a part appears at several levels in the BOM. This is not uncommon for standard components, as demonstrated in Figure 8.7a. A standard support brace denoted as part C in the figure is used at two levels of end-item A and at one level for end-item B. In doing the MRP explosion, we must account for all uses of an item. We could perform multiple processing of the requirements for an individual item, once for each level on which it appears. For now, this would only be a waste of computational effort. However, shortly we examine lot-sizing procedures whereby we may combine net requirements for several consecutive periods into a single production run. In that case, it is important to be aware of all net requirements when planning order releases. The solution to this problem is to process each item only once by delaying processing of each occurrence of an item to the lowest level at which it appears. Figure 8.7b shows the redrawn BOM to place all uses of item C on the same level.

A third point relates to computing gross requirements for a part used by multiple parent items and/or as a spare part with external demand. Note that the stake rivet in Figure 8.5 is such a part because it is used in both the base assembly and top assembly. In computing gross requirements, we must add the gross requirements derived from all parent items and external demand. This is shown in the example on the following page.

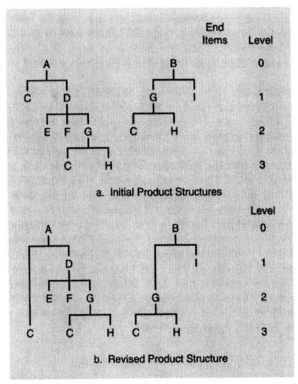

Figure 8.7 BOM Structure for Multiuse Items

EXAMPLE 8.2

A particular component, part C, is used in two final products, A and B. It takes one part C to make a unit of product A; however, each product B requires three Cs. Plans call for starting 100 units of part type A into production each week. Planned order releases over the next four weeks for part type B are 50, 0, 60, and 30 units, respectively. In addition, estimated spare-part demand for part C is estimated at 75 units per period. Find the gross requirements for part type C.

SOLUTION

To find the gross requirements for item C, we must combine planned order releases from A with those from B and external demand. Noting that it takes three units of C per B, the gross requirements in period 1 become $GR_{C1} = POR_{A1} + 3 \cdot POR_{B1} + d_{c1} = 100 + 3(50) + 75 = 325$. Using the same procedure for periods 2 through 4, $GR_{C2} = 175$, $GR_{C3} = 355$ and $GR_{C4} = 265$.

8.2.1   Scrap Losses

The above discussion assumed that order sizes would be determined directly from net requirements. In practice, we may need to plan for scrap losses. If 100 units are needed, and the process usually has a 90% yield factor (meaning that 10% of units are scrapped), then we would need to order $\frac{100}{0.9} = 111$ units on average. Of course, yield losses are typically random. We may want to start extra units to ensure the completion of 100 good items. This becomes a statistical problem, with the best choice being determined by the distribution of actual yield from any run, the value of any extra units produced, and the cost of shortages. It may be possible to start a second batch if evidence indicates that the first batch will have too many defective items. Alternatively, there may be a safety stock of units kept for use if too many are scrapped, or we may have the option of delivering fewer final products if the scrap rate is too high. The economics and customer expectations of the situation determine the best course of action.

8.2.2   Safety Stock and Safety Lead Time

Related to the decision of how many units to start into production because of scrap losses, we may want to guard against variability in demand requirements and/or lead time. In a dynamic environment, requirements can change. If, tomorrow, a customer requests that an order be expedited or increased, we need to have materials available to accommodate the request. If a workstation becomes temporarily overloaded and its cycle time increases, then we have a problem maintaining a schedule for all end items that use the components made in that workstation. These situations can be handled by added safety stock or safety lead time. With safety stock, we order extra items, intending to have a few on hand at all times for quantity variations. The reorder point inventory models, addressed previously, can help determine the amount of safety stock needed. Carlson and Yano [1986] describe an incremental approach for setting safety stock levels in discrete-time MRP systems in which emergency setups are possible. If variability comes primarily in the form of the timing of requirements, then safety lead time should be used. Orders are planned for release prematurely to ensure that parts will be scheduled to arrive before their expected use. Then, if changes in demand create a premature need for the part, we are covered. Likewise, if the workstation takes longer than expected to complete the order, we are covered.

Safety lead time can be implemented by artificially increasing the lead time. If the distribution of lead time is known, then we can select the release date for an order based on ensuring a desired probability of on-time delivery. If we increase the lead-time parameter value by l periods, then orders will arrive l periods early on average. The mate-

Part: $\tau = 2$	0	1	2	3	· · ·
Gross Requirements		12	6	11	
Scheduled Receipts		15			
(Projected) On-Hand	5	8	2		
Net Requirements				9	
Planned Order Releases		9			

Figure 8.8 Initial Material Plan for Safety Stock Illustration

rial plan will show the order quantity being carried in inventory for l periods before being used.

Safety stock is slightly more complex to accommodate in MRP. Two common choices are to: 1) reduce the initial on-hand quantity by the size of the safety stock or 2) increase gross requirements in period 1 by the size of the safety stock. Calculations can then proceed as normal. Either method will work. However, if changes occur because of an increase in requirements or shortages in receipts, then the material plan may indicate a shortage, when in reality, we have extra safety stock on hand. The material planner must be aware of how to interpret the plan. Consider the initial plan shown in Figure 8.8. Desired safety stock of 10 units has been subtracted from the initial on-hand quantity. Thus, whereas it appears that we have five units on hand, we actually have 15. With the current plan, the first order is to be placed in period 1 for nine units. Suppose that an important customer requests a change in an order that increases gross requirements for the part from six to 11 units in period 2. With only two units left on hand in period 2, and a two-week lead time, it appears that we will have a shortage of three units that cannot be met. This revised situation is shown in Figure 8.9. In reality, we can meet this demand and still have seven units of safety stock. With this knowledge, we can respond to the customer that we can handle his request and still deliver on time. We add the extra five units to the order intended to be released in period 1. This will return the safety stock to the desired level of 10 units when the order arrives in period 3.

Part: $\tau = 2$	0	1	2	3	· · ·
Gross Requirements		12	11	11	
Scheduled Receipts		15			
(Projected) On-Hand	5	8			
Net Requirements			3	11	
Planned Order Releases		14			

Figure 8.9 Revised Plan for Safety Stock Illustration

8.3 CAPACITY PLANNING

A plan for meeting demand is meaningless if insufficient capacity exists to execute the plan. Thus, just as we have a hierarchy for planning production with shorter time frames and increasing levels of detail as we move down that hierarchy, we have a parallel planning process for ensuring that adequate capacity exists. A key factor in aggregate planning is trying to match demand as closely as possible given aggregate resource capacity constraints. In an aggregate sense, then, we should be confident that the mid-range aggregate capacity plan is feasible. Similarly, when the aggregate plan is decomposed into a master production schedule, we should ensure that the plan has a chance to succeed. At this step, rough-cut capacity planning is usually executed to ensure that sufficient production hours and materials are available to meet the MPS goals. At the detailed explosion level, capacity requirements planning (CRP) is integrated with MRP to find the resource requirements for every work center in every period. In this section, we discuss the processes of rough-cut capacity planning and the more detailed capacity requirements planning. In both cases, the methods addressed in this text are oriented toward determining the capacity required to execute the production plan. If the derived capacity requirements for a plan exceed available capacity, then manual intervention is required for adding capacity or adjusting the production plan. MRP systems that perform requirements explosion without regard for available capacity are called **infinite loading** systems.

8.3.1 Rough-Cut Capacity Planning

Several approaches exist for **rough-cut capacity planning.** We concentrate on the use of resource profiles for checking the feasibility of the MPS. A resource profile for an end item is a combination of BOM and process plan. The profile first breaks down the end item into each of its component parts and offsets them in time based on their precedence structure and the lead time for each component. Figure 8.10 demonstrates the profile for an end item in a three-level BOM. End item A is made from one part C and one part D. Part C has a two-week lead time, and part D has a one-week lead time but in turn requires parts E and F. Each part is further broken down into the operations required for manufacture if it is produced internally. The standard times per unit for each operation and the work center involved are recorded. When the entire Figure is viewed, we see the profile of resource requirements. The profile indicates the loading placed on each work center in the plant over time to make one end item A. This profile assumes that the standard times are correct and that the timing of part coordination will be as tight as indicated in the schedule. Thus, no consideration of slack time, current inventory levels, or lot-sizing is included.

To compute total resource requirements by time and work center, we add all end items and all periods in the MPS. Let r_{ijk} be the standard time per unit of end item i required in work center j in the kth period before the item is completed. Thus, in Figure 8.10,

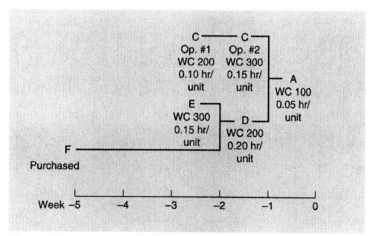

Figure 8.10 Resource Load Profile for Example 8.3

$r_{A,100,0} = 0.05$ hours and $r_{A,200,2} = 0.10$ because each unit of end item A requires 0.05 hours at work center 100 in the period it is produced, and component C requires 0.10 hours in work center 200 two periods before the completion of an A. According to the Figure, the lead time for part type C is two weeks. An order will spend one week in work center 200 and then proceed to work center 300 for the second week. The total load on work center j implied by the MPS in period t is then given by:

$$w_{jt} = \sum_{i=1}^{n} \sum_{k=t}^{T} r_{i,j,k-t} \cdot Q_{ik}$$ (8.8)

where Q_{ik} is the planned production quantity for end item i in period k as listed in the MPS. A basic check of feasibility for the MPS is to ensure that all w_{jt} values are within capacity limits.

EXAMPLE 8.3

A production facility produces two final products, A and B. The resource profile for A is shown in Figure 8.10. Item B is similar but requires two Cs. Each work center has 100 labor hours available per period. Determine the capacity requirements for the following MPS. The MPS represents planned order releases for end items A and B.

End item	Period					
	1	2	3	4	5	6
A	50	75	25	50	200	
B	75		125	100		150

SOLUTION

We have two end items and three work centers (identified as 100, 200, and 300 in Figure 8.10). The first step involves determining the aggregate load profile for each end item. This aggregate profile shows the amount of resource required in each work center, in each previous period for each unit of the end item produced. For our problem, the resource is labor hours. The aggregate profile is shown in Figure 8.11. To complete a unit of A requires 0.05 hours of final assembly time in the last period (lag = 0 periods) as indicated in Figure 8.10. In the previous period, we had to spend 0.2 hours making part type D in work center 200 and 0.15 hours in work center 300 for the second operation of part type C. Two periods ago, we had to do the first operation for part type C in work center 200 and also produce part type E in work center 300. Because part type F is purchased, it does not consume any internal production resource. The profile for end item B is similar except that the contribution from the two operations of part type C are doubled because we require two Cs for each B.

The next step is to apply the aggregate profiles to the planned production quantities for each end item. For each period and each item appearing in the MPS, the load profile for that item is multiplied by the MPS planned quantity, and the resulting resource values are assigned to the work center in the corresponding time-lagged period. For example, to produce 50 units of A in period 1 requires a workload of $50(0.05) = 2.5$ hours to work center 100 in period 1, $50(0.2) = 10$ hours to

	End Item					
	A Period			B Period		
Workcenter	0	−1	−2	0	−1	−2
100	0.05			0.05		
200		0.20	0.10		0.20	0.20
300		0.15	0.15		0.30	0.15

Figure 8.11 Aggregated, Time-Phased Load Profiles for End Items (Hrs/Unit)

Workcenter	−1	0	1	2	3	4	5	6
					Period			
100			6.25	3.75	7.50	7.50	10.00	7.50
200	20.00	32.50	42.50	55.00	50.00	70.00	30.00	
300	18.75	41.25	33.75	63.75	67.50	52.50	45.00	

Figure 8.12 Total Resource Requirements (Hrs.) for MPS of Example 8.3

work center 200 in the previous period (labeled period 0 in the figure), $50(0.15) = 7.5$ hours to work center 300 in period 0, $50(0.1) = 5$ hours to work center 200 in period -1 and $50(0.15) = 7.5$ hours to work center 300 in period -2. The values in Figure 8.12 are the total hour requirements found by adding the workloads implied by all MPS values. Consider work center 200 in period 1. The workload of 42.5 hours is the sum of three orders. The MPS calls for producing 75 units of A in period 2. Each A produced in period 2 requires 0.2 hours in work center 200 in period 1, a workload of 15 hours. We also plan to produce 25 As in period 3 and 125 Bs in period 3. Each A requires 0.1 hours in period 1 at work center 200 for an additional 2.5 hours. Each B requires 0.2 hours for an additional 25 hours. Adding 15 hours $+$ 2.5 hours $+$ 25 hours yields the 42.5 hour total.

We can also obtain these values directly by using Equation (8.8). Consider the busiest work center/period combination, namely, work center 300 in period 3. From Equation (8.8):

$$w_{300,3} = \sum_{i=A,B} \sum_{t=3}^{6} r_{i,300,t-3} \cdot Q_{it}$$

$$= 0.15 \cdot Q_{A4} + 0.15 \cdot Q_{A5} + 0.30 \cdot Q_{B4} + 0.15 \cdot Q_{B5}$$

$$= 0.15 \cdot (50) + 0.15 \cdot (200) + 0.30 \cdot (100) + 0.15 \cdot (0) = 67.5$$

Workload for negative-numbered periods implies the need for on-hand inventory of these items. Otherwise, it is too late to start production of the components and meet the MPS.

8.3.2 Capacity Requirements Planning

Using average load profiles for end items, rough-cut capacity planning provides an efficient means of estimating time-phased workloads by work centers. However, rough-cut planning ignores several details such as accounting for current inventory levels and the impact of lot-sizing decisions. To proceed to the next level of detailed capacity planning requires integration of capacity planning with detailed production plans. **Capacity Requirements Planning** (CRP) performs this task. Essentially, CRP augments key capacitated resources to the MRP explosion process. Planned order releases and open shop orders are combined with data on standard times for setups and variable processing times to determine requirements in each time period for each resource. Stored capacity in the form of item inventories and WIP are allocated to requirements after which remaining net requirements are computed.

CRP processing entails adding the information from the resource profile reports to the BOM explosion. Resources required to produce a part are viewed as extra components for that part. When an order is planned, it produces a gross requirement for the corresponding resources in the same time-phased manner and quantity as stated in the load profile for that item. Planned order releases and scheduled receipts dictate the gross requirements for resources. For open shop orders listed as future scheduled receipts, work already completed forms the on-hand inventory for the resource. This completed work is effectively stored capacity for the work center because it contains work that need not be performed in the future.

EXAMPLE 8.4

Returning to Example 8.3, current on-hand inventory quantities for items C, D, and E are 260, 200, and 100, respectively. There is an open shop order for 110 units of C that is scheduled to arrive next week. Determine the capacity requirements in work center 200 over the planning horizon.

SOLUTION

We begin by determining the type of planned order releases that place a load on work center 200. From Figure 8.10, we see that the first week of production for part type C places a load of 0.10 hours per unit on work center 200. Part type D also requires 0.2 hours of work center 200 time per unit produced during its one-week lead time. The next step is to determine the scheduled receipts and planned order releases for part types C and D. Based on the MPS for items A and B, the material plans for part types C and D are given in Figure 8.13. Gross requirements for C are equal to the MPS plan for A plus twice that for B. Gross requirements for D are the sum of end items A and B releases. After allocating current inventory and scheduled receipts and offsetting net requirements by the lead time, we obtain the planned order releases shown in Figure 8.13. The final step in our analysis involves allocating the effort required to execute the planned order releases for parts C and D. Gross requirements for work center 200 time come from part type C in the week it is released and also from part type D. Thus, in period 1, the gross requirement stems from the release of 180 Cs, and this requires $180(0.10) = 18$ hours. In week 2, we require (25) hours for the release of 250 Cs plus 30 hours for the release of 150 Ds. We do not need to worry about any scheduled receipts in this case because the scheduled receipt for part type C in period 2 should have completed operation 1 last week and no longer require work center 200 time. The full capacity requirements plan for work center 200 is given in Figure 8.14.

Two comments need to be made before leaving this example. First, we choose not to count WIP that had left the work center but that was still in process as on-hand work for work center 200. The open shop order for 110 Cs could alternatively be counted as 11 hours of on-hand capacity for work center 200. If this approach was taken, then the 11-hour requirement imposed on work center 200 by this order and reflected in the scheduled receipt should be counted as a net requirement for work center 200. Adding the on-hand WIP and then subtracting the implied scheduled receipt workload will cancel out, and the net requirements will be the same. Second, the difference between the workload estimates in this example and those in the previous example are attributable to the addition of detailed information here on current inventory status and planned order release timing.

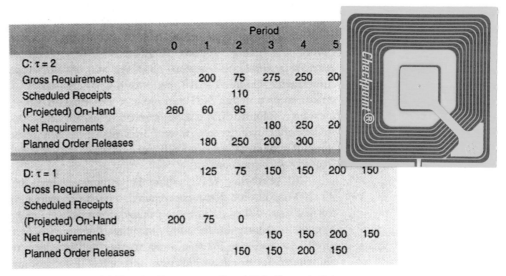

	Period					
	0	1	2	3	4	5
C: τ = 2						
Gross Requirements		200	75	275	250	200
Scheduled Receipts			110			
(Projected) On-Hand	260	60	95			
Net Requirements				180	250	200
Planned Order Releases		180	250	200	300	
D: τ = 1		125	75	150	150	200	150
Gross Requirements						
Scheduled Receipts						
(Projected) On-Hand	200	75	0			
Net Requirements				150	150	200	150
Planned Order Releases			150	150	200	150

Figure 8.13 Material Plan for Components C and D in Example 8.4

	Period						
	0	1	2	3	4	5	6
Gross Requirements		18	55	50	70	30	
(Projected) On-Hand	0						
Net Requirements		18	55	50	70	30	

Figure 8.14 Hours or Capacity Required for WC 200 in Example 8.4

A potential problem can arise in our use of counting completed work as resource inventory. If the stored (completed) work is on parts not currently needed, those parts cannot be used to meet current needs. Counting the stored capacity may make us believe we have sufficient capacity when, in reality, we have a shortage of capacity for executing the high-priority jobs. To illustrate, suppose that we have completed 10 hours of work on a job for part type A and have five hours of work remaining to complete the job. However, the order is not due for two days. Assume that the work center has 20 hours of capacity per day, and an outstanding order due tomorrow morning requires 25 hours for some part other than A. Looking at the report, the planner will see a net requirement of -5 hours for day 1 resulting from the gross requirement of 25 hours, minus 10 hours of inventory on hand, minus 20 hours of capacity. Thus, the report appears to state that everything is acceptable. In reality, we cannot meet the due date by the end of today because we cannot use the 10 hours of stored capacity to meet today's order. Even if we switch from the job in process to the higher-priority job, we will need five hours tomorrow to complete the job.

8.3.3   Load Reports

Load reports are a common means for displaying the workload on a work center resulting from a plan. A sample load report is shown in Figure 8.15. The load report shows the number of hours required because of open orders and planned order releases. Any backlog of uncompleted orders may also be shown. If required workloads exceed the capacity level, then we know we have a problem and orders must be rescheduled or extra capacity found. In the Figure, we see that the current level of backorders is small and can be handled in week 1, but a problem appears likely in week 2.

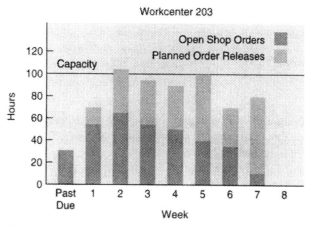

Figure 8.15 Sample Load Report

8.3.4 Incorporating Stochastic Behavior

The preceding capacity planning models assumed that processing requirements and processor availabilities were known. We could thus project the flow of batches using a deterministic view of the world. In actuality, the world is less predictable. Machines break down and temporary backlogs follow idle periods because interarrival and processing times vary. The MRP model assumes a fixed lead time for planning order releases and estimating loads. If shop conditions are random but stationary (average utilization levels are constant over time), we can use historical data to estimate lead times or use queuing models. We noted previously that we would want to pick some upper percentage point of the lead-time distribution when setting the MRP lead-time parameter to avoid late completion of a large number of orders. Historical records showing empirical histograms of time at each work center may be maintained for this purpose. With estimates of the interarrival and service time distributions for batches at workstations, we can use GI/G/1 (general interarrival time, general service time, single server) queuing models to estimate capacity and lead times.

The shop floor functions as a collection of related work centers. Completion of batches at one work center leads to arrivals at the subsequent workstations. The shop therefore becomes a network of interrelated queues. More accurate estimates of lead times can be found by incorporating production sequence information from process plans to relate interdeparture and interarrival times. Shanthikumar and Buzacott [1981] describe one approach for connecting these work centers. This approach is further illustrated in Askin and Standridge [1993].

8.4 LOT-SIZING DECISIONS

During our MRP explosion, we sized order releases to match net requirements in each period, just offset for lead time considerations. This is referred to as the **lot-for-lot** (LFL) lot-sizing policy. The LFL strategy makes sense in make-to-order environments and when the setup cost is smaller than the cost to carry an extra period's worth of demand in inventory. In these situations, there is no reason to accumulate inventory of parts for later use. On the other hand, if setup is a significant factor, and the same part types are used again and again over time, then less frequent production releases can be used to help lower total costs. The simplest approach would be to compute an EOQ or EMQ for each item as we did with reorder point inventory systems and then always plan order releases for this amount. Another common alternative is to determine a regular order frequency or period, such as placing an order every four weeks. The order quantity will then vary based on net requirements in the upcoming weeks. This periodic order quantity (POQ) approach facilitates planning because we know which part types will be produced in each upcoming period. Parts sharing the same production resources can be sequenced based on ease of part type to part type changeovers. Their sequence can be spread over the repeating cycle length to smooth workload from period to period.

EXAMPLE 8.5

Consider the net requirements for a gear housing shown in Figure 8.16. Compute the planned order releases for an EOQ strategy. Each production job costs $100 to set up. Holding costs are estimated at $0.10 per housing per week.

SOLUTION

The first step is to determine the economic order quantity. We use the basic EOQ expression $Q = \sqrt{\dfrac{2AD}{h}}$. To apply this result, we need to estimate weekly demand. Cumulative gross require-

Gear Housing: $\tau = 1$	0	1	2	3	Period 4	5	6	7	8	9	10
Gross Requirements		117	145		175	128	211		74		242
Scheduled Receipts		215									
(Projected) On-Hand	13	111									
Net Requirements			34		175	128	211		74		242
Planned Order Releases											

Figure 8.16 Net Requirements for Example 8.5

ments over the 10-week period given in Figure 8.16 are for 1092 housings. A reasonable estimate then is: $\overline{D} = \dfrac{1{,}092 \, units}{10 \, weeks} = 109.2 \, units/week$. Using the quantity, $Q = \sqrt{\dfrac{2 \, AD}{h}} = \sqrt{\dfrac{2(100)(109.2)}{0.10}} = 467$. We round to the more even number of $Q = 450$ units per order.

We have a positive net requirement for period 2 in Figure 8.16. With a one-week lead time, we must plan an order release for period 1. Figure 8.17a shows a planned order release for 450 units in period 1. This will cause an order to be scheduled for receipt in period 2. (Usually, we would not duplicate the planned order releases and scheduled receipts, but we show both in the Figure to more clearly illustrate the changes in planned inventory position.) These values along with updated on-hand inventory values are shown in italics in Figure 8.17a. The 450 units will last until period 6, at which time we will have a net requirement for 98 additional housings. Because we always order 450 units at a time, an order for this amount is scheduled in period 5. Figure 8.17b shows the updated inventory position that would be expected. Our plan then is to release an order for 450 gear housings in period 1 and in period 5. Note that this plan results in the carrying of significant levels of inventory. Although we present this approach because it is sometimes used in practice, it is clear that the use of the EOQ in some sense ignores the potential of MRP to eliminate unnecessary inventory levels by matching production to actual demand of parent items.

Gear Housing: $\tau = 1$	0	1	2	3	Period 4	5	6	7	8	9	10
Gross Requirements		117	145		175	128	211		74		242
Scheduled Receipts		215	*450*								
On-Hand	13	111	*416*		*241*	*113*					
Net Requirements							98		74		242
Planned Order Releases		450									

Figure 8.17a Modified MRP Inventory Status with First Planned EOQ Order

Gear Housing: $\tau = 1$	0	1	2	3	Period 4	5	6	7	8	9	10
Gross Requirements		117	145		175	128	211		74		242
Scheduled Receipts		215	*450*				*450*				
(Projected) On-Hand	13	111	*416*	*416*	*241*	*113*	*352*	*352*	*278*	*278*	*36*
Net Requirements											
Planned Order Releases		450				450					

Figure 8.17b Final MRP Inventory Status for EOQ Orders

EXAMPLE 8.6

Repeat the previous example using a periodic order quantity policy.

SOLUTION

The POQ policy is to produce the item on a regular cycle based on expected time between orders in an EOQ system. POQ provides the advantage of cyclical and advanced planning. We can sequence part production each period based on setup and other considerations.

The EOQ = 467. Average demand is $\overline{D} = 109.2$. On average, we need to produce every $\frac{EOQ}{\overline{D}} = \frac{467}{109.2} = 4.28$ periods. Rounding to the nearest integer, we should plan to produce this item every four weeks. Each time we produce, we set the order quantity to equal the net requirements over the succeeding four weeks. We have already established that an order must be placed in period 1. With a four-week cycle, this means we will also release an order in weeks 5 and 9. The week 1 order quantity is set equal to net requirements in periods 2 through 5. The week 5 order release must cover net requirements in periods 6 through 9, and the week 9 release covers requirements for periods 10 through 13. Because net requirements are unknown beyond period 10, we will plan this last order based only on period 10 requirements. Our planned order releases are then for $Q = 34 + 175 + 128 = 337$ units in period 1, $Q = 211 + 74 = 285$ units in period 5, and $Q = 242$ units in period 9. This schedule is shown in Figure 8.18. Although inventory is still carried over for some periods because of setup costs, the POQ strategy at least manages to avoid carrying inventory into periods in which a receipt is scheduled. In principle, this should lead to lower costs in the long term.

							Period				
Gear Housing: $\tau = 1$	0	1	2	3	4	5	6	7	8	9	10
Gross Requirements		117	145		175	128	211		74		242
Scheduled Receipts		215	337								242
(Projected) On-Hand	13	111	303	303	128	0	74	74	0		0
Net Requirements											
Planned Order Releases		337				285				242	

Figure 8.18 POQ Orders and Resultant Inventory Status

You may recall that we solved a very similar problem for dynamic reorder point based systems. As an alternative to LFL, EOQ, and POQ, we could use the Silver-Meal heuristic or the Wagner–Whitin (WW) algorithm described in Chapter 6 to combine net requirements into batched planned order releases. At first look, it would seem that the WW procedure would in fact give the optimal solution of how to batch requirements. Unfortunately, however, the WW algorithm can be sensitive to the length of the planning horizon and is seldom used for MRP lot-sizing. WW assumes the product life ends at the end of the planning horizon. Instead, the horizon moves forward each period so that there will always be another T period.

The use of lot-sizes for grouping net requirements into planned order releases seems like a straightforward problem. However, lot-sizing creates significant "nervousness" and workload variability in MRP systems. Nervousness refers to significant changes in order release plans from period to period as the schedule is updated for new data. To understand how this occurs, refer to Figure 8.16. Suppose that a customer asks to delay an order of 40 units from period 2 to period 3. Now the gross requirements are 105 in period 2 and 40 in period 3. The 111 units in stock are now enough to cover period 2 requirements. Using the lot-for-lot approach, we would simply cancel the period 1 release of 34 units and add a planned order release for 40 units in period 2. With EOQ, we would postpone releasing the order for 450 units from period 1 and reschedule it in period 2. This is a much more significant change in planned workload. Our manpower plan could be disrupted, and we may arrive at period 2 and find that we have insufficient capacity to make the 450 units when combined with orders for all other products because we did not plan far enough ahead. The problem is potentially worse if the customer wants units early, and we have to try to move up an order to an earlier period. We suddenly have a request for a large number of raw materials that was not expected.

The use of lot sizes significantly increases the variability of the workload on work centers from period to period (see Askin [1988]). Periods of high production follow periods of low production, the exact opposite of the smooth continuous production flow we desire. This problem worsens if multiple levels of lot-sizing further increase order sizes and the time between orders. The use of rotation cycles in which many products are produced on the same equipment can help alleviate this problem, and thus a POQ policy is reasonable. Beyond this, there is still an issue of dependency between levels. Batching of net requirements into orders results in intermittent planned order releases. This in turn forces large infrequent requirements for the item's components. The setup and holding cost structure for the entire product hierarchy should therefore be considered when setting order sizes. Considerable research has been conducted in the past two decades on determining the optimal method for grouping net requirements into orders for multilevel production systems. Some of this work is described in the next chapter.

8.5 MANAGING CHANGE

As the world evolves, so must our production plans. As time passes from one period to the next, our production schedule rolls forward, and we must update for the new time period. The old period is removed from our planning horizon, and the new period is added. Engineering changes become incorporated in product designs, manufacturing implements new technology, new production orders are released, and old orders are completed in accordance or contradiction with our old plan. In this section, we discuss in more detail some of the issues that must be addressed in a dynamic environment and how we can address those issues when using MRP.

8.5.1 Net Change vs. Regenerative Updates

At a minimum, we will probably want to update the production schedule once each period. As forecasts and the MPS change, so do gross requirements for end items. Even if all events go according to plan, each MPS period moves the planning horizon forward, and new end-item production plans need to be passed down to the MRP system. More than likely, however, during this interval, some outstanding orders have arrived early and some are late; some inventory records have been corrected for discrepancies, some parts have been lost or damaged. ECOs have changed the BOM and thus changed part requirements, and scrap rates on some orders have deviated from their expected values. The shop is in a state of flux and replanning is, therefore, the norm.

The processing of data and updating of operational plans can be performed in either a **net change** or **regeneration mode.** The choice depends largely on the frequency with which inventory transactions are input to the system and at which changes in plans are to be considered. In the regenerative approach, the entire system plan is recreated anew at prescribed intervals. The interval length may be as long as the length of a time bucket. Transactions are stored and then inventory records are updated, and order releases are replanned for the entire product hierarchy, one level at a time. Old plans are no longer rel-

Parent: $\tau = 2$	0	1	2	3	4	5	6	7	8
Gross Requirements		5	25	12	42	0	25	55	0
Scheduled Receipts			30						
(Projected) On-Hand	33	28	33	21					
Net Requirements					21		25	55	
Planned Order Releases			21		25	55			

Component: $\tau = 1$									
Gross Requirements			21		25	55			
Scheduled Receipts									
(Projected) On-Hand	23	23	2	2					
Net Requirements					23	55			
Planned Order Releases				23	55				

Figure 8.19 Initial MRP Plan for Parent Item and Component

evant except that they determine on-hand inventory status and scheduled receipts for orders in process. In the net change approach, we update frequently, perhaps even continuously, as new inventory transactions are reported. Instead of recreating the entire plan, only the differences in status are input to the explosion process. In this section, we address the processing of inventory transactions and the two approaches to replanning.

Regenerative systems are only indirectly concerned with deviations from expected shop performance. Inventory transactions such as receipt or issuance of items are stored and then processed at the start of the regeneration cycle. Then the material requirements explosion is executed. The newly updated files containing current values of on-hand and on-order inventory status are consulted for each part when its turn comes in the hierarchical explosion process. The information is used to fill out the current on hand and scheduled receipt entries that help determine net requirements. The fact that we expected an order to arrive last week with 30 units, but instead, it is still in process and five units have already been discarded, is irrelevant. We need only know that 25 units are now scheduled to arrive next period.

This is not the case for net change operation. When events occur that change inventory status, they are fed to the system and necessary changes are processed. Consider the parent item and component part shown in Figure 8.19. A problem occurs in the work center that produces the parent item. Suppose four out of 30 parent items scheduled to be received in week 2 have to be scrapped and the remaining items must be reworked, delaying receipt a week. The updated plan for the parent item is shown in Figure 8.20. Because planned order releases change for the parent, gross requirements change for the component as given in Figure 8.21a. This change in gross requirements is processed yielding the new plan in Figure 8.21b. Because we added a gross requirement of 4, the 2 units of on-

Parent: $\tau = 2$	0	1	2	3	4	5	6	7	8
Gross Requirements		5	25	12	42	0	25	55	0
Scheduled Receipts				26					
(Projected) On-Hand	33	28	3	17					
Net Requirements					25		25	25	
Planned Order Releases			25		25	55			

Figure 8.20 Updated Plan for Parent Item Due to Scrap Loss

Component	Week								
	0	1	2	3	4	5	6	7	8
Gross Requirements			+4						

Figure 8.21a Net Change in Gross Requirements for Component

Component: $\tau = 1$	Week								
	0	1	2	3	4	5	6	7	8
Gross Requirements			25		25	55			
Scheduled Receipts									
(Projected) On-Hand	23	23							
Net Requirements			2		23	55			
Planned Order Releases		2		23	55				

Figure 8.21b Updated Plan for Component Item Due to Scrap Loss

hand inventory changed to a net requirement of 2 units. This initiated a new order for 2 units planned to be released in period 1. If this component is also a parent for other items, then we would need to next drop down and modify these plans as well based on the change in planned orders for this component.

It should be apparent that net change processing requires more computation than regeneration. For each event, we must trace through the hierarchical explosion process updating all affected material plans. However, net change processing spreads out the computational workload over a continuous time interval instead of batching. This avoids a surge in computation whenever the schedule is regenerated. Net change processing also provides more timely plans.

8.5.2 Order Issuance

The MRP system must handle the process of actually issuing planned order releases. In each period, action must be taken by planners to convert the planned order releases into actual shop orders. When this authorization is given, and the order is released to the shop, then the planned order release must be converted to a scheduled receipt in the material plan. At the same time, the on-hand and net requirement fields for components are updated. This process is demonstrated in Figures 8.22a and 8.22b showing the before and after cases.

To actually release the order, we must have all materials and component parts available. Thus, the material planner must first check to make sure sufficient quantities of all required components are on hand. These units must then be designated for use in this order. Somehow, we must indicate that these units are no longer freely available for use even though they will temporarily still be sitting in the same inventory storage location. These claimed, but as yet unused items, are called **uncashed requisitions.** The necessary quantity must be subtracted from the on-hand stock of the materials and component parts. Also, as the order is released and these component parts and materials are allocated to the order, their gross requirements are adjusted accordingly to reflect that the planned order release of the parent item is removed. The uncashed requisitions are listed separately much in the same manner that the starting on-hand stock is listed. Consider a component item that is used by the part type planned in Figure 8.22. Two of these lower-level parts are needed for each parent item. Figure 8.23 shows the before and after status of this part as the planned order for its parent item is released. By updating the status of the available

Parent Item: $\tau = 2$	Week					
	0	1	2	3	4	5
Gross Requirement		30	15	40	25	· · ·
Scheduled Receipt						
(Projected) On-Hand	45	15	0			
Net Requirements				40	25	
Planned Order Release		40	25			

a. Material Plan Immediately Before Order Release

Parent Item: $\tau = 2$	Week					
	0	1	2	3	4	5
Gross Requirement		30	15	40	25	· · ·
Scheduled Receipt				40		
(Projected) On-Hand	45	15	0	0		
Net Requirements					25	
Planned Order Release			25			

b. Material Plan Immediately After Order Release

Figure 8.22 Illustration of Order Release Execution

on-hand inventory in this manner, we can be sure that we do not try to use the same parts twice. By reducing the gross requirements, we account for the fact that we no longer need to be concerned with planning to acquire components for this parent-item order. When the order for 40 parent items is released to the floor, we reduce the on-hand inventory of the component by 80 units from 110 to 30 in this case. The period 1 gross requirements are reduced by a corresponding 80 units. If the component was used by more than one

$a_{ij} = 2$ Component: $\tau = 1$	Period			
	0	1	2	3
Gross Requirements		80	50	· · ·
Scheduled Receipts				
(Projected) On-Hand	110	30		
Net Requirements			20	
Planned Order Releases		20		

a. Component Item Immediately Before Order Release

Uncashed Requisitions = 80

$a_{ij} = 2$ Component: $\tau = 1$	Period			
	0	1	2	3
Gross Requirements		0	50	· · ·
Scheduled Receipts				
(Projected) On-Hand	30	30		
Net Requirements			20	
Planned Order Releases		20		

b. Component Item Immediately After Order Release

Figure 8.23 Effect on Component of Actual Order Release of the Parent Item

parent, then it is quite possible that some gross requirements destined for other parent items would remain. For simplicity, in our example, we show the case in which all the gross requirements result from a single-parent item planned order.

8.5.3 Order Pegging

When problems occur during the execution of orders, we would like to be able to modify the schedule to reflect the new situation. Shortages of a component will affect the ability to meet production goals for its parent items. Rather than tie up a productive resource waiting for components that are not available, we should reschedule production of the parent item. This in turn affects the schedule for the parent item's parent(s). If the schedule for the parent item changes, then requirements for other components that go into that parent item are also impacted. We see a cascading effect as the component shortage sends ripples throughout the entire web of items to which it is linked in the BOM. Indeed, taking a broader view, if the customer order calls for multiple products, all of which must be delivered together, then delays experienced for one component could conceivably affect the requirements for other items completely unrelated through the BOM structure.

Order pegging allows us to trace the source of requirements upward from a component to end items. This is useful when we must adjust the schedule for a parent item resulting from the lack of availability of one of its components. The process of correcting the material requirements plan commences when we determine that a disruption has occurred in the timing or quantity of a scheduled receipt for a lower-level item. A correction to an on-hand inventory count for the component could initiate a similar process. In general, the MRP system stores a **where–used** file that is essentially an inverse view of the BOM. The where–used file for an item lists all of its parent items. This file lists the possible items that may be directly affected by a component shortage. However, if external demand or multiple-parent items exist, pending shortages cannot be immediately identified with a specific parent item. Thus, for any current material requirements plan, the system maintains a subset of the where–used file for each positive gross requirement on every item. This subset is referred to as a **pegged requirements** file. The pegged requirements file lists each of the parent items and their corresponding order numbers that contributed to that gross requirement. Pegged requirements files, therefore, provide the knowledge needed to identify the affected parent or parents and make the appropriate changes in their production plan. For example, consider a gear, number B3516, that is used to make a variety of old-fashioned cuckoo clocks. Gears are also shipped directly to dealers who use them in repairing clocks. An abbreviated MRP record with the associated where–used file is shown in Figure 8.24. The 60 units required in period 3 are destined for three locations. Being produced that week are two different clock models that use the gear and other gears are being shipped directly to a customer who warehouses spare parts. Production control personnel can decide the best parent item to absorb any delays in obtaining parts for period 3. A production planner will review the information contained in the pegged requirements file and decide the least disruptive or costly way to accommodate the change. Suppose, for example, the problem resulted when 15 of the gears were found to be defective. The gears are being reworked and will be available in one week. Delaying order 10716 would not solve the problem by itself, since that would only reduce the gross requirement by 12 units. If the dealer that generated order number 31789 was not a key customer or was in a hurry, we could delay shipment one week. Otherwise, we could delay order 10831 for parent item A1117 by one week or split it into two orders—one with a gross requirement of 13 for this week and the second with a gross requirement of 15 for next week. To make this decision, we may need to trace up the BOM hierarchy, one level at a time, to find the parent of A1117 and then its parent, and

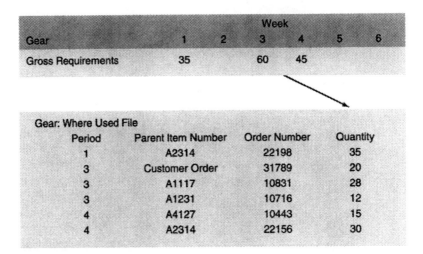

Gear		Week					
	1	2	3	4	5	6	
Gross Requirements	35		60	45			

Gear: Where Used File			
Period	Parent Item Number	Order Number	Quantity
1	A2314	22198	35
3	Customer Order	31789	20
3	A1117	10831	28
3	A1231	10716	12
4	A4127	10443	15
4	A2314	22156	30

Figure 8.24 Illustration of Pegged Requirements for a Commonly Used Gear

so on until we reach the end-item level. At each level, we use the pegged requirements file for the gross requirements to identify which of its parent items should be changed. As we delay the scheduled receipt date for either 15 or all 28 units of A1117, then we may likewise need to modify the planned order releases for one of its parents.

When parent items are replanned as we move up the BOM, we may choose to explode back down through their BOM structure to update gross requirements for all their children. Because one or more planned order releases were delayed for the parent, the gross requirements for its other children are also affected. Whereas this may not be necessary since the due dates are usually pushed back in time, failure to update in this manner will cause the plant to use its resources ineffectively. Some critical orders may sit idle while unnecessary parts are produced. As this becomes known to shop personnel over time, the MRP system tends to lose its credibility, making it difficult to maintain a shop discipline for producing to the schedule.

We have implicitly assumed in this section, that the production scheduler can affect only the timing and quantity of order releases. This may well be the case. However, the scheduler may also be able to affect lead time. Pleas to a vendor to expedite shipment or changes to shop due dates may allow the planner to recover from problems without the need to replan parent. In the next section, we will understand how the use of firm-planned orders can help in this regard as well as how gross requirements are nominally set based on lot-sizing considerations.

We have described a single-level pegging system. Item requirements are tied to planned orders for their direct parent. In using this system, we have to continue moving upward level by level until reaching the end-item level to determine the effect of item shortages. We then have to trace all the way back down the hierarchy to determine the effects of the problem with the original component and to adjust all affected plans. Although not commonly done, it is possible to maintain this entire trace in the where–used file to avoid having to reconstruct this trail each time.

8.5.4 Firm-Planned Orders

As data records are updated, the MRP system will constantly attempt to change the timing and size of planned orders. This nervous condition of constantly changing plans can wreak havoc on purchasing and manufacturing and destroy the credibility of the MRP plan. Suppose that a change in inventory status resulting from a few parts being discarded causes us to want to change an order release for a part from week 3 to week 2. We have to be concerned with whether the corresponding materials and components can be obtained in time. By the very nature of MRP, if we do not require these components until week 3, it is unlikely that they will be available in week 2. However, the MRP system will reschedule the parent item from week 3 to week 2 if allowed to act on its own. The computerized program will assume that the gross requirements imposed on the component parts can be met when it gets to the next level. This, however, may not be possible. To avoid such problems and reduce the nervousness of MRP systems, **firm-planned orders** are used. Several studies have shown that freezing a portion of the planning horizon can be effective in MRP systems (see for example, Blackburn et al. [1986] and Zhao and Lee [1993]). With low-demand uncertainty and nervousness because of lot-sizing rules, it may be best to freeze the entire planning horizon. In general, shorter frozen horizons are preferred as demand uncertainty increases. Reducing the replanning frequency can also reduce nervousness and lower costs. Although the schedule may be tabulated on the basis of a weekly period, it is not necessary to rerun the schedule that often.

The human planner may specify a particular planned order as being "firm." The computer is then prevented from automatically changing the timing or size of that order. If all planned order releases scheduled for the cumulative lead time of a part are made firm, then the system can operate in a calm manner, meeting these requirements (assuming of course that capacity is assured through feasible CRP). On the other hand, firming all orders for the length of the cumulative lead time can be overly restrictive and result in a system that is slow to react to a changing market. In general, some compromise between the extremes of continuous replanning and long firm-order horizons is used.

To demonstrate firm-order planning, consider the following scenario. One component of an electronic keyboard is a set of 36 white keys. The keys are purchased from a vendor and take three weeks for delivery. Keyboards take one week to assemble. Keys are usually ordered once every three weeks. The current inventory status and MRP plan is shown in Figure 8.25. The planned order release of 23 units in period 2 accounts for the net requirements in periods 3, 4, and 5. Suppose that a customer calls and asks for an ex-

		Period						
	0	1	2	3	4	5	6	7
Keyboard: $\tau = 1$								
Gross Requirements		10	10	10	10	10	10	10
Scheduled Receipts		25						
(Projected) On-Hand	2	17	7					
Net Requirements				3	10	10	10	10
Planned Order Releases			23			20		
Key Package: $\tau = 3$								
Gross Requirements			23			20		
Scheduled Receipts			20					
(Projected) On-Hand	8	8	5	5	5			
Net Requirements						15		
Planned Order Releases			15					

Figure 8.25 Initial Inventory Status and Material Plan

			Period					
	0	1	2	3	4	5	6	7
Keyboard: $\tau = 1$								
Gross Requirements		10	10	10	10	25	10	10
Scheduled Receipts		25						
(Projected) On-Hand	2	17	7					
Net Requirements				3	10	25	10	10
Planned Order Releases			38			20		
Key Package: $\tau = 3$								
Gross Requirements			38			20		
Scheduled Receipts			20					
(Projected) On-Hand	8	8						
Net Requirements			10			20		
Planned Order Releases		10[a]	20					

[a] Indicates Order Will Be Late

Figure 8.26 Tentative Material Plan with Shortage Problem for Revised Customer Order

tra 15 keyboards in week five. This causes a problem because if we order the extra keys now, it will take six weeks before they can be converted into finished keyboards. This situation is shown in Figure 8.26. The new net requirement for keys in week 2 cannot be met and the plan will fail. To meet the net requirement of 10 sets of keys in period 2, we would have to release the order three periods earlier. But it is obviously too late to release an order in period 1. Consider the alternative shown in Figure 8.27 instead. We can fix the first order for keyboards at 23 in week 2, and then the MRP system will automatically place a second order for keyboards in period 4 in time to meet all demand. We have previously planned to obtain keys for the order of 23 sets in period 2 thus we are okay. The

			Period					
	0	1	2	3	4	5	6	7
Keyboard: $\tau = 1$								
Gross Requirements		10	10	10	10	25	10	10
Scheduled Receipts		25						
(Projected) On-Hand	2	17	7					
Net Requirements				3	10	25	10	10
Planned Order Releases			23*		35			
Key Package: $\tau = 3$								
Gross Requirements			23		35			
Scheduled Receipts			20					
(Projected) On-Hand	8	8	5	5				
Net Requirements					30			
Planned Order Releases		30						

*Firm Planned Order

Figure 8.27 Update Plan with Firm Planned Order

new period 4 planned order release for keyboards can be met by placing a new order for keys in period 1.

8.6 LIMITATIONS AND REALITY OF MRP

MRP has been used extensively in practice since large scale computing capability first started becoming available in the 1960s. It was felt that MRP would provide the opportunity to take control of production schedules and increase inventory turns.[1] Ideally, by ordering only what was needed, when it was needed, inventory could be minimized and the storage of excess parts eliminated. The vision held that centralized MRP software systems would deliver instructions to all work centers to coordinate and control production activities in an optimal manner. This vision did not come to pass as easily or as effectively as was hoped. What then are the problems implicitly overlooked by MRP? We first describe the assumptions necessary for MRP to operate efficiently, and next, we examine the limitations of the MRP concept.

8.6.1 Environmental Assumptions

MRP makes several assumptions about the production environment and the quality of data. We blend the assumptions and data requirements into a list of necessary technical conditions for success of MRP.

1. Dependent demand. While not a requirement for success, MRP logic is based on the assumption that the gross requirements for one item are determined by the planned schedule for parent items. If this is not the case, the MRP explosion process is unnecessary. We can simply use the basic inventory models studied previously in the text.

2. Feasible master production schedule. One of the most important assumptions is that capacity is adequate to meet the production goals set in the MPS. Many MRP systems use infinite loading and fail to ensure that the MPS plans can be reasonably met. Even if a feasibility check is made in constructing the MPS, it will be made at an aggregate level. This will not ensure that each workstation will have adequate time and production capacity to meet the MPS. Bottleneck workstations may not be able to meet production goals. The capacity requirements planning tools of Section 8.3 are, therefore, essential for maintaining worker confidence in the MRP plan. If insufficient capacity exists, lead-time estimates will be inaccurate and planned orders will not be completed on time. This will disrupt the timing of all related shop orders and our carefully planned coordination of orders will fail to materialize. Job priorities will not be valid, and workers charged with meeting schedules will abandon the plan. The shop will become chaotic as the loudest expediter rules instead of our published plan.

3. Accurate bill of materials. As addressed earlier in this chapter, BOMs evolve over time and serve different purposes for different functional areas. If the BOM is incorrect because of the component parts required or how they are integrated during production, then our production schedule will not be effective and final end-item assembly will be delayed. The required parts will not be available when the plan calls for the release of an order for the parent item.

---

[1]Inventory turns is a common measure used in accounting and finance. Inventory turns are the number of times per year the inventory stock turns over as measured by the ratio of annual value of parts and materials used to the average on-hand inventory value.

4. Accurate and current inventory records. Because MRP allocates current inventory and expected receipts to forecasted demand, our measure of net requirements can be only as good as the accuracy of these data. If the data are incorrect, order sizes will not meet actual net requirements.

5. Known and constant lead times. Order release timing is offset from net requirements based on the lead time. In practice, lead time may be random based on the current level of utilization in the shop. Nevertheless, the order planning process of Equation (8.6) admitted a single lead time per part. This can be one of the major shortcomings of MRP and leads to the need for safety lead time as previously addressed. The safety lead time ends the myth that MRP can precisely match the timing of orders to needs. The related issue of unpredictable yield losses also means that we must order extra parts and carry safety stocks.

6. Employee discipline. For MRP to work, we must keep to the schedule. Parts cannot be cannibalized from one order to serve another because that would put errors into inventory records. Order priorities must be maintained in the production process to ensure the joint replenishment of related components. To the extent that employees do not follow the plan, even for a single order, order priorities and due dates for all shop orders are affected.

7. Batch withdrawals of inventory at known points in time. MRP regards withdrawals of parts from inventory as coming in discrete batches at discrete points in time as being in contrast to the models in traditional inventory theory that typically assume constant withdrawal of parts to meet demand. In actuality, MRP can work in either situation since we plan to have all parts needed for the period available at the start of the period. If withdrawals are continuous, MRP will result in unnecessarily large buffer inventories, but the system will still function.

8.6.2 Structural Limitations

The required assumptions and basic structure of MRP limit its effectiveness. The major factors that limit MRP are as follows.

1. Discrete time buckets. In the MRP explosion, we assume that all the parts to be used in a period have to be available at the start of the period. This is necessary to ensure that a feasible operational schedule will exist wherein each work center can execute its assigned tasks. In actuality, this inventory will be used in some uniform manner throughout the period represented by a time bucket. Unfortunately, when planning MRP order releases, we do not know the sequence of activities in each workstation during the period. On average, this gives us a half-period worth of staging inventory[2] on hand. Much of the benefit of MRP coordination is lost if we must carry half-periods worth of inventory on average. Two choices exist for reducing this inventory. First, we could reduce the length of the time bucket. The amount of inventory is linearly related to the length of a time bucket thus shorter time buckets mean less inventory is needed on hand at the start of each period. Reducing time bucket size, however, requires a similar proportional increase in computation. Moreover, if lead times are variable, it becomes more difficult to predict order completions to the precise period if periods are short. Thus, our plans are less likely to play out as planned if we use short time buckets. Second, we could incorporate a de-

---

[2]By staging, we mean the storage of parts for use by a specific machine.

tailed schedule into the time bucket. This would provide input on the precise time during the period that each raw material would be needed. However, this would require the integration of considerably more decision-making into the MRP process. Even if we have enough computational hardware to sequence the order releases at each work center in each period, for work centers that make parts that reside on different levels of the BOM hierarchy, it is unclear how to create this integration. When the initial workload is assigned to the work center from the higher levels of the explosion, we are unaware of other jobs that will be added later in the explosion process. It is also unlikely that we could predict batch processing times exactly. For both these reasons, the integration of MRP and detailed scheduling has not yet come to pass.

2. Lead time variability. In our MRP explosion, we assumed lead time was known and constant. Order releases were offset from net requirements based on the given value of lead time. Jobs are, therefore, scheduled to arrive just-in-time (where time is measured in periods). Unfortunately, lead time varies in practice. A long run average lead time may exist, but, for any specific order the lead time will depend on the current availability of raw materials, the queue at the required workstations, the availability of labor, and the state of the equipment. Even if the workstation and materials are available, we may complete the operation only to find that the yield is less than expected and thus we have to wait for rework or order additional parts. MRP, however, has planned the entire plant operation assuming that each job will be completed on time and in acceptable condition. We could inflate the expected lead time using perhaps the 99% percentile of the lead-time distribution instead of the mean. Likewise, we could also carry extra inventory of parts in the form of safety stock to guard against this variability. In either case, however, we end up with extra inventory and effectively defeat the purpose of MRP.

3. Infeasible capacity plans—infinite loading. Historically, MRP software has tended to assume infinite capacity. We pretend that workstations can produce whatever quantity is desired. This is not the case in reality, and capacity limits contribute to the lead time variability addressed above. Many software packages present a workload report for each work center after the MRP schedule is computed. However, problems have to be detected, and plans have to be changed manually. The system can then be rerun but this trial and error approach is seldom effective for large systems. The amount of work assigned to a work center can vary drastically from period to period with MRP. This makes the assignment of labor and scheduling of work hours complex and frustrating for both managers and employees. Weeks with high overtime can be followed by droughts of idleness. Upper management becomes disturbed by reports of high overtime followed by low total utilization of resources. The willingness of software vendors and production control personnel to carry data fields showing past due orders as a regular part of planning indicates that the consensus is that we cannot expect to meet all production obligations on time.

4. Accurate data. MRP assumes inventory records are correct. If on-hand quantities are incorrect then net requirements are incorrect leading to incorrect planned order releases. This leads to part shortages, in some cases, with excess inventory in others. This dilemma spawned the job category of cycle counting. Each morning the cycle counters receive a list of part numbers to verify. For each part number, their task is to find and count all the parts of that type on the shop floor and in storage. This total is compared to the electronic inventory record, which is then updated for corrections. Counts differ for many reasons—scrap rates may not be recorded correctly after processing and inspection, the actual number of parts obtained from a batch of raw material may differ from the standard because of material quality, parts may be lost or damaged during transport or storage, and parts may be removed from storage or cannibalized for other uses without permission or feedback to the inventory record file. With MRP trying to produce the exact number of units needed, all such discrepancies lead to part shortages.

5. Hierarchical planning interactions. MRP is part of the production planning and control hierarchy. Morecroft [1983] documents how MRP can increase inventory fluctuations and decrease labor productivity when interfaced with materials ordering, aggregate planning, and labor management decisions in an environment with limited capacity. Suppose that external demand increases. The MRP explosion will cause orders to be released for all parts. Work centers become backlogged and lead times increase. As lead times increase, the MRP system will try to release orders even earlier, thus increasing WIP. With limited capacity, final goods production may not increase significantly and inventory stocks will decrease as we try to fill demand. The aggregate planning function will see inventory dropping and shortages increasing. Overtime will be authorized and authorization to hire workers will be given. The early release of jobs will overstate the demand rate and we will soon increase capacity more than necessary to meet the new average demand level. Then, with excess capacity, the aggregate plan will authorize excess production. Eventually, lead times become shorter than the original values, and finished goods inventory increases. Now order releases are placed at a rate slower than actual demand. Soon, we will decide to keep workers idle and slowly reduce the labor hours available. Thus, in the first half of the cycle inefficiencies occurred resulting from overtime, and in the second half of the cycle we have idle resources. The cycle will be self perpetuating. By slowing the transfer of information and relating orders for materials from one production stage to actual consumption by its successor, traditional reorder point inventory does not cause the entire hierarchy to overreact to this level. The remedy for this problem requires a more sophisticated integration of aggregate planning with MRP and a finite loading approach to MRP.

In summary, MRP is doomed by its attempt to plan detailed actions using an aggregated, deterministic model of shop cycle time. Its efficiency is limited by the length of a time bucket and its effectiveness is limited by instability in forecasts and scrap rates. The implicit assumption of infinite capacity produces infeasible production requests in a limited capacity production environment.

8.7 EXTENSIONS AND INTEGRATION INTO THE FIRM

We have presented push-oriented MRP as a stand-alone system. In actuality, MRP is part of a larger information processing scheme for the firm. It can be implemented as a high-level shop planner that coordinates with a business planning system to ensure adequate materials, finances, and capacity. Likewise, it can feed a kanban or other shop floor control policy to ensure manufacturing plans are executed. In this section, we address extensions of MRP and its overall role in the firm's decision-making system. MRP becomes a vital node not only in the temporal, hierarchical, and horizontal integration of decisions within the firm, but also in the extended network linking customers, producers, and suppliers.

8.7.1 Manufacturing Resources Planning (MRP II)

MRP has been in existence since the early 1960s. Since then, we have learned to appreciate the need for total system integration. We have already addressed integrating capacity planning into MRP. The focus there was on the labor and machine **resources** required for executing the plan. Material Requirements Planning has evolved into Manufacturing Resources Planning (MRP II). MRP II links modules for capacity planning and other pertinent functions to the MRP system. In addition to capacity planning, typical key modules include a bill of material processor, customer order entry, engineering change order management, financial planning, forecasting, inventory and warehouse management, master scheduling, and purchasing. The importance of the BOM and inventory modules are evident as these constitute key inputs into the MRP explosion. Forecasting and master scheduling are needed to initialize the gross requirements at the end-item level. These modules must feed back to MRP and capacity planning to ensure plans can be realized. In general, there must be two-way communication between the MRP II modules to ensure plans are workable. This is referred to as a **closed-loop** system.

A full fledged MRP II system drives all business operations. The interface with finance determines cash and other financial needs over the planning horizon. If planned order releases are known, then the amount of cash required to purchase raw materials can be computed. If resources are embedded via CRP, then labor costs and energy consumption can be estimated for each future period. Simulation capability may even be added to assist in financial planning. Cash flows and inventory levels can be predicted and various production plans tested to determine their performance and required outlays over time. Levels of accounts payable and receivables may be extrapolated over the planning horizon. The simulation may evaluate the impact of limited capacity as some orders become past due, and the schedule integrity deteriorates. Integrating order entry reduces the delivery lead time by reducing the delay in ordering parts and scheduling orders. It also allows marketing to provide more realistic estimates to customers.

8.7.2   Enterprise Resources Planning (ERP)

The next step beyond MRP II moves outside the firm and integrates functions across the supplier–producer–customer network. Electronic data interchange (EDI) may be used to connect partners across the network, reduce communication time and facilitate coordination. We have seen the advantages in efficiency from sharing information across the supply chain. Imagine a system where real-time point-of-sale data at the retail level drives not only replenishment orders to the warehouse, but also production and distribution schedules are driven across the supply chain. As jobs are scheduled, internal resources are encumbered and purchase orders are automatically issued to vendors all the while checking feasibility and projecting financial parameters. The common term for such systems is Enterprise Resource Planning (ERP). Typically, sophisticated information system designs are utilized including relational databases, friendly graphical user interfaces, and client/server architectures. Experience has shown that such systems are expensive and time–consuming to implement, maintain, and operate. The decision problems for specifying production schedules are large and necessarily require heuristic approximations when setting plans. Recent trends seem to move toward the use of Internet-based service bureaus that specialize in providing ERP solutions and managing the system for the manufacturing enterprise. We are rapidly reaching the point in which vendors may be included in the information system creating an extended virtual logistics hierarchy.

8.7.3   Hybrid Systems

MRP and kanban systems are not necessarily incompatible. MRP handles the control of intermittent or low-volume items best. Kanban works very well for mid- to high-demand items such as those components that are used in all end items. In practice, some parts may be only needed when their specific end item is to be produced while other common parts are used at a constant rate. It is feasible to use both approaches simultaneously and to let the parameters for each part determine its inventory control strategy. Items, both finished and component parts with a high but relatively stable demand rate use a pull system. Maximum levels are set for each item at each of their production stages using the rules described in the previous chapter. If large orders are occasionally received for these items, it is possible to handle these large orders separately with a push strategy. Items with intermittent and/or lumpy demand can be controlled by an MRP scheduler. Work centers that produce both JIT and MRP controlled items should have a constant proportion of their time dedicated to the JIT items and should also be operated with finite loading in the MRP planned order release module. Setting aside time for the pull items helps ensure predictable replenishment lead times for those items. Using a finite capacity MRP scheduler will ensure that the dependent-demand items are produced on time. A hybrid approach may also be used when item importance varies. The MRP system is simplified if only relatively expensive, dependent-demand items are included. Lower-cost items can be controlled by kanban or reorder point models. It is also possible to use MRP to ensure the feasibility of the master schedule and to plan material purchases. Batches for end items are then released accordingly, but a kanban system is used on the shop floor to control the flow of parts.

EXAMPLE 8.7

Consider, the finished product family shown in Figure 8.28. Arc labels indicate the number of child items needed per parent item. Numbers in parentheses are average lead times for that part. Finished product assemblies A1 and A2 are produced to a master production schedule. Common parts C and D are kanban controlled with containers that hold 20 units. Lead time is usually no more than one third of a week for the kanban-controlled items. A 10% safety factor has been established by plant management. Part types B and E are specific to each product model and use MRP control. Planned assembly for product A1 is 0, 25, 0, and 30 over the next four periods. For A2, the plan calls for producing 75, 50, 120, and 50 units per week in the next four weeks. Historically, the average demand for A1 and A2 is 50 products per week of each. Part C and part family B are small metal rotational parts and are made in the same fabrication cell. This cell is the key capacity constraint in the plant. The fabrication cell operates ten hours per period, and each unit requires 0.02 hours. The injection molding department manufactures part type D. Capacity is sufficient in injection molding as this area has the most slack capacity in the plant. Part Es are ordered from a vendor. Planned lead times for ordered and MRP-controlled items is set to one week, the smallest time period possible since items are not always available for use in the same week we produce them. Currently, there are 20 B1s, 100 E1s, and 80 B2s on hand. An open order for 170 E2s is due very shortly. Develop a hybrid production schedule for this facility over the next four weeks.

SOLUTION

For the first step, consider the constant-use items C and D. If demand for A1 and A2 average 50 units per week each, then demand for C averages $2(50) + 4(50) = 300$ units per week. Likewise, demand for D averages $1(50) + 2(50) = 150$ units per week. To find the number of kanbans required, recall that $K_i = \dfrac{\tau_i \cdot D_i \cdot (1 + \alpha)}{n_i}$ where n_i is the container size, α is the safety lead-time factor, and K_i is the number of containers for part type i. Using this result, $K_C = \dfrac{(0.33) \cdot (300) \cdot (1.1)}{20} = 6$

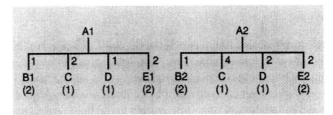

Figure 8.28 Product Family Structure for Hybrid System Example 8.7

and $K_D = \dfrac{(0.33) \cdot (150) \cdot (1.1)}{20} = 3$. Note, that in producing 300 part type Cs per week, we use $(0.02)(300) = 6$ hours per week. This leaves 4 hours of capacity available per week for MRP-controlled items.

The second step involves computing net requirements and planning order releases for parts B1, B2, E1, and E2. Based on the planned assembly of A1 and A2, and the BOM, the requirements for these items are shown in Figure 8.29. Scheduled receipts and on-hand quantities are taken directly from the inventory and shop status data given in the problem statement. The MPS build schedule and a_{ij} coefficients determine the gross requirements.

Finally, we produce a load report for the bottleneck work center. Using planned order releases for items B1 and B2, we obtain the net requirements for the fabrication cell shown at the bottom of Figure 8.29. Because requirements never exceed four hours in any period, it appears that we have adequate capacity, at least for the next several weeks.

			Week		
	0	1	2	3	4
B1: $\tau = 1$					
Gross Requirements		0	25	0	30
Scheduled Receipts					
(Projected) On-Hand	20	20			
Net Requirements			5	0	30
Planned Order Releases		5		30	
B2: $\tau = 1$					
Gross Requirements		75	50	120	50
Scheduled Receipts					
(Projected) On-Hand	80	5			
Net Requirements			45	120	50
Planned Order Releases		45	120	50	
E1: $\tau = 1$					
Gross Requirements		0	50	0	60
Scheduled Receipts					
(Projected) On-Hand	100	100	50	0	
Net Requirements					10
Planned Order Releases				10	
E2: $\tau = 1$					
Gross Requirements		150	100	240	100
Scheduled Receipts		170			
(Projected) On-Hand	0	20			
Net Requirements			80	240	100
Planned Order Releases		80	240	100	
Fabrication Cell (Hours)					
Gross Requirements		1.0	2.4	1.6	
On-Hand					
Net Requirements		1.0	2.4	1.6	

Figure 8.29 Order Releases and Capacity Requirements for Example 8.7

Another hybrid possibility uses MRP to estimate loads and determine the feasibility of the MPS through capacity requirements planning, but lets actual order releases be controlled by kanbans. The kanban system provides very little information for financial or capacity planning. MRP can fill this void. We might even use MRP to ensure that purchased parts are ordered in time if their usage rate varies over time depending on the MPS. As long as the CRP determines the MPS is feasible, we can allow individual jobs to be controlled by desired kanban stocks. This approach uses the best of both worlds. Because it is often easier to predict total business volume than individual end-item demands, we use MRP for general load balancing but use kanbans to ensure our production adjusts rapidly to actual product demand.

8.7.4   MRP Performance Level

Although the ideal of the fully integrated MRP II system is appealing, most companies have failed to implement such an ambitious system. It is customary to refer to MRP users as being a class A, B, C, or D user. At the top, rests the class A user company. Class A users are few, but those that climb this high enjoy the fruits of their labor. In a class A company, a closed-loop MRP system is in full operation. Capacity requirements planning is integrated to ensure feasibility of plans and order priorities are set based on the MRP plan. Feedback from capacity plans, inventory transactions, and shop status factor into plan updates. The MRP system dictates order release, shop floor dispatching, and purchasing. The entire plant agrees to the model. Engineering has bought into the system and works with it for planning changes. Financial planning and general managerial decision-making are based on the plan. At the next level, Class B users have less integration with engineering and business functions, but MRP does serve as the production and inventory control system. CRP helps to maintain credibility, but there may still be a fair amount of expediting since not all functions have accepted the MRP model. Class C users operate in an open-loop manner. MRP is used for ordering parts but not for scheduling. Typically, the class C user will have little if any capacity planning in the data processing system. At the bottom lies the class D user. The MRP records are exploded and made available to departments, but because of an unrealistic MPS and inaccurate inventory records, the plan is unreliable and no real attempt is made to follow the materials plan. As you would expect, the benefits from MRP are highly dependent on the class of usage reached by the firm.

8.8   SUMMARY

Material requirements planning has been implemented in a variety of manufacturing companies for guiding the process of planning and controlling the release of orders to the shop floor. MRP coordinates the release of orders for dependent-demand items in dynamic-demand environments to minimize the carrying of unnecessary inventory. It can be used in both make-to-stock and make-to-order environments with appropriate choices for lot-sizing. MRP decision-making relies on inventory records, bill of material structures, the master production schedule, and lead-time estimates.

MRP operates on a rolling horizon. Each period, inventory status is updated, and the bill of materials for end items are exploded down the product hierarchy one level at a time. At each level, current inventory and open orders are matched against gross requirements over the planning horizon. Order releases are then planned to meet the remaining net requirements. The planned order releases for parent items become the gross requirements at the next lower-level of the hierarchy. Schedule updates can be performed either on a continuous net change basis or by periodic regeneration of the entire production schedule. Production plans for the first few periods can be frozen to minimize nervousness in the schedule and ensure the availability of component parts and materials. If problems do occur, pegging of part requirements to their parent item in which they will be used allows the production control department to detect problems and take corrective action.

Without accurate and current values for these data parameters, MRP systems will fail. Thus, considerable pressure is placed on the manufacturing system to acquire and process shop status data and for all workers to demonstrate the discipline necessary to continue with the plan. Additional requirements for success include stable processes with predictable scrap rates, good demand forecasts, and sufficient capacity to execute MPS production goals. Capacity requirements planning is usually integrated with MRP to ensure this last condition. Load profiles can be used to estimate the feasibility of a master production schedule. Detailed capacity requirements planning can be embedded into the explosion process to ensure sufficient capacity at each resource in each period given standard time data and current inventory status.

Despite the obvious logical appeal of MRP, it has its limits. By trying to eliminate all unnecessary inventory, MRP becomes sensitive to random disturbances in the shop. Minor deviations from predicted shop performance can make plans for all items invalid (not just those directly affected). In addition, the use of finite length time buckets and the separation between MRP planning and actual order scheduling leads to carrying at least a half period's worth of production in inventory on average.

MRP has evolved into more general manufacturing resource planning (MRPII) or even enterprise resource planning (ERP). These approaches attempt to integrate all planning functions of the firm from capacity planning to finance. Although integration into the firm clearly provides benefits, MRP continues to be limited by its use of discrete planning periods and static lead times.

8.9 REFERENCES

ASKIN, R.G. & RAGAVAN, M. The effect of lot-sizing on workload variability. *Journal of Operations Management,* 1989, 4(1), 53–71.

BLACKBURN, J.D., KROPP, D.H., & MILLEN, R.A. A comparison of strategies to dampen nervousness in MRP systems. *Management Science,* 1986, 32(4), 413–429.

CARLSON, R.C. & YANO, C.A. Safety stocks in MRP–Systems with emergency setups for components. *Management Science,* 1986, 32(4), 403–412.

DING, F. & YUEN, M. A modified MRP for a production system with the coexistence of MRP and kanbans. *Journal of Operations Management,* 1991, 10(2), 267–277.

GARDINER, S.C. & BLACKSTONE, J.H. JR. The effects of lot sizing and dispatching on customer service in an MRP environment. *Journal of Operations Management,* 1993, 11(2), 143–160.

JACOBS, F.R. & WHYBARK, D.C. A comparison of reorder point and MRP inventory control logic. *Decision Sciences,* 1992, 23(2), 332–342.

MORECROFT, J.D.W. A systems perspective on MRP. *Decision Sciences,* 1983, 14, 1–18.

ODEN H., LANGENWALTER, G.A., & LUCIER, R.A. *Handbook of material and capacity requirements planning.* New York: McGraw-Hill, Inc., 1993.

ORLICKY, J. *Material requirements planning.* New York NY: McGraw-Hill, Inc., 1975.

PTAK, C.A. *MRP and beyond: A toolbox for integrating people and systems.* Chicago: Irwin Professional Publishing, 1997.

PLOSSL, G. *Orlicky's material requirements planning.* New York: McGraw Hill, Inc., 1994.

SAMITT, M.D. & BARRY, A. Mixed model operations: Solving the manufacturing puzzle. *Industrial Engineering,* 46–50, 1993.

VOLLMANN, T.E., BERRY, W.L. & WHYBARK, D.C. *Manufacturing Planning and Control Systems,* Irwin, Homewood, IL, 1988.

WHYBARK, D.C. & WILLIAMS, J.G. Material requirements planning under uncertainty. *Decision Sciences,* 1976, 7, 595–606.

ZHAO, X. & LEE, T.S. Freezing the master production schedule for material requirements planning systems under demand uncertainty. *Journal of Operations Management,* 1993, 11(2), 185–205.

8.10 PROBLEMS

8.1. What are the data requirements for operating an MRP system?

8.2. What is meant by a push system in contrast to a pull system?

8.3. List the standard modules that comprise an MRP II system.

8.4. What information is contained in a bill of materials, and what problems can develop if different functional areas of the firm are using different versions?

8.5. Describe the fundamental flaws in MRP that prevent it from eliminating excess inventory in practice.

8.6. Disassemble an old hand-held calculator. Construct a bill of materials for the calculator showing all parts and levels.

8.7. Visit an office supply store and sketch a picture of a computer desk with drawers for storage. Construct a BOM for the desk.

8.8. The net requirements for a product for the next twelve weeks are 0, 120, 95, 67, 83, 110, 0, 59, 134, 120, 91, and 108.

Demand averages 100 units per week, each setup costs $150 and holding cost is $0.07 per unit per week. Lead time for the item is one week.

a. Plan order releases over the twelve-week horizon using the EOQ.

b. Plan order releases over the twelve-week horizon using POQ.

8.9. In Example 8.4, determine the capacity requirements for work centers 100 and 300.

8.10. What is order pegging and when is it necessary?

8.11. A company produces combination locks for household use on gates, lockers, and so on. The general parts structure and lead time is shown in Figure 8.30. For one particular model, the MPS calls for assembling 0, 500, 0, 300, 300, and 400 in the next six weeks, respectively. Currently, there are 140 metal bodies, 230 U-bars, 600 lock assemblies, and 347 turn knobs on hand. An open-shop order for 700 locking mechanisms is due to be received at the start of week 2. Using a lot-for-lot decision rule, plan order releases for each item over the horizon.

8.12. Repeat the previous problem using a POQ strategy where the natural cycle for U-bars is three weeks and for all other parts the cycle is two weeks.

8.13. Repeat problem 8.11 for the case that the scrap loss for lock mechanisms is 15% and 5% for U-bars.

8.14. Consider product A shown in Figure 8.31. All lead times are one week unless stated otherwise in the figure. Numerical values indicate the number of child items need to make one parent item. The master production schedule for planned assemblies over the next ten weeks for product A is as follows: 0, 150, 0, 200, 125, 0, 235, 110, 0, and 175. Current on-hand inventory levels are 236 part Ds, 328 part Fs, and 118 parts of type G. In addition, an order for 165 part Cs is in process and should be completed in time for use in period 1. A batch of 500 part type Ds is in the shop and due at the start of week 2. Plan order releases for all items using lot-for-lot scheduling. Assuming the lead times can be kept, do you anticipate any problems in meeting the schedule?

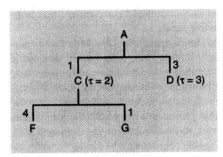

Figure 8.31 Product Structure for Problem 8.12

8.15. Repeat problem 8.14 using a periodic order quantity strategy. Parts C and D are produced in alternating weeks at the same work center. Part types F and G are ordered every four weeks from a vendor.

8.16. In problem 8.14, suppose that we also produce product B shown in Figure 8.32. Currently, there are 635 parts of type H in stock. The MPS for product B calls for the following assembly schedule by week: 0, 0, 300, 0, 250, 125, 125, 0, 400, and 275. Plan all order releases necessary for the next ten weeks to meet the assembly schedule for products A and B. Use lot-for-lot scheduling.

8.17. In the previous problem, subassemblies C and D are made at the same work center. The work center has forty hours of time available per week. It takes two hours to set up either product. Variable processing time is 0.1 hour per unit of C and 0.005 hours per unit of D. Develop a capacity requirements plan for the work center. Part type C visits the work center in its first week of production. Part type D visits the work center in its second week of processing.

8.18. Describe the problems that can occur when attempting to implement an MRP explosion schedule that was derived by infinite loading (i.e., no capacity requirements planning).

8.19. An end-item widget is assembled from 4 subassemblies of type A36, 2 B11s and 1 C35. Each A36 is made from a B22 and 2 B19s. Each B22 is made using a C7 and a purchased part number D19. All A type parts are made in the A department, B type parts are made in the B department, and C type parts are

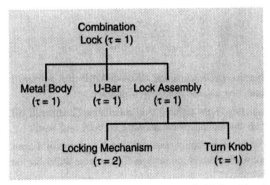

Figure 8.30 BOM Structure for Household Combination Lock

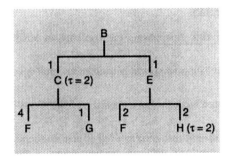

Figure 8.32 Product Structure for Problem 7.14

End Product: τ = 2		Period			
		1	2	3	4
Gross Requirements		20	18	25	40
Scheduled Receipts			10		
On-Hand	34				
Net Requirements					
Planned Order Releases					

Figure 8.33 Initial Inventory and Gross Requirements for Problem 8.21

made in the C department. Lead times are all one week. Each A36 requires 0.5 hours of department time. Each B22 requires 2.5 hours of department time. The other part types all take one hour of department resource per unit. Develop a load profile for a widget.

8.20. The load profile for a particular product places a workload of 0.2 hours per unit in the cut-off department and 1.5 hours per unit in the machining department during the first week. In the second week, there is a load of 1 hour per unit in the assembly department. Currently, there is an open-shop order for 200 units prepared to start week 2 of its route. Planned order releases call for 50 units in week 1 and 125 units in week 3. Find the anticipated workload in cut-off, machining, and assembly for the next four weeks.

8.21. Production of a product takes two weeks and requires operations in two workstations with some activities being performed in parallel. In the first work center, capacity is based on weight of product processed. This work center is used only the first week of production and each unit of product requires the processing or 2,000 lbs. of material. Capacity in the second workstation is based on labor hours. Each unit of end product produced requires 0.5 hours in the first week of processing and one hour in the second week. Gross requirements and current inventory conditions are given in Figure 8.33.

a. Find on-hand quantities, net requirements and planned order releases for the end product.

b. Using CRP, determine the weight of product that will be processed in the first work center and the labor hours required in the second work center, each period, if this plan is executed.

8.22. A production controller has been alerted to a material shortage for component AZ478. The component has a two-week lead time but the current MRP record shows:

Period	0	1	2	3	4
Gross reqts			50		
Sch. receipts/on hand	35				
Net reqts			15		
POR	(15)				

The pegged requirements file for the 50 units required in period 3 indicate a source of 25 units destined for product A and 25 units for product B. Both products have a one-week lead time. The material records for these items are:

A	0	1	Period 2	3	4
Net reqts				10	15
POR			25		

B					
Net reqts				25	
POR			25		

Show how the pegged requirements file can be used to reschedule order releases so as to avoid all shortages.

Chapter 9

Multistage Models

In Chapter 6, we considered production batching and ordering models for individual items. The items were treated independently unless related through some system level constraint such as storage space. In this chapter, we develop lot-sizing models for items that go through several stages of production. The assembly of final products depends on the availability of subassemblies. The production of the subassemblies depends on the availability of their component parts. To optimize systems with dependent items, we must develop replenishment policies that exploit the coordinated nature of the items.

9.1 MULTISTAGE PRODUCT STRUCTURES

We consider three types of multistage structures (Figure 9.1). In each case, a stage refers to a set of production operations/work centers that remove materials or parts from a controlled storage area, add value, and then store the output. Thus, a transfer line in which material moves directly from workstation to workstation in a continuous manner without being checked into and out of a storage area each time is considered a single stage.

The simplest type of multistage structure is the **serial structure (Figure 9.1a).** In a serial structure, materials enter the first stage and progressively pass through a sequence of production stages until final products exit at the last stage. Inventory can be stored between stages. An example of a serial structure is a metal fabrication and coating process. The material starts in raw form and proceeds through a sequence of operations such as grinding, milling, polishing, coating, and packaging. Material is transferred through the plant in batches. Each batch requires a setup at each process, and batches may be held in inventory.

The second system is the **assembly structure (Figure 9.1b)** in which each production stage has at most one successor but may have several predecessor stages. For example, consider a facility that assembles computers. Wafer fabs are producing and packaging the memory and processing chips. Simultaneously, another production facility produces raw circuit cards. The cards are then fed to circuit card assembly areas that populate, solder, and test the boards. Meanwhile, other areas are producing the power units, CD ROM drives, hard drives, fans, cases, and other components in the bill of materials. Each of the components of the finished product are stored and brought to the assembly area as needed.

The key aspect of the assembly structure is that each production stage has a unique successor. By contrast, in the third structure, **the general structure (Figure 9.1c),** each stage can have multiple successors and multiple predecessors. The general structure is quite common when multiple products are produced that share some of the same components. Suppose that you produce 1/4 hp. submersible pumps. These pumps are used as part of several models of residential fountains, hot tubs, small boat options, and are sold

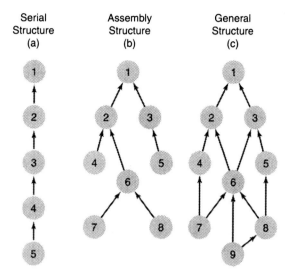

Figure 9.1 Product Structures

directly from hardware stores. When setting times and quantities for production orders, all uses must be considered, and the pumps have several successor locations.

Even if the production system is serial or assembly, when we include distribution systems, with shipments to warehouses followed by delivery to retail outlets, most systems adopt a general structure. Nevertheless, the serial and assembly systems provide a useful model for portions of the overall system and lead to some useful results for understanding how to coordinate inventory plans.

To formalize definitions, define the following notation concerning multistage structures:

$i = 1 \ldots N$ stages where stage 1 denotes the end product and the stage numbers increase as we move further down the bill of materials structure

$a(i)$ = the set of stages that are **immediate** successors of stage i—$j \in a(i)$ if j is "after" stage i in the production process

$b(i)$ = the set of stages that are **immediate** predecessors of stage i—$j \in b(i)$ if j is "before" stage i in the production process

$A(i)$ = the set of all stages that are successors of stage i

$B(i)$ = the set of all stages that are predecessors of stage i

r_{ij} = number of stage i units required per unit of **immediate** successor stage j

ρ_{ij} = number of stage i units required per unit of successor stage j

A serial structure is such that $a(i)$ and $b(i)$ contain at most one element, $A(i)$ are all stages from stage i to the end-product stage 1, and $B(i)$ are all stages from stage i back to the lowest-level stage N. Assembly structure stages have unique successors, therefore, $a(i)$ contains one element, $b(i)$ is the set of components/subassemblies that immediately precede stage i in assembly, $A(i)$ is the unique path of stages from i to the end item, and $B(i)$ resembles an assembly structure in which stage i is the end item. For serial and assembly structures, ρ_{ij} is easily computed by taking the product of r_{ij} values on the unique path from item i to item j. General structures are such that $a(i)$ and $b(i)$ have multiple elements and $A(i)$ and $B(i)$ can be general structures also. Finally, it is easy to show that $i \in B(j)$, if and only if, $j \in A(i)$.

<table>
<tr><td>**EXAMPLE 9.1**</td><td></td></tr>
</table>

For the product structure below, compute the following:

$$a(4) \qquad\qquad b(2) \qquad\qquad A(9) \qquad\qquad B(3) \qquad\qquad \rho_{51}$$

Assume that $r_{52} = 3$, $r_{21} = 2$, $r_{73} = 3$, and $r_{ij} = 1$ for all other (i, j) pairs.

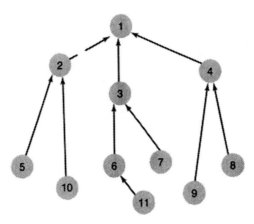

SOLUTION

This is an assembly structure, which simplifies calculations somewhat:

$a(4)$ = the immediate successor of stage 4 is stage 1.

$b(2)$ = the immediate predecessors of stage 2 are stage 5 and stage 10.

$A(9)$ = all successors of stage 9. This is the path from stage 9 to the end item thus the stages in the set are stage 4 and stage 1.

$B(3)$ = all predecessors of stage 3. The stages in this set are stage 6, stage 7, and stage 11.

ρ_{51} = number of units of stage 5 used in each stage 1 item produced. There are three units of stage 5 in each stage 2 unit and two units of stage 2 in each stage 1 unit thus there are six units of stage 5 in each stage 1 unit.

9.2 TYPES OF INVENTORY

Inventory of a specific stage in the system held at that stage location, (finished goods or work-in-process) is called **installation inventory. We use the notation I_i to denote installation inventory.** When we considered inventory in single-stage models, our inventory was always installation inventory. Finished goods inventory is always installation inventory. Installation inventory is the traditional method of describing inventory levels and estimating inventory costs.

When dealing with multistage systems, it is often convenient to look at inventory in a slightly different manner than installation inventory. We next discuss a method for computing inventory levels and costs that will provide a mathematical structure that facilitates solving large mathematical programming models for determining production decisions.

Echelon Inventory of stage i is a measure of how much of stage i production is still in the system. Echelon inventory was first introduced when discussing distribution problems in Chapter 4. Let E_i denote the echelon inventory of stage i. Stage i production can be stored at stage i, and that is the installation inventory level. Also, stage i production is contained in production at stages $j \in A(i)$ because these stages use i in their production (ρ_{ij} is the amount of i per unit of j). When work-in-process moves between stages within a facility, the echelon inventory value does not change, although the location at which it is stored changes. Echelon inventory decreases when demand is satisfied—units leave the system and these units include items from all stages in the bill of materials. Echelon inventory of i increases only when we produce units of stage i.

We can compute echelon inventory as a function of installation inventory of all successor stages or recursively based on immediate successor stages. Then,

$$E_i = I_i + \sum_{k \in A(i)} \rho_{ik} I_k \qquad (9.1)$$

or

$$E_i = I_i + \sum_{k \in a(i)} r_{ik} E_k \qquad (9.2)$$

EXAMPLE 9.2

Echelon Inventory and Installation Inventory Computations: Consider the three-stage structure depicted below. Demand for C, the end item, is one pallet per day. The lot sizes are depicted below each stage, therefore, A is produced once every eight days, B is produced once every four days and C is produced once every two days. For example, assume that the batch production is instantaneous, or equivalently, the lot is delivered in one shipment. Compute the installation and echelon stock over time.

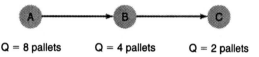

Q = 8 pallets   Q = 4 pallets   Q = 2 pallets

SOLUTION

Installation inventory is simple to graph. We start with stage C. Production occurs every two days and one pallet is demanded every day. Because production arrives in one shipment, we can time this to correspond to a point when inventory is 0 (just-in-time arrival). Therefore, in a production period, the batch of two arrives and one is immediately demanded. The next period, the second pallet is demanded, and then the cycle repeats. The following graph depicts the pattern:

A similar pattern occurs for stage B. Here, the production cycle is four days long, and there are two pallets withdrawn each odd period because the production of C triggers a demand of B. The pattern is:

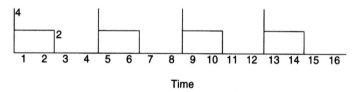

Finally, we arrive at stage C at which production is every eight days and demand occurs in four pallet orders corresponding to the production points of stage B. The pattern is:

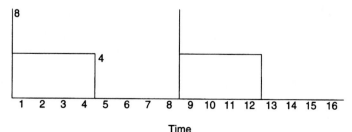

To depict echelon inventory, we begin with stage C. Because it is an end item, echelon inventory is equal to installation inventory and the profile is:

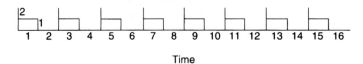

Time

Next, we move to stage B where echelon inventory is the installation inventory of stage B plus the echelon inventory of the successor of B—stage C. The result is:

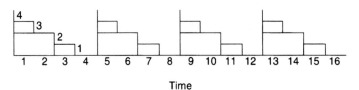

Time

Finally, we obtain the echelon inventory of stage A by adding the installation inventory of A to the echelon inventory of its successor—stage B. The result is:

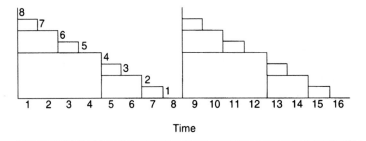

Time

One can see the possible advantages of using echelon inventory. When the production batches are sized to be **integer multiples of production batches at the immediate successor stage,** then the echelon inventory pattern resembles the traditional saw-tooth pattern of many continuous time inventory models and calculating inventory costs is easier. In this case, the echelon inventory level at a stage depends only on the production decision at that stage and no other stages. Installation inventory is different in that the level depends on the decision at the stage as well as the decision at the immediate successor stages. The pattern can be more difficult to use in optimization models when we are solving problems on assembly structures. When the integer multiple idea does not hold, then it can be easier to use installation inventory in models.

9.3 INVENTORY COSTS

The cost for holding a unit of **product i installation inventory for one time period is** denoted h_i and generally is a percentage of the total dollars invested in the item. In multistage systems, successor items have a higher installation holding cost. As more operations are performed on an item and more parts are combined, more dollars are invested in that item. A finished unit has the highest installation holding cost of any item in its bill of materials.

Define e_i, **the echelon holding cost at stage** i**,** to be the value added at stage i. This includes materials, labor, and machining done **only** in stage i. This is simple to compute based on installation holding costs:

$$e_i = h_i - \sum_{k \in b(i)} h_k \, r_{ki} \qquad (9.3)$$

Alternatively, we can define h_i as the sum of echelon holding costs for all stages that are predecessors of stage i plus the value added at stage i. That is:

$$h_i = \sum_{k \in B(i)} e_k \rho_{ki} + e_i, \qquad (9.4)$$

or we can use [Equation (9.2)] and understand that:

$$h_i = \sum_{k \in b(i)} h_k r_{ki} + e_i. \qquad (9.5)$$

Here, installation holding cost at stage i is the sum, over all immediate successors, of the successor's installation holding cost multiplied by the number of the successor in i, plus the value added at stage i. To use this definition, $h_i = e_i$ for the lowest-level stages.

We can show that the total system inventory cost using echelon inventory costs is equivalent to the total system inventory cost using installation inventory costs. We assume that $r_{ij} = 1$ for all (i, j) pairs to simplify the notation, however, the result still holds in the more general setting.

Total installation inventory cost:

$$= \sum_{i=1}^{N} h_i I_i = \sum_{i=1}^{N} I_i \left(\sum_{k \in B(i)} e_k + e_i \right) = \sum_{i=1}^{N} \sum_{k \in B(i)} I_i e_k + \sum_{i=1}^{N} I_i e_i \qquad (9.6)$$

We know $k \in B(i)$ if and only if $i \in A(k)$, therefore,

$$\sum_{i=1}^{N} \sum_{k \in B(i)} I_i e_k + \sum_{i=1}^{N} I_i e_i = \sum_{k=1}^{N} \sum_{i \in A(i)} I_i e_k + \sum_{k=1}^{N} I_k e_k = \sum_{k=1}^{N} e_k \left(\sum_{i \in A(k)} I_i + I_k \right) = \sum_{k=1}^{N} e_k E_k \qquad (9.7)$$

The third equality follows from the definition of echelon inventory at stage k as the sum of installation inventory at k and all its successor stages $A(k)$.

9.4   LINEAR MODELS FOR LOW SETUP TIME (NONBOTTLENECK) FACILITY

We will begin with a discussion of a general multistage system where setup times are short enough to be ignored or require nonbottleneck resources and thus need not be explicitly considered in setting production schedules. In this case, production coordination for a multistage system can be modeled as a linear program. The model resembles those used for dynamic aggregate planning models, except now we use the master production schedule to define production goals and schedule for shorter time periods such as a day or a week.

We assume that minimum lead times are known for all components. The minimum lead time is the time required to actually produce and transport a batch of material through its process route if adequate capacity exists at each work center encountered. We use finite loading of work centers to ensure adequate capacity exists. Part batches will be released early only when necessary to balance workload over time. Likewise, we attempt to release work to the shop to minimize the shortages of finished products where a shortage implies a failure to meet the master production schedule. Production costs are assumed to be stationary, and because the total demand is a constant, production costs will be constant regardless of the production plan. The model includes the possibility of multiple-end items and external demand at all stages. Shortage costs are assessed only on external demand and inventory costs are accrued at all stages. We use the following notation:

Data parameters:

$k = 1 \ldots K$ resources

$i = 1 \ldots N$ stages

$t = 1 \ldots T$ time periods

J is the set of end items: $i \in J$ is a stage that is an end item

π_j = shortage cost per unit per period for end-item $j \in J$

h_j = holding cost per period per unit of end-item $j \in J$

p_{ik} = the number of units of resource k used per unit of stage i produced

P_{kt} = the amount of resource k available in period t

r_{il} = the number of units of stage i needed per unit of stage l, (requirements data from the bill of materials, note that l may be an end item)

τ_i = minimum lead time for stage i

D_{it} = external demand for stage i in period t

Decision variables:

X_{it} = the number of units of stage i begun in period t

I_{it}^+ = on-hand inventory of stage i at the end-of-period t

I_{it}^- = on-hand backorders of stage i at the end-of-period t.

The production scheduling problem denoted **model LP** is:

$$Minimize \sum_{t=1}^{T} \left(\sum_{i=1}^{N} h_i I_{it}^+ + \sum_{i=1}^{N} \pi_i I_{it}^- \right), \tag{9.8}$$

$$subject \ to: \sum_{i=1}^{N} p_{ik} X_{it} \le P_{kt}, \ k = 1, \ldots, K; \ t = 1, \ldots, T \tag{9.9}$$

$$I_{it}^+ - I_{it}^- = I_{i,t-1}^+ - I_{i,t-1}^- + X_{i,t-\tau_i} - D_{it} \ for \ i \in J \ and \ t = 1, \ldots, T \tag{9.10}$$

$$I_{it}^+ - I_{it}^- = I_{i,t-1}^+ - I_{i,t-1}^- + X_{i,t-\tau_i} - \sum_{l \ne i} r_{il} X_{lt} - D_{it} \ for \ stages$$

$$i = 1 \ldots N, \ i \notin J \ and \ t = 1, \ldots, T \tag{9.11}$$

$$X_{it} \ge 0, I_{it}^+ \ge 0, 0 \le I_{it}^- \le \sum_{l=1}^{t} D_{il} \tag{9.12}$$

The objective function [Equation (9.8)] minimizes the sum of inventory holding costs for all stages plus shortage costs for finished goods not produced in time to satisfy the master production schedule. Backorders are limited to external demand for component items and this is done in constraint set [Equation (9.12)]. If holding cost is not crucial and our objective is to simply find a feasible solution, we can remove the on-hand inventory terms from the expression. The constraint set [Equation (9.9)] limits production in each period based on the availability of resources. Constraint sets [Equation (9.10) and Equation (9.11)] are inventory balance constraints, one constraint for each part type in each period. For end items the constraint equates end-of-period net inventory to beginning of period net inventory plus production started a lead time number of periods previously minus the external demand for this period (as specified in the master production schedule). The inventory balance equations for lower-level stages account for items removed as components for successor stages ($r_{ij}X_{jt}$, the dependent demand) and also any potential spare parts external demand for this stage as specified in the MPS.

Limiting backorders is necessary to ensure the integrity of the solution. We must ensure that enough component parts are produced to meet the planned production schedule for successor stages. Suppose that each end item 1 requires a part type 2. Without these restrictions, the solution might be to produce item 1 in a period for which part 2 was in a backorder position. This would of course not be possible because the part type 2 is necessary to produce the product.

EXAMPLE 9.3

LP Formulation: Lamps are made from a shade, a base assembly and an electrical wiring/fixture assembly. The electrical wiring assembly goes into the base and then a shade is attached to the base/wiring assembly component. Finally, the lamp is tested in quality control. The lead time for each operation is one day. Skilled electricians construct the electrical wiring assembly and do the testing. This is the key resource and we have 200 electrician hours per day. The lamp requires 1.5 electrician hours per unit for assembly and 1.2 hours per unit for testing. The master production schedule calls for 50 lamps of model 1 in day 4 and 100 lamps in day 6. Backorders of the end item cost $5 per occurrence, and external demand exists only for the end item. Inventory holding costs are $1 per unit per period of wiring assembly, $1.25 per unit per period of the base/wiring assembly component, $1.5 per unit per period of the attached shade assembly, and $2 per unit per period of the tested finished product. Write out the linear program model for this problem.

SOLUTION

The product structure for this problem is a serial system that looks like:

Stage labels:

    stage 1–test product

    stage 2–attach shade

    stage 3–attach base to assembly

    stage 4–construct assembly

We use the variables defined for the model where the time period is one day and $J = \{1\}$. The model is:

$$\text{Minimize} \sum_{t=1}^{T} 1I_{4t}^{+} + 1.25I_{3t}^{+} + 1.5I_{2t}^{+} + 2I_{1t}^{+} + 5I_{1t}^{-}$$

Subject to:

$$1.5X_{4t} + 1.2X_{1t} \le 200 \text{ hours} \qquad \text{for } t = 1 \dots 6$$

$$I_{1t}^{+} - I_{1t}^{-} = I_{1,t-1}^{+} - I_{1,t-1}^{-} + X_{1,t-1} - D_{1t} \qquad \text{for } t = 1 \dots 6$$

$$D_{14} = 50,\, D_{16} = 100,\, D_{11} = D_{12} = D_{13} = D_{15} = 0$$

$$I_{it}^{+} - I_{it}^{-} = I_{i,t-1}^{+} - I_{i,t-1}^{-} + X_{i,t-1} - r_{i,i-1}X_{i-1,t} \qquad \text{for } i = 2, 3, 4 \text{ and } t = 1 \dots 6$$

$$r_{i,i-1} = 1 \qquad \text{for } i = 2, 3, 4$$

$$X_{it} \ge 0,\, I_{it}^{+} \ge 0,\, I_{it}^{-} = 0 \text{ for } i = 2, 3, \text{ and } 4,\, I_{1t}^{-} \le \sum_{l=1}^{t} D_{1t}$$

The model follows Equation (9.8) through Equation (9.12) for the specific problem. We eliminate the backorder variables for components by setting variables directly to 0. The values for demand and for the requirements matrix are set separately, however, we could have included them with the constraints if we had used a less compact style of writing the formulation.

Linear programming is an extremely flexible modeling framework and model LP can be extended in many directions. We could easily add overtime and undertime considerations, dynamic production unit costs, subcontracting possibilities, and a range of other situations that add reality to the model. The drawback is that as we add more reality, more data parameters must be estimated and the solution may be harder to validate. In the next section, we consider models that include a setup cost each time a batch of stage i is produced. This situation cannot be modeled accurately with linear programming because the setup decision is discrete—either on or off.

9.5 CONTINUOUS TIME MODELS FOR STATIONARY DATA

We begin by examining the case of constant demand. In this case, it is tempting to consider a coordinated solution that produces a constant batch size at a fixed time interval for each stage.

9.5.1 Assembly Structures—Lot Aggregation

In an assembly system, each stage has a unique successor. This system was modeled in Crowston, Wagner, and Williams [1975], and we present their approach. We make the following assumptions:

Item 1 is the unique end product

End product demand is deterministic and occurs at a constant rate. There is no external demand for production from other stages.

One unit of stage i is required in each unit of stage $a(i)$

Backorders are not permitted

Production is instantaneous and lead time is deterministic

There are no capacity constraints

The only constraints that link stages are the parts requirements

Production at a stage occurs at a constant frequency and order quantity

The costs considered in the model are inventory holding costs and stage setup costs. Stage production costs are stationary and hence total to a constant because demand is fixed when added over the time horizon. Define the following notation:

Q_i = batch size at stage i

e_i = echelon inventory holding cost at stage i

A_i = setup cost at stage i

D = demand rate

Similar to the EOQ model, the relevant cost function for a single stage is:

$$f_i(Q_i) = \frac{DA_i}{Q_i} + \frac{(Q_i - 1)*e_i}{2} \tag{9.13}$$

The first term represents the setup cost per unit time and the second term represents the echelon inventory holding cost per unit time (recall that echelon inventory has the sawtooth pattern in these dependent-demand systems). The maximum inventory level is $Q_i -$ 1 since the first item moves immediately through the system. To obtain the total cost of the system, we sum over all stages to obtain **model IR (Integer-Ratio):**

$$F = \sum_{i=1}^{N} f_i(Q_i) = \sum_{i=1}^{N} \left[\frac{DA_i}{Q_i} + \frac{(Q_i - 1)*e_i}{2} \right] \tag{9.14}$$

To solve this model, it is helpful to first determine a structure for the optimal solution. Because we assume that both the production rate is instantaneous, and that we process batches after equal spans of time (so the batch size is constant), the following result holds:

Proposition 9.1 Integer Ratio Property: There exists an optimal solution where the ratio $Q_i/Q_{a(i)}$ is an integer for all i.
If we let $k_i = Q_i/Q_{a(i)}$ and $K_i = Q_i/Q_1$, then the cost model can be transformed into the following model:

$$F = \sum_{i=1}^{N} f_i(Q_i) = \sum_{i=1}^{N} \frac{DA_i}{K_iQ_1} + \frac{(K_iQ_1 - 1)*e_i}{2} \tag{9.15}$$

Now, the problem has one Q variable and a vector of $(N - 1)$ ratio values K_i. If the K_i values are given, then Q_1 is easily solved by differentiation. The result is:

$$Q_1 = \left[\frac{2*D*\sum_{i=1}^{N} A_i/K_i}{\sum_{i=1}^{N} e_i K_i} \right]^{1/2} \tag{9.16}$$

The structure of the optimal solution depends heavily on the assumptions. Under finite production rates, Jensen and Kahn [1972] and Szendrovits [1981] show that Proposition 9.1 is not necessary in an optimal solution (it's not even feasible, when the production rate at a stage equals the demand rate)! Finally, Williams [1982] shows that without the assumption of equal spans of time between production batches at a stage, the proposition does not hold, and the cost equation with echelon holding costs is not valid. Despite these theoretical and practical drawbacks, the idea that the batch size of an item should be an integer multiple of its immediate successor item had and continues to have many supporters. This approach facilitates both planning and implementation.

One of the key ideas that can be derived from the integer ratio property is the concept of nesting. A production plan is **"nested"** if whenever you produce at stage i, then you also produce simultaneously at the immediate successor of i. It is easy to understand that if a production plan is nested, then whenever you produce in stage i, then you produce simultaneously along the entire path from i to the end item. When $Q_i/Q_{a(i)}$ is an integer for all i, then every batch of stage i covers the requirements for an integer number of setups of stage $a(i)$, and therefore it is feasible to operate in a nested fashion.

Solving for the optimal K_i values is not a trivial task. Dynamic programming can be used; however, the state space expands rapidly with the number of stages and the potential number of Q_1 values. Let $T_i(Q_i)$ be the optimal cost at stage i and all stages $j \in B(i)$ when Q_i is given. Then, one possible dynamic programming recursion is:

$$T_i(Q_i) = f_i(Q_i) + \sum_{j \in b(i)} \min_{l \in I} T_k(l*Q_i) \tag{9.17}$$

where I is the set of positive integers. Here, in stage i, we must evaluate $T_i(Q_i)$ for all possible Q_i values. For each value, we have an immediate cost at stage i, and then we must evaluate the optimal integer multiple for each stage in the immediate predecessor set of i. Evaluation starts at the lowest level of the product structure, stages with no predecessors, and then continues up the structure to the end item. One can improve the computational efficiency if you can eliminate some stages from consideration a priori.

There is a great deal of work that has extended model IR. Schwarz and Schrage [1975] extended the model to finite production rate systems by adjusting the holding cost coefficients. They considered an integer programming based solution approach that used a clever product structure compression scheme to solve bounding problems. Although the integer ratio property does not hold under finite production rates, Schwarz and Schrage only looked for solutions that satisfied the property. In addition, they considered the idea of myopic policies as effective heuristics. **A myopic policy** only considers two stages at a time when finding solutions. Under myopic policies, one can derive heuristic k_i values for each i by **finding the k_i value that is the smallest integer satisfying:**

$$k_i(k_i + 1) \geq \frac{A_i e_{a(i)}}{A_{a(i)} e_i} \tag{9.18}$$

Once you have all the k_i values, then it is a simple matter to compute K_i then Q_1 using Equation(9.16). The right side of the inequality, $\dfrac{A_i e_{a(i)}}{A_{a(i)} e_i}$ is called the **myopia ratio,** and the holding cost terms are adjusted for the finite production rate situation. The myopia ratio considers the relative size of the setup to holding cost ratio between an item and its immediate successor. This relationship is combined with the square root relationship of the EOQ to suggest appropriate lot sizes. Schwarz and Schrage found that when this ratio is large, then solving the inequality for each i produces solutions close to the optimal solutions. When the ratios are small, then stage i has higher echelon holding costs and/or lower setup costs relative to stage $a(i)$. These two properties suggest that stage i should produce in lower batch sizes than stage $a(i)$, thereby invalidating the integer multiple property.

9.5.2 Assembly Structures—Lot Splitting

Moily [1986] considers the same model information as the Schwarz and Schrage model; however, instead of using the integer ratio property, a lot-splitting policy is used for a finite production rate system. Here, the assumption is that the optimal solution follows the property that:

$$Q_{a(i)}/Q_i \text{ is an integer for all } i.$$

Therefore, the largest lot size is at the end-item level, and the lot sizes decrease as we go down the structure. To implement these splitting policies, the system must be able to produce stage i components fast enough to satisfy production demand at stage $a(i)$, and individual units are transferred between stages (as opposed to batch sized transfers in the lot-aggregation model).

The model is stated in terms of installation inventory and solution is by branch and bound, implicit enumeration, and heuristics. A depiction of the installation inventory levels for a three-stage serial system is given in Figure 9.2. Here, assume that $Q_1/Q_2 = 2$ and $Q_2/Q_3 = 1$. The batches at stage 1 follow a standard finite production inventory pattern. The second batch starts immediately after the first batch is depleted. The "dotted" lines denote the times when production at stage 1 shuts down and the inventory depletes. For stage 2, we produce two batches for each batch at stage 1. The batches are timed to

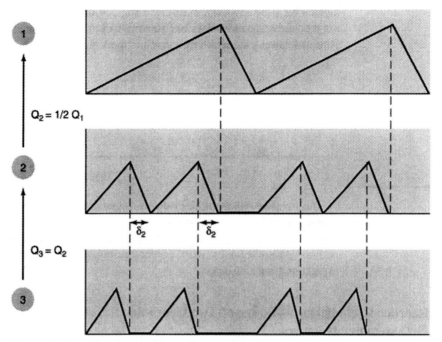

Figure 9.2 Lot Splitting, Installation Inventory Levels

allow a continual feeding of components to stage 1. The idle time between the pair of batches at stage 2 is:

$$\frac{\dfrac{Q_{a(i)}}{P_{a(i)}} - \dfrac{Q_i}{P_i}}{2},$$

and this is given by δ_2 in Figure 9.2. Stage 2 starts production again at the start time of the second batch at stage 1. Finally, stage 3 produces 1 batch for each batch of stage 2, and that batch must be completed before the batch at stage 2 finishes processing. It is clear in this method that the production rate must be increasing as you move down the structure from end items to parts. Also, when using this approach, the batches and the inventory levels get larger as we move closer to the end item.

Solutions on small problems suggest that **lot splitting can save money when the problem yields myopia ratio values less than 1.** This occurs when setup costs increase and echelon inventory costs decrease as you get closer to the end item. Higher setup as we move closer to the end item possibly occurs when the assembly costs have a high setup because of part changeovers. Having low echelon holding costs close to the end item requires low value added and often implies low setup costs because setup cost is part of the value added. Because the myopia ratio is a ratio of products of terms, it is difficult to quantify a priori whether or not lot splitting is economical.

Instead of lot splitting another approach is to simply slow machines down (if this is indeed technologically feasible without incurring additional setup or operating costs). To insure that components are fed to the next stage in a timely manner, the production rate at i must be greater than or equal to the production rate at stage $a(i)$. Because we slow down stage i, we obtain a lower "inventory triangle area" with the same number of setups and hence a lower cost. Consider, again, a three-stage serial structure as in Figure 9.2 in which we slow down the machine at stage 3. Comparing the installation inventory graph at the third stage now looks like (the original production rate graph is the dotted lines and the new lower production rate graph is the solid lines):

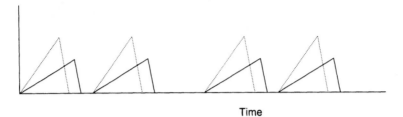

Time

As we slow the machines down to match the production rate at the next stage, the triangles get smaller and smaller until the inventory at stage 3 vanishes. Therefore, the number of setups (and hence the setup cost) remains the same and the inventory decreases, enabling a lower cost system. The end result is a balanced, continuous flow system with single-unit transfer batches.

9.5.3 Reorder Cycle Policies—Power-of-Two Heuristics for Assembly Structures

Reconsider heuristic policies for the instantaneous production, lot aggregation problem on assembly structures. The approach is to consider the problem of determining reorder intervals as opposed to determining reorder quantities. Define a **"power-of-two policy"** for each stage i to be a production policy that is structured as follows:

$$T_i = 2^{k_i} \beta \text{ for all stages } i \text{ and for some } 1 \le \beta < 2,$$

where T_i is the production cycle time of stage i (orders are placed for stage i every T_i time units), k_i is an integer, and β is the base period.

An important result states that near optimal lot sizes can be found by considering only "powers-of-two" for relative production frequencies [Roundy (1986)]. To give some insight on why this might be, consider the single-stage EOQ problem. If we parameterize our decision by the cycle time, $T = Q \cdot D$, we obtain the model Minimize Cost/Time $\equiv$ $Z(T) = \dfrac{A}{T} + \dfrac{h \cdot T \cdot D}{2}$ and this has an optimal solution $T^* = \sqrt{\dfrac{2 \cdot A}{h \cdot D}}$. Now suppose that we use some other, nonoptimal cycle time $T = \alpha T^*$. With a little algebra, we can show that the ratio of the true optimal objective value to the objective value under the nonoptimal cycle time is:

$$\frac{Z(T)}{Z(T^*)} = \frac{\dfrac{A}{\alpha T^*} + \dfrac{h\alpha D T^*}{2}}{\sqrt{2 \cdot A \cdot h \cdot D}} = \frac{\alpha^{-1} + \alpha}{2}.$$

Now, suppose that we have some convenient cycle length, T_L. T_L may be a day, a week, or some other natural period. We restrict the cycle length of a product to $T = 2^k T_L$ for some non-negative integer $k = 0, 1, 2, \ldots$ (ensures that $T^* \ge T_L$ and is a convenient operating policy). Because the cost function $Z(T)$ is convex in T, we choose the smallest k satisfying $Z(2^{k+1} \cdot T_L) \ge Z(2^k \cdot T_L)$ as our "optimal power-of-two policy." Combining this with the result for $\dfrac{Z(T)}{Z(T^*)}$, we have $(\alpha_1 + \alpha_1^{-1}) \ge (\alpha_2 + \alpha_2^{-1})$ where $\alpha_1 = \dfrac{2^{k+1} T_L}{T^*}$

and $\alpha_2 = \dfrac{2^k T_L}{T^*}$. After rearranging terms, $2^k T_L \geq \dfrac{T^*}{\sqrt{2}}$. Likewise, using the relationship

that $Z(2^{k-1} T_L) \geq Z(2^k\ T_L)$ because we stopped searching at k, we can show that

$2^k T_L \leq \sqrt{2} T^*$. Combining these, $\dfrac{T^*}{\sqrt{2}} \leq 2^k \cdot T_L \leq \sqrt{2} T^*$ for the optimal power-of-two

choice of k. Finally, the $\dfrac{Z(T)}{Z(T^*)}$ relation tells us that because $\dfrac{Z(\sqrt{2} T^*)}{Z(T^*)} = \dfrac{Z\!\left(\dfrac{T^*}{\sqrt{2}}\right)}{Z(T^*)} =$

1.06, then restricting ourselves to "powers-of-two" ensures that we are at most 6% from the optimal cost solution. This general result can be generalized to multistage systems, which is addressed in section 9.5.4.

The work of Jackson, Maxwell, Muckstadt, and Roundy at Cornell University in the 1980s (see end-of-chapter references) led to an efficient heuristic for static, multistage assembly systems. Consider the multistage lot-sizing problem **(denoted Problem LS)**:

$$Minimize \sum_{i=1}^{N} \frac{A_i}{T_i} + \frac{e_i D T}{2}$$

Subject to:

$$T_i \geq T_{a(i)} \ \forall i$$

The objective is simply the sum of the individual stage objectives (setup and echelon holding cost) and the constraint set ensures that batches increase as we move down the assembly structure.

Define a cluster as a set of connected stages in the assembly tree. The root of the cluster is the highest level (parent) stage in the cluster. We divide the stages of the assembly tree into clusters with similar production cycles. Then, we show that a solution $T_j^*, j = 1, \ldots, n$ solves problem LS, if and only if, there exists a set of clusters C^j, each with root j such that:

a. $\bigcup\limits_j C^j = \{1, 2, \ldots, N\}$;

b. $T_i^* = T^*(j) \ \forall i \in C^j, \ \forall j$;

c. $T^*(j) = \left[\dfrac{2 \sum\limits_{i \in C^j} A_i}{D \sum\limits_{i \in C^j} e} \right]^{1/2}$;

d. $T^*(j) \geq T^*(k)$ when a stage of cluster k comes after j (i.e., k is higher in the assembly tree); and

e. $\dfrac{2A_i}{D \cdot e_i} < \dfrac{2 \sum\limits_{r \in C^j} A_r}{D \sum\limits_{r \in C^j} e_r} \ \forall i \in C^j, i \neq j.$

Condition a. requires that all production stages are included in the solution. Condition b. ensures that all nodes in the cluster have the same cycle length and that cycle length is defined in c. by aggregating the nodes in the cluster. Condition d. requires that cycle times naturally increase as we go down the assembly tree (similar to the idea of Crowston, Wagner, and Williams). Finally, condition e. is a rule for forming clusters. Except for the root node representing the cluster, all members of the cluster must have a shorter natural cycle than the cluster. If this were not the case, there may be benefit in dividing the cluster.

These results lead to an effective heuristic for finding high quality powers-of-two policies. First, we form clusters, then find the natural cycle for each as follows (we denote this as the **multistage lot-sizing heuristic**):

Step 1: (**Form clusters**). Begin with each stage forming a 1–stage cluster. Select the cluster with smallest $\dfrac{\sum\limits_{i \in C^j} A_i}{D \sum\limits_{i \in C^j} e_i}$ for which it is not yet shown that property d holds.

If

$$\frac{\sum\limits_{i \in C^j} A_i}{D \cdot \sum\limits_{i \in C^j} e_i} \geq \frac{\sum\limits_{i \in C^k} A_i}{D \cdot \sum\limits_{i \in C^k} e_i}$$

for $k = a(C^j)$, then consider property d is established for cluster j.
If

$$\frac{\sum\limits_{i \in C^j} A_i}{D \cdot \sum\limits_{i \in C^j} e_i} < \frac{\sum\limits_{i \in C^k} A_i}{D \cdot \sum\limits_{i \in C^k} e_i}$$

then combine clusters j and k. Continue until all clusters satisfy condition d.

Step 2: (**Find cluster cycle lengths**). For each cluster j find $T^*(j) = \left[\dfrac{2 \sum\limits_{i \in C^j} A_i}{D \sum\limits_{i \in C^j} e_i} \right]^{1/2}$.

Step 3: (**Integrate cluster multipliers**). For each cluster j, find the integer l in which $T^*(j) \approx 2^l \cdot T_L$ and readjust $T^*(j) = 2^l T_L$

If T_L is not predefined by a physical period such as a day, then the optimal T_L can be found by optimizing the inventory cost objective over T with the fixed 2^l ratios for each cluster. Using the optimal base cycle can improve the heuristic from being no worse than 6% above optimal to no worse than 2% above optimal.

9.5.4 General Structures

We conclude this section with a short discussion concerning general product structures. Here, there may be multiple paths from each stage to the end item. This multiplicity causes difficulty in that if we produce at a stage, we do not necessarily produce at all immediate successors to the stage. Therefore, production may occur at a stage although there is positive echelon inventory of the stage in the system. The simplistic regular inventory level pictures simply do not exist for these structures.

A successful line of research on this structure has been to develop effective heuristics similar to the power-of-two policies considered for assembly structures in section 9.5.3. The seminal works of Roundy [1986] and Maxwell and Muckstadt [1985] both consider the idea of developing heuristics to set reorder intervals for each stage. This approach is different than the approach of Crowston, Wagner, and Williams, which focuses on determining batch sizes.

Recall that the power-of-two policy has the following form:

$$T_i = 2^{k_i} \beta \text{ for all stages } i \text{ and for some } 1 \le \beta < 2,$$

where T_i is the production cycle time of stage i (orders are placed for stage i every T_i time units), k_i is an integer, and β is the base period. The Roundy method allows k_i to be both positive or negative, whereas the Maxwell and Muckstadt method considers only $k_i \ge 0$. The final restriction on these policies is that an order at a stage is placed only when the installation inventory of that product is 0 (**called the "zero-inventory" rule**).

A stationary policy is a production plan in which each stage uses a fixed order quantity and a fixed production cycle time. A nested policy is a production plan in which each stage sets up an order each time one of its immediate predecessors sets up an order. **Stationary Nested Policies** are both stationary and nested, and these are exactly the types of plans that are generated when using the methods of Crowston, Wagner, and Williams, Schwarz and Schrage, and Maxwell and Muckstadt. The lot-splitting ideas of Moily yield stationary but not nested policies (actually, these policies are "inverse nested" in that you produce at stage i whenever stage $a(i)$ is producing). Although stationary nested policies often perform well, it is possible that they sometimes perform poorly.

It should be evident from the results for assembly systems, that considering only nested policies or considering only lot-splitting policies may not be the most effective strategy. We saw that when the myopia ratio decreases, lot splitting becomes better and when the myopia ratio increases, lot-aggregation policies become better. The key is the cost structure, and in particular, the ratio of setup cost to echelon holding cost.

To illustrate the impact of multiple immediate successors, consider the product structure in Figure 9.3. Assume that we chose a policy with $\underline{T}_4 = 1$ week, $T_2 = 1/2$ week, and $T_3 = 2$ weeks. Assume that the end item setup cost is very small thus we produce in batches of one to meet demand of one unit per week. Therefore, stages 2 and 3 also see a demand of one unit per week. Using these cycle times, stage 2 produces a half unit every half week, and stage 3 produces two units every two weeks. Now, the plan at stage 4 is complex. Stage 4 sees a demand of two units per week on average. Every week, there is a demand of one unit to supply the requirements of stage 2. Also, every even numbered week stage 4 must supply stage 3 with two units. Therefore, the net result is that on odd weeks, stage 4 produces one unit and on even weeks, stage 4 produces three units. It is clear that this policy is not nested (we don't produce at stage 3 every time we produce at stage 4), and it is not stationary (the production quantity is not fixed and changes with each batch).

The Roundy method transforms the general structure into a mathematically tractable structure by enumerating all of the routes from items to end products. For example, if we take the structure in Figure 9.3, we can split stage 4 into the two routes to stage 1. Figure 9.4 (called a "Hasse Diagram") results. In Figure 9.4, we see all routes. Stage 4 can

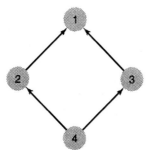

Figure 9.3 General Structure

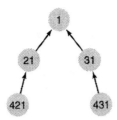

Figure 9.4 Hasse Diagram of Figure 9.3

take either $4 \to 2 \to 1$ or $4 \to 3 \to 1$. The amount of time that the unit spends in the system (and hence must be counted and costed in inventory) depends on the path taken. Stages 2 and 3 have only one route each, they must go directly to stage 1.

This route structure yields an effective method for computing the average cost of a power-of-two policy, and once we have such a function, we can consider ways to find low cost solutions. The key to computing average cost is the realization that echelon inventory along a route follows a saw-tooth pattern (we have seen similar insight in the Crowston, Wagner, and Williams work).

Consider path $\mathbf{421 \to 21 \to 1}$ in Figure 9.4. To understand the echelon inventory picture, assume that demand for end item 1 is 1 unit per week. Also, let $T_4 = 8$ weeks, $T_2 = 1$ week, and $T_1 = 2$ weeks. The installation and echelon inventory levels are in Figure 9.5. Route 1 is rather simple and is the familiar saw-tooth pattern when demand is one unit per week, and we produce every week (Figure 9.5a). Route 21 represents the production of stage 2 and stage 1. Here, we have a cycle at stage 2 every week; however, because of the zero-inventory production rule, we do not produce at stage 2 unless installation inventory is zero. Therefore, every time we actually produce at stage 2, it is immediately transformed into stage 1 production and has already been counted. Therefore, there is no installation inventory of items on route 21. Route 421 represents the production of stage 4, 2, and 1. Because $T_4 = 8$ and demand = 1 per week; we produce eight units at stage 4 once every eight weeks. This inventory is depleted periodically based on the production frequencies at stages 2 and 1. The demand of two units every other week is depicted in Figure 9.5b. The echelon inventory of units on route 421 is simply the sum of the route installation inventories on 421, 21, and 1. This is depicted in Figure 9.5c, and we can see the familiar pattern.

The key here is that the height and base of the triangle depend on **the maximum cycle time on the route from a stage to the end item.** We must compute echelon inventory for all routes in the network to find the system inventory cost. The setup cost is a simple function of the cycle times, so the total cost of the system is the sum of the echelon holding costs over all routes plus the sum of the setup costs over all stages.

Besides the clever modeling done in this problem, the Roundy method shows that power-of-two policies are extremely effective. **Roundy shows that any optimal k_i power-of-two policy is at worst 94% effective, and if we select β correctly, then the policy is at worst 98% effective.** The term 94% effective implies that the average cost of a power-of-two policy is no more than 6% above the average cost of the optimal policy.

Besides low cost, these policies are easy to implement on the shop floor and have a rather natural appeal. The approach to proving solution quality started an entire avenue of research in developing production plans with guaranteed performance. There have been many extensions of this work including models for finite production rate systems, systems with backorders, and special structure systems.

Figure 9.5a Route 1 Installation Inventory:

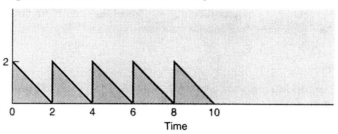

Figure 9.5b Route 421 Installation Inventory

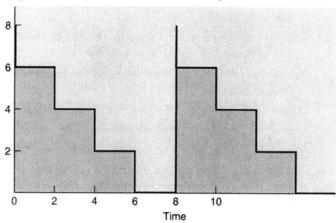

Figure 9.5c Route 421 Echelon Inventory

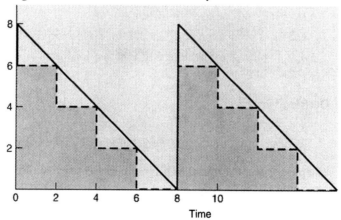

Figure 9.5 Installation and Echelon Inventory along a Path

9.6   DISCRETE TIME MODELS FOR NON-STATIONARY DATA

We have been considering the case of constant demand. With constant demand, it makes intuitive sense to use a simple constant lot-size (over time) policy. This will not be the case where forecasts indicate changing demand over time. Therefore, we next consider problems where the demands and costs can change over time. This adds more realism to the problem structure, however it also detracts from our ability to find optimal solutions quickly. We consider the three structures depicted in Figure 9.1, serial structures, assembly structures, and general structures.

9.6.1 Serial Structures

We consider the serial system as discussed in Love [1972]. Here, the end product is denoted stage 1, its immediate predecessor is denoted stage 2, and the other stages are numbered consecutively with the final stage being stage N. The system must produce to meet demand over $t = 1 \ldots T$ time periods. We make the following assumptions:

Production costs at a stage in a period can be modeled with a concave function of the production level.

Inventory holding costs at a stage in a period can be modeled with a concave function of the end of period inventory level.

Production is moved between stages in batches, and production is assumed to be instantaneous.

All demand must be met on time, and demand only occurs at the end product.

Production capacity is unbounded at every stage in every period.

Only 1 unit of each stage is required in its successor stage.

The first two assumptions are more general than those used in the previous section. Linear holding costs fit the concave model, and production costs that include setup plus a linear component are also concave.

The following notation will be used:

Data Parameters:

D_t = demand at stage 1 in time t

$C_{it}(X_{it})$ = production cost function for stage i in period t

$h_{it}(I_{it})$ = inventory holding cost function for installation inventory held at stage i at the end of period t

Decision Variables:

X_{it} = production at stage i in time period t

I_{it} = installation inventory of units of stage i at the end of time t

The model, denoted **model SER,** minimizes cost while meeting demand:

$$\text{Minimize} \sum_{i=1}^{N} \sum_{t=1}^{T} [C_{it}(X_{it}) + h_{it}(I_{it})] \tag{9.19}$$

Subject to:

$$I_{i,t-1} + X_{it} - X_{i-1,t} = I_{it} \quad \text{for } i = 2 \ldots N \text{ and } t = 1 \ldots T \tag{9.20}$$

$$I_{1,t-1} + X_{1t} - D_t = I_{1t} \quad \text{for } t = 1 \ldots T \tag{9.21}$$

$$X_{it} \geq 0, I_{it} \geq 0 \tag{9.22}$$

The objective [Equation (9.19)] minimizes the production cost plus the holding cost. Constraint set [Equation (9.20)] is the inventory balance equation for all stages other than the end item. Here, the inventory coming into a period, plus the production in a period, minus the production in the successor stage must equal the ending inventory. Constraint set [Equation (9.21)], is the inventory balance for the end item and differs from the previous set in that external demand replaces the successor production level. Finally, all variables must be non-negative.

Solving this model requires using results from nonlinear programming on minimizing concave functions over linear constraint sets. For these problems, an optimal solution can always be found at an extreme point of the feasible set [Zangwill (1966)]. Love [1972] characterizes the extreme points of model SER using the following propositions:

Proposition 9.2 There exists an optimal solution to model SER where:

$$X_{it} * I_{it-1} = 0$$

This proposition states that if you have inventory at stage i at the end of period $t - 1$, then there is an optimal solution in which you are not producing in stage i in period t. This is identical to the Wagner–Whitin result for the single-stage problem.

Proposition 9.3 If $X_{it} + I_{it-1} > 0$, then these units must satisfy an integer number of periods worth of end-item demand. Whenever we produce, we produce for an integer number of period's worth of demand regardless of the stage of production.

Figure 9.3 depicts a network representation of model SER for five stages and four periods. At the top left of the figure, an input equal to the total demand over the horizon is entered. The vertical arcs represent production at the various stages in each period while the horizontal arcs represent inventory held at each stage between periods. For example, assume that we produce only in periods 1 and 3 for stage 5. The production amount in period 1 flows on the arc labeled X_{51}, the remaining flow $\left(\sum_{t=1}^{T} D_t - X_{51} \right)$ flows over the top row of the network, and then the production in period 3 flows on the arc labeled X_{53}. There is no other flow on the top row. Once stage 5 is completed, those units would be used in stage 4 production or be held in inventory at stage 5 (flow on arcs I_{5t}). Eventually, the material will flow out of the bottom of the network to satisfy the demand in each period.

Proposition 9.2 has the following interesting interpretation when used in conjunction with Figure 9.3 (or similarly drawn figures for other problem examples), there exists an optimal solution in which there is positive flow on at most one arc into every node. Again, this is identical to the results for extreme points for the single-stage problem and follows from the theory of network problems.

Next, we redefine the idea of nested production for this periodic problem. A production plan is said to be **"nested schedule"** if:

$$X_{i-1,t} = 0 \Rightarrow X_{it} = 0.$$

This idea states that if you do not produce in a stage in a period, then you do not produce at any of its predecessors in that period. Another way to state the concept is to use the contrapositive. A production plan is nested if:

$$X_{it} > 0 \Rightarrow X_{i-1,t} > 0.$$

The insight here is that if you produce at a stage in a period, then you will produce in every successor stage in that period. The following result shows when there exists an optimal production plan that is nested:

Proposition 9.4 Assume that the production cost function $C_{it}(X_{it})$ is a nonincreasing function of time for all stages i, and that the holding cost function $h_{it}(I_{it})$ is a nonincreasing function of i for all time periods t. Then, there exists an optimal production plan that is nested.

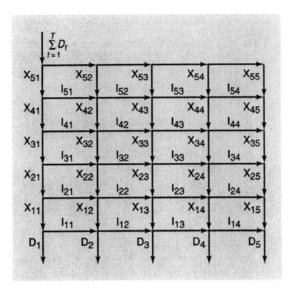

Figure 9.6 Production Network for the Serial System

The conditions of Proposition 9.4 are generally easy to satisfy. Production costs are often stationary or decreasing in time. If we consider the time value of money (money in the future is worth less than money today) or situations where the production process is improving over time, then costs can be expected to decrease. Installation holding costs decrease as we move down the product structure from end items to components and parts as long as we base holding costs on the value added structure. This assumption on holding costs is equivalent to having positive echelon holding costs at each stage.

Love used Propositions 9.2 and 9.3 to develop a dynamic programming formulation for finding an optimal nested production plan. The formulation was based on a structure similar to Figure 9.6. At each node, there is a decision to produce or to hold inventory (only one of these can be positive). If you decide to produce, then you must determine the size of the production run (an integer number of periods worth of demand). The formulation is appropriate for small problems and works best when the demand values are small.

9.6.2 Assembly Structures

We now present a mathematical programming model that is a discrete time analog to model IR (section 9.5.1). Like model IR, we use echelon inventory to help us find computational solutions. We assume that unit production costs are stationary and hence not included in the model.

Data Parameters:

r_{ij} = number of stage i units required per unit of immediate successor stage j

ρ_{i1} = number of stage i units required per unit of end item

A_{it} = setup cost of production at stage i in period t

h_{it} = installation holding cost per unit of stage i installation inventory on hand at the end-of-period t

e_{it} = echelon holding cost per unit of stage i echelon inventory on hand at the end-of-period t

D_t = demand in units of stage 1 end product in time t

Decision Variables:

I_{it} = installation inventory of stage i on hand at the end-of-period t

E_{it} = echelon inventory of stage i on hand at the end-of-period t

X_{it} = production at stage i in time period t

δ_{it} = 1 if $X_{it} > 0$, otherwise 0

The model developed by Afentakis, Karmarkar, and Gavish [1984] attempts to minimize production, setup, and holding cost while meeting demand. We call this **model ASI.**

$$\text{Minimize} \sum_{t=1}^{T} \sum_{i=1}^{N} (A\delta_{it} + h_{it} I_{it}) \tag{9.23}$$

Subject to: $X_{it} \leq M\delta_{it}$ for each (i, t) pair   (9.24)

$\quad\quad\quad\quad\;\; I_{1t-1} + X_{1t} - D_t = I_{1t}$ for each t (9.25)

$\quad\quad\quad\quad\;\; I_{it-1} + X_{it} - r_{ia(i)}X_{a(i),t} = I_{it}$ for each (i, t) pair   (9.26)

$\quad\quad\quad\quad\;\; X_{it} \geq 0, I_{it} \geq 0, \delta_{it} \in \{0, 1\}$ (9.27)

The objective [Equation (9.23)] totals the setup and holding costs. Constraint set [Equation (9.24)] ensures that setup cost is accrued when production is positive. Constraint set [Equation (9.25)] is the inventory balance constraint set for the end item. Finally, constraint set [Equation (9.26)] is the inventory balance for intermediate stages. Here, the demand is based on the production at the immediate successor stage.

Model ASI is a compact way to formulate the problem, but it offers little in the way of solution methodology. The problem is a large integer programming problem with little decomposability. Relaxation approaches have not proven to be effective on the model, and that suggests a reformulation is in order.

We next consider an alternate formulation based on echelon inventory (denoted **model ASE**):

$$\text{Minimize} \sum_{t=1}^{T} \sum_{i=1}^{N} (A\delta_{it} + e_{it}E_{it}) \tag{9.28}$$

Subject to: $X_{it} \leq M\delta_{it}$ for each (i, t) pair   (9.29)

$\quad\quad\quad\quad\;\; E_{it-1} + X_{1t} - \rho_{i1}D_t = E_{1t}$ for each (i, t) pair   (9.30)

$\quad\quad\quad\quad\;\; r_{ia(i)} E_{a(i),t} \leq E_{it}$ for each (i, t) pair   (9.31)

$\quad\quad\quad\quad\;\; X_{it} \geq 0, E_{it} \geq 0, \delta_{it} \in \{0, 1\}$ (9.32)

The objective [Equation (9.28)] totals the setup and echelon holding costs. Constraint set [Equation (9.29)] ensures that setup cost is accrued when production is positive. Constraint set [Equation (9.30)] is the echelon inventory balance constraint set for all items. Echelon inventory increases with production of stage i and reduces by ρ_{i1} per unit of end item demand. Constraint set [Equation (9.31)] ensures that there is enough echelon inventory of stage i on hand to produce the completed amount of the successor item, $a(i)$.

Using the cost equivalence of installation inventory and echelon inventory and the definitions of echelon inventory, Afentakis, Karmarkar, and Gavish show that models ASI and ASE are equivalent in that the production plan optimal to one is also optimal to the other. They developed the following properties of an optimal solution for both models.

9.6.2.1 Optimal Solution to ASE

Model ASE can be solved using a Branch and Bound-Lagrangian relaxation approach, and we will consider that next. Using the structure of model ASE, the following property is easily demonstrated:

Proposition 9.5 Nesting: There exists an optimal solution to model ASE with the property if $X_{it} > 0$ then $X_{a(i),t} > 0$.

Sequentially applying the nesting result, we deduce

Proposition 9.6 Path Production: Suppose that $X_{it} > 0$ at optimality. Then, there exists an optimal solution such that $X_{jt} > 0$ for all $j \in A(i)$ (all stages on the path from stage i to the end item).

Proposition 9.7 Echelon Inventory Level: There exists an optimal solution where $X_{it} \cdot E_{i,t-1} = 0$ for all (i, t) pairs.

These three structural results are similar to results from model IR in the stationary demand case and similar to results from model SER in the serial system nonstationary demand case. Our cost model implicitly assumes production and demand occur instantaneously in the period. Therefore, issues such as lot splitting do not arise.

To find bounds for the branch and bound scheme, we use Lagrangian relaxation and relax constraint set [Equation (9.31)] concerning the link between echelon inventories at i and $a(i)$. Assume that λ_{it} denotes the Lagrange multiplier on constraint (i, t) in Equation (9.31), and assume that these values are known. The relaxed model is:

$$f(\lambda) = \text{Minimize} \sum_{t=1}^{T} \sum_{i=1}^{N} (A\delta_{it} + e_{it}E_{it}) - \sum_{t=1}^{T} \sum_{i=1}^{N} \lambda_{it} (E_{it} - r_{i,a(i)} E_{a(i),t}) \quad (9.33)$$

Subject to:
$$X_{it} \leq M\delta_{it} \qquad \text{for each } (i, t) \text{ pair} \quad (9.34)$$

$$E_{it-1} + X_{1t} - \rho_{i1}D_t = E_{1t} \qquad \text{for each } (i, t) \text{ pair} \quad (9.35)$$

$$X_{it} \geq 0, E_{it} \geq 0, \delta_{it} \in \{0, 1\} \quad (9.36)$$

By regrouping terms, the objective of the relaxed model can be transformed into a form that is separable by stage:

$$f(\lambda) = \text{Minimize} \sum_{i=1}^{N} \left[\sum_{t=1}^{T} \left(A\delta_{it} + E_{it} \left(e_{it} + \sum_{k \in b(i)} r_{ki}\lambda_{kt} + \lambda_{it} \right) \right) \right] \quad (9.37)$$

Using Equation (9.37) as the objective and Equations (9.34) and (9.35) as the constraints for a given set of λ_{it} values, solving the relaxed problem is similar to solving N independent Wagner–Whitin problems (see Chapter 6). The inventory cost for stage i in time t is:

$$e_{it} + \sum_{k \in b(i)} r_{ki}\lambda_{kt} - \lambda_{it}$$

and demand is based on end-item demand and the number of units of stage i used in the end item. Note, that it is possible to have negative inventory cost in this model, and hence the subproblems are not equivalent to Wagner–Whitin problems. However, because an optimal solution to model ASE has $X_{it} E_{i,t-1} = 0$ for all (i, t) pairs, we can use shortest path methods to solve the subproblems.

To find the appropriate λ_{it} values, our goal is to have the tightest lower bound to model ASE. Therefore, we should solve the following problem:

$$\text{Maximize } f(\vec{\lambda}) \qquad (9.38)$$

$$\text{Subject to: } \vec{\lambda} \geq 0 \qquad (9.39)$$

The difficulty with this method is that $f(\vec{\lambda})$ is not differentiable in $\vec{\lambda}$, and hence subgradient optimization must be used. Afentakis et. al. [1984] had excellent success with the subgradient optimization as well as the branch and bound approach and quickly solved for the optimal production plans for problems with up to 50 stages and 18 time periods.

9.6.2.2   Heuristic Solution to ASE—Level by Level Lot Sizing with Cost Adjustments

A first cut approach to solving a multistage production planning problem is to decompose the problem into a sequence of single-stage problems. In Chapter 6, we considered methods for developing single-stage production plans based on periodic demand, holding costs, setup costs, and variable production costs. Both the Wagner–Whitin algorithm and the Silver–Meal heuristic are example methods. In Chapter 8, we used this decomposition process to perform the MRP explosion using lot-for-lot scheduling, EOQ ordering, and POQ ordering as the single-stage planning methods.

The process to construct an optimal plan for the multi-stage structure is difficult since decisions at the successor levels impact the demand at the predecessor levels. If you make decisions optimally for a single stage without considering the impact on predecessor stages, then it is highly unlikely that your decisions will be optimal for the overall structure. For example, when you decide to set up and produce a batch of stage i, you are putting a demand on all of the predecessor stages $B(i)$ in that there must be sufficient inventory of these stages to produce the batch of i. The decomposition approach is said to be "myopic."

Blackburn and Millen [1982] suggested a way to deal with the stage dependencies. Their approach is to decompose to a sequence of single-stage problems. However, first they adjust the costs on each stage to account for decisions made later in the process. The adjustment process is based on insight gained from the work by Crowston, Wagner, and Williams [1973] and Schwarz and Schrage [1975] on stationary demand problems. Recall that the key idea in these papers is that to solve the planning problem, you only need to determine the k_i values, the ratio of the batch size at i relative to the batch size at its successor $a(i)$. Using the appropriate notation, $k_i = Q_i/Q_{a(i)}$, and only integer k_i values need be considered.

Coming back to our multistage problem, if you decide to set up and produce a batch at stage i, then because of the $k_{b(i)}$ values, you are essentially going to produce a fraction of a batch at stage $b(i)$ (for all immediate predecessors of i). For example, consider two stages in series:

If $k_2 = 3$, then for every three times that you set up and produce stage 1, you set up and produce stage 2. Therefore, when you decide to set up and produce stage 1, you are "spending" one of your three allowed setups before incurring the cost of producing at stage 2.

One can account for this in the single-stage decomposition by adjusting the setup cost at stage 1 to account for the future cost. In this example, an appropriate adjustment might be:

$$\hat{A}_1 = A_1 + A_2/3$$

Or more generally,

$$\hat{A}_i = A_i + \sum_{j \in b(i)} \frac{\hat{A}_j}{k_j} \tag{9.40}$$

where $\hat{A}_i$ is the adjusted value of the setup cost at stage i and stage i has one or more immediate predecessors.

Besides adjusting the setup cost, Blackburn and Millen suggest adjusting the echelon holding costs. Again, borrowing from Crowston, Wagner, and Williams, the echelon inventory level increases as we move down the structure away from the end item. When we decide to produce a batch at stage i, then we must increase the echelon inventory of all predecessor stages. Hence, deciding to hold a unit of echelon inventory at stage i actually costs more than e_i because we must also hold echelon inventory at all predecessor stages. Thus, the appropriate adjustment to echelon holding cost is:

$$\hat{e}_i = e_i + \sum_{j \in b(i)} \hat{e}_j * k_j \tag{9.41}$$

where $\hat{e}_i$ is the adjusted value of the echelon holding cost at stage i, and stage i has one or more immediate predecessors.

The adjustment process is recursive. We start the process at the lowest level, the part level, namely $i = N$. We then work backward up the structure until we get to the end item, $i = 1$. Although both adjustment formulas consider only immediate predecessors, when we do the process in this recursive manner and incorporate adjusted cost values as we go, we are actually taking into consideration the impact on **all** predecessors of stage i.

The final issue in using the Blackburn and Millen approach is the method of computing the k_i values for each stage. We discuss three of the five methods that they propose. Note that for $i = 1$, the end item, $k_1 = 1$ by definition.

Continuous–k method: Here, we use the first order condition that the first partial derivative of Equation (9.15) with respect to each k_i must vanish at optimality. After calculating algebraically, we obtain:

$$k_i = \left[\frac{A_i}{A_{a(i)}} * \frac{e_{a(i)}}{e_i} \right]^{1/2} \quad \text{for } i = 1 \ldots N - 1 \tag{9.42}$$

Although this estimate of k_i suggests that only adjacent stages are used (and hence this is myopic), Blackburn and Millen state that this formula is equivalent to using updated values for the setup and echelon holding costs for stage i.

System Myopic Integer–k method: Here, we use the results from Schwarz and Schrage they estimate the optimal value of k_i to be the minimum value k_i that satisfies the following inequality:

$$k_i(k_i + 1) \geq \left[\frac{A_i}{A_{a(i)}} * \frac{e_{a(i)}}{e_i} \right] \quad \text{for } i = 1 \ldots N - 1 \tag{9.43}$$

We note that this method only uses data from successive stages in the product structure, and hence it is myopic.

Integer–k Method: Here, we try to bring a more global view into the system myopic method and use the updated cost structure when available:

$$k_i(k_i + 1) \geq \left[\frac{\hat{A}_i}{A_{a(i)}} * \frac{e_{a(i)}}{\hat{e}_i} \right] \qquad \text{for } i = 1 \dots N - 1 \qquad (9.44)$$

Once the cost adjustments are complete, the method proceeds with solving for the production quantities using the single-stage decomposition. Results from their experiment on small structures suggest that the integer-k method and a variation of the continuous-k method perform best. As the number of levels in the product structure increases, it becomes more important that adjustments be made to consider the effects of current decisions on future decisions. We conclude this section with an example demonstrating the cost adjustments for a small product structure.

EXAMPLE 9.4

Consider the following product structure with setup and echelon holding costs listed below. Solve for the updated cost structure using the continuous-k method, the system myopic integer-k method, and the integer-k method.

Stage	e_i	A_i
1	1	20
2	2	30
3	1	20
4	1	10
5	1	20

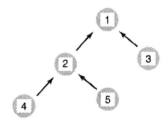

SOLUTION

We summarize the results in the following Table and explain the entries below.

Stage	A_i	e_i	Continuous-k $\hat{A}_i$	$\hat{e}_i$	Myopic Integer-k $\hat{A}_i$	$\hat{e}_i$	Integer-k $\hat{A}_i$	$\hat{e}_i$
1	20	1	108.79	5.44	100	6	100	6
2	30	2	59.57	3.97	60	4	60	4
3	20	1	20	1	20	1	20	1
4	10	1	10	1	10	1	10	1
5	20	1	20	1	20	1	20	1

For all methods, the results for stages 3, 4, and 5 are the same and do not change because these stages all have no predecessors. We first compute k_3, k_4, and k_5 using the continuous-k method.

$$k_5 = \left(\frac{A_5 e_2}{A_2 e_5} \right)^{1/2} = \left(\frac{20 * 2}{30 * 1} \right)^{1/2} = 1.155$$

$$k_4 = \left(\frac{A_4 e_2}{A_2 e_4} \right)^{1/2} = \left(\frac{10 * 2}{30 * 1} \right)^{1/2} = 0.816$$

$$k_3 = \left(\frac{A_3 e_1}{A_1 e_3} \right)^{1/2} = \left(\frac{20 * 1}{20 * 1} \right)^{1/2} = 1$$

We next compute k_2, and then update all of the cost values (note that order does not matter here because we are not using updated cost values):

$$k_2 = \left(\frac{A_2 e_1}{A_1 e_2} \right)^{1/2} = \left(\frac{30 * 1}{20 * 2} \right)^{1/2} = 0.866$$

The updated setup and echelon holding cost calculations for stages 2 and 1 are:

$$\hat{A}_2 = A_2 + \sum_{j \in b(2)} \frac{\hat{A}_j}{k_j} = 30 + \frac{10}{.816} + \frac{20}{1.155} = 59.57$$

$$\hat{e}_2 = e_2 + \sum_{j \in b(2)} \hat{e}_j * k_j = 2 + 1 * .816 + 1 * 1.155 = 3.971$$

$$\hat{A}_1 = A_1 + \sum_{j \in b(1)} \frac{\hat{A}_j}{k_j} = 20 + \frac{59.57}{.866} + \frac{20}{1} = 108.79$$

$$\hat{e}_1 = e_1 + \sum_{j \in b(1)} \hat{e}_j * k_j = 1 + 3.971 * .866 + 1 * 1 = 5.439$$

For the myopic integer-k method, we have to compute the right side values that limit k_i, $\left[\dfrac{A_i}{A_{a(i)}} * \dfrac{e_{a(i)}}{e_i} \right]$, for each i.

Stage	$\left[\dfrac{A_i}{A_{a(i)}} * \dfrac{e_{a(i)}}{e_i} \right]$	Computed k_i value
2	.75	1
3	1	1
4	.67	1
5	1.33	1

Because all of the $k_i = 1$, we simply add setup and holding costs along the structure to arrive at the adjusted costs:

$$\hat{A}_2 = A_2 + \sum_{j \in b(2)} \frac{\hat{A}_j}{k_j} = 30 + \frac{10}{1} + \frac{20}{1} = 60$$

$$\hat{e}_2 = e_2 + \sum_{j \in b(2)} \hat{e}_j * k_j = 2 + 1 * 1 + 1 * 1 = 4$$

$$\hat{A}_1 = A_1 + \sum_{j \in b(1)} \frac{\hat{A}_j}{k_j} = 20 + \frac{60}{1} + \frac{20}{1} = 100$$

$$\hat{e}_1 = e_1 + \sum_{j \in b(1)} \hat{e}_j * k_j = 1 + 4 * 1 + 1 * 1 = 6$$

The calculations for the integer-k method are similar, and the only difference is that the costs are updated during the process. Because of the structure, the k_i values are the same (nothing changes when computing k_2). Because the k_i values remain the same, the adjusted costs are also the same.

Given the adjusted costs, we now solve a standard Wagner–Whitin problem over the planning horizon for each stage beginning at the end-item level. The adjusted costs $\hat{e}_j$, $\hat{A}_j$ are used for holding and setup costs.

9.6.2.3 Heuristic Solution to AKG–e—Parallel Heuristic

We next address another heuristic for solving model ASE. In the previous section, we considered a level-by-level heuristic, in which we solved a single-stage problem over the entire horizon and then fed those decisions to the immediate predecessors as dependent demand. Eventually, we fed decisions down to all N stages and completed a production plan. In this section, we consider an alternate approach, [Afentakis (1987)] that considers the entire structure in parallel and solves a lot-sizing problem with an increasingly longer horizon until we have covered T periods.

The basic ideas of the heuristic can be described in the following steps:

For $t = 0$ to $T - 1$ periods,
   Find a heuristic solution to the $t + 1$-period problem

1. Construct a set of candidate schedules for each stage in the structure that covers the demand for periods 1 through $t + 1$

2. From the candidate set for each stage, select a single schedule that has low cost and meets production needs for the immediate successor stage.

9.6.2.3.1   Schedule Construction for the $t + 1$-Period Problem

The schedule construction process uses schedules selected for smaller horizon problems to construct schedules for the $t + 1$-period problem. We can start with a one-period problem, and this is easy to solve—simply set up and produce in every stage that does not have sufficient inventory to meet the demand in the period. Now, solutions to the one-period problem can be used to generate schedules for the two-period problem. The process continues with schedule selection, and then these selections and past selections are used to generate the next set of schedules for a one-period longer problem. Eventually, we get to generate schedules that cover the entire horizon.

We define the following notation:

$S_l(i) = l^{th}$ candidate schedule for the ith stage for the $t + 1$-period problem

$C_l(i) =$ cost of the lth candidate schedule for the ith stage for the $t + 1$-period problem

$s*(i, t) =$ candidate schedule selected for the ith stage for the t-period problem

Each schedule $S_l(i)$ comprises the periods where i will be set up, and production will be positive. The production quantity for setup in period k is simply the end-item demand for the periods from k until the period preceding the next setup.

Assume that we have generated and selected schedules for problems of size 1 through t and want to generate schedules for period $t + 1$. Consider schedule $s*(i, t)$ selected for stage i in the problem for t, and denote the last setup period in that schedule as period j^*. Then, for each stage i, construct $n_i = t + 2 - j^*$ schedules, each with the format:

$$S_l(i) = [s*(i, j^* + l - 2), 1, 0, 0, ..., 0]$$

where l goes from 1 to n_i.

The key idea behind the method is that we take the selected schedules computed by the algorithm in time periods 1 through $j^* + l - 2$. Then, we schedule a final setup in period $j^* + l - 1$ to cover demand in periods $j^* + l - 1$ through $t + 1$. If j^* is far in the past (much less than $t + 1$) then, we generate many candidate schedules because there are many possibilities for scheduling the last setup. If j^* is close to $t + 1$, than in past iterations, the selection was to produce the last setup close to time period t, therefore, borrowing planning horizon results from the single-stage problem, there is no need to check schedules in which the final setup is substantially earlier. In any case, because $j^* \leq t$, we always generate at least two schedules for each stage.

When $l = 1$, we take the selected schedule for stage i with a horizon of $j^* - 1$, add a setup in period j^* and append 0's up to the $t + 1^{st}$ position. As l increases, we move the last setup closer and closer to $t + 1$, and we use selected schedules from problems with longer and longer horizons. This construction process must be done for each stage for the horizon of $t + 1$ periods. The following example demonstrates the schedule construction method.

EXAMPLE 9.5

We are solving for the production plan for a six-period problem. For stage 3, we have selected the following partial schedules for periods 1 through 5.

$$s^*(3, 1) = [1]$$
$$s^*(3, 2) = [1, 0]$$
$$s^*(3, 3) = [1, 0, 1]$$
$$s^*(3, 4) = [1, 0, 1, 0]$$
$$s^*(3, 5) = [1, 0, 1, 0, 0]$$

Generate the set of candidate schedules for stage 3 for the six-period problem.

SOLUTION

First, we must determine the cardinality of the set of schedules. We look at the selected schedule for period 5 and see that the last setup was in period 3. Therefore, we generate $5 + 2 - 3 = 4$ schedules for the six-period problem. Using the construction formula $S_l(3) = [s^*(3, j^* + l - 2), 1, 0, 0, ..., 0]$ for $l = 1 ... 4$, we obtain:

$$S_1(3) = [s^*(3, 3 + 1 - 2), 1, 0, ..., 0] = [s^*(3, 2), 1, 0, ..., 0] = [1, 0, 1, 0, 0, 0]$$
$$S_2(3) = [s^*(3, 3 + 2 - 2), 1, 0, ..., 0] = [s^*(3, 3), 1, 0, ..., 0] = [1, 0, 1, 1, 0, 0]$$
$$S_3(3) = [s^*(3, 3 + 3 - 2), 1, 0, ..., 0] = [s^*(3, 4), 1, 0, ..., 0] = [1, 0, 1, 0, 1, 0]$$
$$S_4(3) = [s^*(3, 3 + 4 - 2), 1, 0, ..., 0] = [s^*(3, 5), 1, 0, ..., 0] = [1, 0, 1, 0, 0, 1]$$

Computing the cost of the schedule is a rather simple matter. Once the set of setup periods is known, then the production quantities are deterministic because it is best to produce when echelon inventory levels are 0. For a $t + 1$ period problem, if the last setup is in period k, then the final setup covers demand in periods k to $t + 1$. The next to last setup covers demand in the period of that setup through period $k - 1$. We can proceed backward to find all production quantities. Once the quantities are known, the setup and unit production costs are easily computed. The holding cost is based on echelon inventory cost and the end-of-period echelon inventory. For a $t + 1$ period problem, if the last setup for stage i is in period k, then the echelon inventory levels for period $j = k$ through t is:

$$E_{ij} = \sum_{l=j+1}^{t+1} D_l$$

A similar formula holds for all other batches, only the starting and ending periods for the batches change.

EXAMPLE 9.6

Consider the schedules in the previous example, and again focus on stage 3. Assume that only one of stage 3 is required in each end product. Also, assume that the periodic costs for stage 3 and external demands on the end item are:

Period	End item demand	Echelon holding cost/(unit*period)	Setup cost	Production cost per unit
1	20	1	50	5
2	30	1	60	5
3	20	1	40	5
4	15	1	50	5
5	40	1	50	5
6	20	1	60	5

For each of the schedules in the previous example, estimate the total cost.

SOLUTION

We compute the cost of the first schedule and leave the rest as a homework exercise. The schedule is [1, 0, 1, 0, 0, 0], thus we set up and produce in period 3 to meet demand in periods 3, 4, 5, and 6. Next, we set up in period 1 to meet demands in periods 1 and 2. We can compute the costs of each batch separately. For the batch in period 1, we produce $20 + 30 = 50$ units. We spend \$50 on setup, $5 * 50 = \$250$ on unit cost, and at the end of period 1, we have the demand for period 2 (30 units at a cost of $30 * 1 = \$30$) in inventory. Therefore, the total cost of the batch starting in period 1 is \$330. A similar analysis for the batch starting in period 3 yields \$40 for setup, $(20 + 15 + 40 + 20) * 5 = \$475$ in unit costs, and $75 * 1 + 60 * 1 + 20 * 1 = \$155$ echelon inventory costs for a total of \$670 for the batch starting in period 3. Therefore, the total cost of the schedule is $\$330 + \$670 = \$1000$.

9.6.2.3.2   Schedule Selection for the $t + 1$-Period Problem

In schedule selection, we want to find low-cost solutions that are feasible production plans. Because the capacity is assumed to be infinite, if the schedules selected are nested, then the company can produce components in time for successor production. Therefore, if in a selected schedule there is a setup in stage i in period t, then there must be a setup in $a(i)$ in period t also.

Define the **set $F(i, k, l)$ as the set of schedules that are feasible to be selected for stage k where k is an immediate predecessor of i, if schedule $S_l(i)$ is selected.** You may only select a schedule that is in $F(i, k, l)$ for stage k if $S_l(i)$ is selected for stage i. By the nesting property, schedules $F(i, k, l)$ can only schedule production in the periods where $S_l(i)$ has scheduled production. To set up the model to make selections, define the following notation:

$X_{il} = 1$ if the lth candidate schedule $S_l(i)$ is selected for stage i, 0 otherwise;

$C_l(i) = $ cost of the lth candidate schedule for stage i.

The model (**SEL**) to solve is:

$$\text{Minimize} \sum_{i=1}^{N} \sum_{l=1}^{n_i} C_l(i)X_{il} \qquad (9.45)$$

Subject to: $\quad \sum_{l=1}^{n_i} X_{il} = 1 \qquad \text{for } i = 1 \ldots N \qquad (9.46)$

$$X_{il} \leq \sum_{s_j(k)\in F(i,k,l)} X_{kj} \qquad \text{for } i = 1 \ldots N, l = 1 \ldots n_i, k \in b(i) \qquad (9.47)$$

$$X_{ij} \in \{0, 1\} \qquad \text{for } i = 1 \ldots N, l = 1 \ldots n_i \qquad (9.48)$$

The objective [Equation (9.45)] represents the total cost of the selected schedules. Constraint set [Equation (9.46)] ensures that one schedule is selected for each stage. Constraint set [Equation (9.47)] is a logical constraint that ensures if you select schedule $S_l(i)$, then for each of its immediate successors you must select an appropriate schedule that has the nesting property. Note that if you select a schedule at stage $k \in b(i)$ that is compatible with $S_l(i)$, then you may or may not select schedule $S_l(i)$. Finally, constraint set [Equation (9.48)] defines the binary nature of the schedule selection process.

Solving this model is not difficult, and Afentakis suggests a dynamic programming algorithm (also, standard integer programming software can often be used because there are few integer variables). Assume that the stages are numbered/ordered so that predecessors have larger stage numbers than successors. Also, denote stage 0 as the successor to stage 1 (the end item), and let there be one candidate schedule for this stage $S_1(0) = [1, 1, 1, \ldots 1]$ with cost $C_1(0) = \$0$. Because this schedule has setup in every period, it puts no constraints on stage 1 for nesting properties. The method with explanation is:

Step 1: Initialize $i = N + 1$ (we start at the bottom of the structure—the parts).

Step 2: Set $i = i - 1$, $l = 1$, and $j = a(i)$ (we consider a stage and its successor. At this point, we are going to update the costs of the candidate schedules for the successor stage, j. If $i = 0$, then stop (we are at the end item and are finished).

Step 3: Recompute $C_l(j) = C_l(j) + \text{minimum}\{C_l(i):, S_j(i) \in F(j, i, l)\}$ (update the cost of candidate schedule l at stage j by adding the lowest cost schedule at stage i over all schedules at i that are feasible given that we choose schedule l at stage j. In this manner, we accumulate costs as we move up the structure, always selecting the lowest accumulated cost to get all the way through the rest of the structure. This is identical to the optimality principal in dynamic programming. Also, note that the index labels in the F set have been switched. The set is always **stage, predecessor, schedule number at stage** and here, j is the stage, i is the predecessor, and l is one of the possible schedules at stage j).

Step 4: If $l < n_j$, then set $l = l + 1$ and go to step 3, otherwise go to step 2 to move to the next stage in the structure.

EXAMPLE 9.7

Schedule selection: Consider solving a five-period problem for the product structure below. Listed next to the product structure is a Table detailing the candidate schedules for each stage (determined in the previous step of the heuristic), and their costs. Solve for the selected schedule for each stage using the dynamic programming method.

Stage	Candidate schedules	Schedule costs
1	[1, 0, 1, 1, 1]	75
	[1, 0, 1, 1, 0]	84
2	[1, 0, 1, 0, 1]	80
	[1, 0, 1, 1, 0]	75
	[1, 0, 1, 0, 0]	90
3	[1, 0, 1, 1, 1]	65
	[1, 0, 1, 1, 0]	55
4	[1, 0, 1, 0, 0]	45
	[1, 0, 1, 1, 0]	50
	[1, 0, 1, 0, 1]	40

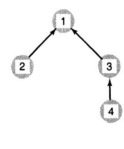

SOLUTION

A Table of the relevant calculations is given below. We start at the bottom of the structure and work toward the top. The Table is organized by (i, l, j) values. We list $S_l(j)$ and then the set of schedules that are feasible $F(j, i, l)$. For each schedule in $F(j, i, l)$, we add its cost to $C_l(j)$ and then find the minimum total over all schedules in $F(j, i, l)$. These values are in **bold** in each section of the Table. Once the calculations are completed, we have to trace back to find the solution.

i	l	$j = a(i)$	$S_l(j)$	$F(j, i, l)$	$C_l(j) + C_j(i)$	Min value
5	—	—	—	—	—	—
4	1	3	[1,0,1,1,1]	[1,0,1,0,0]	65 + 45	
				[1,0,1,1,0]	65 + 50	
				[1,0,1,0,1]	**65 + 40**	**105**
4	2	3	[1,0,1,1,0]	[1,0,1,0,0]	**55 + 45**	**100**
				[1,0,1,1,0]	55 + 50	
3	1	1	[1,0,1,1,1]	[1,0,1,1,1]	75 + 105	
				[1,0,1,1,0]	**75 + 100**	**175**
3	2	1	[1,0,1,1,0]	[1,0,1,1,0]	**84 + 100**	**184**
2	1	1	[1,0,1,1,1]	[1,0,1,0,0]	175 + 90	
				[1,0,1,1,0]	**175 + 75**	**250**
				[1,0,1,0,1]	175 + 80	
2	2	1	[1,0,1,1,0]	[1,0,1,1,0]	**184 + 75**	**259**
				[1,0,1,0,0]	184 + 90	
1	1	0	[1,1,1,1,1]	[1,0,1,1,1]	**0 + 250**	**250**
				[1,0,1,1,0]	0 + 259	

We start with the bottom section. The minimum occurs for the schedule [1,0,1,1,1] with a value of 250. Therefore, for stage 1, the selection is [1,0,1,1,1] and its individual cost is 75. For stage 2, selection is the minimum cost schedule using [1,0,1,1,1] for $S_l(k)$ because stage 1 is the successor to stage 2. Therefore, the minimum occurs at schedule [1,0,1,1,0] and its individual cost is 75. For stage 3, the selection is the minimum cost schedule using [1,0,1,1,1] for $S_l(k)$ because stage 1 is the successor to stage 3. Thus, the minimum occurs using [1,0,1,1,0] and its individual cost is 55. Finally, for stage 4, the selection is the minimum cost schedule using [1,0,1,1,0] for $S_l(k)$ because that is the selection at stage 3—the successor to stage 4. Thus, the selection at stage 4 is [1,0,1,0,0] and its individual cost is 45. Totaling all costs, we obtain 75 + 75 + 55 + 45 = 250 (as expected from the value for $C_1(0)$). Also, it is easy to understand that the nesting property holds throughout the structure.

Computationally, the heuristic requires reasonable work even if the product structure is large, and we have to repeat steps 1 and 2 for each of the T periods. For example, after completing the schedule selection step (Example 9.7), we would then increase t by 1 and move to the next period to generate a new set of candidate schedules for each stage.

Numerical results on example problems suggest that the heuristic works extremely well. Afentakis compared the heuristic results with the optimal solution generated from branch and bound on problems with varied number of stages, levels, setup and holding cost structures, and demand patterns. The heuristic generally was within 1% of the optimal solution and required a fraction of the time required by branch and bound. For more insight on the computational experiment, see Afentakis [1987].

9.6.3 General Product Structures

General structures have the property that many stages may share common components. This is illustrated in Figure 9.1c. Using common components or modules normally has significant advantages. Few parts and modules need to be designed, less tooling is needed, and fewer part types need to be kept in inventory. These advantages usually outweigh the difficulty in trying to coordinate schedules for use by all successors. At a minimum, we could treat each occurrence of a component as a separate item and use our assembly system scheduling algorithms. Whenever batch production times coincided with supplying more than one successor, we reap the savings of reduced setups. At other times, we still avoid the cost of design and tooling and are no worse off in setup than if we had used separate components.

We present two approaches for the general structure problem. The first is a direct extension of model ASE. The approach is to take a general structure and transform it to an equivalent assembly structure with constraints to ensure that setup costs are counted correctly. The model has the drawback that it requires assumptions of infinite production capacity and deterministic lead times. We do not directly include the lead time in the model, therefore, the time subscript for any stage is actually the current time t, offset by the lead time τ_i. The second model is similar to the aggregate planning models developed previously. The model has a large amount of flexibility, however, solution is difficult because of its size.

9.6.3.1 Afentakis and Gavish Model

Afentakis and Gavish extended model ASE to consider general structures. Here, there is no longer a unique path from stage i to the end item. Therefore, there must be a slight adjustment to the structure of an optimal solution. Using the same notation as was used for the assembly structure and the same assumption on stationary production costs, **model GE** is:

$$\text{Minimize} \sum_{t=1}^{T} \sum_{i=1}^{N} (A\delta_{it} + e_{it}E_{it}) \tag{9.49}$$

Subject to:

$$X_{it} \leq M\delta_{it} \qquad \text{for each } (i, t) \text{ pair} \tag{9.50}$$

$$E_{it-1} + X_{1t} - \rho_{i1}D_t = E_{1t} \qquad \text{for each } (i, t) \text{ pair} \tag{9.51}$$

$$\sum_{k \in a(i)} r_{ik} E_{kt} \leq E_{it} \qquad \text{for each } (i, t) \text{ pair} \tag{9.52}$$

$$X_{it} \geq 0, \, E_{it} \geq 0, \, \delta_{it} \in \{0,1\} \tag{9.53}$$

The objective, [Equation (9.49)], totals the setup and echelon holding costs. Constraint set [Equation (9.50)] ensures that setup cost is accrued when production is positive. Constraint set [Equation (9.51)] is the echelon inventory balance constraint set for all the items. Echelon inventory increases with production of stage i and reduces with end item demand. Constraint set [Equation (9.52)] ensures that there is enough echelon inventory of stage i on hand to produce the completed amount of all successor items in the set $a(i)$.

Afentakis and Gavish prove the following result for an optimal solution to GE:

Proposition 9.8 There is an optimal solution to GE with the property that if $X_{it} > 0$, then for some $k \in a(i)$, $X_{kt} > 0$.

Even with this proposition, GE is not easily solved directly. Instead, the general structure model is transformed into an equivalent assembly structure model with a constraint that ensures that setup costs are accrued correctly. The key to the transformation is that in a general structure, there are many immediate successors to stage i, and hence there are many paths from i to the end item. If we enumerate these paths and splits stages appropriately, then we can transform the structure into an assembly structure. Let G be the original bill of material product structure. The construction process is:

Step 1   **(Node construction):** For each stage i compute the number of paths from i to stage 1. Denote this value as P_{i1} and place P_{i1} nodes in the new product structure G'. Label these nodes $(i, 1)$, $(i, 2)$, ...(i, P_{i1}).

Step 2   **(Edge construction):** Place an edge between nodes (i, k) and (j, l) in G', if $j \in a(i)$ and node k for i and node l for j are on the same path in G. This edge has weight r_{ij}.

The following example demonstrates the construction process.

EXAMPLE 9.8   **Structure Transformations:** Convert the following general structure into an assembly structure:

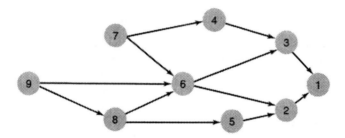

SOLUTION   The structure below is the solution. Stage 6 has two paths to stage 1, stage 7 has three paths to stage 1, stage 8 has three paths to stage 1, and finally, stage 9 has five paths to stage 1. The number of paths determines the number of stages in G'. Next, we link up the nodes so that the paths are realized. From the diagram, we can see that each stage has one immediate successor and the resulting structure is an assembly structure.

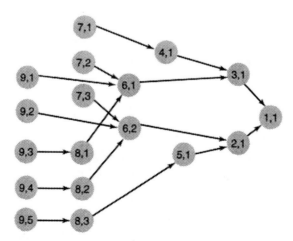

This process yields an assembly structure, however, the number of nodes in G' grows rapidly when the original structure is highly connected. Also, when any node of stage i $((i, 1), (i, 2), \ldots (i, P_{i1}))$ produces in a period, then a setup cost must incur, however if more than one node in the set produces, then only one setup cost is incurred.

The following proposition extends Proposition 9.8 to the transformed product structure:

Proposition 9.9: An optimal batch X_{it} can be decomposed into P_{i1} batches in which each batch size is either 0 or covers an integer number of periods of demand for the end item (including multiplicity requirements using ρ_{i1}).

The significance of the proposition is that when you produce at a stage, you are producing for an integer number of the paths worth of demand. Once you get to the path level, then the results for an assembly structure hold, and you produce for an integer number of periods of demand (produce when echelon inventory is 0).

Afentakis and Gavish use Lagrangean relaxation in conjunction with a branch and bound procedure to solve GE. Their results are encouraging for sparse product structures with few paths.

9.6.3.2   Integer Programming Approaches

We can formulate the general material requirements planning problem as a mathematical program model similar to the models for aggregate planning. In accord with our lower-level decision relative to aggregate planning, time periods in the model will represent shorter periods such as a day or week. Instead of aggregate production, we plan each end item and each component separately. Likewise, each resource such as a work center or labor grade that is limited must be modeled separately instead of aggregated. The material planning model, **denoted model MPM,** is formulated using installation inventory. Define the following notation:

Data Parameters:

$k = 1 \ldots K$ resources

$i = 1 \ldots N$ stages

$t = 1 \ldots T$ periods

A_i = setup cost of production at stage i (stationary costs)

s_i = setup time required for stage i

h_{it} = installation holding cost per unit of stage i installation inventory on hand at the end-of-period t

p_{ik} = production time required in resource k for one unit of stage i production

P_{kt} = production time capacity for resource k in period t

D_{it} = external demand for stage i production in period t

τ_i = lead time periods for stage i production

o_{kt} = cost per unit of overtime for resource k in period t

u_{kt} = cost per unit of under time for resource k in period t

r_{ij} = number of stage i units required per unit of immediate successor stage j

Decision Variables:

O_{kt} = overtime allocated to resource k in period t

U_{kt} = under time allocated to resource k in period t

I_{it} = installation inventory of stage i on hand at the end-of-period t

X_{it} = production at stage i in time period t

δ_{it} = 1 if $X_{it} > 0$, otherwise 0

$$\text{Minimize cost} = \sum_{t=1}^{T} \sum_{j=1}^{N} (h_i \cdot I_{it} + A_i \cdot \delta_{it}) + \sum_{t=1}^{T} \sum_{k=1}^{K} (o_{kt} \cdot O_{kt} + u_{kt} \cdot U_{kt}) \qquad (9.54)$$

Subject to: $I_{i,t-1} + X_{i,t-\tau_i} - D_{it} - \displaystyle\sum_{l \in a(i)} r_{il} \cdot X_{lt} = I_{it},\ i = 1, \ldots, N; t = 1, \ldots, T \qquad (9.55)$

$$\sum_{i=1}^{N} (s_i \cdot \delta_{it} + p_{ik} \cdot X_{it}) + U_{kt} - O_{kt} = P_{kt},\ k = 1, \ldots, K; t = 1, \ldots, T \qquad (9.56)$$

$$X_{it} \leq M \cdot \delta_{it},\ i = 1, \ldots, N; t = 1, \ldots, T \qquad (9.57)$$

$$X_{it} \geq 0,\ I_{it} \geq 0,\ O_{kt} \geq 0,\ U_{kt} \geq 0\ \delta_{jt} \in \{0, 1\} \qquad (9.58)$$

The objective function, [Equation (9.54)] minimizes the setup, processing, overtime, and under time costs over the horizon. Setup and processing are summed over the N product stages. Overtime and under time are accumulated for each of the K resources. The first constraint set [Equation (9.55)] provides inventory balance. Ending inventory each period equals initial inventory plus production started τ_i periods ago minus units consumed either directly by product demand or for use in production of a successor item. Next, constraint set [Equation (9.56)] balances the capacity for each resource k in each period. As always, the left side of the equation totals the resource required for the planned production as a function of the decision variables. The right side of the equation contains the amount of resource available for use, P_{kt}. This equation form assumes the amount of resource available in a period is independent of the amount consumed in other periods. The constraint set [Equation (9.57)] ensures that a setup be performed if we produce a positive amount of the item in the period. In practice, time may be continuous and it may be possible to save a setup by starting a period by producing the same product that was produced last in the previous period. We assume that this is not the case, however. Model MPM will try to fit the entire batch into one period or the other when possible. M is an upper bound on the maximum production of stage i product per period. This could be set equal to the total horizon demand or another value based on an upper bound on overtime and regular time production.

The major difficulty with MPM is its size. We have NT binary variables, $2T(N + K)$ continuous variables and $T(2N + K)$ constraints. For a full size production facility with 10,000 products, 50 periods and 100 resources, this translates to 500,000 binary variables, over one million continuous variables and over one million constraints. Building and managing the data for a model this size is a formidable but possible task. However, resolving this model on a regular basis as each inventory transaction occurs is not possible. This explains why MRP decision models are still relatively simplistic. Nonetheless, we address some basic strategies for reducing the size of this problem.

The formulation above assumes deterministic lead times following production starts. The two complicating factors are the binary variables needed to model setups and the resource constraints caused by limited capacity. If capacity is infinite and setup is free, then the optimal solution would be to use a lot-for-lot (LFL) strategy and could be found using the standard MRP explosion process. One strategy to reduce the size of the problem is to predetermine that certain stages should follow a LFL strategy and effectively remove these from the decision model. First, identify those stages that have a small setup time and cost and are produced on a nonbottleneck work center. These items are candidates for elimination by automatically assigning an LFL strategy.

For those items that are successors to the items produced at the bottleneck work center, we use the normal MRP explosion process to set production quantities in each period. Let CNR_{it} be the cumulative net requirements for stage i in period t. Starting at the end-item level and working down the structure, we can compute these cumulative net requirements using:

$$CNR_{it} = \max\left[0, \sum_{v=1}^{t}\left(D_{iv} + \sum_{k\in a(i)} r_{ik} X_{kv}\right) - I_{i0} - \sum_{v=1}^{\min(t,\tau_i)} X_{i,v-\tau_i}\right] \qquad (9.59)$$

The first summation in Equation (9.59) accounts for direct demand for the item as an end item or spare part plus units required for production starts of immediate successor parent items. From this total, we subtract initial inventory and scheduled receipts up through period t. At each stage, planned production quantities are then formed using a LFL strategy. Production is set equal to the increase in cumulative net requirements from one period to the next, offset by the lead time. Thus,

$$X_{i,t-\tau_i} = CRN_{it} - CNR_{i,t-1} \; ; \; t = 1, ..., T \qquad (9.60)$$

Billington et al. [1983] addresses additional ways to compress the product structure by eliminating other items that should follow a LFL strategy. In many cases, it is possible to substantially reduce the size of the material planning model.

EXAMPLE 9.9

Consider the product structure shown below.

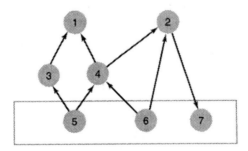

Items within a set of dotted lines are produced in the same capacity constrained work center. Assume all setups are small and that $r_{ij} = 1$ for all (i, j) pairs. Demand over the next five weeks for end item 1 is (0, 100, 75, 80, 60) and (0, 140, 160, 110, 90) for end item 2. All previously planned orders will arrive in period 1, and they total (110, 170, 30, 40, 300, 250, 250) for each of the seven stages, respectively. Lead times are one period for all stages. Reduce the product structure and determine the production schedule for the eliminated items.

SOLUTION

All items above the bottlenecks can be eliminated. Because setup cost is small, we should use lot-for-lot scheduling. Therefore, we have to use MRP explosion on items 1, 2, 3, and 4—all using lot-for-lot scheduling. We first compute the CNR_{it} values using Equation (9.59) for stages 1 and 2, the end items.

$$CNR_{it} = \max\left[0, \sum_{v=1}^{t}\left(D_{iv} + \sum_{k\in a(i)} R_{ik} X_{kv}\right) - I_{i0} - \sum_{v=1}^{\min(t,\tau_i)} X_{i,v-\tau_i}\right]$$

CNR_{it}	1	2	3	4	5
1	Max(0, −110) = 0	0	65	145	205
2	Max(0, −140) = 0	0	130	240	330

For stages 1 and 2, we can now compute the lot sizes using Equation (9.60) with a lead time of one week.

$$X_{i,t-t_i} = CRN_{it} - CNR_{i,t-1}; \quad t = 1, \dots, T$$

Stage/time	1	2	3	4	5
1	0	65	80	60	0
2	0	130	110	90	0

We can now take these decisions and explode back to stages 3 and 4. Stage 3 has only stage 1 as a predecessor whereas stage 4 has both stages 1 and 2 as predecessors.

CNR_{it}	1	2	3	4	5
3	Max(0, −30) = 0	65 − 30 = 35	115	175	175
4	Max(0, −40) = 0	195 − 40 = 155	345	495	495

Finally, we compute production decisions for these stages.

Stage/time	1	2	3	4	5
3	35	80	60	0	0
4	155	190	150	0	0

These values along with the decisions at stages 1 and 2 generate net requirements for stages 5, 6, and 7. For this problem, finding a production schedule for stages 5, 6, and 7 reduces to a single stage, multiproduct, constrained lot-sizing problem.

9.7   APPLICATION

We have considered a great deal of material in this chapter, much of it based on simplified models with extensive data requirements. Understanding and applying the material in this chapter is not easy; however, the cost benefits can be great. A fair question would be:

How much of this material is applicable and how would it be applied?

Much of the early work had little direct application; however, it served as insightful work that led to more applicable directions. It takes time for a body of knowledge to mature to application, and today, as computers become more powerful, multistage production planning models are used to give insight into supply chain management (order cycle times and vendor selection based on costs and response times), and production control (cycle-time computation). The work on powers-of-two policies has been especially significant in that it shows that simple, easily-implemented policies can have excellent quality. Some of the chapter references contain data from industry cases.

The data requirements for these models can often be found in company accounting systems and market forecasting systems. The key parameters are the setup and holding costs and the periodic or stationary demand rates. One can estimate setup cost from standard labor times and labor rates, and/or material usage and material cost. Holding costs can be estimated using the value added amount for each process (for example, added labor cost plus added material cost plus added material handling cost). Finally, demand forecasting problems can certainly be tackled using statistical methods and data from past periods. Certainly, advertising strategies and promotions affect demand, and these events must be coordinated with production control.

9.8 SUMMARY

This chapter has briefly covered multistage production scheduling problems. We first discussed different types of structures and then considered two equivalent approaches for defining inventory levels—installation and echelon inventory. Models for making decisions in these environments may use the approach that makes model solution easier.

We next split the presentation into three major sections. First, we considered problems in which setup cost was not a factor. Here, linear programming was used to model a variety of system issues. Model solution can be done using standard mathematical programming software. Next, we considered problems in which setup was important and demand was stationary. We presented models and solution approaches that depend on the specific product structure. As the structure becomes more complex, and there are multiple paths from a stage to an end product, heuristics must be used. For a variety of problems settings, recent research has improved upon these heuristics, and they now have excellent performance guarantees. Finally, we considered problems where setup is important and demand varies with time. Here, we developed mathematical programming models and solution approaches. Similarly to the stationary demand case, more complex structures generally require heuristics or extensive computational work.

The topics in this chapter are a small portion of the research on multistage lot sizing. Most of the material is from the 1970s and 1980s. Research on these topics has continued to present; however, it is critical to understand the early work before starting on more advanced papers. Interested students are encouraged to look at the more theoretical operations research literature as well as the more applied production planning and scheduling literature. To effectively solve complex production planning problems, it is necessary to understand the key objectives, constraints, and cost structures of the actual system, as well as the theoretical topics of mathematical modeling, discrete optimization, and heuristic development.

9.9 REFERENCES

AFENTAKIS, P. A parallel heuristic algorithm for lot-sizing in multistage production systems. *IIE Transactions*, 1987, 19(1), 34–42.

AFENTAKIS, P., & GAVISH, B. Optimal lot-sizing algorithms for complex product structures. *Operations Research*, 1986, 34, 237–249.

AFENTAKIS, P., GAVISH B., & KARMARKAR, U. Computationally efficient optimal solutions for the lot-sizing problem in multistage assembly systems. *Management Science*, 1984, 30, 222–239.

BILLINGTON, P.J., McCLAIN J.O., & THOMAS, L.J. Mathematical programming approaches to capacity constrained MRP systems: review, formulation and problem reduction. *Management Science*, 1983, 29(10), 1126–1141.

BLACKBURN, J., & MILLEN, R. Improved heuristics for multi-stage requirements planning systems. *Management Science*, 1982, 28, 44–56.

CROWSTON, W.B., WAGNER, M., & WILLIAMS, J.F. Economic lot size determination in multistage assembly systems. *Management Science*, 1973, 19(5), 517–527.

JENSEN, P., & KAHN, H. Scheduling in a multi-stage production system with setup and inventory costs. *IIE Transactions*, 1972, 4, 126–133.

LOVE, S. A facilities in series inventory model with nested schedules. *Management Science*, 1972, 18, 327–338.

MAXWELL, W., & MUCKSTADT, J. Establishing consistent and realistic reorder intervals in production distribution systems. *Operations Research*, 1985, 33, 1316–1341.

MUCKSTADT, J., & ROUNDY, R. Analysis of multistage production systems. In: GRAVES, R.K. & ZIPKIN, eds; *Logistics of production and inventory, handbooks in operation research and management science*, 1993, 59–131.

MOILY, J. Optimal and heuristic procedures for component lot-splitting in multistage manufacturing systems. *Management Science*, 1983, 29, 113–125.

ROUNDY, R. A 98% effective lot-size rule for a multi product, multi stage production/inventory system. *Mathematics of Operations Research*, 1986, 11, 699–727.

SCHWARZ, L., & SCHRAGE, L. Optimal and system myopic policies for multi-echelon production/inventory assembly systems. *Management Science*, 1975, 21, 1285–1294.

SZENDROVITS, A. Comments on the optimality in optimal and system myopic policies for multi-echelon production/inventory assembly systems. *Management Science,* 1981, 27, 1081–1087.

VEINOTT, A.F., Jr. Minimum concave cost solution of leontief substitution models of multifacility inventory systems. *Operations Research,* 1996, 17(2), 262–291.

WILLIAMS, J. On the optimality of integer lot size ratios in economic lot size determination in multi-stage assembly systems. *Management Science,* 1982, 28, 1341–1349.

ZANGWILL, W.I. A deterministic multiproduct multifacility production and inventory model. *Operations Research,* 1966, 14, 486–507.

9.10 PROBLEMS

9.1. Show that for a product structure, $i \in B(j)$, if and only if, $j \in A(i)$.

9.2. Show that for an assembly structure with one end item, there is a unique path between any stage i and any other stage j.

9.3. Show that the number of stage i items used in each item of stage 1 is the product of the requirements values of the stages along the path from i to 1.

9.4. Describe the difference between installation inventory and echelon inventory levels.

9.5. Consider the product structure below, and assume that all r_{ij} values are 1. Suppose that the installation inventory levels and costs are as follows:

Stage	1	2	3	4	5
Installation Inventory	25	10	5	45	60
Installation Inventory Cost	$22	$6	$10	$5	$1

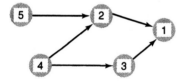

a. Find the echelon costs and echelon inventory levels for each stage. This can be effectively done on spreadsheet software.

b. Show that total installation inventory costs and total echelon inventory costs are the same here. This can be effectively done on spreadsheet software.

9.6. Extend Equations (9.6) and (9.7) to the case when $r_{i,a(i)} >$ 1. That is, prove that total echelon inventory cost is equal to total installation inventory cost for the case when more than one item i can be used in each immediate successor $a(i)$.

9.7. Show that model ASI is equivalent to model ASE. That is, show that any production plan to model ASI is optimal, if and only if, it is an optimal solution to model ASE.

9.8. Prove that the relaxed problem for ASE is a relaxed problem for $\lambda_{ij} \geq 0$. That is,

a) Show that any solution feasible to ASE is also feasible to the relaxed problem.

b) Show that for any feasible solution to ASE, the objective of the relaxed problem is no greater than the objective of ASE.

9.9. Describe how you could modify Model ASI to incorporate a yield loss during production. Let y_i be the yield proportion for stage i production.

9.10. Consider the Afentakis parallel heuristic described in section 9.6.2.3. Compute the cost of each of the schedules generated in Example 9.6 using the data of Example 9.7.

9.11. Consider a single stage in a multistage assembly structure with a single end item. Using the Afentakis parallel heuristic, we have solved the problem for the one, two, three, four, and five-period problems. Generate the set of candidate schedules for the sixth period. The previous solutions are:

$$t = 1 \quad [1]$$
$$t = 2 \quad [1,0]$$
$$t = 3 \quad [1,1,0]$$
$$t = 4 \quad [1,1,0,0]$$
$$t = 5 \quad [1,1,1,0,0]$$

9.12. Using the following data, compute the costs of the schedules generated in problem 11. Assume that the unit production cost is stationary. This can be effectively done using spreadsheet software.

Period	Demand	Echelon holding cost	Setup cost
1	20	1	40
2	30	2	60
3	10	1	40
4	0	1.5	50
5	15	1	40
6	20	2	50

9.13. Consider solving a six-period problem for the product structure below. Listed next to the product structure is a table detailing the candidate schedules for each stage, and their costs. Solve for the selected schedule for each stage using the dynamic programming method proposed in the Afentakis parallel heuristic. (If you have access to mathematical programming software, or a spreadsheet "solver.") Solve for the selected schedule for each stage using mathematical programming software.

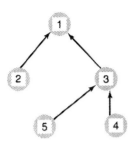

Stage	Candidate schedules	Schedule costs
1	[1, 0, 1, 1, 0, 1]	75
	[1, 0, 1, 1, 1, 0]	84
	[1, 0, 1, 1, 0, 0]	90
2	[1, 0, 1, 0, 1, 1]	80
	[1, 0, 1, 0, 1, 0]	75
	[1, 0, 1, 1, 0, 0]	90
	[1, 0, 1, 0, 0, 0]	100
3	[1, 0, 1, 1, 1, 1]	65
	[1, 0, 1, 1, 1, 0]	55
4	[1, 0, 1, 0, 0, 0]	45
	[1, 0, 1, 1, 0, 0]	50
	[1, 0, 1, 0, 1, 0]	40
	[1, 0, 1, 0, 1, 1]	55
5	[1, 0, 1, 0, 0, 1]	80
	[1, 0, 1, 0, 1, 0]	75

9.14. For the Blackburn and Millen level-by-level heuristic described in section 9.6.2.2, derive the continuous-K formula for setting k_i.

9.15. For the following data set, use the Blackburn and Millen level-by-level heuristic to adjust the setup and holding costs using the:

a. Continuous-K method

b. Integer-K method

c. System myopic-K method

Stage	Setup cost	Echelon holding
1	20	2
2	15	1
3	18	1
4	15	2
5	15	0.5
6	20	0.5
7	10	1

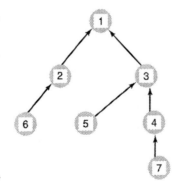

9.16. Using the results of problem 15, solve for the lot sizes at each stage using the following single-stage techniques:

a. Lot-for-lot scheduling (all three cost adjustment techniques)

b. Wagner-Whitin method (all three cost adjustment techniques)

c. Silver-Meal heuristic (all three cost adjustment techniques)

9.17. For the following general structure, construct the generated assembly structure for the method of Afentakis and Gavish.

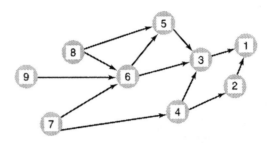

9.18. Consider a serial two-stage system where stage 1 is the end product and stage 2 is the part. Let:

d_{it} = the external demand for stage i in time t,

r_{21} be the number of units of stage 2 used in each unit of stage 1

A_{it} = the setup cost of stage i in time t,

h_{it} = the installation inventory cost per unit of inventory on hand at stage i at the end-of-period t, and

c_{it} = variable production cost per unit at stage i in period t.

Assume that there are no initial inventories for either item, lead time is zero periods, and $d_{11} > 0$. Consider the following heurisitic for finding the minimum cost production plan:

Step 1: Solve a Wagner–Whitin problem for stage 1 using the data d_{1t}, A_{1t}, h_{1t}, and c_{it}. If X_{1t} is the production quantity of stage 1 in time t, then $r_{21} X_{1t}$ is the demand imposed on stage 2 by the schedule of stage 1.

Step 2: Set $\hat{d}_{2t} = d_{2t} + r_{21}X_{1t}$. Solve a Wagner–Whitin problem for stage 2 using the data $\hat{d}_{2t}, A_{2t}, h_{2t}$, and c_{2t}. From the solution to this problem, determine γ_{2t}, the marginal cost of increasing demand in stage 2 in period t by one unit (the dual variable in the demand constraint for this Wagner–Whitin problem).

Step 3: Set $\hat{c}_{1t} = c_{1t} + r_{21}\gamma_{2t}$. Solve a Wagner–Whitin problem for stage 1 using the data d_{1t}, A_{1t}, h_{1t}, and $\hat{c}_{1t}$. If there is no change from the previous optimal solution for stage 1, then stop. Otherwise, return to step 2.

a. Explain why the heuristics of Blackburn and Millen, and Afentakis are called "single-pass" heuristics, whereas this is called a "multi-pass" heuristic.

b. Devise a method for estimating γ_{2t}.

c. How might you extend this method to more complex structures such as larger serial systems and assembly systems?

Chapter 10

Lean Manufacturing and the Just-in-Time Philosophy

Kanbans and the "pull" approach to controlling material flow constitute one of the key aspects of a just-in-time (JIT) production system. Beyond kanbans however, JIT encompasses a variety of practices that can improve the performance of a production system. These practices are important for push as well as pull systems. The goal in both cases should be the establishment of a competitive, **lean manufacturing** system. In this chapter, we describe a set of techniques for achieving JIT, or equivalently, lean production.

10.1 LEAN, JUST-IN-TIME PRODUCTION SYSTEMS

Just-in-time (JIT) is a philosophy for optimizing the performance of a manufacturing system. The origins of JIT stem from the work of Taiichi Ohno at Toyota Motor Company. After World War II, Japan needed to rebuild its manufacturing industry. In the United States, the call was for mass production to satisfy the needs of a large populace that had saved and sacrificed during the war. The Japanese market was much smaller, and investment capital was scarce. With smaller production volumes per part, there was a need for more flexible systems that could produce many different items on the same equipment. After years of experimenting, Toyota settled on an effective strategy based on:

1. Kanban-based pull production;
2. Elimination of waste as a guiding philosophy;
3. Faith in the value and importance of quality;
4. "Kaizen" or continuous improvement as a day to day operating strategy;
5. Belief in the value and utilization of human resources;
6. Emphasis on reducing setup times for machines;
7. Integration of suppliers and material acquisition into the corporate planning process; and
8. Efficient, cellular layouts with balanced material flow.

The kanban system developed as a production control system that could be implemented without sophisticated information technology. Quality at the source, or "do it right the first time" was essential in Japan where limited capacity and resources made it unacceptable to tolerate waste. Striving to prosper on limited resources necessitates productive use of all resources and elimination of waste. Continuous improvement recognized the large gap between current and optimal practices as well as an appreciation for the future improvements that would be made possible through technological advances. In addition, this spirit allowed workers to take pride in their accomplishment. Note, that pride in workmanship and recognizing the value of the human served to help rebuild a sense of self worth following the demoralizing loss Japan experienced in WWII.

Pull-oriented kanban systems are sometimes erroneously equated to JIT. While the pull approach is sufficiently important to merit its own chapter (see Chapter 7), in this text on production systems, it represents only one aspect of the broader JIT philosophy. It is the combination of all these factors that transform the shop into a lean facility. We can address each aspect separately, but it should be kept in mind that they are dependent on each other. A kanban system, without superior quality or low setups, will fail to achieve the anticipated benefits. In the following sections, we elaborate on the basic tenets of lean, JIT production. In the organization of this chapter, these tenets are divided into techniques for improving the production environment, quality engineering, and improving material flow. However, this division is somewhat arbitrary, and the concepts are best viewed as one integral philosophy.

10.2   IMPROVING THE PRODUCTION ENVIRONMENT

A variety of activities can be employed to improve the general production environment. The presence of these programs or conditions make products flow through the system quicker and in a more predictable manner. In this section, we address several areas where investments in improving the production environment can have large payoffs in system performance.

10.2.1   Eliminating Waste

Beginning with Frederick Taylor's *The Principles of Scientific Management* in 1911, managers in the United States applied scientific reduction to production systems. Systems were dissected and specialists ordained to optimize each piece of the system. The intent was to eliminate waste thereby permitting better wages for workers, higher profits for owners, and a better quality of life for consumers. This effort quickly produced significant improvements to product quality and worker productivity. The emphasis on science led to the EOQ and other basic models. However, in some cases, we eventually became sidetracked, the small pieces became the system, and the science became the objective. The models that evolved are correct but in a sense are suboptimal and at times divert attention from viewing the total system. For example, is the primary lesson of the EOQ that we must train everyone to optimally tradeoff inventory and holding cost? Or should the lesson be that we must work to minimize setup cost? Sadly, many heard the first lesson while overlooking the second and more important lesson. Along a similar line, is the lesson learned from stochastic models that we should collect data and model the lead-time demand distribution to find the optimal safety stock? Or, do we work to reduce the causes of variability? Instead of emphasizing the science of system pieces, Japan chose to develop a system that optimally connected the pieces. Instead of minimizing inventory cost for a given setup cost, the philosophy developed in Japan to reduce the setup cost directly. Instead of tracking production statistics and trying to inspect out defective products, the Japanese strategy revolved around eliminating the production of defective items. The JIT system design of T. Ohno eloquently articulates the common sense, hands-on approach to effective operation. Others followed, including K. Suzaki, who noted seven types of waste commonly found in industry. His sources of waste to be eliminated include:

1. Waste from overproduction. Production costs money. There is no reason to produce items that cannot be sold. Reducing cycle time and matching production schedules for parts that are assembled together reduce system costs. Traditionally, supervisors were judged by the quantity of production. The thought was that resource utilization was to be maximized. This leads to waste. Machines and humans should only be busy when they have useful tasks to accomplish. Why waste materials and energy producing unnecessary products? In addition to the wasted cost of materials, machines will wear out sooner and be unavailable for productive tasks. We should balance the flow of parts toward final assembly to meet demand instead of working to keep resources busy.

2. Waste of motion. Motion consumes time and energy. Eliminate those motions that do not add value. This objective should guide workplace design, process planning, writing of detailed job procedures, and material handling. It encompasses everything from describing detailed hand motions in assembly to selection of machines and design of fixtures to reduce the number of setups and intermachine transfers for parts.

3. Transportation waste. In a well-designed system, work and storage areas are positioned whereby material moves are short. Tooling is kept near its point-of-use, and materials are only moved as necessary to obtain proper orientation for feeding into production operations.

4. Processing waste. Nonvalue added operations should be eliminated. Changing product designs can often reduce the number of parts in a final product. Changing part designs, limiting functionally unnecessary tolerances, and rethinking process plans can often eliminate processing steps.

5. Wasted time (queuing). Small lot sizes with planned order release and coordinated production of dependent items can reduce cycle time and inventory costs. A well-balanced system with coordinated order processing can achieve full capacity output without excessive WIP and throughput time. This has been repeatedly demonstrated by synchronous assembly lines throughout this century. Parts normally mate at final assembly or shipping. Their production priorities are therefore linked, and this knowledge should be utilized when scheduling shop operations.

6. Defective products. Defective products incur cost, deplete resources, and negatively impact customer perception. Quality engineering strategies should be employed for designing products and production systems, monitoring those systems, and improving process yield over time. Several of these strategies are discussed later in this chapter.

7. Excess inventory. We have seen the myriad of costs associated with inventory—space, obsolescence, damage, opportunity cost, lagged defect detection, and handling. Excess inventory should be eliminated. This can be accomplished through design activities such as the use of modular components. Modular components that are combined into many different products reduce the number of items stored and take advantage of economies of scale in item usage. Administrative activities such as the negotiation of long-term contracts with suppliers to ensure a steady stream of high-quality parts also contribute by eliminating the need for safety stock to cover market-based material shortages.

The Japanese term for nonvalue–added activity is **muda.** The pursuit of continuous improvement through identifying and eliminating muda is called **kaizen.** In the United States, many companies have adopted the practice of holding kaizen events. In a typical implementation, a multidisciplinary team of a trained facilitator, managers, engineers, and line workers will come together for three days to focus on improving an area of the plant. The kaizen process begins with questioning the current methods. A typical strategy involves asking "why" five times to discover the root cause or motivation for an action. For example, "Why do we store those items?" followed by "Why is it necessary to produce large batches?" followed by "Why is the setup time so long?" and so on until options for improvement that can be easily accomplished and will not inflict complications elsewhere are identified. The golden rule of kaizen is to utilize everyone's knowledge to identify and implement improvements quickly and without significant cost. Thereafter, the gains must be maintained while new improvement opportunities are pursued. This activity meshes with the traditional industrial engineering credo of "simplify, combine, eliminate."

10.2.2   Employee Cross-Training and Job Rotation

Cross-training and rotation of employees enhances worker flexibility and enthusiasm. Under cross-training, workers are trained over time to perform a variety of tasks within their work area. Once trained, workers are often rotated through the positions. On a basic level, this prevents boredom, brings fresh perspective and ideas for improvements, provides a context for communication between workers in the same team, brings a view of the big picture to all workers to ensure they perform their individual tasks cognizant of the overall objective, maintains enthusiasm (job enrichment), minimizes fatigue, and minimizes repetitive stress injuries. Manufacturing flexibility increases as well. It is possible to move cross-trained workers between workstations to deal with temporary bottlenecks caused by changes in product mix or time lost because of machine breakdowns. U-shaped manufacturing cells (see Figure 10.1) can be used with varying numbers of workers to match production rates to demand. The U shape results in short walking distances between

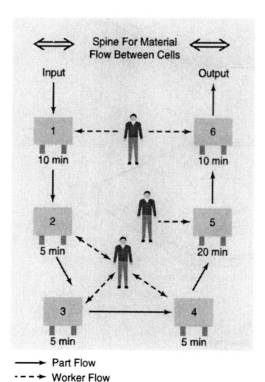

---→ Part Flow
- - -➤ Worker Flow

Figure 10.1 U-Shaped Cell with 3 workers for 3 Parts/Hr.

machines and good visibility. Thus, each worker can easily see what is happening through-out the cell and adjust their actions accordingly. This may mean helping out a coworker or adjusting the production rate of a machine. Workers can also operate more than one machine at a time, perhaps only being responsible for loading and unloading parts from several machines. As demand changes, workers are added to or removed from the cell. Processing times for the five operations required to complete a particular part are shown in Figure 10.2. Total processing time is 24 minutes. With two workers, the task assign-ment shown in Figure 10.2a has a cycle time of 12 minutes. This is the rate at which the two worker cells can produce parts. If demand increases, a third worker can be added (Figure 10.2b). The maximum cycle time for the three workers is nine minutes, and thus we will produce one unit every nine minutes.

Too much cross-training can be detrimental. Cross-trained workers may be tempted to stray from their primary assignment as a queue piles up nearby. This is costly if the worker abandons a long-term bottleneck processor to help out where a temporary queue has backed up. In the long run, this lost production at the bottleneck will reduce capac-ity. In general, workers should stay in their normal orbit of machine(s) unless idle from starvation or blockage, or needed for repair. When they are moved to other locations for temporary help, they should keep an eye on the status of their primary workplace and re-turn when needed. For bottleneck operations, it is usually best to keep workers there at

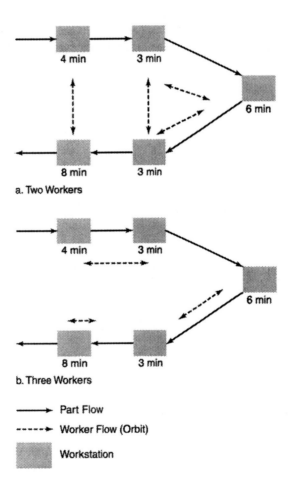

Figure 10.2 U-Shaped Cells

all times except for the rare occasions when the bottleneck is completely starved for work. Otherwise, we can develop a pattern in which workers follow each other from temporary bottleneck to temporary bottleneck, getting in each other's way and destroying any chance for continuous, smooth production. The key is to maintain a steady, balanced stream of parts flowing through all workstations. Avoid the bubble effect of hectic spurts moving from workstation to workstation. Such spurts serve only to reduce total throughput and encourage quality to then be sacrificed to short-term production goals.

10.2.3   Employee Empowerment and Involvement

Humans come with brains and bodies. It would be foolish to hire from the best educated population in world history only to use the body and disengage the brain. Employees should be assigned the responsibility for ensuring the process runs smoothly and for improving it over time. Coinciding with responsibility for achieving quality and productivity improvement goals, employees must be empowered to make changes. This includes minor investment and procedural changes to support continuous improvement, but also the authority to stop and correct a production system that is not operating properly. Too often management dwells on the cost of idle resources when a process is taken down for repair but ignores the cost of making bad product on an on-going basis. Sometimes referred to as the "hidden factory," it is not uncommon to find 10% or more of the resources (cost) in a facility dedicated to repairing defective product. Workers should be charged with identifying production problems as early and close to the problem source as possible. This avoids on-going waste while defective product continues to be made. One sure road to disaster in a competitive environment is to adopt the all too common managerial philosophy, "just ship it, we'll fix it later." There may be no "later" if customers find more reliable suppliers elsewhere. A common approach to empowering workers on a production line is to provide each with a switch to stop the line. In hitting the switch, a red light signals a line stoppage and the location of the problem to other workers. Nearby workers assist the signaler in solving the problem and acquiring additional expertise if necessary. Instead of casting blame, the identifier and solver of the problem are rewarded. Minor problems and situations that cause a short, temporary delay at the workstation can be signaled with a yellow light. The Japanese refer to this warning system as **jikoda** and the set of lights for broadcasting the warnings as **andon.** In some plants, every workstation will have a bank of three lights—green, yellow, and red. One light will be lit at all times indicating the status of the workstation. Additional boards may be hung in prominent locations to display the workstations currently exhibiting problems.

Standard operating procedures represent an important aspect of performance improvement. Standard procedures help ensure consistency in productivity and quality. Employees should be involved in developing the standard methods. Involving employees in the development of standards and procedures allows the worker to buy into the results. Such standards are more likely to be feasible and more likely to be followed. The worker can then be given the responsibility for following these procedures and meeting expectations. Workers take pride in knowing that if they do their job according to the standard procedures, their output will meet the customer's expectations. In an effort to provide this improved workplace environment of worker pride and dedication, the workplace should be kept organized and clean. The 5S concept of sort, straighten, sanitize, sweep, and sustain is sometimes used to summarize this policy. The organized workplace reduces the chances of misplacing, contaminating, or otherwise jeopardizing the successful manufacture of the product. A tidy workplace brightens the atmosphere and conveys the feeling that the system is under control and sloth and sloppiness are not to be tolerated.

10.2.4 JIT Purchasing

Managing the selection, certification, and commerce with suppliers constitutes a critical aspect of JIT production. The entire nature of the relationship between suppliers and customers changes from competition to cooperation under JIT. The major categories of changes can be summarized as:

1. Sole sourcing versus multiple supply sources. Traditionally, companies would maintain databases of qualified suppliers for each item. Each time a part was to be ordered, purchasing agents would call for price and delivery quotes. Thus, the customer was rarely a significant portion of the supplier's business and incoming product quality varied. JIT purchasing favors negotiation of long-term contracts with a single supplier. The fate of the two companies becomes linked thus encouraging cooperation instead of competition over terms. It is now in the long-term interest of the supplier to provide high-quality product on time at a fair price to ensure the success of their customer. It is not uncommon for personnel to move freely back and forth between supplier and customer to ensure communication of product requirements and to troubleshoot problems. The long-term, high-volume arrangement allows for the use of packaging designed to be compatible with the customer's material handling and production system. Often times, this will be in the form of pallets, containers, or carts that enter directly into the production system on arrival. This customization might not be economically feasible for an occasional customer.

2. Frequent delivery of small lots versus quantity discounts. The traditional approach has been occasional deliveries of large quantities of parts. These parts went through sampling inspection to ensure compliance with requirements. Inspection is now eliminated as the supplier is certified for quality and has, over time, proven competent at meeting requirements. Otherwise, a different supplier is found. In addition, by receiving small lots just-in-time, parts proceed quickly into production and quality problems are discovered soon after receipt. In a truly ambitious system, parts are delivered several times a day.

3. Flexible ordering versus paperwork. Long-term contracts call for a steady flow of product at a specified rate plus or minus 10%. The customer can change the delivery quantity within this range on very short notice without significant paperwork and administrative hassle. This allows the customer to adjust production rates to demand, but does require some reserve capacity on the part of the supplier. To further simplify procedures while simultaneously improving performance, electronic data exchange (EDI) of order entry, quality reports, and shipping lists can be established. Information is thus transmitted faster, more accurately, and at lower cost.

4. Vendor owned and managed inventories. If suppliers are known, responsibility for maintaining a ready supply of input parts and materials can be turned over to the vendor. The vendor stores these supplies at the customer's site. Only when the customer extracts an item for use in production does the customer take ownership and get charged. The supplier assumes additional responsibility but can also integrate their own delivery and production schedules. In addition, the supplier can establish their own link to collect immediate point-of-sale information for planning replenishments. As we have seen, such information can reduce the need for large warehouse safety stocks and the volatility of the bull-whip effect.

10.2.5 Impact of Reducing Variability

In the evolutionary and revolutionary progress of technology, variability provides benefits by creating genetic mutations and outlying events that lead to fundamental changes.

However, when trying to operate a thoughtfully designed production system, variability impedes efficiency. A central theme of the recent quality revolution focuses on eliminating variability from production processes. When material supplies, process yields, or machine availability are unpredictable, managers feel compelled to build large safety stocks of inventory in order to ensure meeting customer expectations. Customers pay for products and services, thus the manager's first responsibility dictates delivering the desired products on time. The efficiency of the production and delivery process will usually be secondary.

We have already seen how variability in lead time creates a need for safety stock. To further demonstrate the pernicious impact of variability in production processes, we call on the basic models of queuing theory for exponential and deterministic service times. In both cases, we assume a Poisson arrival process, i.e. the time between the arrival of customer orders follows an exponential distribution. Let the mean time between order arrivals be λ^{-1} and the mean time to service an order be μ^{-1}. The average utilization of the facility is given by:

$$Utilization \equiv \rho = \frac{E \text{ (number of orders arriving per time)}}{E \text{ (number of orders that can be produced per time)}} = \frac{\lambda}{\mu}.$$

For a single server queuing system with exponential service time, the average number of orders at the work station (N) is given by[1]:

$$N = \frac{\rho}{1 - \rho}, \tag{10.1}$$

and the average time spent in processing and waiting at the workstation is T where:

$$T = \frac{1}{\mu(1 - \rho)} \tag{10.2}$$

On the other hand, if the service time is deterministic and always equal to μ^{-1}, the corresponding averages are:

$$N^D = \frac{\rho(2 - \rho)}{2(1 - \rho)} \tag{10.3}$$

and

$$T^D = \frac{2 - \rho}{2\mu(1 - \rho)} \tag{10.4}$$

A comparison of Equation (10.1) with Equation (10.3) and Equation (10.2) with Equation (10.4) shows that by eliminating variability in service time, we adjust the WIP level and cycle time by a factor of $\frac{2 - \rho}{2}$. For $0 < \rho < 1$, this ratio varies from 1 to 1/2. Thus, in heavily utilized systems ($\rho \approx 1$), inventory and throughput time are cut in half by eliminating variability in the processing times. This assumes that average processing time stays the same. Improvements are even larger if by eliminating variability because of machine breakdowns, then we also increase the capacity (μ) of the processor. If we can eliminate the variability in arrival rate and thereby pace the system, we may be able to eliminate

---

[1]This is the standard *M/M/*1 queuing model for which the three parameters *M, M*, and 1 indicate exponential interarrival time, exponential service time, and a single server, respectfully. Exponential service times may be because of variability in processing activities directly, or it may serve as an approximate model for the case of occasional breakdowns in the process that yield long delays between successive job completions.

waiting time completely. This is seen in synchronous assembly line systems. In this case,

$$T = \frac{1}{\mu} \text{ and } N = \rho, \text{ a reduction of } \frac{1}{1 - \rho} \text{ relative to the } M/M/1 \text{ situation.}$$

To complete the picture, we provide an approximation for a workstation (or system) with a general intermediate level of variability. We refer to this as a general interarrival and general service ($GI/G/1$) process. To present the results, we must define the squared coefficient of variation for the arrival (C_a^2) and service (C_s^2) processes. The squared coefficient of variation is always the variance divided by the square of the mean. For any random variable X, $C_X^2 = \frac{V(X)}{E^2(X)}$. For a $GI/G/1$ queuing process, we can use the approximation:

$$T^G = \mu^{-1} + \frac{\rho^2(1 + C_s^2)(C_a^2 + \rho^2 C_s^2)}{2\lambda(1 + \rho^2 C_s^2)(1 - \rho)} \tag{10.5}$$

The first term in Equation (10.5) is processing time, the second term is expected queuing time waiting for the server to become available. Given T^G, we can always relate flow time to WIP level by Little's Law,

$$N^G = \lambda \cdot T^G \tag{10.6}$$

EXAMPLE 10.1

A particular machine is capable of producing ten jobs per day on average. Jobs are considerably shorter but occasional breakdowns, most short but some long, give the appearance of an exponential processing time for the machine. The industrial engineer wants to determine if it is worth investing in a major overhaul of the machine. She estimates that the modified machine could average 14 jobs per day and processing time would have a standard deviation of 0.02 days. Determine the effect of the overhaul. Jobs arrive equally spaced at the rate of nine per day. The subsequent production process has been plagued by the inconsistent output of parts from this machine.

SOLUTION

Under the current system, we have $C_s^2 = 1$ (exponential service) and $C_a^2 = 0$ (constant interarrival times). Thus,

$$T^G = \mu^{-1} + \frac{\rho^2(1 + C_s^2)(C_a^2 + \rho^2 C_s^2)}{2\lambda(1 + \rho^2 C_s^2)(1 - \rho)} = 0.1 + \frac{0.9^2(1 + 1)(0 + 0.9^2 \cdot 1^2)}{2(9)(1 + 0.9^2 \cdot 1^2)(1 - 0.9)} = 0.50 \text{ days.}$$

If we overhaul the machine, we increase μ to 14 and reduce $C_s^2 = \frac{0.02^2}{(1/14)^2} = 0.0784$. Utilization has dropped from $\rho = 0.9$ to $\rho = 9/14 = 0.643$. For the new system,

$$T^G_{new} = \frac{1}{14} + \frac{0.643^2(1 + 0.0784)(0 + 0.643^2 \cdot 0.0784)}{2(9)(1 + 0.643^2 \cdot 0.0784)(1 - 0.643)} = 0.14 \text{ days.}$$

The mean time through the workstation has been reduced by just over 70%. At the same time, capacity has increased attributable to less down time. We leave it to the truly curious reader to find the equations for throughput time variability for a $GI/G/1$ queue and further explore the impact of this change.

EXAMPLE 10.2

The wire bonding stage in a electronics assembly system receives ten jobs per day. The station is capable of completing an average of 12 jobs per day when operating at capacity. The variance of processing time is 0.005 days^2. Currently jobs arrive at random according to a Poisson process. An engineer is considering coordinating the production stages to ensure that jobs will arrive evenly spaced, once every hour (the system operates for ten hours per day). Determine the impact of the change on average WIP levels at the stage and average production time.

SOLUTION

Using a day as the time period, we are given that $\lambda = 10$ and $\mu = 12$. Thus, $\rho = \dfrac{\lambda}{\mu} = \dfrac{5}{6}$. The service process has $C_s^2 = \dfrac{0.005}{(1/12)^2} = 0.72$. Consider, first, exponential arrivals for which $C_a^2 = 1$.

Then, $T^G = \mu^{-1} + \dfrac{\rho^2(1 + C_s^2)(C_a^2 + \rho^2 C_s^2)}{2\lambda(1 + \rho^2 C_s^2)(1 - \rho)} = \dfrac{1}{12} + \dfrac{0.694(1.72)[1 + 0.694(0.72)]}{2(10)[1 + 0.694(0.72)](1 - 0.833)}$

$= 0.44$ days and

$N^G = \lambda \cdot T^G = 10(0.44) = 4.4$ jobs. For evenly spaced interarrivals, $C_a^2 = 0$ and

$T^G = \mu^{-1} + \dfrac{\rho^2(1 + C_s^2)(C_a^2 + \rho^2 C_s^2)}{2\lambda(1 + \rho^2 C_s^2)(1 - \rho)} = \dfrac{1}{12} + \dfrac{0.694(1.72)[0 + 0.694(0.72)]}{2(10)[1 + 0.694(0.72)](1 - 0.833)} = 0.20$ days.

By controlling the input process, we have reduced this portion of the lead time by 55%. Average WIP at the workstation will drop to two jobs.

The output variability at one stage impacts the performance at subsequent workstations. It is important to recognize the dependency between workstations and parts. **With infinite buffer space between workstations, the effective capacity of a workstation is determined by its own mean interarrival and processing times.** However, the distribution of time a part spends at the workstation and whether the arrival rate matches capacity does depend on the surrounding workstations. **With limited or no buffer space between workstations, the effective capacity of a workstation may be affected by the characteristics of surrounding workstations.** To help visualize these relationships, consider the case of two workstations in series. Parts must pass through both workstations during processing as illustrated in Figure 10.3. Suppose that both processes are always operational and have deterministic processing times of t_1 and t_2. If we dispatch parts to the first machine at a spacing of $t \geq \max(t_1, t_2)$, then each processor will be able to keep the pace, and parts will flow through the system at the rate t^{-1} units per time. The Figure shows the progression of the first six items through both workstations for the case $t_1 = 8$, $t_2 = 10$. In the tabular portion of the Figure, S_{ij} indicates the starting time of item j at workstation i and C_{ij} is the time the item completes processing at that workstation. For the deterministic case, the bottleneck resource (machine 2) stays busy and paces the line. Parts never wait. If variability exists in processing times or machines are unavailable at times, then the outcome changes significantly. Because we cannot depend on a regular supply of output from the first process, we must either purchase a faster machine (extra capacity) for workstation 2, or we must carry a buffer of inventory between the processors. A faster machine at workstation 2 allows us to make up for the time workstation 2 is starved. A buffer of parts between the workstations helps to ensure that parts are available to keep workstation 2 busy. The longer the potential delay for stage 1 output, the larger the buffer we need to ensure a constant supply for stage 2. Figure 10.4 illustrates with happens when processing times are variable and no buffer exists. For the Figure, $t = t_2 = 10 > E(t_1) = 8$. $E(t_1) = 8$ results from alternating times of 8, 12, and 4. It would appear at first glance that demand can be met. Mean processing time is no longer than the mean time between arrivals at each workstation. Now, without a buffer, workstation 1 cannot start a new item until the previous item completes at workstation 2. Thus, $S_{i1} = \max(R_i, S_{i-1,2})$ where R_i is the arrival or release time for part i. Likewise, note that $S_{i2} =$

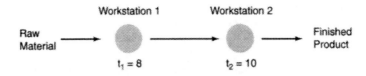

Item i	t_1	t_2	Release Time	S_{i1}	C_{i1}	S_{i2}	C_{i2}
a	8	10	0	0	8	8	18
b	8	10	10	10	18	18	28
c	8	10	20	20	28	28	38
d	8	10	30	30	38	38	48
e	8	10	40	40	48	48	58
f	8	10	50	50	58	58	68

Workstation

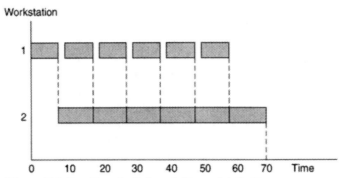

Figure 10.3 Part Flow for Balanced, Deterministic Line

$\max(C_{i1}, C_{i-1,2})$. Workstation 2 cannot begin processing a part until that part has been completed at the first workstation, and workstation 2 has completed the previous part. On those occasions when t_1 exceeds ten, such as for the second and fifth item, the second workstation is forced to wait. In the constant time case, we had $S_{b2} = 18$. With the varying times, machine 2 cannot start the second unit until time 22, i.e. $S_{b2} = 22$. This lost idle time can never be made up. Thereafter (you will have a chance to show this in the end-of-chapter exercises), in each cycle of three units, workstation 2 will be forced to wait for two time units and capacity is reduced. If we insist on releasing units every ten time units, then a queue will gradually build up in front of the line making it appear that workstation 1 is the bottleneck. In reality, workstation 1 has sufficient capacity, but its own variability causes it to be blocked at times by workstation 2. In our example, workstation 2 has the lowest capacity when considered in isolation, making it a natural bottleneck. Capacity is further reduced by the occasional starvation of this workstation whenever workstation 1 requires a long production time. The bottleneck thus results from the interaction between the workstations.

Another problem exists if parts with variable cycle times have to be mated to produce a finished product. Let's consider a product that is assembled from n different parts. Let the lead time for part i be τ_i. If lead times are deterministic, we simply release item

Item i	t_1	t_2	Release Time	S_{i1}	C_{i1}	S_{i2}	C_{i2}
a	8	10	0	0	8	8	18
b	12	10	10	10	22	22	32
c	4	10	20	22	26	32	42
d	8	10	30	32	40	42	52
e	12	10	40	42	54	54	64
f	4	10	50	54	58	64	74

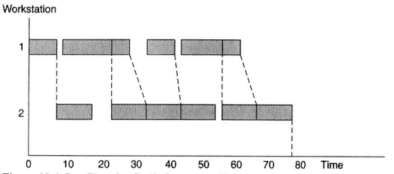

Figure 10.4 Part Flow for Cyclic Processing Time

i into production τ_i time units before final assembly is scheduled. The finished inventory of parts waiting to be assembled is zero. Now, suppose that the τ_i are independent random variables with density functions $f_i(\tau)$. Assembly cannot begin until all parts are available. Assuming the $f_i(\tau)$ are symmetric, the probability we could begin assembly on time if each part was released at its mean lead time prior to scheduled assembly is only 0.5^n, a paltry chance of less than one in a thousand if $n = 10$. To ensure a service level probability of α percent for starting on time, we would need to release each item i at a time equal to $F_i^{-1}(\alpha^{1/n})$ before scheduled assembly.[2] The difference $F_i^{-1}(\alpha^{1/n}) - E(\tau_i)$ for each item represents slack time for which we must store the part on average before its use in assembly.

EXAMPLE 10.3

A product is assembled from five subassemblies. The lead time for subassembly i is normally distributed with mean μ_i and variance σ_i^2. If we want to be 95% sure that assembly can begin on time, how long before assembly is scheduled should each subassembly be started? Specifically, what does

---

[2]This assumes route times are independent with an equal allocation of delay probability to each item. We could play with the probabilities and release some items earlier and some later to also achieve the same service level.

this imply for a particular rotor housing subassembly that has a mean lead time of five days and standard deviation of two days?

SOLUTION

To have a 95% chance that all subassemblies are available on time, each must have a β probability of on-time availability where β satisfies $\beta^5 = 0.95$. Thus, $5 \cdot \ln \beta = \ln 0.95$ and $\beta \approx 0.99$. For a standard normal distribution, $\Phi(2.33) = 0.99$, and we must release subassembly i at time $\mu_i + 2.33 \cdot \sigma_i$ prior to assembly. For the rotor housing subassembly, we must release the order $\mu_i + 2.33 \cdot \sigma = 5 + 2.33 \cdot (2) = 9.66$ days before it is due. We then expect to carry the completed housing in inventory for 4.66 days. In addition, we must now schedule assembly further in advance, which means we must rely on even longer-term forecasts of demand.

10.2.6   Techniques for Mistake-Proofing Processes

Eliminating opportunities for production errors offers one of the most cost effective investments for a production system. Producing defective parts clearly represents waste. Instead, we should attempt to design processes to avoid producing defective items. And, if defects occasionally happen, we should detect them as soon as possible before we waste more resources processing the defective item through subsequent production steps. The Japanese term **poka-yoke** refers to efforts at "mistake proofing" processes. Poka-yoke takes a variety of forms, but the basic tenets are that: 1) 100% of units should be inspected; 2) defects should be identified as close to the source of the defect as possible; 3) upon detecting a defect, production should be halted immediately and corrective action should be taken to avoid repeating that defect; and 4) processes should be designed to avoid producing defects. Several techniques have proven useful for poka-yoke. These include:

1. Checklists and worker source inspection. It may be helpful to assign each worker the responsibility to check his own work. However, it is doubtful that this will eliminate all defects. Humans have a tendency to approve of their own work. Likewise, any error in judgment or training concerning what constitutes an acceptable item will affect both production and inspection. It can help to provide a checklist for the worker. For each unit produced, the worker must certify that the checklist has been fully executed. This may help eliminate the worker who occasionally forgets to perform some task in the operation sequence. Even better, the checking process could be automated. For example, suppose that a worker is to insert components into an assembly. As assembly begins on an item, the worker must push a button or scan a barcode to indicate that a new unit is in production. An indicator light at each part feeder that is to be used to make this product then lights up red. When the proper number of parts are removed from the feeder, the light turns green. Only when all lights are green and no loose parts remain is the worker sure that the proper components have been inserted. As another example, suppose that a worker performs a visual source inspection of surface finish. The worker may be asked to compare each product unit to a set of standard templates and indicate the closest match. The templates provide a training and calibration guide to ensure consistency. To ensure the check is performed, the worker may be required to input the outcome for each part into a shop floor control/labor reporting system before the worker can receive his next work instruction.

2. Successive check system. Source inspections seldom catch all defects. In a successive check system, each worker is charged with inspecting incoming parts before beginning their operation. In particular, the worker must approve the performance of the previous worker. This provides a more impartial inspection and also serves to identify defects before wasting more resources on the item. It is important that defects are immediately communicated to the previous worker. In addition to stopping the production of consecutive defective items, the immediate feedback serves to train the previous worker on what constitutes acceptable quality. Immediate feedback also removes any possible motivation to pass on shoddy work because it will be immediately detected by a peer and returned.

3. Mistake-proof part and fixture design. Parts should be designed whereby they can only fit into a tooling fixture in the proper orientation. Nearly symmetrical parts can be reshaped to exaggerate their asymmetries thus making it easier to detect misalignment. Nonfunctional slots or holes can be drilled into parts as an initial step in their processing. These features are incorporated into part holding fixtures to ensure the parts are properly aligned (Figure 10.5). Four identical circuit boards are shown mounted in a fixture. The polarity of parts mounted by different machines must be compatible. The boards could be rectangular, but instead one corner is angled, and a matching feature is added to the fixture for all machines to ensure consistent orientation. Likewise, the part feeders should be mistake-proofed to ensure that the parts are oriented correctly. The parts could be designed with an asymmetrical feature to ensure proper loading into the part feeder, or the parts could be given a special mark during manufacture that can be read and checked for ID and orientation as the part is being mounted.

4. Integrated machine gaging. Integrate input and output gages into production operations to ensure that quality parts enter and leave a production operation. The word "gage" is used in a general sense here. For ensuring that dimensions on mechanical parts are within specifications, we could add minimum dimension and maximum dimension gages to a machine whereby parts cannot be loaded or removed unless they meet speci-

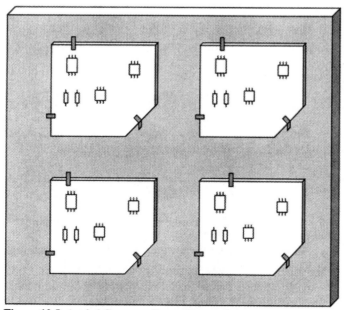

Figure 10.5 Angled Corner to Ensure Proper Orientation

fications. If a defective part is produced, the process must be stopped to manually remove the defective part. At the same time, the part should be examined and the process adjusted to eliminate the source of the defect. Instead of physically stopping the machine, a system may be set up whereby parts being removed by a machine are automatically diverted to one defect bin if they are too large to fit through one gage and diverted to a second defect bin if they are small enough to fit through a second gage. Only parts meeting specifications are then passed onto the next operation. Although we used dimension as the quality measure in this example, clearly thickness, weight, reflectiveness, resistance, or other measures could be used as appropriate.

10.2.7 Economics of Setup Time Reduction

The direct and indirect impact of setup has been a recurring theme throughout the text. Reducing setup time and cost may well be the most productive investment that can be made to improve production system performance, particularly if product variety or sales are currently limited by setup requirements. In this section, we attempt to quantify the direct savings from setup time reduction. We then elaborate on the intangible and second order effects that also occur throughout the firm. To facilitate our discussion, we assume that setup cost (A) is directly related to setup time (S). To parameterize our findings, we examine the impact if setup cost (time) reduces by $100(1 - \alpha)$ percent.

Suppose that after a setup reduction program, the new $A' = \alpha A$, and the new $S' = \alpha S$. Using the basic EMQ result of $Q^* = \sqrt{\dfrac{2AD}{i[C + 2pwD(M + v/2)]}}$, we have that:

$$\frac{Q'^*}{Q^*} = \frac{\sqrt{\dfrac{2\alpha AD}{i[C + 2pwD(M + v/2)]}}}{\sqrt{\dfrac{2AD}{i[C + 2pwD(M + v/2)]}}} = \sqrt{\alpha} \tag{10.7}$$

If we reduce setup cost by 20% such that $\alpha = 0.8$, then the new batch size is 89% of the old batch size. The number of setups per time is given by D/Q, and:

$$\text{Setups/Time} = \frac{D/Q'^*}{D/Q^*} = \frac{Q^*}{Q'^*} = \frac{1}{\sqrt{\alpha}} \tag{10.8}$$

The processing time for a batch is given by $S + pQ$. Assuming unit processing time, p, stays the same, batch processing time changes by:

$$\frac{S' + pQ'^*}{S + pQ^*} = \frac{\alpha \cdot S + p \cdot \sqrt{\alpha} \cdot Q^*}{S + p \cdot Q^*} < 1, \tag{10.9}$$

for $0 < \alpha < 1$. In other words, if we reduce setup time, we can reduce both batch sizes and batch processing time.

We have seen in earlier chapters that for both deterministic and exponential processing time, shop throughput time is proportional to batch processing time. Reductions in setup cost translate into smaller batch sizes that in turn reduce lead time. Assuming total production stays the same, Little's Law tells us that WIP will be reduced by the same percentage as flow time. In reality, the impact will be larger in a stochastic environment because flow time is also related to utilization and utilization will also be reduced. Total setup time, SD/Q, will be affected as:

$$\frac{S'D/Q'^*}{SD/Q^*} = \frac{S' \cdot Q^*}{S \cdot Q'^*} = \frac{\alpha}{\sqrt{\alpha}} = \sqrt{\alpha} \qquad (10.10)$$

Total unit variable processing time remains unchanged by setup reduction. Thus, the reduction in total setup time translates directly into a reduction in system utilization, or equivalently, an increase in system capacity. Thus, not only do we have the option of producing more batches that flow through the system more quickly while meeting the current demand rate, we will have the opportunity to increase total production without purchasing new capacity.

EXAMPLE 10.4

A work center produces several similar part types. Each setup costs $100 and takes ¼ day (two hours). Batch size is set using the EMQ. Demand occurs at a constant rate of 1,000 units per year. Processing time is also deterministic with a capacity of 4,000 units per year. Production batches flow through the system with only minimal waiting time. The carrying cost rate is 40%/year, material cost is negligible, and the completed item is worth $40. Determine the impact on batch size, batch throughput time, number of setups, and total relevant inventory cost for a $100 (1 - \alpha)\%$ reduction in setup time and cost. The plant operates 2,000 hours per year.

SOLUTION

For this deterministic system, prior to setup reduction we have:

$$Q^* = \sqrt{\frac{2AD}{i[C + 2pwD(M + v/2)]}} = \sqrt{\frac{2(100)(1,000)}{0.4[40 + 2(0.00025)(1)(1,000)(40/2)]}} = 100.$$

Ignoring the negligible waiting time, the throughput time per batch comes from setup and processing time.

Therefore, Throughput Time $= 2$ hrs. $+ \dfrac{100 \text{ (units)} \cdot 2,000 \text{ (hrs/yr)}}{4,000 \text{ (units/yr)}} = 52$ hrs.

The number of setups per year is $D/Q^* = 10$.
Total relevant annual inventory cost is

$$Cost/Year = \frac{AD}{Q} + \frac{hQ}{2} + i(S + pQ)wD\left(M + \frac{v}{2}\right) \qquad (10.11)$$

$$= \frac{(100)(1,000)}{100} + \frac{0.4(40)(100)}{2} + 0.4\left(\frac{2}{2,000} + \frac{100}{4,000}\right)(1)(1,000)\left(\frac{40}{2}\right) = \$2,008.$$

Now, consider the system after setup reduction. Figure 10.6 shows the evolution of costs, order frequency, and batch size as setup is reduced. For any value of α, we have from Equation (10.7) that $Q' = \sqrt{\alpha} \cdot Q^*$. The new throughput time $= 2$ hrs. $+ \dfrac{Q'}{2 \text{ units/hr}} = 2 + \dfrac{\sqrt{\alpha} \cdot (100)}{2} = (2 + 50 \cdot \sqrt{\alpha})$ hrs. The new number of setups is $\dfrac{D}{\sqrt{\alpha} \cdot Q}$.

Finally, replacing Q by Q' in Equation (10.11) we have

$$Cost/Year = \frac{AD}{\sqrt{\alpha} \cdot Q} + \frac{h\sqrt{\alpha}}{2} Q + i(S + p\sqrt{\alpha}Q)wD\left(M + \frac{v}{2}\right).$$

Using $Q = 100$, we obtain the curves shown in Figure 10.6.

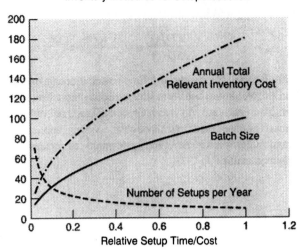

Figure 10.6 Inventory Measures vs. Setup Time and Cost Reduction

In addition to those savings quantified above, a variety of more intangible benefits come with reduced setup. As lead time is reduced, many improvements are possible. Forecast time horizons can be reduced leading to lower costs and more accurate forecasts. This in turn reduces safety stock inventory requirements. With shorter forecast horizons and lead times, the rate of stock obsolescence from unforeseen changes in customer demand declines. Recall the stochastic demand inventory models of Chapter 6. For the continuous review model, the amount of safety stock carried (or expected inventory on hand when an order arrived) was proportional to the standard deviation of lead-time demand. Likewise, in the single-period news vendor problem, the amount of product scrapped increased with the standard deviation of the demand forecast. Because it is more difficult to forecast the farther we look into the future, the standard deviation of forecast error increases with lead time. It then follows that the cost of holding safety stock or scrapping product that cannot be sold increases proportionally with lead time.

Significant reductions in lead time may permit a switch from make-to-stock to make-to-order operation virtually eliminating finished goods cycle inventories. As inventory levels drop, so does the need for storage space. With short production runs, it may be possible to avoid periodic quality monitoring checks. Statistical process control checks are usually performed on a periodic basis during long runs. These checks are designed to detect shifts in a process during operation. With a foolproof setup and short run, the machine may not have time to drift and will be less likely to go out of control during the run. Thus, monitoring and failure costs are reduced. Quicker setups provide the option to select some combination of increased setups and production volume. More setups translate into increased ability to customize products as well as savings in WIP, inventory obsolescence, cycle time, and lead time. Increased capacity effectively lowers production cost per unit and enhances competitiveness.

EXAMPLE 10.5

Suppose that the demand in Example 10.4 was actually random with a standard deviation of ten units per week and safety stock is set for a 98% fill rate for make-to-stock items. If the reduction in setup time in Example 10.4 was sufficient to allow change from a make-to-stock to a make-to-order system and still meet customer delivery expectations, compute the savings in safety stock cost. Assume the company uses a periodic review system.

SOLUTION The initial lot size of 100 units with ten setups per year indicate a lot being produced once every five weeks. This makes the cycle + lead time equal to about six weeks (five weeks plus 52 hours). For independent weekly demands, the variance over this period will be $6(10^2) = 600$. The standard deviation of demand over this period is thus $600^{1/2} = 24.5$. For a 98% fill rate, the normal distribution indicates we would need to order up to 2.06 standard deviations to the right of the mean. In other words, our safety stock = $(2.06)(24.5) = 50$ units. Safety stock will be eliminated under the make-to-order system. The savings then is $iC = (0.4)(40) = \$16$/per unit of safety stock per year. With a safety stock of 50 units, this translates into an additional \$800 per year savings.

10.2.8   Technology of Setup Time Reduction

In many cases, the desire to reduce setups, a little creativity, and common sense are sufficient to obtain significant reductions. Assigning setup reduction responsibility to teams of production line workers often generates a variety of improvement ideas. In addition, a variety of technological innovations have been developed to enable setup time reduction. Basic strategies for reducing downtime for setups include:

1. Design parts for manufacturability. Design parts so as to minimize the need for operations and the accompanying tooling. Nonfunctional part dimensions or arbitrary tolerances may be assigned during product design that complicate setup and processing. Educating part designers about the effect of these decisions can simplify the manufacturing process and reduce the number of part setups and tooling changes required. Suppose that a tolerance of ± 0.010 is sufficient for a particular hole diameter or part length, but the designer indicates ± 0.001 out of habit. Suddenly, we need an extra tool and operation to ream the hole after drilling, or finish the length after sawing.

Designing features that can be produced with standard tools can avoid the need for tooling changeovers on machines. Odd-shaped grooves, unusual hole sizes or chamfer angles are all examples of features that increase manufacturing complexity but can often be eliminated by redesign. Suppose that the tool magazine on a machining center or the feeder slots on a part sequencer can hold M tools of which m are standard.[3] If the family of parts made by the machine requires, at most, a collection of M tools, then tooling changeovers are eliminated (except for replacements resulting from wear). On the other hand, if a part requires more than $M - m$ nonstandard tools, then multiple tooling changeovers are required each time that part is run.

When dimensioning a milled slot or cavity in a part, simply matching the corner radius of the cavity to that of the milling tool can eliminate the need for adding an extra tool and using it on each part. These and similar points are addressed in detail in Hoffman [1996].

2. Develop standard methods. Standard procedures can be developed for setup and changeover that eliminate unnecessary steps and idle time. Standard industrial engineering procedures can be very effective here. In many cases, setup, because it is an intermittent activity, receives only scant documentation. It may be helpful to begin by videotaping setup activities and then charting each activity in a traditional man-machine format. The operator and engineer can study the tape or chart to look for improved methods that minimize wasted time and effort.

---

[3]On many machines, it is common to allocate some space to standard tools that are required by many parts and remain on the machine at all times. This simplifies the production of process plans for parts and ultimately reduces the frequency that these high-use tools are loaded and unloaded.

3. Divide setup activities into internal and external tasks. Internal tasks are those that require the machine to be down while external tasks are those that can be performed with the machine running. Obtaining tooling, reading blueprints, and prepositioning fixtures for easy roll onto the machine are external activities. Consider, for example the process of loading a large fixture or die onto a machine. We could move the fixture in by some arbitrary material handling technique such as a forklift or overhead crane. We could then raise or lower the fixture and carefully position it onto the machine. Alternatively, we could design all fixtures with a common size and roll them in on a standard cart that is set to the bed height of the machine. The cart is prepositioned so that changeover can be done by quickly and easily rolling the fixture from the cart onto the machine. In general, we should move as many activities as possible from internal setup to external setup, that is, prepare the machine in advance as much as possible. This will minimize downtime during changeovers. Activities such as reviewing setup instructions, retrieving and presetting tools in the tool magazine of a machining center, obtaining and loading part feeders, and preheating dies for injection molding machines can be performed offline.

4. Design procedures to perform setup tasks in parallel. This may require setup teams of workers that travel between machines or simply require support from other workers in the vicinity of the machine. While one worker is changing tooling, another can be loading part feeders, and another can be downloading instructions to the machine or setting stops. By performing internal setup tasks in parallel, total idle time of the machine for setups is reduced. If setup tasks require that workers frequently move about from one location to another, such as one side of a machine to load tools and another side to input tool data into the controller, then groups of workers can eliminate time spent moving between locations.

5. Utilize family tooling to minimize the need for setups. Family tooling can often be developed for a set of similar parts. A generic family fixture is constructed with the flexibility to hold any part in the family, perhaps needing only an adjustable setting or quick insert that is peculiar to that part type. Tool magazines can be set up to produce any part within a family without the need for tool changeovers. This can apply to standard turret lathes holding a few tools, large rotating tool magazines holding over 100 tools for a CNC machining center, or component placement machines that hold large numbers of part feeders for assembling circuit cards. Tool heads may also be able to be configured off-line and quickly exchanged when necessary. This technique can be used for mounting electrical components as well as in machining. Easily exchangeable, modular fixtures can also be designed for holding circuit cards of various sizes when drilling holes or mounting components.

Modular fixturing systems have been developed in recent years for quick assembly of fixtures for low-volume parts. A base plate with a grid pattern is used. Standard clamps and locators can be attached anywhere on the grid to fit a specific part. Not only can the fixture be assembled quickly, but the standard grid provides a coordinate system for locating reference points. Fixtures can be assembled and stored or disassembled after use depending on the production frequency for the part.

6. Locally stored tools and tooling kits. If part types and their operations are assigned to specific machines, then tooling can be kept at the appropriate work center instead of in a centralized area. This reduces the time spent finding and transporting tooling. In addition to the time spent transporting the tooling to the work area, it is easier to develop methods such as standard height carts or conveyors for quickly loading tooling onto machines. All of the fixtures, inserts and tools needed to set up a particular part type can be stored together instead of the traditional approach of one rack of fixtures, a second for clamps, a third for cutting tools, and so on. The clamps used for a particular die can be kept with the die instead of in a separate location. Significant advancements have been made in rapid exchange of heavy dies for presses. Dies can be stored locally, and using roller conveyors, easily pushed onto a cart that delivers the die directly to the press at the correct height for loading. You may be able to add a carousel with a roller conveyor top surrounding the press whereby dies can be pushed quickly into place, perhaps even automatically loaded and unloaded. Similar improvements have been made in changeovers for injection molding machines.

An even better step, when possible, would be to locate all four fixtures permanently within the work envelope of the machine. Part size may prevent this, but consider the machine shown in Figure 10.7. The machine produces 15 different part numbers using four different fixtures. Instead of maintaining a single fixture on the machine at a time as shown in Figure 10.7a, requiring a changeover whenever the next part uses a different fixture, we can modify the tool bed as shown in Figure 10.7b. When starting a new part, it is simply inserted into the proper fixture.

7. Use standard sized intermediate workholders. Even if tooling does not need to be the same size, it may be expedient to design the tooling with similar characteristics.

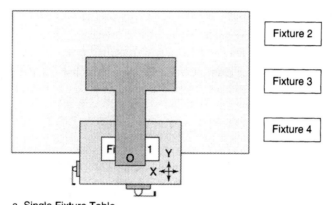

a. Single Fixture Table

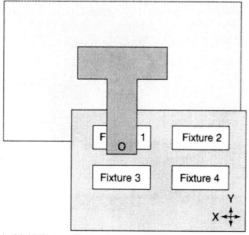

b. Multi-Fixture Table

Figure 10.7 On-line Fixtures

You probably have CDs that vary in length from 30 to 70 minutes. Yet, they are all the same size and fit into the same CD-ROM drive. Imagine if you had to modify the CD-ROM drive for each CD. A significant impact in industry has been obtained by designing flexible fixtures that adapt to multiple parts. By changing the size of the die base or adding spacers, all dies can be made to be exchangeable without the need to adjust the shut height on a press. Dies can be made to fit on the base plate in exactly the same position. Stop blocks and guides may be added to the machine to ensure that each fixture can be quickly placed in the correct position. The semiconductor manufacturing industry has adopted a standard for handling wafers. The next generation of heterogeneous fab equipment will all receive wafers in a standard carrier and at a standard height and orientation to facilitate automated material handling.

8. Eliminate adjustments. When setting up a machine, it is common to run off one or more parts and wait until these parts pass inspection before starting the remainder of the batch. Machine adjustment is a basic part of setup that should be minimized. The setup procedure should be developed whereby the need for machine adjustment is eliminated. On-line measurement can reduce the delay, but foolproofing the setup process and eliminating the need for these test parts is optimal. The addition of intelligent features produces autonomous control (referred to as "Jidoka" in Japan) in many cases. Machines start and stop automatically, and diagnose themselves, transmitting signals to operators when problems occur. Digital readouts of positions and settings can eliminate the guess work that leads to iterative adjustments. Limit switches can be added to aid in aligning tooling or setting stroke lengths for use with specific parts.

9. Use power clamps. Many machines still use manual clamps and long screws that must be set during changeovers. Switching to push button activated pneumatic or hydraulic clamps reduces the time to secure fixtures. Simple activities such as sawing excess threads off of a screw can reduce setup time. Monden [1983] describes several quick fastening techniques for reducing internal setup time. Typically, to bolt two items together, the bolt is placed through a hole in both and a nut tightened. Tightening the nut requires multiple turns on a wrench. The need to completely remove the nut to separate items can be eliminated. In one approach, a U-shaped washer sits between the nut and work piece. The opening in the washer is larger than the diameter of the bolt. The diameter of the entire washer exceeds the size of the nut as does the diameter of the bolt hole in the work piece. The washer is placed directly under the nut. The nut is either tightened or loosened with one turn. When loose, the washer can be inserted or removed between the nut and the top piece. As soon as the washer is removed, the work piece can be lifted off the bolt. Thus, one turn on the wrench replaces complete removal. The setup is shown in Figure 10.8.

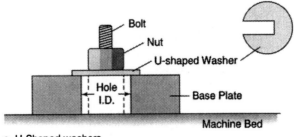

a. U-Shaped washers

Figure 10.8 Quick Fastening Methods

10.3   QUALITY ENGINEERING

JIT relies on the availability and predictable performance of suppliers, production processes, and demand. The field of quality engineering has developed a variety of tools and management strategies for ensuring that products and processes meet target specifications. Quality engineering assumes that key quality characteristics can be defined, and measures of these characteristics can be observed. The key characteristic may be as simple as the length of some part dimension or the percent impurities in a chemical solution. Once the key characteristic(s) are defined, systems are established for controlling the mean value of the characteristic(s) to match the target specification and for minimizing the variation of the observed characteristic to ensure a continual close fit between the target and actual performance.

Quality systems require activity at all hierarchical levels of the firm. At the executive level, there must be a clear statement of support for a quality improvement program backed up by the allocation of resources and the consistency of decision-making. Employees and suppliers must sense that quality is truly demanded and rewarded. The executive level must commission multidisciplinary teams of middle managers to redesign procedures and improve quality. The middle management team must define significant problem areas and then empower teams of workers under their direction to solve these problems. Each of these task teams needs the time, personnel, training, and financial resources to re-engineer their assigned activity. The goal is continuous improvement. Most of this improvement will be gradual in form resulting from sequential elimination of sources of variation. Occasionally, new paradigms such as switching from gears to quartz sensors, or manual to automated control, or functional layout to cells, will provide quantum leaps in performance. Throughout all this is the necessity to involve all employees, prioritize the needs for improvement and monitor performance to ensure that gains are maintained. Thorough discussion of the design of quality systems and use of quantitative techniques would require too much time and space for this text. The interested reader should consult other texts such as Montgomery [1996]. In the following subsections, we provide a brief introduction to some of the useful techniques. We begin with statistical process control for describing the capability and monitoring the performance of existing processes. This is followed by a model for adjusting processes in response to sample data, a brief overview of experimental design, the value of product grading, and the basics of total productive maintenance programs.

10.3.1   Statistical Process Control

When it is not practical or possible to physically eliminate all defects or immediately detect defective items via poka-yoke, it may be possible to monitor a process and either predict when it is about to go out of control or to quickly detect once it has degraded. Although statistical process control (SPC) may not catch all defective items, the procedures are often quite effective at reducing the number of defective items leaving a workstation to a few parts per million.

We present SPC as a two-step procedure. The first step involves determining the capability of a process. The second step involves implementing and operating a control system to ensure that the process remains in control. The word "control" here refers to a statistical inference that the process is performing at the desired mean level with the only variation from part to part coming from the natural inherent repeatability of the process.

10.3.1.1   Determining Process Capability

Process capability measures the inherent repeatability of a process over time (minute to minute) or from part to part in a high-volume production environment. In determining capability, we must separate **assignable causes** of variation from **natural** (inherent) **causes.** Assignable causes are those sources of variation that can be attributed to specific causes and can be eliminated by careful management without fundamentally changing the process. Consider, for example, a punch press used to form a part by pressing a sheet metal blank through a series of successive dies. If we make a high proportion of defective items every morning because the machine is started off cold, then this is an assignable cause that can be eliminated by preheating the dies. Likewise, if untrained workers occasionally set up the machine incorrectly or blanks with thicknesses outside of the engineering specifications are occasionally purchased leading to defective parts, then these are assignable causes easily correctable by management. However, even when the machine is properly maintained and the raw material appears good, successive parts will not be identical. This results from the limits of the technology and may be caused by local stresses or minor undetectable thickness variations in the sheet metal blanks. There may be factors that could be corrected, but it would not be economical to do so. For example, we might be able to reduce part variation slightly by putting a better voltage controller on the machine or cleaning the dies more frequently. If the cost is high and the improvement minor, we may decide that the current process is acceptable for our market and lump the variation caused by these potentially assignable causes into the natural variation category.

Process capability is measured by the repeatability of a process during a very small interval of time. Data are collected indicating a "snapshot" of the process when it is believed to be operating in control. A statistical description of the process at this snapshot is used to describe process capability. This description is usually produced by measuring the key characteristic(s) of the products made during this snapshot. Suppose that the target value for the key quality characteristic is labeled T. The mean of the process at this snapshot is denoted by μ, and the standard deviation of the process is denoted by σ. Unfortunately, the mean and standard deviation are not knowable in general. Instead, we use the sample of process output taken during the snapshot and measure the key characteristic(s) for each of the n items in the sample. Let the value for the ith sample be X_i. (We will assume a single key characteristic, otherwise X_i is a vector of measurements.) We then estimate the mean by the sample mean:

$$\hat{\mu} \equiv \overline{X} = n^{-1} \cdot \sum_{i=1}^{n} X_i \qquad (10.12)$$

and the standard deviation by the sample standard deviation,

$$s \equiv \sqrt{\frac{\sum_{i=1}^{n}(X_i - \overline{X})^2}{n-1}} \qquad (10.13)$$

The difference between the mean and target value indicate whether the process is properly centered. The standard deviation indicates how much the process output will naturally vary from the mean. Several summary indicators have developed into popular measures of process capability. The basis for these is the historical reliance on the normal

Table 10.1   PPM Defective for Normally Distributed Processes
                in Control

C_p	Defective parts per million	
	Two-sided specifications	One-sided specifications
0.50	133,614.	66,807.
0.80	16,400.	8,200.
1.00	2,700.	1,350.
1.20	316.	158.
1.40	27.	13.5
1.50	6.8	3.4
1.60	1.80	0.8
1.80	0.066	0.033
2.00	0.002	0.001
2.50	0.000006	0.000003

distribution as a model for statistical behavior in many situations in which multiple factors aggregate to produce random behavior.[4]

Consider, a product with two-sided specifications. For example, the product is defective if a particular dimension is too long or too short. Let the engineering specifications for the product be given by U for the upper specification and L for the lower specification. The first process capability measure is C_p given by:

$$C_p = \frac{U - L}{6 \cdot \sigma} \quad \text{or} \quad \hat{C}_p = \frac{U - L}{6 \cdot s} \tag{10.14}$$

C_p can be used to estimate the number of defective parts per million produced by a process when it is in control. For a process with only upper or lower specifications, we can replace C_p by:

$$C_{pu} = \frac{U - \mu}{3\sigma} \quad \text{or} \quad C_{pl} = \frac{\mu - L}{3\sigma}, \tag{10.15}$$

respectively. For a process that **is centered, in-control, and with normally distributed random deviations,** Table 10.1 gives the number of defective parts per million. Unfortunately, this Table is often misused to overestimate the capability of a process. If the process is not centered, or wanders out of control, or s does not accurately estimate σ, or the random deviations are non-normal, then the Table will not accurately indicate ppm defective.

EXAMPLE 10.6

A particular dimension is specified by a design engineer as 3.600 ± 0.010 mm. The process has historically performed with a standard deviation of 0.002 mm. Assuming that the process is in control, estimate process capability by C_p, and also estimate the ppm defective if the random deviations are normally distributed.

---

[4]Those with a basic background in statistical theory should be able to identify the central limit theorem as justification for this stance. However, note that some real-world phenomena can produce other random distributions, and the assumption of normality should be checked on a case-by-case basis.

SOLUTION

Using Equation (10.14), we estimate process capability by:

$$C_p = \frac{U - L}{6\sigma} = \frac{3.610 - 3.590}{6(0.002)} = 1.67.$$

Interpolating in Table 10.1, we may conclude that there will be only a few ppm defective if the deviations are normally distributed and the process is in control.

One of the key assumptions behind C_p concerned the process being currently centered between the upper and lower specifications. If this is not valid, we can use the measure C_{pk}. C_{pk} incorporates the effect of the process mean by using the measure:

$$C_{pk} = \min\left\{\frac{U - \mu}{3\sigma}, \frac{\mu - L}{3\sigma}\right\}. \tag{10.16}$$

To indicate if a low C_{pk} problem lies with process bias or excess variability, we can check a measure of bias such as $\frac{\mu - T}{\sigma}$ where T is the centered target $T = \frac{U + L}{2}$. This is a relative measure, but any nonzero bias indicates an opportunity for recentering, and values with magnitude in excess of one suggest attention unless the bias is justified by other operational considerations such as cost or safety.

EXAMPLE 10.7

Suppose that in Example 10.6 the actual process mean is 3.604mm. Estimate process capability.

SOLUTION

We see that the process is performing at 3.604 mm on average instead of the target 3.600. Using Equation (10.16),

$$C_{pk} = \min\left\{\frac{U - \mu}{3\sigma}, \frac{\mu - L}{3\sigma}\right\} = \min\left\{\frac{3.610 - 3.604}{3(0.002)}, \frac{3.604 - 3.590}{3(0.002)}\right\}$$

$$= \min\{1, 2.33\} = 1.$$

This is a more accurate measure of process capability and indicates several orders of magnitude more defective ppm than if the process were centered at 3.600 mm.

10.3.1.2 Control Charts for Monitoring Statistical Control

Process capability studies are based on a statistical analysis of observed data. If the data is collected at a single point in time then the study only provides a snapshot picture of the process. Such studies indicate how the process varies over a short interval of time. For a process to be in control, it must sustain its consistency over longer periods of time. It is preferable, therefore, to collect the data in small subgroups over a set of successive periods. We can then test to see that the means of the subgroups are equal, and the process variance is holding steady. Control charts provide a simple means for monitoring process performance over time to check whether the process remains in a state of statistical control.

For a process producing a large number of product units, we can take small periodic samples. Each sample should be taken under homogeneous conditions—approximately same point in time, same operator, same machine, same raw material batch, and so on. These small samples are referred to as **rational subgroups**. Each provides a snapshot view of a particular process at a specific point in time. The change in observed quality characteristic values between subgroups indicates changes in the process. The key concept behind control charts stems from the fact that rational subgroups should look like purely random picks from the same distribution. If a subgroup is observed and its mean or variance is significantly different (statistically speaking) than the values for past subgroups then this indicates a high probability that the process has shifted and needs to be brought back into control. Thus, we are essentially using a dynamic hypothesis-testing procedure. We hypothesize that the underlying process produces product with a stationary distribution and we then check the rational samples periodically to see if the data indicates convincing evidence that does not support this hypothesis.

A wide variety of statistical process control charts have been proposed for monitoring various types of processes and detecting particular types of shifts in process performance. We confine our discussion to the basic **X-bar** and **s charts** for monitoring quality characteristics that are measured by a continuous variable and **p charts** for monitoring qualitative variables indicating whether products are acceptable or defective. A wide selection of other chart types can be found in texts devoted entirely to statistical quality control such as Montgomery [1996]. This includes special charts for the cases of monitoring multiple-quality characteristics (multivariate charts), quickly detecting small shifts in the process (exponentially-weighted moving average charts), charts for use with short production runs or small lot sizes (Q charts), and charts for use when product specifications are much wider than normal process variation (acceptance control charts) among others.

10.3.1.2.1   X-bar and S Charts for Variables Data

Suppose that the key quality characteristic of a process is a measurable variable such as length, thickness, or electrical resistance. We will collect our rational subgroup samples and look for deviations from our hypothesized values. Typically, we monitor the distribution mean by using the sample mean $\overline{X}$, and the distribution spread using the sample standard deviation s. Because the variance of a sample mean on n observations, $\left(V(\overline{X}) = \dfrac{\sigma_X^2}{n}\right)$ depends on the process variance (σ_X^2), we usually check the process variance for each subgroup first. If that seems to have remained the same, i.e. the dispersion is under control, then we can test the process mean. We describe the X-bar and s charts for the case that we have m historical samples (subgroups) each of size n to describe the process, and we believe all these samples were collected when the process was performing as expected with only random variation. In other words, when the process was "in-control." Let X_{ij} be the jth observation in the ith subgroup. Under these conditions, the grand mean of the process can be estimated by:

$$\overline{\overline{X}} = \frac{\sum\limits_{i=1}^{m}\sum\limits_{j=1}^{n} X_{ij}}{n \cdot m}. \tag{10.17}$$

The standard deviation of the process can be estimated by the average of the standard deviations for the subgroups or,

$$\overline{s} = \frac{\sum\limits_{i=1}^{m}\left(\dfrac{\sum\limits_{j=1}^{n}(X_{ij} - \overline{X}_i)^2}{n-1}\right)^{1/2}}{m} \quad \text{where } \overline{X}_i = \frac{\sum\limits_{j=1}^{n} X_{ij}}{n}. \tag{10.18}$$

Note that in Equation (10.18), the deviation for each observation was taken from its subgroup mean and not the grand mean of Equation (10.17). This is to guard against mis-

Table 10.2 X-bar and S Control Chart Constants

n	c_4	A_2	A_3	D_3	D_4	B_3	B_4	d_2
2	0.7979	1.8880	2.659	0	3.267	0	3.267	1.128
3	0.8862	1.023	1.954	0	2.575	0	2.568	1.693
4	0.9213	0.729	1.628	0	2.282	0	2.266	2.059
5	0.9400	0.577	1.427	0	2.115	0	2.089	2.326
6	0.9515	0.483	1.287	0	2.004	0.030	1.970	2.534
10	0.9727	0.308	0.975	0.223	1.777	0.284	1.716	3.078
15	0.9823	0.223	0.789	0.347	1.653	0.428	1.572	3.472
20	0.9869	0.180	0.680	0.415	1.585	0.510	1.490	3.735

leading computation resulting from changes in the process mean over time. The safest estimate of the process mean at any time is the subgroup sample mean.

Control charts are constructed by plotting sample results of the statistic of interest (such as the sample mean or standard deviation) for each subgroup on a chart and comparing those values to the critical values from a test of hypothesis that the process is in control. The vertical axis contains the scale for the sample statistic. Time in the form of "sample number" forms the horizontal axis. The maximum and minimum critical values for the hypothesis are indicated as the upper control limit and lower control limit, respectively. A tradition with control charts is to consider any plotted values which differ by more than three standard deviations from its mean to be significant evidence of a change in the process.[5] This rule of thumb stems from the normal distribution for which there is a 99.73% chance that any observed value will lie within three standard deviations of the mean. Thus, if we find a value more than 3 σ from the mean, it is more reasonable to conclude that the process has changed rather than that we observed such an unusual value by chance. The values for each subgroup are plotted on a historical chart and compared to the reference control limits at $\pm 3\sigma$ from the mean. Consider a chart for monitoring $\overline{X}$. Let $\overline{X}_i$ be the sample average for subgroup i. Based on all our historical data, if the process is in control $E(\overline{X}_i) \approx \overline{\overline{X}}$. The $V(\overline{X}_i) = \dfrac{\sigma^2}{n}$. Thus, $\sigma_{\overline{X}_i} = \dfrac{\sigma}{\sqrt{n}}$. However, because of the nonlinearity of the square root transformation, while $E(s^2) = \sigma^2$, $E(s) \neq \sigma$. For small sample sizes, we need to make the adjustment $E(s) = c_4 \cdot \sigma$ where c_4 is shown in Table 10.2. To summarize, we should have individual snapshot sample averages revolve around the center line $\overline{\overline{X}}$ within the reasonable range of variation

$$\text{Upper Control Limit} = \overline{\overline{X}} + \frac{3 \cdot \overline{s}}{c_4 \cdot \sqrt{n}} = \overline{\overline{X}} + A_3 \cdot \overline{s} \qquad (10.19)$$

and

$$\text{Lower Control Limit} = \overline{\overline{X}} - \frac{3 \cdot \overline{s}}{c_4 \cdot \sqrt{n}} = \overline{\overline{X}} - A_3 \cdot \overline{s} \qquad (10.20)$$

---

[5]Other values than 3 can certainly be used. In the economic design of control charts, the cost of Type I and Type II errors are used to determine the number of standard deviations to choose. Other charts use rules such as detecting two out of three consecutive values beyond two standard deviations to try to more quickly detect processes that are shifting. A variety of rules have been proposed, but the user should be careful not to use too many rules or else false positive results will be detected too often.

A_3 values are shown in Table 10.2. Any plotted point outside the control limits indicates the process may have shifted. In addition, any nonrandom pattern such as eight increasing (or decreasing) values in a row or several consecutive values near a control limit, indicate a possible shift and justify investigation of process status.

To check the process standard deviation, we make use of the result that the standard deviation of the sample standard deviation is $\sigma_s = \sigma\sqrt{1 - c_4^2}$. Once again, using the result that $\sigma \approx \dfrac{s}{c_4}$, we can develop the control limits for the sample standard deviations from each rational subgroup as

Upper Control Limit =

$$\bar{s} + 3 \cdot \left(\frac{\bar{s}}{c_4}\right) \cdot \sqrt{1 - c_4^2} = \left(1 + \frac{3 \cdot \sqrt{1 - c_4^2}}{c_4}\right) \cdot \bar{s} = B_4 \cdot \qquad (10.21)$$

and

Lower Control Limit =

$$\bar{s} - 3 \cdot \left(\frac{\bar{s}}{c_4}\right) \cdot \sqrt{1 - c_4^2} = \left(1 - \frac{3 \cdot \sqrt{1 - c_4^2}}{c_4}\right) \cdot \bar{s} = B_3 \cdot \bar{s} \qquad (10.22)$$

where B_3 and B_4 are tabulated in Table 10.2.

For small subgroup sizes, the sample range of subgroup i, given by:

$$R_i = \max_j\{X_{ij}\} - \min_j\{X_{ij}\} \qquad (10.23)$$

contains almost as much information as the sample standard deviation. In addition, the sample range is easier to compute which may be an issue when training shop personnel. The range and standard deviation are both measures of the spread in these data. Their relationship has been tabulated and for a given subgroup size n, $E(R_i) = d_2 \cdot \sigma$. Values of d_2 are shown in Table 10.2. Knowledge of this relationship allows us to replace $\bar{s}$ by $\bar{R}$ in Equations (10.19) to (10.22). The new expressions are:

$$UCL_R = D_4 \cdot \bar{R} \qquad (10.24)$$

$$LCL_R = D_3 \cdot \bar{R} \qquad (10.25)$$

for the R chart, and

$$UCL_{\bar{x}} = \bar{\bar{X}} + A_2 \cdot \bar{R} \qquad (10.26)$$

$$LCL_{\bar{x}} = \bar{\bar{X}} - A_2 \cdot \bar{R} \qquad (10.27)$$

for the X-bar chart. The parameters A_2, D_3 and D_4 are provided in Table 10.2.

EXAMPLE 10.8

Nine volt batteries are produced in high volume for the consumer market. Each hour, a sample of three batteries is taken. Results for individual samples over the last 15 hours are shown in Table 10.3. Construct a control chart, and determine if the process is in control. What is the natural process variability?

SOLUTION

We begin with the R chart to test whether the process variance is in control. Results are shown in Table 10.4. The sample range for the first subgroup is:

$$R_1 = \max\{X_{11}, X_{12}, X_{13}\} - \min\{X_{11}, X_{12}, X_{13}\} = 8.95 - 8.75 = 0.20.$$

Table 10.3	Historical Voltage Data		
	Observed		
Sample	voltages		
1	8.75	8.95	8.93
2	8.79	9.03	9.00
3	9.30	8.91	9.01
4	8.84	9.02	8.94
5	8.91	8.98	9.05
6	8.76	9.00	8.89
7	8.99	8.85	8.97
8	9.01	9.22	8.95
9	8.96	8.84	8.93
10	8.93	8.88	9.01
11	9.03	8.86	9.10
12	8.95	9.00	8.92
13	8.83	9.07	9.02
14	9.00	8.78	9.11
15	9.06	8.94	8.88

Table 10.4	Sample Results for Example 10.9	
Sample	$\overline{X}_i$	R_i
1	8.877	0.20
2	8.940	0.24
3	9.073	0.39
4	8.933	0.18
5	8.980	0.14
6	8.883	0.24
7	8.937	0.14
8	9.060	0.27
9	8.910	0.12
10	8.940	0.13
11	8.997	0.24
12	8.957	0.08
13	8.973	0.24
14	8.963	0.33
15	8.960	0.18
Average	8.959	0.208

Computing the ranges for each of the 15 subgroups, we can find the average range for samples of size 3 by:

$$\overline{R} = \frac{\sum_{i=1}^{15} R_i}{15} = \frac{3.12}{15} = 0.208.$$

The control limits are $UCL_{\overline{R}} = D_4 \cdot \overline{R} = 2.575 \cdot (0.208) = 0.536$ and $LCL_{\overline{R}} = D_3 \cdot \overline{R} = 0$. All sample ranges fall within the control limits in Figure 10.9. Because process variability is in control, we can check the process mean. The X-bar chart is shown in Figure 10.10. The sample process mean is given by $\overline{\overline{X}} = \frac{\sum_{i=1}^{15} \sum_{j=1}^{3} X_{ij}}{(15) \cdot (3)} = \frac{403.15}{45} = 8.959$. The control limits are at $UCL \equiv \overline{\overline{X}} + A_2\overline{R} = 8.959 + 1.023 \cdot (0.208) = 9.172$ and $LCL = \overline{\overline{X}} - A_2\overline{R} = 8.959 - 1.023 \cdot (0.208) = 8.746$. Figure 10.10 shows that the sample mean for each of the 15 rational subgroups is within the control limits and there does not appear to be an unusual pattern. Therefore, we can conclude that the process seems to be in control. The process variability is described by $\hat{\sigma} = \frac{\overline{R}}{d_2} = \frac{0.208}{1.693} = 0.123$. Unfortunately, the mean is not centered precisely at nine volts.

The basic test for each new rational sample concentrates on determining whether the plotted statistic falls within the control limits. The chart can also be used as a visual aid for determining many types of nonrandom behavior. Figure 10.11 shows several patterns that might be observed in practice. Figure 10.11a shows a process that is drifting linearly over time. Perhaps we have not yet obtained an "out-of-control" point, but this seems imminent. Tool wear or sensitivity to changing ambient environmental conditions may produce such trends. Figure 10.11b shows oscillation often found when multiple processes are mistakenly plotted on the same graph. For example, the facility may operate two shifts.

R Chart for Voltages

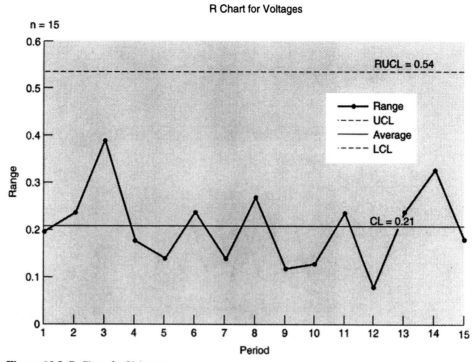

Figure 10.9 R Chart for Voltages

XBar Chart for Voltages

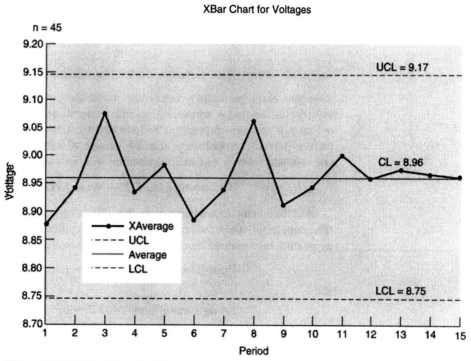

Figure 10.10 X Bar Chart for Voltages

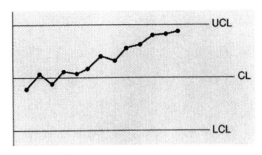

a. Linear Drift

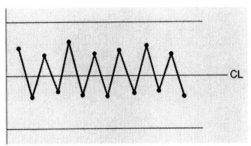

b. Oscillation from a Mixed Process

Figure 10.11 Examples of Nonrandom Behavior

In the Figure, product from shift 1 is always under specification and shift 2 output is always over specification. Instead of concluding that we are "correct on the average," the Figure should lead the manufacturing engineers to adjust operations on both shifts to yield a continuously on-target output. Looking at the consistency of every second plotted value, we see that variability is small, and we could control the process with much tighter limits if the two shifts are calibrated separately.

10.3.1.2.2 Proportion Defective Charts for Attribute Data

Some processes, particularly newer ones in the early stages of development, have high-yield losses. Instead of measuring specific traits, we may use a gaging system that only indicates if the item is good or bad (defective). If the process has a nominal process fraction defective proportion p, then the number of defective items found in a sample of size n should follow a binomial distribution with probability density function $b(x; n, p) = \binom{n}{x} p^x (1 - p)^{n-x}$. The binomial has $E(X) = np$ and $V(X) = np(1 - p)$. The proportion of defective items in a sample, X/n, has $E(X/n) = p$ and $V(X/n) = p(1 - p)/n$. If we take samples of size n, we can then check the stability of the process fraction defective by plotting on a control chart with centerline at p and control limits

$$\text{Upper Control Limit} \equiv UCL = p + 3\sqrt{p(1 - p)/n} \qquad (10.28)$$

and

$$\text{Lower Control Limit} \equiv LCL = p - 3\sqrt{p(1 - p)/n} \qquad (10.29)$$

If p is unknown, we can estimate it by the average of a large set of historical samples taken when we believe the process is performing normally. We use $\bar{p} = \dfrac{\sum_{i=1}^{m} X_i}{m \cdot n}$ where X_i

is the number of defective items found among the n items in sample i. We then replace p in Equations (10.28) and (10.29) with $\bar{p}$.

EXAMPLE 10.9

A process operating on the edge of technological feasibility has an historical defect rate of 25%. Samples of 20 units are to be taken twice a shift to determine if the process is functioning normally. Set up a control chart for monitoring the process by testing for defective items in each sample.

SOLUTION

We will use a p chart. The center line will be at $p = 0.25$. The control limits will be at

$$UCL = \bar{p} + 3\sqrt{\frac{\bar{p}(1 - \bar{p})}{n}} = 0.25 + 3\sqrt{\frac{0.25(0.75)}{20}} = 0.54$$

and

$$LCL = 0.25 - 3\sqrt{\frac{0.25(0.75)}{20}} = -0.04.$$

Because we cannot, in practice, have a negative number of defective items in the sample, we effectively use $LCL = 0$. At each sampling point, we collect and test 20 items. The number of defective items in the ith sample is labeled X_i. There is evidence of a change in the process only if the proportion of defective items in sample i, X_i/n, exceeds 0.54. With $n = 20$, this translates to the condition $X_i \geq 11$.

Proportion defective charts are only appropriate when sample sizes are large and the percent defective is significant. Otherwise $np < 1$, and the number of defects in a sample will generally be zero. In an effective process, p is small, and it may make more sense to monitor the number of consecutive units produced or tested between defects. If each item represents a Bernoulli trial, then the random number of units until a defect is detected, X, will have a geometric distribution with probability mass function $f(x; p) = p \cdot (1 - p)^x$.

10.3.2   Process Adjustments During Setup

When setting up a process to produce a batch of Q parts, it is customary to include a quality check to approve the setup prior to starting production of the batch. This check may involve running and testing one or more parts. In this section, we describe an efficient mechanism for planning setup adjustments. Beyond the initial setup, this process can be integrated into the production of the entire lot.

Let us assume that the parts being produced have a key quality characteristic X with target value T. This characteristic might be length, thickness, voltage, brightness, or other measure. Suppose that the initial machine setup has an expected value for the process mean equal to the target, but the mean setting has a variance of σ_s^2 about the target. We will let S_0 denote the initial set value. Also, let σ_x^2 describe the inherent variance of unit to unit values about the set point when the process is operating. Note, that we have two sources of variation about the target. The first is any systematic error in the setup (σ_s^2). The second is the natural variation of the process (σ_x^2). We want to minimize the deviation between the machine setting and the desired target. The squared difference between the true setting and the target value, referred to as the **squared bias,** and given by

$$Bias^2 = [X - T]^2$$

serves as the measure of interest. For the initial setup, $X = S_0$ and, if $E(S_0) = T$, $E(Bias^2) = \sigma_s^2$. However, we may be able to improve on this by obtaining information from a small sample run of k units, $k < Q$. Using these first units, we estimate bias error in the initial setup by comparing the quality characteristic on these units to the target. If initial measurements show that the bias is large, then adjustment may be in order. It would make sense to adjust the process based on the negative of the observed deviations from the target, i.e. if X_i is the measured value for the ith unit in the initial sample, we modify

the current settings by moving $\dfrac{-\sum\limits_{i=1}^{k}(X_i - T)}{k}$. In practice, we might hedge our bets and only

modify the initial setup by some α proportion of this deviation, that is by $\dfrac{-\alpha\sum\limits_{i=1}^{k}(X_i - T)}{k}$,

$0 \le \alpha \le 1$. Each sample value, X_i, is composed of the initial setting plus a random deviation, thus $X_i = S_0 + \varepsilon_i$ where ε_i is a random variable with mean 0 and variance σ_x^2. The bias error in the process after adjustment is therefore:

$$
\begin{aligned}
Bias &= S_0 - \alpha\,\frac{\sum\limits_{i=1}^{k}(X_i - T)}{k} - T \\[2em]
&= S_0 - T - \alpha\,\frac{\sum\limits_{i=1}^{k}(S_0 + \varepsilon_i - T)}{k} \\[2em]
&= (1 - \alpha)(S_0 - T) - \alpha\,\frac{\sum\limits_{i=1}^{k}\varepsilon_i}{k}
\end{aligned}
\tag{10.30}
$$

The expected squared bias of the resulting setting is therefore:

$$
E(Bias^2) = (1 - \alpha)^2 E^2(S_0 - T) + \alpha^2\,\frac{\sigma_x^2}{k} - 2\alpha(1 - \alpha)E\left[(S_0 - T)\left(\frac{\sum \varepsilon_i}{k}\right)\right] \tag{10.31}
$$

$$
= (1 - \alpha)^2 \sigma_s^2 + \alpha^2\,\frac{\sigma_x^2}{k}
$$

because $(S_0 - T)$ and ε_i are independent. The best α is that which minimizes the expected squared bias. To minimize the bias, we differentiate Expression [Equation (10.31)] with respect to α and set to zero. This yields:

$$
\alpha^* = \frac{k}{k + \dfrac{\sigma_x^2}{\sigma_s^2}} \tag{10.32}
$$

or an adjustment of:

$$
Adjustment = -\frac{\sum\limits_{i=1}^{k}(X_i - T)}{k + \dfrac{\sigma_x^2}{\sigma_s^2}}. \tag{10.33}
$$

We leave it to the reader to take the second derivative and show that Equation (10.32) actually does minimize the expected squared bias.

EXAMPLE 10.10

Environmental conditions such as temperature, humidity, and impurities in the chamber affect the machine used to deposit a thin coating on metallic disks. The target coating thickness is 0.0010 mm but the standard deviation of coating thickness on initial setup is 0.0003 mm. The standard deviation of part-to-part difference for a given setup is 0.0001 mm. An initial sample of four disks had an average coating thickness of 0.0013 mm. How should the machine be adjusted?

SOLUTION

The total deviation of the four parts is $\sum_{i=1}^{4}(X_i - T) = 4(0.0013 - 0.0010) = 0.0012$ *mm*. Using Equation (10.33), the desired adjustment in coating thickness is:

$$Adjustment = -\frac{\sum_{i=1}^{k}(X_i - T)}{k + \frac{\sigma_x^2}{\sigma_s^2}} = \frac{-0.0012}{4 + \frac{0.0001^2}{0.0003^2}} = -0.00029 mm.$$

It is not necessary to restrict ourselves to a single adjustment. We may want to periodically adjust the settings as we proceed through the batch. It can be shown that after k observations, the best choice is to readjust the process by:

$$\frac{-\sum_{i=1}^{k}(X_i - T)}{k + m + \frac{\sigma_x^2}{\sigma_s^2}} \tag{10.34}$$

where m is the number of units checked from the beginning up through the last adjustment. Thus, if we make an adjustment every k units, for the first adjustment, $m = 0$, for the second $m = k$, for the third $m = 2k$, and so on.

10.3.3 Design of Experiments for Process Optimization

The better we understand a process, the easier it is to optimize the parameter settings and control the key factors. Unfortunately, most processes have multiple variables (measures) that could be used to control the process, and we are never sure exactly how these interact and determine process performance. We may have years of observational experience, but unless we used controlled experiments to collect and then carefully analyze the data, we probably have mistaken impressions because of correlation between variables, occasional random events that left lasting impressions, a myopic view of only part of the process, and shifts in process performance over time. The field of experimental design and analysis offers a scientific approach to planning experiments and analyzing data for accurately modeling the effect of the key controllable variables that determine process performance. There are many uses for experimental design. For our purposes, we can group these into the following categories:

1. Identifying influential variables and sources of variation. What causes output of a process to vary? A process may have 1, 10, 100 or more parameters that can affect product outcome. What proportion of the variation in the process is caused by each of these potential sources? The question is which of these are the vital few that have the most impact. These variables, sometimes referred to as experimental factors, need to be understood and carefully controlled to the optimal settings. Efficient screening experiments are available to help identify these key variables from a large collection of possible variables. The proportion of total variation observed in the response variable attributed to each factor indicates the importance of that factor.

2. Identifying the impact of variables on process performance. Once the vital few variables are identified, we would like to develop a detailed understanding of their individual impact on process performance and the interaction of these variables. This understanding can then be used to set the optimal target values for each of these variables.

The topic of experimental design is much too complex to address adequately in this text. Further details can be found in a variety of textbooks such as Montgomery [1991].

10.3.4 Process Improvement vs. Product Grading

In some cases, particularly with new technology that is necessary to provide a competitive advantage, it is not possible to iron out the major sources of variability prior to implementation. It may also be the case that product of varying quality still has value. The microprocessor chip industry offers a prime example of this environment. You can buy a 233MHz or 300MHz chip at different prices. Some chemical processes have similar characteristics whereby product price is based on the purity or percentage of a key element in the product. Warehouse outlets that sell clothing "seconds" at reduced prices represent a low-tech example of this practice. In each of these cases, the variation of the key product characteristic serves as the metric for grading the product. The various grades are sold to different customers at different prices.

10.3.5 Total Productive Maintenance

We have seen the impact that can be achieved by reducing variability. A **total productive maintenance** (TPM) program can reduce variance in processor availability. TPM works by taking a proactive approach to identifying key machines, and developing inspection and maintenance schedules to prevent breakdowns. Ideally, a TPM program will reduce the frequency and duration of down times resulting from machine breakdowns. The prime candidates for TPM policies are bottleneck processes and those with both expensive parts and long down times resulting from failures. Breakdowns in bottleneck processes limit production and thus are costly. Long down times, even at nonbottlenecks, require carrying large safety stocks to avoid starvation of subsequent processes. If output products are expensive, then inventory investment cost will be high.

Differences in **mean time to failure** (MTTF) and **mean time to repair** (MTTR) for machines help quantify the impact of a TPM program. To implement a TPM program, MTTF and MTTR measures should be computed for all key resources. Using the criteria mentioned above, namely, bottleneck processes, and costly safety stocks, candidates for TPM should be identified. Ishikawa (also called fishbone or cause and effect) diagrams are then constructed for each class of problem to indicate possible problem causes. Figure 10.12 shows a fishbone diagram for a low-volume circuit card production area. The diagram uses major ribs emanating from a central spine to represent major categories of problems such as worn parts, operator misuse, or contamination. Minor bones then spread out from each rib to further specify the possible sources of problems in each category. The process can continue for as many levels of detail as necessary to identify potential problem sources. Expert opinion can then be used to prioritize the frequency of each cause, and the diagram can be used to search for causes of problems and to suggest improvement projects. If historical data exists, the actual causes of problems may be listed in a Pareto diagram as shown in Figure 10.13. The Pareto diagram rank orders the potential

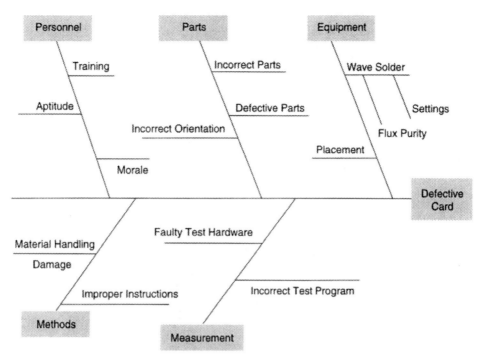

Figure 10.12 Ishikawa (Fishbone) Diagram for Circuit Card

sources of breakdowns by frequency of occurrence. Maintenance and improvement plans can then be targeted to the vital few problem sources shown in the Figure. For maintenance problems, a plan of scheduled maintenance activities and allocation of responsibility for these activities should be developed for each key machine.

We divide proactive maintenance programs into three classes—**operator maintenance, preventive maintenance,** and **predictive maintenance.** Involving machine operators in simple maintenance constitutes the most basic aspect and first line of defense for a total maintenance program. Routine cleaning activities, including changing of consumable parts such as filters, should be performed by machine operators. This prevents the need to wait for a maintenance specialist and serves to include machine operators in taking ownership of maintenance responsibility. The line workers are most closely associated with the machine and, therefore, are best situated to detect problems early. It is important to instill in operators the mindset that noticing such problems and taking responsibility for the availability and performance of their machine falls within their job description. The operators should have the authority to call for help from the maintenance crew when problems occur or when the operator detects indications of impending problems. Changes in sound or operating temperature, and an increase in output variance may indicate that it is time for a thorough checkup.

Preventive maintenance is often performed by trained maintenance personnel. These activities may involve breaking down a machine to change worn out belts or bearings, or to change oil at a prescribed frequency. Preventive maintenance actions should be based on statistical distributions for MTTF. These distributions are estimated from historical data and may be based on chronological time or machine usage. Only when the item has an increasing failure rate should preventive maintenance be considered. Consider, for example, a light bulb that has an exponentially distributed life. Would there ever come a time when it makes sense to proactively replace that bulb to avoid a burn out? The answer is

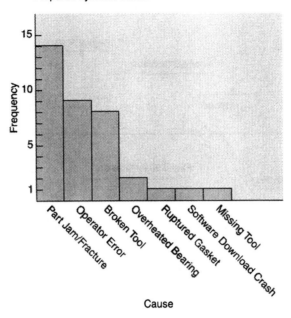

Breakdown Report

Machine: VMC 100
Period: May, 1999
Prepared by: John Brown

Figure 10.13 Pareto Diagram for Machine Breakdowns

no. If the mean life is λ^{-1}, the distribution of life time X for an exponentially distributed random variable is $f(x) = \lambda e^{-\lambda x}$, $x > 0$. Given that the bulb has burned for time t, the residual life expectancy has the same distribution as a new bulb. This is the well-known **memoryless property of the exponential.** It results from the constant hazard or failure rate $h(t)$ defined as $h(t) = \dfrac{f(t)}{1 - F(t)}$. $F(t)$ is the cumulative distribution function of the time to failure (TTF). The hazard rate at time t is the failure time density divided by the probability the item has survived to time t. For the exponential, $h(t) = \dfrac{\lambda e^{-\lambda t}}{1 - \displaystyle\int_0^t \lambda e^{-\lambda x} dx} =$

$\dfrac{\lambda e^{-\lambda t}}{1 + e^{-\lambda x}\big|_0^t} = \dfrac{\lambda e^{-\lambda t}}{e^{-\lambda t}} = \lambda$, a constant. If, on the other hand, an item has an increasing failure rate, preventive maintenance in the form of planned replacement may be cost effective.

Consider the following maintenance policy. A system undergoes preventive maintenance each time it has been operating for T time units without failure since the last preventive maintenance. If the system fails at some point, it is repaired immediately. We refer to this policy as **age replacement.**[6] After a repair or maintenance activity, we assume the system restarts in new condition. Times to failure, repair and maintenance time can all be random variables, but we assume they are independent. Failure time has density

---

[6]Other policies are also prevalent. In **block replacement,** preventive maintenance occurs at every multiple of T time units, regardless of whether a failure and repair have occurred since the last maintenance event. This allows advance scheduling of maintenance checks. Other policies consider options such as minimal repairs in addition to "good-as-new" replacements and inspection intervals that define the points at which failures can be detected and repaired.

$f(t)$, the mean repair time is μ_R and the mean time to perform preventive maintenance is μ_P. Our objective will be to maximize the proportion of time the system is operating, i.e., neither in repair nor maintenance. We assume maintenance is quicker on average than repair, otherwise it would be better to wait for failures. This system can be modeled as a renewal process. The system "renews" each time a repair or maintenance activity finishes. System availability is given by the average proportion of a renewal cycle for which the system is operating. This can be represented by the ratio of operating time per cycle to total cycle time or:

$$A(T) = \frac{\text{Mean Up Time}}{\text{Mean Cycle Time}}$$

$$= \frac{T \cdot (1 - F(T)) + \int_0^T t \cdot f(t)\, dt}{\mu_P \cdot (1 - F(T)) + \mu_R \cdot F(T) + T \cdot (1 - F(T)) + \int_0^T t \cdot f(t)\, dt} \quad (10.35)$$

Expected up time in the numerator weights the two cases of operation until maintenance with probability $(1 - F(T))$ and operation until failure if failure occurs before T. In some cases, it may be easier to represent the mean up time by the equivalent expression $\int_0^T (1 - F(t)) \cdot dt$. This expression recognizes that for any $0 \leq t \leq T$, the probability the system is still up is equal to one minus the probability it has failed. We can therefore integrate this survival probability over the possible up times.

The denominator adds expected down time because of maintenance with probability $(1 - F(T))$ and owing to repair with probability $F(T)$. Although we could differentiate Equation (10.35) to find the optimal value to T, it is perhaps more informative and easier to plot $A(T)$ versus T in most cases.

EXAMPLE 10.11

The time (days) to failure for a component has been shown to have a Weibull distribution with shape parameter $v = 2$ and scale parameter $\beta = 100$. Thus, $f_w(t; v, \beta) = \frac{v}{\beta}\left(\frac{t}{\beta}\right)^{v-1} \exp\{-(t/\beta)^v\}, t \geq 0)$. Mean repair time is four days as the work order must be scheduled, tests run to determine if other parts in the system were affected, and then replacement parts ordered. If scheduled on a routine basis, preventive maintenance can replace the part in 0.2 days. How often should preventive maintenance be performed to maximize system availability?

SOLUTION

For the Weibull, $F_w(t; v, \beta) = 1 - \exp\{-(t/\beta)^v\} = 1 - e^{-(0.01t)^2}$. Thus, for a maintenance interval of T, Equation (10.35) then yields

$$A(T) = \frac{\int_0^T e^{-(0.01t)^2}\, dt}{0.2 \cdot e^{-(0.01T)^2} + 4 \cdot (1 - e^{-(0.01T)^2}) + \int_0^T e^{-(0.01t)^2}\, dt}.$$

Availability is plotted in Figure 10.14. Although MTTF for this Weibull is 88.6 days, availability is maximized by preventive maintenance replacement every 25 days. (In practice, we might choose to perform maintenance once per month.) In addition to increased long-term availability, preventive maintenance will improve the predictability of the hour-to-hour availability of the machine and simplify operations planning.

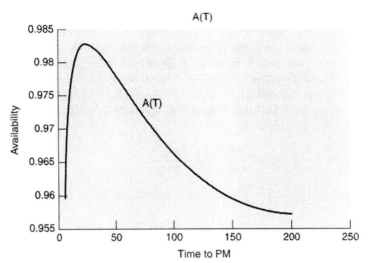

Figure 10.14 Sample Impact of Preventive Maintenance

Equation (10.35) can be extended to account for the costs of repair and replacement. If we let c_1 be the expected cost to replace a failed item and c_2 the cost of a preventive maintenance action (repair or replacement to as good as new). The cost per time is then given by:

$$E\left(\frac{\text{cost}}{\text{time}}\right) = \frac{c_1 \cdot F(T) + c_2 \cdot (1 - F(T))}{\mu_P \cdot (1 - F(T)) + \mu_R \cdot F(T) + T \cdot (1 - F(T)) + \int_0^T t \cdot f(t)\, dt}. \quad (10.36)$$

Given parameter values, Equation (10.36) can be searched over T to find the optimal maintenance interval. Note that our model does not insist that the item be replaced at each interval T. The assumption is that the item is "as good as new" after maintenance. A good cleaning or other adjustment action may accomplish this in some cases.

Predictive maintenance and scheduled inspections are based on correlations between observable conditions and breakdowns. For example, the maintenance department may periodically analyze the condition of oil for impurities that might indicate a developing problem. Based on historical data, we may know that impurity levels beyond a certain point indicate excessive wear on some part and foretells a catastrophic breakdown in the near future. Thus, when the inspection detects this level of impurities, a major machine overhaul begins. This approach makes sense when indicators are highly correlated and the down time caused by the machine overhaul is significantly less than the down time resulting from an imminent machine breakdown if no action is taken.

10.4 IMPROVING PRODUCT FLOW

Lean manufacturing depends on maintaining a steady flow of small batches of product moving through the facility. For the lean system to respond quickly to external customer demand without incurring excessive inventory costs, response to internal requests for replenishment of parts must likewise occur quickly. The organization of the shop floor and the allocation of resources to tasks can have a significant impact on job throughput time. As a general rule, small, balanced serial flow lines are more efficient than large process-oriented shops. In this section, we address how to evaluate the flow patterns in a shop, divide the machines into groups to serve part families, balance the workload within groups, and manage the dual resource constraints of workers and machines.

10.4.1   Process Flow Charts

The process of improving product flow begins with understanding the current flow patterns. Process flow charts depict the sequence of operations, inspections and storage events as material moves through the facility. Information on the distance material moves between each pair of successive activities should be included. Information on standard processing times, setup times and transfer batch sizes at each process stage facilitates understanding the product flow. An example of a process flow chart appears in Figure 10.15.

Item: HDIP Cover	Item Number: 27A-34672		
Drawing Number: 0143629	Approved By:	Date:	
Batch Size:	Annual Quantity:	Cost/Unit:	

Symbol	Description of Activity	Distance Moved (ft.)	Set Up Time (Min)	Unit Op. Time (Min)
	To Receiving			
	Inspection			
	To Raw Material Storage			
	To Material Storage			
	To Drill Press			
	Drill Top and Side Holes		12.0	8.0
	To CNC HMC	30		
	HMC Operations		1.0	29.0
	To Deburr	50		
	Tumble			16.5
	To Inspection Table	200		
	Inspection			14.0
	To Storage	125		
	Store			
	To Assembly	300		
	Await Assembly			
	Assembly		0.6	12.3
	To Inspection	10		
	Inspection			125.0
	To Shipping	100		
	Await Rest of Order			
	Assemble Order and Ship			25.0

Figure 10.15 Process Flow Chart for Instrument Package Cover

The figure summarizes shop data, listing the moves and operations required for a particular product. The chart may be embellished by superimposing the moves on a schematic diagram of the facility layout. In this form, the extent of material handling is more readily captured. After construction, flow charts should be studied for improvement possibilities. Unnecessary operations or moves can be eliminated. Opportunities to improve product flow by moving machines should be considered. If technological constraints on operation sequences are only partially constrained, the distance traveled by parts might be reduced by changing the order in which operations are performed.

10.4.2 Forming Manufacturing Cells

In many cases, dividing the production floor into manufacturing cells can improve productivity and quality. A cell is a collection of machines or processes constructed to produce a family of products. Cells come in many forms—large and small, serial and random flow patterns, automated and manual, part-oriented and product-oriented. A small automated cell, which comprises a material handling robot and a pair of CNC machining centers may machine a family of similar castings into a family of similar finished parts such as nozzles or mounting plates. Such a cell appears in Figure 10.16. At the other end of the spectrum, we may have a labor-intensive cell of ten or more workers with the machines required to produce a family of similar products. This could be a cell dedicated to cutting and sewing textile products such as shirts or to fabricating the parts and then assembling small motors. A schematic of a manual cell appears in Figure 10.17. Workers

Figure 10.16 Automated Cell with One Horizontal Machining Center, Transport Robot, Loading Station and 7 Pallets. (Courtesy of Mazak Corporation.)

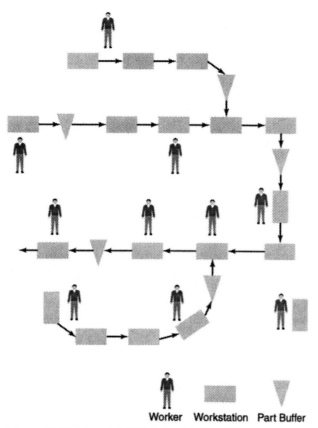

Worker   Workstation   Part Buffer

Figure 10.17 Manual Cell Showing Workers, Workstations and Product Flow

form a team that runs the main line and the feeder lines that produce components and sub-assemblies. The team may be responsible for activities such as production scheduling, quality assurance, preparing part orders, and machine maintenance as well as part fabrication and assembly. The common thread between the two cell types stems from both cells being constructed using heterogeneous processes that combine to produce a set of related items.

In the first example, a part-oriented cell manufactures a set of parts that are used in different finished products but have similar geometry and processing sequence. Each of the parts in the family follows a similar machining sequence. Machines can potentially be set up such that any part (nozzle or mounting plate) can be produced with only minimal changeover. This permits single-unit or small lot production. By placing the machines next to each other, interoperation moves are short. The small lot production and rapid transport serves to minimize inventory and to detect quality problems. Rapid response time to demand changes means we can rely on relatively accurate short-term demand forecasts instead of looking far into the future.

The product-oriented cell described for shirt or motor production ensures that all the pieces will be produced just-in-time for use in final assembly. Emphasis here is on tying production of parts to the end product in which they will be used. The individual parts may be dissimilar, but they are combined into a common finished product. Cellular operation allows the production of the individual parts to be coordinated in time. Machines in the cell may receive low utilization but workers are cross-trained to produce the vari-

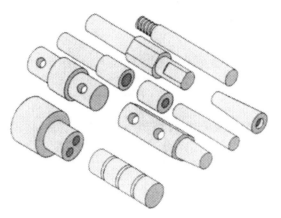

Figure 10.18 Part Family of Parts Machined from Cylindrical Stock by Turning with Some Milling and Drilling. (Reprinted from M. Groover, *Fundamentals of Modern Manufacturing*.)

ety of parts needed for the final product. In this manner, we at least avoid wasted production. Through close coordination in the cell, proper job priorities can be maintained. Product cells are useful when many parts are purchased and only a relatively small number of parts (those that fit the company's core competencies) are produced internally. These cells work particularly well if the parts being produced internally can share flexible production capacity.

Many procedures exist for forming part families and machine cells. The cell formation process should consider standard and alternative operation sequences, feasible machine alternatives for operations, utilizations of machines, correlation between part demands, and ability to share setups. Some companies choose to assign codes to parts that capture information on design and manufacturing similarities. A coding system designed for prismatic mechanical parts might indicate part dimensions, main shape, and some categorization of external and internal features such as threads, chamfers or hole patterns. Secondary information such as tolerances and surface finish may also be included. Parts with the same code could be manufactured using the same process plan and thus would naturally form a small part family as shown in Figure 10.18. Similar code values with sufficient total demand can be grouped into a family to justify a set of dedicated machines.

Typically, an analysis of production flow will easily lead to candidate families. The 0-1 part-machine incidence matrix forms a useful starting point for forming cells. Fig-

Machine Type	Part Type									
	1	2	3	4	5	6	7	8	9	10
A		1	1		1		1			
B	1			1						1
C	1			1		1				1
D		1				1		1	1	
E		1				1		1	1	
F	1			1						1
G			1		1		1			
H		1							1	

Figure 10.19 Initial Listing of Parts and Machines

ure 10.19 illustrates such a matrix for a set of ten parts that are made on eight machine types. A "1" in element ij of the matrix indicates that the part represented by column j uses the machine represented by row i during processing. Only major equipment types need to be represented. Inexpensive operations that can be easily provided to all cells need not be listed as machine types. View the rows and columns of the 0–1 incidence matrix as binary numbers. By iteratively sorting the rows and columns based on this binary representation, we can convert the unstructured matrix of Figure 10.19 to Figure 10.20a and then Figure 10.20b. We repeat the sorting procedure until the row or column order converges (Figure 10.20c). The resulting ordering is not unique. It may depend on the arbitrary initial ordering of rows and columns, and it may not produce clearly separable cells, but it is easy to see in Figure 10.20c that there are three natural cells. The only issue is whether to replicate machine type C in cells 1 and 2 and machine type A in cells 2 and 3. This decision depends on the economics of redundant machines versus the disruption of moving part types between multiple cells when processing.

The basic binary ordering procedure fails to consider a variety of important practical issues such as the cost of duplicating machines in multiple cells, and possible alternative process plans for each part. Nevertheless, it forms a useful starting point for solving these more difficult issues. A wide array of techniques exist for looking beyond the basics of the part-machine incidence matrix when forming cells. An overview can be found in Selim et al. [1998].

10.4.3 Capacity Balancing

When distinct processors are used for each of the operations required to manufacture parts, it becomes relatively easy to select the number of processors of each type needed. We begin with the minimum number needed to meet demand, and then consider the trade off between cost and performance as additional processors of each type are considered to ensure machine availability and reduce queuing. If the standard processing time for operation i is p_i, and the demand for this operation is D_i items per period, then we require $p_i \cdot D_i$ processor time per period. If the processor is available T time per period on average, then the number of processors for operation i, N_i, has a lower bound given by:

$$N_i \geq \left\lceil \frac{p_i \cdot D_i}{T} \right\rceil. \tag{10.37}$$

EXAMPLE 10.12

The aggregate production plan for a product sets production at 3,000 units per week. The third operation for this product requires a special sealing machine. The sealer can seal six units at a time with a standard operation time of 5.40 minutes. The facility operates two 8-hour shifts, five days per week and the expectation based on experience and vendor statements is that the sealer will have a 70% availability. The plant manager plans for 80% utilization of capacity. The yield from this process is expected to be 90% and an additional 4% of items will fail at final inspection. How many sealers will be needed?

SOLUTION

We start by computing the expected time needed per week. Generalizing the $p_i D_i$ term in Equation (10.37) we have:

$$\frac{Required\ Time}{Week} = \frac{(3,000\,units/week) \cdot (5.4\,min/operation)}{(6\,units/operation) \cdot (0.90) \cdot (0.96)} = 3,125\,min/week.$$

Machine Type	Part Type									
	1	2	3	4	5	6	7	8	9	10
C	1			1		1				1
B	1			1						1
F	1			1						1
A		1	1		1		1			
D		1				1		1	1	
E		1				1		1	1	
H		1							1	
G			1		1		1			

a. Part-Machine Incidence Matrix After First Row Sort

Machine Type	Part Type									
	1	4	10	6	3	5	7	2	9	8
C	1	1	1	1						
B	1	1	1							
F	1	1	1							
A					1	1	1	1		
D				1				1	1	1
E				1				1	1	1
H								1	1	
G					1	1	1			

b. Part-Machine Incidence Matrix After First Row and Column Sort

Machine Type	Part Type									
	1	4	10	6	2	9	8	3	5	7
C	1	1	1	1						
B	1	1	1							
F	1	1	1							
D				1	1	1	1			
E				1	1	1	1			
H					1	1				
A					1			1	1	1
G								1	1	1

c. Convergent Ordering After Three Row and Column Sorts

Figure 10.20 Binary Ordering of Part-Machine Incidence Matrix

The denominator recognizes that we need to start $\dfrac{1}{(0.90) \cdot (0.96)}$ units through the sealing operation to obtain one good final unit because of poor quality. Available time per machine is given by:

$$\frac{Available\ Time}{Machine\ Week} = (16\,hrs/day) \cdot (5\,days/week) \cdot (60\,min/hr) \cdot (0.70) \cdot (0.80) = 2,688\,min/week.$$

This means we need at least $N = \left\lceil \dfrac{3,125}{2,688} \right\rceil = 2$ sealers.

When operations are primarily manual or require the same type of processor, the issue becomes how to allocate the required production tasks to each processor. The division of n operations (tasks) into workstations to meet demand at the minimal cost defines the **assembly line balancing** (ALB) problem. Assembly line balancing has been widely applied for assembling automobiles, appliances, computers, and other consumer products. The desired production rate defines the time between successive production starts. We refer to this as the allowable cycle time, c. Every c time units, a new unit enters the production line, and a completed item departs the line. Each workstation has a set of assigned tasks. Each unit of product moves through the line, stopping at each workstation for c time units to have its required tasks performed. Given the time to perform each task, and possible restrictions on the order in which tasks are performed, the problem becomes assigning tasks to workstations whereby all workstations can complete their assigned tasks in at most c time, and the fewest possible number of workstations are used. We explore methods for line balancing in the next chapter.

An important corollary to capacity balancing comes from the concept of **takt time.** For every product, the demand rate determines the frequency at which the product should begin (and complete) production. All of the constituent parts of the final product should also enter production at the same frequency. This rhythmic pacing of the interval between successive releases of a product to the system in order to precisely meet demand produces several benefits. Finished goods inventories are not accumulated. Differences in processing time requirements between the products at a workstation are smoothed out by interspersing the product starts. Thus, each workstation completes items at a fairly constant rate. This is illustrated in the following example.

EXAMPLE 10.13

Two products are produced on the same two-stage process. Stage 1 takes four minutes for product A and one minute for product B. Stage 2 takes three minutes for each product. Both products have demand that averages five units per half hour. Determine the output rate of parts and WIP levels if we produce in lots of five units versus interspersed mixed product releases.

SOLUTION

As shown in Figure 10.21, if we produce the products with the sequence A, B, A, B, ..., releasing a unit every three minutes, then units will be completed at stage 1 and passed onto the next workstation at minutes, 4, 5, 10, 11, 16, 17, 21, 22, 26, 27. For each machine, the first graph indicates the flow of parts from the time they are released until that item is completed. The second graph shows total number of parts at the workstation as a function of time. We note that the release rate from the first workstation is not completely smooth, but it allows stage 2 to maintain a fairly constant production rate with minimal WIP. Suppose that instead, we release five As consecutively and then five Bs each half hour but maintain the steady three minute inter-release time. The flow is shown in Figure 10.22. Units A complete stage 1 at minutes 4, 8, 12, 16, 20, and units B at 21, 22, 23, 24, 25. We note a busy period followed by a slack period at stage 1. Stage 2 sees a slow period followed by a busy period. Total waiting time per half hour increases from five to 19 minutes at stage 1 and ten to 26 minutes at stage 2.

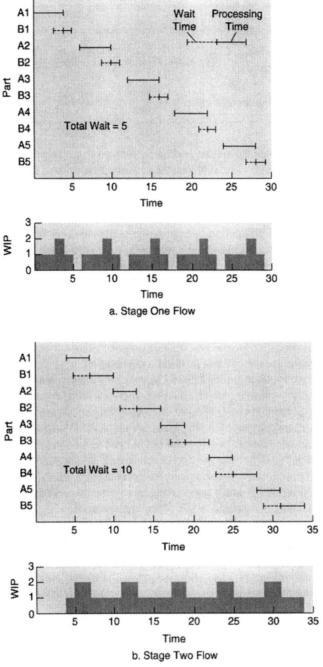

Figure 10.21 Work-in-Process for Smooth Releases and Mixed Order

Batch sequencing, in this example, still took advantage of paced releases to the first machine. If we had released a batch of five part As at time 0 and five part Bs at time 15, which would be a fairly common procedure in many firms, the waiting time and WIP would have been significantly greater. We ask the reader to examine this scenario in the end-of-chapter problems, and explore the impact of batch releases further in the next chapter.

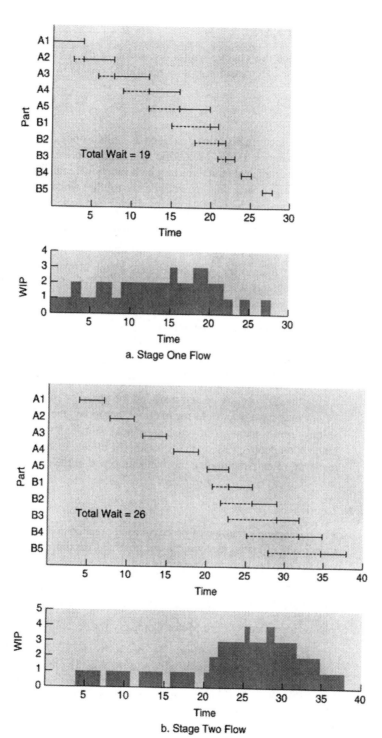

Figure 10.22 Work-in-Process for Smooth Releases

10.5   THE TRANSITION TO LEAN

We conclude this chapter with a discussion of the relative merits of lean manufacturing and how we can transition from a traditional system to a lean system.

10.5.1   Was Taylor Lean?

Lean manufacturing is often described as a philosophy in opposition to the historic teachings of Frederick Taylor's scientific management principles. In the sense that Taylor believed in hierarchical structures in which functional managers make plans and define procedures, which are then passed down to the workers for physical execution, this interpretation may be valid. However, Taylor also believed in minimizing waste costs and aggressively utilizing available technology to facilitate production. For Taylor, that meant the slide rule and the rigid mechanization available in the early 1900s. Product customization was not a key issue for customers, and the rate of technological innovation was slower than today. A large segment of the available workforce consisted of uneducated immigrants who often spoke limited English. To Taylor, the worker was like a machine in the sense that both were designed to perform a limited set of tasks well. Lean manufacturing recognizes that today's customers expect customization and that modern information and flexible manufacturing technologies allow this to happen. At the same time, lean manufacturing recognizes that ultimately, the modern worker is the most flexible machine. Properly educated and motivated, the modern line worker can often solve problems and quickly adapt to new situations. We have thus witnessed a fundamental shift during the 20th century centered around the speed of customer and technological change. Taylor may have been right for his time, but today, the overhead cost and time delay resulting from a manager-driven hierarchical decision-making and training system would hinder competitiveness in many cases. In effect, today's line worker is the first line manager, often directing the mechanization within their work sphere.

10.5.2   From Mass to Lean

The production paradigm of process-based organization, economic order quantities, large unit handling loads, receiving inspection, and maximizing equipment utilization permeates the psyche of many operations and finance managers. Converting to lean thinking of product cell layouts, single-unit mixed model production, continuous material flow, mistake-proofing, and balanced production represents revolutionary change. Such change only comes through strong leadership, education, and time. Liker [1997] describes the five phases identified by Ford Motor Company for becoming lean.

1. Process stabilization. We begin by improving the production environment. Processes must become predictable and reliable. The techniques of total productive maintenance, total quality control, poka-yoke, setup time reduction, development of standard procedures, and organizing/cleaning of the workplace all contribute to this objective. Employees are trained in lean thinking and employee involvement are expectations in this phase.

2. Continuous flow. The second phase attempts to reduce WIP inventories and batch sizes. The mentality of running machines as fast as possible begins to fade. Parts flow in small or even single-unit quantities between adjacent workstations.

3. Synchronous production. Weekly production schedules are now not only produced but followed. The former daily production meetings to review machine and material availability and revise the published schedule are no longer necessary and are eliminated. All processes are producing parts in concert whereby parts enter final assembly operations in the proper sequence. Likewise, suppliers have been integrated into lean behavior with frequent deliveries of the appropriate quantity and type of parts to point-of-use workstations.

4. Pull authorization. Production authorization occurs by the pull of parts from successor workstations. Kanbans, either physical or electronic, dictate production.

5. Balanced (level) production. Finally, all processes produce at a constant level, continuous rate. Every part type is made daily, and parts flow through the system in a steady and continual manner—materials transform into products.

A number of companies have reported significant improvements from implementing lean techniques. Liker [1997] includes case studies for several companies. As an example, Garden State Tanning, a supplier of leather to Toyota, reported a 50% drop in inventory and labor. Labor savings resulted from modifying a batch-oriented operation to a flow line with balanced workloads and workstations connected by conveyor. Production lead time went from 60 days to nine days, and quality defects went from over 1% to under 0.05%.

10.5.3   Push vs. Pull Control

Push systems such as MRP control throughput but allow WIP and cycle time to vary. Kanban and CONWIP pull systems control WIP at a level intended to produce the desired average throughput. Push systems rely on accurate and timely demand forecast and shop execution data to coordinate workstation actions. Pull systems simplify coordination through physical linkage. We could smooth production in MRP by forcing this on the master production schedule, or purposefully vary WIP levels in pull systems to match throughput to demand forecasts; however, the underlying concepts call for throughput-driven decisions during push operation and WIP control when using pull approaches. There are several other differences between push and pull systems that are worth noting when striving for lean.

Pull systems can be modeled as closed queuing networks. The amount of WIP is kept constant or at least bounded. Push systems resemble open networks. Arriving jobs are dispatched to the shop floor and proceed as fast as possible through the system, i.e. work centers are active whenever their input buffers are not empty. Spearman and Zazanis [1992] used this representation to compare push and pull systems. Several basic observations are instructive. First, pull systems with fixed levels of WIP require lower average WIP (and hence cycle time) than push systems to achieve the same throughput. However, the above statement does not include the time jobs spend waiting to enter the shop. If the material supply process cannot be tied to the pull chain, and expensive parts must be queued outside the pull system waiting their turn to be released, then these advantages of the pull system are diminished. Nonetheless, the pull system would still require less space for accommodating fluctuations in WIP levels and exhibit less congestion.

The second observation is that pull systems are more robust to errors in setting operating parameters. If demand is underestimated in a push system, shortages occur. If production plans overestimate demand, final inventory costs increase. A major breakdown occurs if jobs are released to the shop floor at a rate that exceeds capacity. Output will not be able to increase beyond resource capacity but WIP levels will continue to grow until the mistake is noted and corrective action taken. The resultant congestion may actually reduce throughput. The pull system will automatically match demand, and assuming basic rules are applied to adjust the production quota and number of kanbans as conditions change, the pull system can also adapt to demand changes.

In MRP push systems, one would typically freeze the short-term production schedule for a few weeks representing the cumulative lead time for producing end items from parts. Once items are released to the shop floor, the quantity and timing of open orders is fixed. The schedule may be firm for even longer periods of time to incorporate the ordering of raw materials and external parts. Frozen and firm schedules mandate the use of precise demand forecasts or large end-item safety stocks. With shorter lead times and reliance on actual customer demands to set final assembly schedules, pull systems avoid the need to rely on precise forecasts. Instead, pull systems assume production will be relatively constant and utilize their innate robustness to minor variations. In addition, pull systems may have more shallow (fewer levels) in the bill of materials (BOM). The BOM for a push system will include a level for every production stage, potentially every operation in a process sequence, to accommodate detailed capacity requirements planning. The BOM for a pull system need only list the major control levels at which controlled storages of items are kept. If a work cell is constructed to create a complete part (or product) with the part (product) flowing through multiple processing and assembly operations in the cell, there still only needs to be one level in the pull-system BOM for the cell.

A final observation relates to the simplicity of pull systems. Production workers are automatically empowered and do not need to wait to be told what to do. In addition, fewer workers are needed to create and monitor production plans.

10.6 SUMMARY

Kanban-control pull systems are one aspect of just-in-time (JIT) production. Elimination of waste in its many forms highlights the objective of lean or JIT production. Variability in machine availability and process yields limits effectiveness by requiring safety stocks and safety lead time. Reducing variability through better technology, training, and methods improvement can have as much impact on improving performance as increasing capacity. Minimizing setup time provides a plethora of advantages from increasing capacity and reducing throughput times to allowing greater product customization. Empowerment and cross-training of workers can also be an effective means for improving performance. Empowered, independent workers can often devise creative approaches to problems. Teams can be configured to attack specific problems or improve operations in individual areas. Cross-training allows workers to be brought to bear on bottlenecks as needed and to reassign workers as demand patterns change. The purchasing strategy should support the objectives of improving production operations. A variety of techniques are available for supporting setup time reduction including using similar product designs and process plans, localized storage of family tooling, standardized methods, and fast-clamping systems. More activities can be performed during external setup to reduce down time. Family fixtures can be designed to speed up the changeover process between similar part types.

A commitment to quality and the use of practices to back up that commitment can be an important aspect of reducing waste and improving performance. Poka-yoke, or mistake-proofing techniques should be considered to prevent the possibility of mistakes. On-line gauges can ensure that parts are within specifications when made. Simply adding keys such as checklists or indicator lights can often reduce error rates measurably. Incorporating a successive check policy wherein each worker begins by inspecting the performance of the prior worker serves to improve worker performance by providing immediate feedback and reduces the investment of resources into defective parts by catching errors quickly. Statistical process control charts and other data analysis techniques can also be used to identify quality problems as soon as they appear. Processes can be periodically adjusted to correct for any systematic deviation from the target performance. On-line maintenance, preventive maintenance, and predictive maintenance all have roles to play in reducing variability and improving worker morale. If the inherent variability in a process is significant, product grading can be used to match output to the customer price-function trade-off curve.

Improving the flow of material through the production facility can also provide benefits of easier scheduling and faster throughput. Flow charts provide a simple yet useful means for capturing the state of the material flow system and serve as a source of information when looking for system improvements. Whenever possible, workstations should be arranged so that parts need only travel a short distance between operations. Excess moves should be eliminated. Cellular manufacturing provides a technique for partitioning the factory into smaller independent subfactories, each dedicated to a family of similar items. Setup and move times are reduced along with labor and material handling cost. U-shaped cells can be used to increase flexibility in adjusting the production rate to changing demand rates and to improve communication between workers. The release of parts to the shop should create a steady pace of production. Parts that feed into the same finished product should have coordinated releases to minimize queuing time and smooth workloads throughout the shop.

The process of converting to lean manufacturing requires a cultural change beginning with a commitment from top management. As the system is brought under control, improvements can be gradually implemented until achieving a balanced, continuous flow process that responds rapidly to customer requests through coordinated production with minimal inventory investment. Workers should be viewed as important sources of knowledge and flexible capacity.

10.7 REFERENCES

ASKIN, R.G., & STANDRIDGE, C. *Modeling and analysis of manufacturing systems.* New York: John Wiley & Sons, Inc., 1993.

BLACK, J.T. *The design of the factory with a future.* New York: McGraw Hill, Inc., 1991.

BURBIDGE, J.L. *The introduction of group technology.* New York: John Wiley & Sons, 1975.

CHENG, T.C.E., & PODOLSKY, S. *Just in time manufacturing: an introduction,* New York: Chapman & Hall, Inc., 1996.

FORD, H. *My life, my work.* New York: Garden City Publishing Co., 1922.

GERTSBAKH, I.B. *Statistical reliability theory.* New York: Marcel Dekker, Inc., 1989.

GHOSH, S., & GAGNON, R.J. A comprehensive literature review and analysis of the design, balancing and scheduling of assembly systems, *International Journal of Production Research,* 1989, 27(4), 637–670.

HOFFMAN, E.G. *Setup reduction through effective workholding.* New York: Industrial Press, Inc., 1996.

IMAI, M. *KAIZEN, The key to Japan's competitive success.* New York: McGraw-Hill, Inc., 1989.

KING, J.R. "Machine-component grouping in production flow analysis", *International Journal of Production Research,* 1980, 18 (2), 213–232.

LIKER, J.K. ed., *Becoming lean: inside stories of U.S. manufacturers,* Portland, Oregon: Productivity Press, 1997.

MONDEN, Y. *Toyota production system.* Atlanta, GA: Industrial Engineering and Management Press, 1983.

MONTGOMERY, D.C. *Design and analysis of experiments.* New York: John Wiley & Sons, Inc., 1991.

MONTGOMERY, D.C. *Introduction to statistical quality control.* New York: John Wiley & Sons, Inc., 1996.

SELIM, H., ASKIN, R.G. & VAKHARIA, A. "Cell formation in group technology: Review, evaluation, and direction for future research, *International Journal of Computers and Industrial Engineering,* 1998, 34(1), 3–20.

SHINGO, S. *Zero quality control: Source inspection and the poka-yoke system.* Cambridge, MA: Productivity Press, 1986.

SHINGO, S. *A Study of the Toyota production system from an industrial engineering viewpoint.* Portland, Oregon: Productivity Press, 1989.

SPEARMAN, M.L., & ZAZANIS, M.A. Push and pull production systems: Issues and comparisons, *Operations Research,* 1992, 40(3), 521–532.

STEUDEL, H.J., & DESRUELLE, P. *Manufacturing in the nineties: How to become a mean, lean, world-class competitor.* New York: Van Nostrand Reinhold, 1992.

SURESH, N.C., & J.M. KAY, eds. *Group technology and cellular manufacturing.* Boston, MA: Kluwer Academic Publishers; 1998.

SUZAKI, K. *The new manufacturing challenge.* New York: Free Press, 1987.

TAYLOR, F.W. *The principles of scientific management.* New York: Harper & Row, 1911.

TRIETSCH, D. The harmonic rule for process setup adjustment with quadratic loss, *Journal of Quality Technology,* 1998, 30(1), 75–84.

ZACKS, S. *Introduction to reliability analysis.* New York: Springer-Verlag, 1992.

10.8 PROBLEMS

10.1. Describe the seven types of waste that can be costly to a production system.

10.2. List five techniques for reducing setup time.

10.3. What are the advantages of pull production over push production in trying to establish a lean production system? Is push always better?

10.4. Explain how converting setup activities from internal to external can improve the performance of a production system.

10.5. What are the potential advantages and possible pitfalls from employee cross-training on a production line?

10.6. Describe the natural progression of changes you may see in a manufacturing system as it is gradually converted to a lean system.

10.7. What is the difference between a part-oriented and a product-oriented manufacturing cell? What factors should be used to determine which approach you should use for defining cells?

10.8. If setup times are reduced by 50%, what percentage change in the frequency of setups should occur? What impact will this have on capacity utilization and product customization opportunities? Consider two cases: 1) setup time is proportional to setup cost; 2) setup cost is zero but total time allowed for setups is fixed.

10.9. A production system produces an average of eight jobs per day. This is considered 80% of shop capacity. Jobs arrive randomly according to a Poisson process, but all jobs take the same amount of time to produce.

a. Find the average flow time and WIP level in the shop.

b. Suppose that the arrival process was changed so that arrivals were equally spaced in time. What effect would this have on average flow time and WIP level?

10.10. Consider a process that has exponential arrivals at a rate of 20 per time. Current capacity is 25 jobs per day but this includes machine down time. Actual service time is deterministic but when the repair time from infrequent breakdowns is added, the service time looks exponential. It is estimated that a preventive maintenance program could be developed that would increase capacity to 30 jobs per day and prevent all breakdowns. (Maintenance would be done between shifts.) Determine the impact on flow time of implementing the preventive maintenance plan.

10.11. Suppose that in Figure 10.4, the next three arriving parts had processing times of 8, 12, and 4, respectively at workstation 1 and 10 at workstation 2. Find the start and completion times of each of these parts at both workstations. At what points in time is workstation 1 blocked and workstation 2 starved?

10.12. Determine the impact on cycle time, flow time, average inventory, process time, batch size, number of setups and total setup time for a 75% reduction in setup time and cost. Assume processing times are deterministic, and the system uses an order release policy that maintains two jobs in the system for each machine.

10.13. Determine the impact on cycle time, flow time, average inventory, process time, batch size, number of setups and total setup time for a 25% reduction in setup time and cost. Assume that the overall mix of products has exponential arrival and service times.

10.14. In example 10.5, how much less safety stock would be needed, and, therefore, how much less would the savings be if we were originally using a continuous review system instead of a period review system?

10.15. Increased lead times result in increased safety stock and/or scrap costs. Using the news vendor problem from Chapter 6, show that if the standard deviation of demand uncertainty is proportional to the length of the acquisition lead time, then the expected cost of scrapped product is also proportional to the acquisition lead time when demand is normally distributed.

10.16. Using the stochastic, continuous review model from Chapter 6, show that if the standard deviation of demand uncertainty is proportional to the length of the replenishment lead time, then the expected cost of safety stock is also proportional to the replenishment lead time when demand is normally distributed.

10.17. Every 15 minutes a sample of three cans is removed from a high-speed canning line, and can weights are measured. The process is thought to be in control during the past week. Eighteen samples from this period are shown in Table 10.5. Construct X-bar and R charts. Is there any evidence that the process has gone out of control during this period?

10.18. A process has a mean surface roughness of 3.56 and a standard deviation of 0.35. The standard is to have a roughness of less than 4.00.

a. Find the capability of this process.

b. Estimate the percent defective if the process is normally distributed.

c. Estimate the percent defective if the process is exponentially distributed.

10.19. Using the data in Table 10.5, find the value of C_p and C_{pk}. The process has a lower specification of $11.30 - 0.05$ oz.

Table 10.5   Can Weights in Oz.

Time	Can 1	Can 2	Can 3
1:05	11.27	11.26	11.31
1:22	11.40	11.27	11.27
1:33	11.32	11.25	11.29
1:48	11.26	11.31	11.25
2:01	11.28	11.29	11.30
2:19	11.34	11.27	11.34
2:30	11.29	11.34	11.27
2:47	11.30	11.32	11.27
3:03	11.32	11.28	11.35
3:16	11.25	11.30	11.32
3:39	11.27	11.31	11.27
3:44	11.31	11.34	11.28
4:00	11.27	11.26	11.25
4:19	11.29	11.28	11.26
4:26	11.33	11.25	11.35
4:49	11.25	11.27	11.31
5:03	11.35	11.31	11.33
5:17	11.28	11.28	11.26

10.20. A school lunch program wants to include bananas in its box lunches. Lunches are sold at a fixed price, and a standard must be met in order to receive reimbursement from a federal lunch support program. Go to your local food store. Examine 15 bunches of bananas. Treat each bunch as a rational subgroup and record the size of each banana. Construct X-bar and s control chart limits for bananas. Is the process constant from bunch to bunch? What other quality measures should be considered?

10.21. A process has a historical defect rate of 15%. Find the upper and lower control limits for proportion defective chart using rational subgroups of 40 units.

10.22. Consider a process with a percent defective of 10%. What sample size would be needed to establish a control chart with $\pm 3\sigma$ control limits with a total width of at most 0.10 (i.e. UCL $-$ LCL $= 0.10$)?

10.23. The first three units made after a machine was set up exceeded the target value by 0.12, 0.09, and 0.05, respectfully. Experience shows that the initial setup is unbiased but has a variance of 0.01. Once set up, units are produced with a variance of 0.0004. Would you adjust the machine? If so, how?

10.24. In section 10.3.2, we implicitly assumed that there was no measurement error in the initial sample of items checked and that we could adjust the machine precisely by any amount desired. Address the implications of these assumptions for a process. Are these assumptions realistic? If they do not apply, how would you modify the adjustment process? For example, suppose that you could only adjust the machine in discrete increments. Suppose that there was a measurement error in X_i with variance σ_m^2.

10.25. Show that Equation (10.34) is appropriate for sequential adjustments to a setup.

10.26. A machine has a Weibull time to failure with parameters $\nu = 4$, and $\beta = 2000$. Failures cause 20 hours of down time. A four-hour preventive maintenance overhaul can renew the ma-

chine. Plot machine availability as a function of the time interval between preventive maintenance.

10.27. Suppose in the previous problem that the cost of a maintenance technician and lost production during down time was $\$50t + \$200t^2$ where t is the length of the down time. What preventive maintenance schedule would you suggest?

10.28. A U-shaped production line has seven stages with unit processing times of 4.0, 3.4, 5.4, 7.6, 1.8, 3.7 and 2.5 minutes, respectively. Find the minimum number of workers required and the assignment of processing steps to workers required to produce six units per hour. Workers are not allowed to cross paths, that is, workers must have nonoverlapping work areas.

10.29. For the U-shaped line described in the previous problem, what is the maximum production rate that can be achieved with two workers? What allocation of operations to workers achieves this production rate?

10.30. Production for a product is planned at 5,000 units per week. The standard operation time for one particular operation is 2.45 minutes. The facility operates 40 hours per week. The required machine is expected to have a 70% availability, and the plant manager plans for 80% utilization of capacity. The yield from this process is expected to be 99%, and an additional 2% of items will fail at final inspection.

a. How many machines are needed?

b. Suppose that there was a fallout of 10% in the operation that precedes this operation. How would that affect your decision? Does it matter whether those defects are found immediately or not until final inspection for the product?

10.31. Construct a process flow chart detailing the operations and moves you make in preparing your favorite breakfast at home. Could you improve the process?

10.32. Consider the process routes for the 10 parts shown in Figure 10.23. Reorder the matrix to find a natural grouping of part families.

Machine	\multicolumn{10}{c}{Part}									
	1	2	3	4	5	6	7	8	9	10
A	1		1					1		
B		1			1					
C			1	1		1		1	1	1
D				1		1			1	
E	1		1				1	1		1
F		1					1			
G						1				
H		1			1		1			

Figure 10.23 Process Routes for Problem 10.30

10.33. A company manufactures eight major parts. Given the following set of processes used for each part, divide the parts into part-oriented cells.

Part type	Processes required
1	A, F, H, K
2	B, C, G
3	B, C, E, L
4	B, E, G
5	F, H
6	A, K, M
7	C, E, G, L
8	F, K, M

10.34. Using the WWW or your local library, find a description of the MICLASS or Opitz part coding system. Using a CAD system, draw a nozzle for a garden hose and assign it a part code.

10.35. Consider Example 10.13. Suppose that all five As had been released at time 0 and all five Bs had been released at time 15. Construct modified graphs similar to Figure 10.21 showing the flow times for each item and WIP as a function of time. Calculate the total wait times.

10.9 CASE STUDY *Dream Desk Company*

Dream Desk Company is a major supplier of office desks for home and business. The company has been in existence since 1875. After serving an apprenticeship as a cabinet maker in the east, George Dreamer had a violent disagreement with the shop owner and quickly decided it was best to move west. He opened up his own shop for making custom desks for homesteaders, ranchers, and small businessmen who could not afford expensive imported desks. The company is still run by the Dreamer family and has maintained a niche in the middle-priced home and small business office furniture market. Dreamer furniture has a good reputation for value. This reputation has been built on sturdy construction including tongue-and-groove corners, good customer service, and strong regional marketing.

The company has three basic grades of desks, each with several styles. The grades are solid oak construction, solid pine construction, and oak veneer. All together, there are fourteen standard material-style desk combinations. Two styles, the contemporary and frontier, have accounted for 65% of sales in the past two years. The oak veneer-frontier model represents approximately 25% of sales. Pine-frontier and oak-contemporary each account for another 10% of sales. In addition to the standard styles, customers occasionally request a custom order of special design. Custom orders account for approximately 20% of sales and 40% of profits. Last year, Dream Desk sold 81,450 desks for a revenue of $15.7 million. Orders vary in size from five to 500 desks, but most orders are for 50 desks or less. Sales are seasonal with about 70% of deliveries requested for August through November, but Dream Desk has always tried to maintain a constant workforce. Overtime is used during peak seasons and machine maintenance, skills cross-training, and development of new models is scheduled for the slow season. The demand for desks also follows the 3 to 5 year business cycle. Currently, we are entering an upswing and the local market is expected to grow 20% over the next 18 months. Dream Desk

can profit from this market growth if prices can be held in line and quick delivery can be promised.

Dream Desk operates a 250,000 sq. ft. manufacturing facility in Casa Petite, TX. The plant currently employs 160 full-time workers of which 125 work in production. Recent pressure from increased wood prices and more aggressive marketing from eastern manufacturers has made it clear to Jim Dreamer, Plant Manager, that the company needs to reduce its manufacturing cost and delivery lead time. Dreamer has considered developing a new production and inventory control system with some of the features of the just-in-time system he gained knowledge of at a recent seminar.

The Production Process All desks go through similar production sequences. The desk is essentially comprised of a frame and drawers. Drawers are assembled from a front, handle, two sides, back, and bottom. Veneer drawers have a two-piece front. Each piece is rough cut from stock and then one or two edges are sanded to provide a reference surface for fixturing on machines. Other edges are then sawed and sanded. During this stage, the patented Dreamer top (rounded tops to drawer sides and engraved pictograph on the top of the drawer front) is added. The priceless Dream trademark seal is then applied to the left side of the drawer. For drawers with a lock, the hole is punched and the top of the door grooved during drawer part construction. Historically, drawers were made by sending them through a series of saws, drills, routers, sanders. Parts then come together at an assembly bench for fitting and gluing. A new $1 million automated system has been installed that can fabricate parts automatically with rapid changeover between styles. The previous arrangement was a process layout with three saws, three drills, four routers, six sanders, and six assembly stations. These machines still exist, and, with the exception of assembly, require approximately 30 minutes to changeover between part types. Over time, the market has demanded more of the six-drawer desks, and this department had become the bottleneck in production. The new system increases capacity about 80% while requiring only one additional operator. (It was felt that the new system would also improve quality and reduce the need for overtime.)

Frames are made in a separate department. Legs are cut from bar stock and sanded. The contemporary styles require a turning operation on one of the three engine lathes. The other styles require a shaping operation after cutting. Both six and eight leg versions of desks (drawers on one side or both sides) are produced. Desk tops, sides, and backs are cut and sanded. In each case, the reference surface is created first. The contemporary styles have elaborate detail on the sides and require extensive router time. The veneer models have the details engraved onto the veneer coverings which are then glued onto the plywood frames. Both manual and automated gluing stations are available with the manual stations being used for short runs because setup is quicker. All together, the frame department has 20 first shift direct-line workers with five saw operators, six routers, four gluing workstations, five shapers, and two lathe operators.

All parts and frames visit the paint department. Drawers are hung on a conveyor and frames ride on pallets. The first step is an air spray to remove dust and particles. Two coats of finish are then added followed by a clear lacquer coat. Parts require two hours to dry between each coat. The paint department requires 24 workers when operating at full strength. Turnover has been a problem in the paint department. As a result, Dream Desk has recently increased the wages in this area by $0.20/hour. Temporarily, this has solved the turnover problem.

Material handling is mainly conducted by workers pushing carts. A cart can hold 40 drawer parts (sides, backs, bottoms, or fronts), 40 desk legs, or ten desk parts (sides, tops, backs). Frames are moved on flat carts, one frame per cart or on overhead conveyors. Up to 12 drawers can be stacked on a flat cart for transport.

The final step is an assembly line. Here, the table top is hand waxed, caps are added to the bottom of legs, locks are inserted and screwed tight, drawers are added to the frame, the finish is checked for blemishes, and then the desk is boxed and moved to the shipping warehouse. Any required fit adjustments and cosmetic repairs are performed by line workers. At full speed, the line takes 12 line workers and completes one desk every 90 seconds.

Dream Desk has historically hired unskilled workers. As experience and training are acquired, wages and responsibility increase. Currently, workers have an average loaded cost of $14 per hr. On average, a desk requires four hours of labor. The plant usually operates a full first shift and a partial second shift, both shifts operate five days per week. The plant shuts down for the first two weeks of July. During this time, major machine overhauls and changes in machine layouts may be implemented.

Project Scope   As plant industrial engineer, you are charged with putting together a team to design and implement a new production and inventory control system for this plant. The main objective is to reduce cost and WIP inventory. You should determine how the facility will be arranged and the procedure for controlling the flow of orders through the shop. For example, will you use a process layout, cellular layout, or product layout? What material handling load size will you use for moving parts between workstations (i.e., move one part at a time to the adjacent machine or move an entire batch at a time)? You should specify which items should be kept in inventory and which items should only be produced when ordered. You must specify production batch sizes for items made-to-stock.

The first step is to list the products and parts via a bill of materials. Document this in a flow chart. You may then decide which parts will be made-to-order and which will be made-to-stock. Using an economic criterion, select appropriate batch sizes for make-to-stock items. Document the procedure for selecting the part to produce at each stage and for authorizing production. Likewise, document the material handling system, including part transfer quantities and move equipment that will be used. You may assume ample warehousing space is available.

Chapter 11

Shop Scheduling

To effectively run a production system, the manager must be able to make decisions on a fine time scale. This chapter deals with decisions related to the detailed scheduling of machines in the shop as well as shop loading on a daily or shift level. The models in the previous chapters tended to involve mid-term decisions (weekly or monthly planning). Even if these production plans are well formulated and lead to low-cost systems that meet demand, it is critical to be able to schedule well to meet the goals of the production plan. For example, consider the following situational questions:

The shop houses five machines and each has a buffer of jobs in process waiting to be completed. Each job takes a different routing through the shop and processing times for each job operation on each machine as well as the job due date are known. The current operation on machine 4 is complete. Which operation should be scheduled next on machine 4 to ensure that the due dates of all jobs are met?

Your shop houses eight machines and each has a buffer of jobs in process waiting to be completed. Each job takes the same routing through the shop and processing times for each job operation on each machine are known. You believe that you should use the same sequence of jobs on each machine. How should you sequence the jobs on the first machine in the routing to ensure that all the jobs are done as quickly as possible?

Today, your shop has five working machines and four workers. Each worker has a different set of skills for doing required operations. Each job that must be completed during the day has a set of required operations. Each operation has a set of machines that can be used for completion, and only trained operators can complete operations successfully. How do you assign operations and machines to workers to ensure that all jobs are completed during the day, no worker is overworked, and no machine must run above capacity?

The problems described above are difficult optimization problems, and this is typical of the large majority of multiple machine scheduling problems. To gain a more thorough understanding of how difficult these problems are, one must study elements of computational complexity and integer programming, both of which are beyond the scope of this text (there are excellent references on these subjects, and many are listed in the chapter references). Here, the approach will be to solve simple problems to optimality and suggest heuristic strategies for complex problems. Mathematical programming formulations will often be used to formulate models and used as a basis for heuristic solution.

To simplify the presentation and save repetition, the problems considered in this chapter have the following basic characteristics unless otherwise noted:

All data are assumed to be deterministic.

The material handling system is not considered a critical resource, and travel times from machine to machine are not considered.

Each machine can process only one job/operation at any one time and each job/operation can be processed on only one machine at a time.

Setup times for jobs/operations are known values, and do not depend on the sequence of the jobs/operations.

Once an operation starts on a machine, the operation runs until completion. Preemption—stopping an operation before completion and putting another operation on the machine, is not permitted.

11.1   SCHEDULING SYSTEM REQUIREMENTS, GOALS AND MEASURES OF PERFORMANCE

The key requirement of any scheduling system is that it must help company management make decisions about the short-term scheduling of operations, machines, and labor. The system must often be used in real time and hence must be linked to a real-time database that accurately describes the state of the production system. The system must be easy to use, generate feasible schedules, and consider objectives that include customer as well as company concerns. A schematic of such a system is depicted in Figure 11.1.

A large part of the system is the database containing information about jobs and operations in progress, near-term jobs that must be started to meet due dates, employees' skills and availabilitie's, machine capabilities, and tooling capabilities. These data are then used in a decision-making process to arrive at a schedule for the time period under consideration.

The system can be used in many situations. For example, at the start of each day, the manager can run the system and determine the work for the day. Also, if a machine fails, then the manager can re-run the system and see how the schedule should be changed to help alleviate the problem. If a new job arrives in the shop because of an expedited order or rush order from a customer, the manager can run the system and determine the impact of the new job on the performance of the existing jobs. The system can be used to determine if there are sufficient resources to complete all jobs and to help the company predict short-term delivery dates for customers. As can clearly be seen, a well-designed scheduling system helps manage operations and provides critical data for running and gauging the effectiveness of the production operations.

The purpose of this chapter is to present decision-making methods for scheduling operations on machines (see the upper right block in Figure 11.1). At this point, it is necessary to become specific as to the types of systems considered, the terminology and notation used, and the objectives of the system.

Assume that the system is to schedule a set of N **jobs** (indexed $j = 1 \ldots N$) on a set of M **machines** (indexed $i = 1 \ldots M$). Each job has a **routing**, an ordered list of the machines that the job visits. An operation is performed when each job visits each machine. A job may visit a machine more than once (these are **termed "re-entrant shops"**); however, we will concentrate on cases in which each job visits each machine once, at most. Let P_j be the set of operations in the routing for job j, t_{kj} is the processing time for operation k of job j. For each job, we define d_j as the due date of job j and r_j as the release

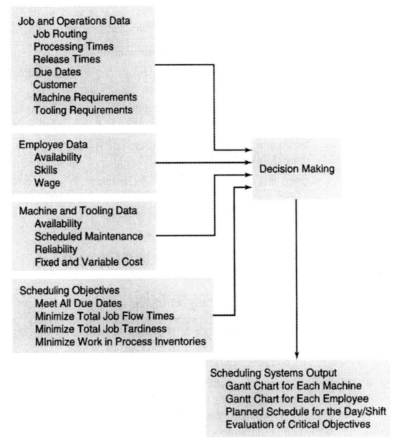

Figure 11.1 Scheduling System Schematic

date for job j (date at which material for the start of processing of job j is first available, and hence we can begin processing of job j on or after r_j). When all jobs have the same routing and visit each machine only one time, the system is termed a **"flow shop,"** and when jobs have different routings, the system is termed a **"job shop."** Generalizations of the flow shop include the **"flexible flow shop,"** in which each job takes the same route through the shop; however, there can be multiple machines of the same type and only one is used by a job, and the **"re-entrant flow shop,"** in which jobs may visit each machine multiple times. Both systems will be addressed in Chapter 12. Figure 11.2 depicts the different types of systems.

Routing sheets for two jobs and the associated data are given in Figure 11.3. Note that the jobs follow different routings, have different numbers of operations, require different times on each machine, and are really two separate entities in the system. The key in scheduling is how to allocate the resources that these two entities must share—the machines and operators.

As operations are completed, the job gets closer to completion. Although the individual operation times are important, the key completion time for the customer is when the final operation is complete thus the job is complete. Define C_j as the completion date for job j. Once the completion date is set, one can easily compute measures of performance for the job. Define the **job's lateness as $L_j = C_j - d_j$**. Note that lateness can be positive or negative under this definition. Because negative lateness (early) has a different

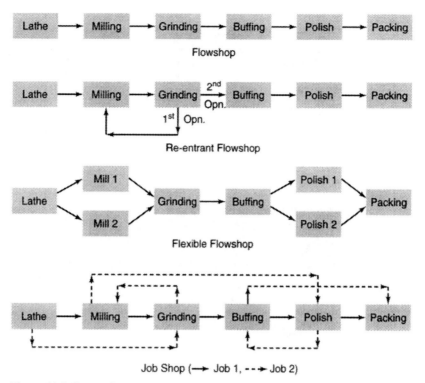

Figure 11.2 System Layouts

cost than positive lateness, the **job's tardiness is defined by** $T_j = max(0, C_j - d_j)$. For a given schedule for a particular time horizon, **NT denotes number of tardy jobs.** The time a job spends in the system, the **job's flow time** F_j, is the difference between its completion and its earliest possible start date or $F_j = C_j - r_j$. By Little's Law, flow time will relate directly to work-in-process inventory.

Job lateness, tardiness, flow time, completion time, and *NT* are termed **"regular measures."** They are regular in the sense that they cannot be improved by intentionally inserting idle time into the schedule. Inserted idle time occurs when a job operation is intentionally delayed although it is feasible to begin processing (the machine is available and all previous operations in the routing have been completed). The structure of regular measures is critical in developing strategies for finding optimal solutions because only the sequence of the jobs on the machines is important. The idea of time tabling, or determining operation starting times, is not critical because each operation will start as early as possible depending on the operation and job sequence.

After computing job performance measures, system objectives are easily defined. For example, one goal of the system is to find the schedule that minimizes **total system lateness.** This is simply:

$$\text{Total system lateness} = \sum_{j=1}^{N} L_j \qquad (11.1)$$

We could similarly use tardiness, flow time, or even simply completion in the above relation and **minimize total system tardiness, flow time, or completion time.** Equivalent criteria would be to optimize **average system lateness, tardiness, flow time, or com-**

Routing Sheet for Job 01101

Job 01101 Smith Mfg Released 5/25/99 Due 6/21/99	Routing - Operations					
Operation	1	2	3	4	5	6
Machine	Lathe	Mill	Polishing	Heat-Treat	Inspect	Package
Time (min)	10	4	21	45	8	5

Routing Sheet for Job 01102

Job 01102 Howard Mfg Released 5/23/99 Due 6/22/99	Routing - Operations				
Operation	1	2	3	4	5
Machine	Lathe	Mill	Buffing	Test	Package
Time (min)	12	6	30	12	3

Figure 11.3

pletion because the optimal solution does not change when the objective is scaled by a constant (the average value is simply the total value divided by N, the number of jobs).

The total lateness criteria will trade off poor performance on a single job against the benefit for the entire system. This approach has the problem that a small subset of jobs can be extremely late although the total lateness is as small as possible. Therefore, another goal of the system is to balance performance across all jobs and find a schedule that **minimizes the maximum lateness (or tardiness, or flow time)** over all jobs:

$$\text{Maximum lateness} = \underset{j=1}{\overset{N}{Max}} L_j \qquad (11.2)$$

Another goal of the system is to find the schedule that minimizes the number of tardy jobs—**NT**. This objective is particularly relevant when the company has many jobs with similar cost and revenue structures, and the revenue for each job depends strongly on meeting the due date. This objective has a problem similar to the total tardiness criteria in that once a job is tardy, there is no incentive to expedite the job, and it therefore yields schedules with a few jobs that are extremely late.

When the system has multiple jobs and multiple machines, a typical goal of the system is to get all work done as fast as is possible. The system goal is to minimize this total time and is termed the **makespan.** For any schedule, the makespan is easily evaluated by computing the largest completion time over all jobs:

$$\text{Makespan} = \underset{j=1}{\overset{N}{Max}} C_j. \qquad (11.3)$$

It has been implicitly assumed that all of the jobs have equal weight for the company. This is not generally valid because each job has a different cost and revenue structure, and jobs may come from different customers—some of which may be more important than others. An easy adjustment to the above criteria is made by adding in multiplicative weights for each job. Define w_j as the **importance weight of job j** (if $w_k > w_1$, then job k is more important than job l). Then the **weighted lateness of job j is defined as:**

$$\textbf{Weighted lateness} = w_j\, L_j \qquad (11.4)$$

This idea can easily be extended to weighted job tardiness, weighted job flow-time, and weighted completion. The system criteria using these new job performance measures are direct extensions of the unweighted criteria (traditionally, NT is not extended to the weighted case; however, this could be done using integer programming approaches). A summary of the job and system criteria is given in Table 11.1.

Depicting a schedule is often done through the use of a Gantt chart. Each row in the chart represents a resource in the system (machine, operator, or tooling, for example). The rows consist of blocks representing job operation processing. The length of each block is proportional to the length of the processing time, and time is the dimension of the horizontal axis. Often, the blocks are labeled with only the job number and it is often convenient to note the key operation and job completion times on the chart.

Figure 11.4 contains a routing sheet and a Gantt chart for one possible schedule for a five-machine, five-job example in which all jobs are released now. The schedule depicted is feasible because:

Each machine has only one job active at any one time.

Each job is processed on one machine, at most, at any one time.

Job operations are processed according to their routing sequence.

Additional key points concerning the chart include:

Idle time may be necessary because an operation may not be started until the predecessor operation in the routing is completed (job 3 on machine 3, for example)

Table 11.1   Job and System Criteria

Job performance	Minimize total performance	Minimize weighted performance	Minimize max performance	Minimize max weighted performance
Lateness	$\sum_{j=1}^{N} L_j$	$\sum_{j=1}^{N} w_j L_j$	$\underset{j=1}{\overset{N}{Max}}\, L_j$	$\underset{j=1}{\overset{N}{Max}}\, w_j L_j$
$L_j = C_j - d_j$				
Tardiness	$\sum_{j=1}^{N} T_j$	$\sum_{j=1}^{N} w_j T_j$	$\underset{j=1}{\overset{N}{Max}}\, T_j$	$\underset{j=1}{\overset{N}{Max}}\, w_j T_j$
$T_j = max\,(0,\ C_j - d_j)$				
Flow time	$\sum_{j=1}^{N} F_j$	$\sum_{j=1}^{N} w_j F_j$	$\underset{j=1}{\overset{N}{Max}}\, F_j$	$\underset{j=1}{\overset{N}{Max}}\, w_j F_j$
$F_j = C_j - r_j$				
Tardy or not	NT	N/A	N/A	N/A
Completion	$\sum_{j=1}^{N} C_j$	$\sum_{j=1}^{N} w_j C_j$	$\underset{j=1}{\overset{N}{Max}}\, C_j$	N/A
C_j			Makespan	

N/A = Not Applicable

Job		Operation 1		Operation 2		Operation 3		Operation 4		Operation 5	
Number	Due	Machine	Time	Machine	Time	Machine	Time	Machine	Time	Machine	Time
1	17	1	2	2	4	4	3	5	3	NA	NA
2	18	1	4	3	2	2	6	4	2	5	3
3	19	2	1	5	4	1	3	3	4	2	2
4	17	2	4	4	2	1	2	3	5	NA	NA
5	20	4	5	5	3	1	7	NA	NA	NA	NA

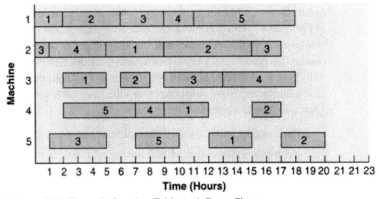

Figure 11.4 Example Routing Table and Gantt Chart

The criteria for the schedule are easily computed:

> makespan is 20 hours because job 2 is completed last, and it completes at time 20;
>
> job completion times for the schedule—job 1 completes at time 15, job 2 at time 20, job 3 at time 17, job 4 at time 18, and job 5 at time 18;
>
> job tardiness—jobs 1, 3, and 5 are not tardy, job 2 completes at time 20 and is 2 days tardy, and job 4 completes at time 18 and is 1 day tardy, and
>
> *NT* is 2.

There are many excellent reference texts devoted entirely to job sequencing and scheduling. For example, Conway, Maxwell, and Miller [1967], Baker [1974], French [1982], Pinedo [1995], and Sule [1997] all provide in-depth coverage of some of the topics covered in the remainder of this chapter. In addition, professional journals such as the *International Journal of Production Research, IIE Transactions, Operations Research, Management Science,* and the *European Journal of Operational Research* contain significant material on the latest results in scheduling research.

11.2 ORDER RELEASE STRATEGIES

The section considers the general problem of determining the set of jobs and operations that should be released to the shop during a standard time period (shift, day, week...). Generally, a company has many orders available to the shop, each has a particular due date and each requires different resources. Management must determine which jobs to start, or equivalently the job release dates, so that materials and parts can be acquired in a timely manner, machines are not over capacitated, employees can perform the required tasks, and jobs can be completed in a timely manner relative to their due dates.

This section considers two basic approaches, dispatching rules and mathematical programming. Dispatching rules are simple decision rules that consider the urgency of each job relative to the available resources. They build a set of jobs to release to the shop in a sequential manner adding jobs based on simple decision rules and simple capacity computations. Mathematical programming approaches consider the entire system simultaneously by optimizing a single criterion subject to constraints on machine and labor capacity.

11.2.1   Dispatching Approaches

Assume that the planning horizon consists of a set D of working days, and the production planner must decide on the day to release each job to the system. To simplify the discussion, assume that the m machine times are the only critical resources, and each machine i has U_{it} hours available during day t. Define the set of jobs to be released in day t as R_t. When deciding the jobs to process during a day, the following job and shop characteristics all impact the decision:

Job due date - d_j

Job importance - w_j

Individual operation processing times for each job on each machine (measured in days) - t_{ij}

Processing time available on each machine on each day - U_{it}

The general format for dispatching rules is as follows:

Step 1 (Initialization): Initialize each R_t to the empty set and compute U_{it} for each machine in each day (remember to consider jobs already released to the shop)

Step 2 (Selection of job): If no jobs remain to be scheduled, then stop with the current R_t sets. Otherwise, locate the job, $j*$, with the highest priority index.

Step 3 (Scheduling of job): Schedule $j*$ as early as possible so that no machine capacity constraints are violated. If $j*$ can be scheduled, add $j*$ to the appropriate R_t set and adjust the U_{it} values to account for $j*$ being in the schedule starting on day t. Eliminate job $j*$ from the list of jobs to schedule. Return to step 2.

The general format requires a specification of the priority index criterion used in step 2 and many simple rules are readily available from the basic data set including:

Earliest Due Date: The job with the closest due date has the highest priority index.

Shortest Processing Time: The job with the smallest total processing time, $\sum\limits_{i=1}^{M} t_{ij}$, has the highest priority index.

Shortest Weighted Processing Time: The job with the smallest ratio of total processing time to job importance, $\sum\limits_{i=1}^{M} t_{ij}/w_j$, has the highest priority index.

VIP: The job with the largest importance factor w_j has the highest priority index.

Minimum Slack Time: The job with the smallest slack time, $d_j - \sum_{i=1}^{M} t_{ij} - current$ $time$, has the highest priority index (note that if slack is negative, then the job must be late).

Minimum Percentage Slack Time: The job with the smallest percentage slack time, $\left(d_j - \sum_{j=1}^{N} t_{ij} - current\ time\right)/\sum_{i=1}^{M} t_{ij}$, has the highest priority index.

Critical Ratio: The job with the largest critical ratio value has the highest priority.

Critical ratio is

$$\left(\sum_{i=1}^{M} t_{ij}\right)/(d_j - current\ time),$$

If the job is late, then it has the highest critical ratio by caveat.

EXAMPLE 11.1

For the following data set with five jobs and four machines, use the VIP method priority generator to construct the order releases using three batches. The U_{it} values represent availability of machine hours for each batch.

Job	w_j	d_j	t_{1j}	t_{2j}	t_{3j}	t_{4j}
1	1	1	2	4	2	5
2	2	1	2	3	6	7
3	1	2	3	5	4	4
4	3	2	3	5	4	8
5	2	3	4	2	6	8
U_{it}	—	—	8	10	8	12

SOLUTION

First, consider the VIP selection rule. Breaking ties arbitrarily, the jobs should be considered in the sequence of highest to lowest w_j values:

$$(4, 2, 5, 1, 3)$$

Job 4 can be placed in R_1 because there is available time on each machine, and it is first in the list. Adjust U_{i1} for all i by subtracting t_{i4} from each current U_{i1} value. The resulting values are:

$$U_{i1} = (8 - 3, 10 - 5, 8 - 4, 12 - 8) = (5, 5, 4, 4)$$

Now, consider job 2. It cannot be placed in R_1 because there is insufficient time available on machines 3 and 4 for this job. Therefore, it is placed in R_2, and U_{i2} must be adjusted to all i.

$$U_{i2} = (8 - 2, 10 - 3, 8 - 6, 12 - 7) = (6, 7, 2, 5)$$

Now, consider job 3. Its processing will fit in the remaining capacity in batch 1, thus it is included in R_1. The remaining capacity for batch 1 is:

$$U_{i1} = (5 - 3, 5 - 5, 4 - 4, 4 - 4) = (2, 0, 0, 0)$$

All jobs use all machines, therefore, batch 1 cannot be used further because there is no remaining processing time on machines 2, 3, and 4. Continuing on for jobs 1 and 3, the results are listed in the following Table:

Batch	R_t	U_{1t}	U_{2t}	U_{3t}	U_{4t}
1	{4, 3}	2	0	0	0
2	{2, 1}	4	3	0	0
3	{3}	5	5	4	8

Different selection index rules could lead to different R_t sets. Note that there is still time remaining in the 3rd batch, so additional jobs could be allocated.

The selection index used should make job selections that are in-line with the company's criteria for scheduling. If due dates are critical, then critical ratio, slack time or early due date methods should be used. If low inventory is important, then processing time criteria should be used. If customer priorities are critical, then VIP or weighted processing time criteria should be used.

Once the selection rule is chosen and the R_t sets are computed, then it is a simple matter to set the release dates for all jobs. These release dates are then used in developing detailed schedules for each day.

11.2.2   Mathematical Programming Approaches

The problem with dispatching strategies is that they are myopic, that is, they make decisions that are based on past decisions and do not take future decisions into consideration. Scheduling a particular job for release today may preclude many other jobs from being released. Therefore, in step 2 of the dispatching process, great care should be taken when making selections. One alternative to using sequential myopic decisions is to consider and solve a model of the entire system at one time. Integer programming models and solution techniques can be used to effectively assign release dates.

The first model considers only the selection of the N jobs that should be scheduled over the time horizon of T periods. The model has multiple time periods, and it is assumed that all operations of the job are completed in the same time period. Each job has an importance factor that must be set externally and drives the decision-making process. Define the following notation:

$\delta_{jt} = 1$ if job j is released in period t, 0 otherwise

$w_{jt} = $ importance weight of job j when released in period t

$t_{ij} = $ time required on machine i for job j

$U_{it} = $ time available in period t on machine i

The model (R1) to select the release times for the jobs is:

$$\text{Maximize} \quad \sum_{j=1}^{N} \sum_{t=1}^{T} w_{jt} \delta_{jt} \tag{11.5}$$

$$\text{Subject to:} \quad \sum_{t=1}^{T} \delta_{jt} \leq 1 \qquad \text{for } j = 1 \dots N \tag{11.6}$$

$$\sum_{j=1}^{N} \delta_{jt} t_{ij} \leq U_{it} \qquad \text{for } i = 1 \dots M, t = 1 \dots T \tag{11.7}$$

$$\delta_{jt} \in \{0, 1\} \qquad \text{for } j = 1 \dots N, t = 1 \dots T \tag{11.8}$$

[Equation (11.5)] represents the objective of maximizing the system importance of selected jobs. Constraint set [Equation (11.6)] ensures that a job is released in one period at most. If $\delta_{jt} = 0$ for all $t = 1 \dots T$, then the job is not released in the study horizon. Constraint set [Equation (11.7)] ensures that no machine is overloaded during the horizon and Constraint set [Equation (11.8)] ensures that the decision variables take on only allowable values.

Model R1 can be refined down to the operations level, and operations for a job may be scheduled in different time periods. Each operation is linked to a specific machine. The decision variable notation changes slightly, and constraints must be added to ensure that the routing structure holds (implicitly assumed to be serial structure). Define the following new notation:

$\delta_{kjt} = 1$ if operation k of job j is done in period t, 0 otherwise

$\delta_j = 1$ job j is released to the shop, 0 otherwise

$t_{kj} = $ time required for operation k of job j

$L_j = $ index value of the last operation in the routing of job j

$N_i = $ set of (job, operation) pairs (j, k) that use machine i

The enhanced selection model (R2) to select the release times for the jobs and processing times for operations is:

$$\text{Maximize} \quad \sum_{j=1}^{N} \sum_{t=1}^{T} w_{jt} \delta_{L_j t} \tag{11.9}$$

$$\text{Subject to:} \quad \sum_{t=1}^{T} \delta_{kjt} = \delta_j \qquad \text{for } k = 1 \dots L_j, j = 1 \dots N \tag{11.10}$$

$$\sum_{t=1}^{T} t\, \delta_{kjt} \leq \sum_{t=1}^{T} t\, \delta_{ljt} \qquad \text{for any operation } k \text{ before } l \tag{11.11}$$

$$\text{in the routing for job } j = 1 \dots N$$

$$\sum_{(j,k) \in N_i} \delta_{kjt} t_{kj} \leq U_{it} \qquad \text{for } i = 1 \dots M, t = 1 \dots T \tag{11.12}$$

$$\delta_{kjt} \in \{0, 1\}\ \delta_j \in \{0, 1\} \quad \text{for } j = 1 \dots N, t = 1 \dots T \tag{11.13}$$

Equation (11.9) represents the objective of maximizing the system importance of selected jobs and is based on when the last operation for the job is scheduled. Constraint set [Equation (11.10)] ensures that each operation for each job is done in one period at most only if the job is released during the horizon. Constraint set [Equation (11.11)] ensures that operations early in the routing must be scheduled simultaneously or before operations later in the routing of each job. Note, that the functions used in [Equation (11.11)] simply transform the decision indicators into the time periods for which they engage. Constraint set [Equation (11.12)] ensures that no machine is overloaded during the horizon and Constraint set [Equation (11.13)] ensures that the decision variables take on only allowable values.

Solutions of models R1 and R2 can be found using integer programming techniques such as branch and bound or heuristics (such as dispatching rules) can be formulated. Model R2 has significantly more decisions than model R1, and hence it generally will be more difficult to solve optimally. R2 also has significantly more constraints because of the routing constraints.

As an alternative to loading specific resources, aggregate measures of WIP can be used to guide order releases. Strategies of this type were addressed in Chapter 7.

11.3 BOTTLENECK SCHEDULING

Once the release times of the jobs are known, detailed scheduling can begin. The difficulty of optimally scheduling multiple operations on multiple machines cannot be understated, hence schedulers generally concentrate their efforts on carefully scheduling only critical processes in the system. These critical processes are called **"bottlenecks"** or **"capacity constrained resources (CCRs)."** Once the CCR is scheduled, then the other machines can be scheduled to effectively feed material and WIP to the CCR and use WIP from the CCR in later operations. This methodology is generally given the name **"Theory of Constraints"** and is addressed in the next subsection.

11.3.1 The Theory of Constraints (TOC)

TOC is a management philosophy that helps firms become and stay profitable by maximizing the production throughput of the factory, and minimizing all relevant costs such as inventory cost, direct costs, overhead costs, and capital costs. The key idea in the philosophy is that there are key areas, CCRs, within the factory that limit the output of the entire facility. If you can manage these CCRs well, then you can maximize output and determine where to invest future capital dollars.

The basis for TOC comes from the shop floor control system OPT and is explained in a series of books by Eliahu Goldratt and his associates: The Goal [1987], The Race [1986], The Haystack Syndrome [1990], and The Theory of Constraints [1990]. OPT can be characterized by nine rules that guide management of the factory:

1. Balance flow in the factory, not capacity.
2. Constraints determine the utilization of the nonCCR processes. Maximizing utilization of nonCCR processes only adds to costs, idle time, and inventory.
3. Activity is not equal to utilization. Don't waste time producing if the output cannot fit within the schedule at the CCR. You are only creating inventory.
4. An hour lost at the CCR is an hour lost in the entire factory.
5. An hour saved at a nonCCR is not immediately important.
6. The CCRs govern throughput and inventory. The production rate of the CCR determines the requirements of processes that feed material to the CCR and the maximum amount of material available to processes after the CCR.
7. The transfer batch to the next process should not always equal the process batch. It may be advantageous to throughput time to move material in smaller batches.
8. Process batches should be variable, not fixed. The process batch size depends on the state of the system and current estimates of costs.
9. The system and the CCR should be scheduled by examining all constraints simultaneously and not by considering only one constraint.

A good product is one that has a high profit contribution and uses little of the CCR resources [Bakke and Hellberg (1991)].

Whereas MRP may assume infinite production capacity, TOC creates capacitated schedules only for the CCR operations. Upstream and downstream operations are scheduled based on the schedule at the CCR. In this way, TOC simplifies system scheduling by decomposing the problem into smaller problems, some of which are easy (nonCCR processes) and some of which are harder (CCR processes).

MRP assumes that all "timing" data (setup times, processing times, handling times, lead times) are deterministic, while TOC makes no such assumptions. The idea of TOC is similar to the PERT and CPM project management techniques in that management must find the critical path through the factory and then manage resources on the critical path. Lead times are a result of the TOC schedule and hence are variable and cannot be predicted a priori. To start the process of scheduling, TOC sets lead times to three times operation processing time; however, TOC uses lot splitting (see Section 12.3.2 for details) to increase overlapping operations. TOC expedites processing or material orders by using overtime and express shipping when two thirds of the lead time has passed.

TOC is designed so that order batches can be split on the production floor. The advantage of splitting batches is that material can be moved quicker and processing expedited. The drawback is that setup time may be increased because of more setup processes. Therefore, TOC splits batches only at the nonCCR processes because these processes can afford additional setup time. Batches at the CCR are not split and actually may be combined to save setup time and increase machine availability.

Inventory in TOC systems is designed to ensure that the CCR is never starved and never blocked. This is contrasted to MRP or JIT in which both have inventory throughout the plant. Because setup cost is often traded against inventory cost, TOC is different than JIT in that TOC does not seek to reduce setup cost everywhere—only at the CCR. Setup reduction is a goal when such reduction increases throughput at the CCR.

11.3.2 Identification of Bottlenecks

Simple checks can help identify bottlenecks or CCRs in the system. One can walk through the factory and see the CCR because inventory often accumulates in front of this operation. Another method is to take the routings, demand, and operation processing times and compute machine utilization. This method works when setup time is negligible and variability is low.

EXAMPLE 11.2

Consider a four-machine shop with three jobs. Routing, processing time, and demand data are given below:

Job	Demand	Routing with machine processing times in minutes				
1	300	M1–3	M2–4	M3–5	M4–1	
2	200	M2–2	M3–1	M4–4	M1–2	M4–1
3	400	M3–2	M4–1	M2–2		

If the time available on each machine is 3,000 minutes, which machine is the bottleneck?

SOLUTION

First, the total time required on each machine must be computed. We obtain:

$$M1 \text{ time} = 300 * 3 + 200 * 2 = 1300 \text{ minutes}$$
$$M2 \text{ time} = 300 * 4 + 200 * 2 + 400 * 2 = 2400 \text{ minutes}$$
$$M3 \text{ time} = 300 * 5 + 200 * 1 + 400 * 2 = 2500 \text{ minutes}$$
$$M4 \text{ time} = 300 * 1 + 200 * 4 + 200 * 1 + 400 * 1 = 1700 \text{ minutes}$$

Because M3 has the most required time, and each machine has the same availability, then it will be the bottleneck. Note, that if demand changes slightly, then M2 may become the bottleneck.

When setups are relevant, then they must be factored into the computation of workload. Batch sizes or equivalently the number of setups must be estimated. TOC assumes that the CCRs remain stable, thus if the demand is fluctuating or the labor and processing times are highly variable, then it is difficult to determine a CCR. Often a company will take some percentage of their highest loaded processes and designate them as CCRs. Ronen and Starr [1990] report using the highest 20% of the processes.

11.3.3   Forward and Backward Scheduling

In TOC, one first schedules the CCR, and then uses this schedule to set the schedules for nonCCR processes that feed the CCR and take material from the CCR. In this backward and forward manner, a schedule for the entire system is generated. The schedule at the CCR can be looked at similarly to the master production schedule in MRP. The key difference is that in MRP, demand drives the schedule; however, in TOC the CCR drives the schedule, and its capacity must be set to ensure demand can be met in a timely manner.

To create the schedule at the CCR, one can use as input the current MRP schedule for the CCR operation and add in additional jobs that must be performed (it may be necessary to add overtime to complete these additional orders). In TOC systems, **three buffers** are created to guard against variability: the first buffer is at the end product area and guards the customer against production shortages; the second buffer is at the raw material area and guards the company against raw material and purchased part shortages; and the third buffer is a time buffer before the CCR area. For the time buffer, the planning lead times to the CCR area are increased beyond that which is expected to ensure that there is a high probability that the CCR area is not starved for parts. A workable heuristic for buffer setting is to have inventory equal to three times the lead-time demand (using an estimate of lead time for product, materials, and time to get WIP from the start to the CCR). For instance, if the expected flow time is four hours from release to the buffer, we may keep 12 hours of work in that pipeline. Another rule of thumb is to adjust buffers until 90% of the parts can be processed without the use of any expediting measures.

The word **"rope"** is used to represent the time it takes material to get from the first operation to the CCR, when only processing time is considered. The rope is generally an underestimate of actual lead time to the CCR because material handling time and other delays may be in the system. (This is one of the reasons why the planning lead time for the CCR was inflated). Given the CCR schedule, one moves backward and schedules operations along the rope so that WIP arrives at the CCR in time for processing. It will always look like the first operation is late because of the underestimate of lead time, therefore, operations along the rope always use the scheduling rule to work whenever WIP is available for processing. Similarly to JIT, TOC is a pull system upstream from the CCR.

The word **"drum"** is used to represent the pace of the CCR. If the market demand is the system constraint (your plant can easily meet demand on time and no process is tightly constrained), then the demand rate is the drum. When the demand rate increases or capacity is reduced, then one of the processes becomes the drum and sends WIP to processes after the CCR. The drum determines the schedule for processes after the CCR and these are easily scheduled given the rate, or beat, of the drum. The schedule for these operations is to simply produce whenever you have work. In contrast to JIT, TOC is a push system downstream from the CCR.

The following simple example shows how the drum, buffer, and rope work to control the system.

EXAMPLE 11.3

Consider a four-machine series system with a single product that visits each machine once. The CCR for the system is machine 3, and the batch processing times for all operations are detailed below:

M1 time	M2 time	M3 time	M4 time
2 periods	1 period	3 periods	2 periods

The CCR, M3, release times have been scheduled, and the following Table lists the release dates of the batches:

Release Periods for M3	5	8	11	14

Develop a plan for the nonCCR machines in the system. Assume that the planned releases include the time buffer at the CCR.

SOLUTION

The schedule for M3 is given, thus we need to back-schedule the batches for M2 first. Because the processing time is one period, we arrive at the following release periods:

Release Periods for M2	4	7	10	12

Now, use the release periods for M2 with the processing time for M1 to arrive at the release times for M1:

Release Periods for M1	2	5	9	10

Now, we forward-schedule M4 using the M3 release dates and the processing time for M3:

Release Periods for M4	8	11	14	17

and this yields the following output for the system:

Completion Periods for M4	10	13	16	19

In work by Newman and Sridharan [1992], 185 companies were surveyed and 5% were using OPT- or TOC-based production systems. Of these, half had less than spectacular results and these occurred in systems with significant variability in job routing (job shops). The process industries performed best, and this is not surprising because these have a stable flow through the system.

The next three sections of this chapter address specific methods to schedule the CCR. Different structures are considered such as when the CCR is a single machine, when the CCR is a flow line, and when the CCR is a job shop. Different criteria are also considered. Algorithms, models, and heuristics to help lend insight into more complex problems are explained and demonstrated.

11.4 SINGLE-MACHINE SCHEDULING

At this point, consider systems consisting of a single machine or CCR processes that are a single machine. Assume that there are N jobs pending processing, and each job must be processed once on the machine. To simplify notation, let t_j be the processing (plus setup) time of the job. Recall that r_j is the job release time, w_j is the job importance, and d_j is the job due date. The goal is to find a sequence of the jobs that minimizes a particular objective in which no job starts before its release time r_j.

The section is organized around the two basic categories of objectives, flow time related criteria and lateness/tardiness related criteria. In each subsection, we first consider the **static shop** case where $r_j = 0$ for all j (all jobs are released at time 0), and then the **dynamic shop** where $r_j \neq 0$ for some j values (jobs arrive to the system over time). When considering the dynamic shop, we assume that all jobs can be processed with a **"preempt resume"** mode of operation. That is, any time spent on the job before preemption counts toward the completion of the job. The alternative case of **preempt repeat** usually requires integer programming approaches and can be computationally difficult when the number of jobs is large.

11.4.1   Flow Time Related Criteria

Consider, first, the case where $w_j = 1$ for all j. Then, the criterion to minimize total flow time is:

$$\text{Minimize} \sum_{j=1}^{N} F_j. \tag{11.15}$$

Using the definition of flow time for a job $F_j = C_j - r_j$, the criterion is easily transformed to a criterion relating solely to completion times:

$$\sum_{j=1}^{N} F_j = \sum_{j=1}^{N} (C_j - r_j) = \sum_{j=1}^{N} C_j - \sum_{j=1}^{N} r_j \tag{11.16}$$

The term $\sum_{j=1}^{N} r_j$ is a constant and does not depend on the sequence of jobs. Therefore, a sequence that minimizes total flow time must also minimize the sum of all completion times of all jobs and vice versa.

Before stating the scheduling rule, it is important to develop intuition on the structure of the problem. The job that is done first, **denoted by the subscript [1]** ([j] always denotes the jth job in the sequence as opposed to the jth job in the data list) will be completed at time $t_{[1]}$ **and will delay the completion of all other jobs in the system.** The second job [2] will be completed at time $t_{[1]} + t_{[2]}$ and will delay all jobs in the system except itself and the first job. Continuing on, each job in the sequence delays only jobs that come after it in the sequence. This is depicted in Figure 11.5 in which the height of

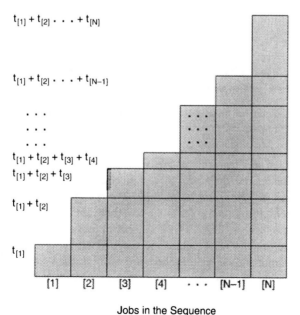

Jobs in the Sequence

Figure 11.5 Flow Time Delays

each column represents the completion time for the job in the sequence. To minimize the sum of the completion times, we must minimize the total area on Figure 11.5. It should be intuitive that we want to minimize the delays imposed on the jobs. Therefore, it seems that the best sequence would have the shortest processing time job first, the second shortest processing time second, and the longest processing time job last. This is indeed true and is stated in the following rule:

SPT Rule For the problem of minimizing system flow time under a single-machine static system in which all jobs are equally important, the optimal solution is to order the jobs so that:

$$t_{[1]} \leq t_{[2]} \leq t_{[3]} \leq \cdots \leq t_{[N]} \tag{11.17}$$

Ties can be broken arbitrarily.

This is called the **"SPT rule"** because the jobs are in shortest processing time order. The following example demonstrates an application of the rule.

EXAMPLE 11.4

Consider a static single-machine system with five jobs and processing times as follows (time in hours):

Job	1	2	3	4	5
t_j	10	12	8	15	12

How should the jobs be sequenced to minimize total flow time, and what is the optimal objective value?

SOLUTION

Following Equation (11.17), job 3 should be first because it has the shortest processing time of eight hours. Job 1 should be second. There is a tie for third between jobs 2 and 5. We can break ties arbitrarily thus, we place job 2 in the third position and job 5 in the fourth position. Job 4 has the longest processing time and therefore is placed fifth. The completion times are given in the following Table:

Sequence	[1]	[2]	[3]	[4]	[5]
Job	3	1	2	5	4
t_j	8	10	12	12	15
C_j	8	18	30	42	57

The total flow time for these five jobs is the sum of the completion times or 155.

Note that one can look at the rows in Figure 11.5 and consider system inventory. Whereas the columns represent flow time of each job, the rows represent number of jobs in the system between completion times. For example, from time 0 to $t_{[1]}$, there are N jobs in the system. From time $t_{[1]}$ to $t_{[1]} + t_{[2]}$, there are $N - 1$ jobs in the system. If you multiply the time interval length by the number of jobs in the system during the interval, then, this is a measure of time weighted inventory. Minimizing the area under the rows is equivalent to minimizing the time average inventory in the system. Therefore, SPT is the best approach for minimizing inventory—get short jobs out of the system quickly to minimize inventory.

A simple extension of the SPT rule can be made when the importance of jobs varies. Now, the criterion is:

$$\text{Minimize} \sum_{j=1}^{N} w_j F_j \qquad (11.18)$$

Similarly to the unweighted case, this function can be transformed to a function of completion times and constants. Intuition is that short jobs should be early in the sequence, and high weighted jobs (important jobs) should be early in the sequence. The ideas are summarized in the following rule:

WSPT Rule For the problem of minimizing total weighted flow time under a single-machine static system, the optimal solution is to order the jobs so that:

$$t_{[1]}/w_{[1]} \leq t_{[2]}/w_{[2]} \leq t_{[3]}/w_{[3]} \leq \cdots \leq t_{[N]}/w_{[N]} \qquad (11.19)$$

Ties can be broken arbitrarily.

This is called the **"WSPT rule"** because the jobs are in **w**eighted **s**hortest **p**rocessing **t**ime order in which each processing time is adjusted by the job importance. The ratio has the desired bias in that low processing times and/or high weights can make the ratio low. The following example demonstrates an application of the rule.

EXAMPLE 11.5

Consider a static single-machine system with five jobs and processing times and weights as follows (time in hours):

Job	1	2	3	4	5
t_j	10	12	8	15	12
w_j	2	1	1	2	3
t_j/w_j	5	12	8	7.5	4

How should the jobs be sequenced to minimize total weighted flow time, and what is the optimal objective value?

SOLUTION

Following Equation (11.19), job 5 should be first since it has the lowest ratio. Job 1 should be second, job 4 should be third, job 3 should be fourth, and therefore job 2 is placed fifth. The weighted completion times are given in the following Table:

Sequence	[1]	[2]	[3]	[4]	[5]
Job	5	1	4	3	2
t_j	12	10	15	8	12
C_j	12	22	37	45	57
$w_j * C_j$	36	44	74	45	57

The total weighted flow time time for these five jobs is the sum of the weighted completion times or 256. Note, that the optimal sequence in this case was in the order of the importance weights (VIP sequencing), but this is specific to this problem data and not optimal in general.

Both SPT and WSPT can be extended to dynamic systems as long as preempt resume processing holds. The basic idea is that the machine should always be busy processing the job with the shortest remaining processing time (equal weight case) or the smallest ratio of remaining processing time/job importance (unequal weight case). Each point in time when a job completes processing or when a new job arrives to the system, the schedule must be re-evaluated to ensure that the correct job is being processed. The following example illustrates the **DSPT (dynamic shortest processing time)** sequencing rule.

EXAMPLE 11.6

Consider a dynamic single-machine system with five jobs with equal weights. Processing times and release times are (time in hours):

Job	1 .	2	3	4	5
t_j	10	12	10	15	12
r_j	0	11	14	0	25

How should the jobs be sequenced to minimize total weighted flow time, and what is the optimal objective value?

SOLUTION

At time 0, jobs 1 and 4 are released to start. Because job 1 has the shortest processing time (10 versus 15), job 1 starts on the machine. The next event in the system is that job 1 completes at time 10. Now, job 4 is placed on the machine and begins processing. At hour 11, job 2 arrives. It has a processing time of 12 hours and job 4, currently on the machine, has 14 hours of processing remaining (start with 15 hours and one hour has elapsed since it was started). Because $12 < 14$, we preempt job 4 and place job 2 on the machine. The next event is at hour 14, and job 3 arrives. The remaining processing time of job 2 is nine hours and the processing time of job 3 is ten hours. Therefore, we keep processing job 2 ($9 < 10$). The next event is at hour 23, and job 2 completes processing. At this time, there are two jobs available: job 4 with remaining processing time of 14 hours and job 3 with a remaining processing time of ten hours. Because $10 < 14$, we place job 3 on the machine and start processing. Job 5 arrives at hour 25 signaling the next event. The remaining time of job 3 is 8 hours (started at hour 23, and it is now hour 25), and this is smaller than the processing time for job 5. Therefore, job 3 remains on the machine. Because all jobs have now arrived to the system, only job completion events can occur. Job 3 finishes at hour 33, and we choose job 5 next because its remaining processing time (ten hours) is less than that of job 4 (11 hours). Job 5 completes at hour 43, and we put job 4 back on the machine to complete its 11 remaining hours of processing. It completes at hour 54. A summary of the job completion times and the job flow times follows:

Job	1	2	3	4	5
r_j	0	11	14	0	25
C_j	10	23	33	54	43
F_j	10	12	19	54	18

The total flow time for the system is 113.

All of the rules considered for flow time when jobs have equal importance favor short jobs at the expense of longer jobs. This may result in one job remaining in the system for an unacceptable length of time. The final result in this subsection tries to remedy this drawback by considering the criteria:

$$\text{Minimize } \underset{j=1}{\overset{N}{\text{Max}}} \, F_j \qquad (11.20)$$

Note that this criterion is only interesting in the dynamic case because if $r_j = 0$ for all j, all schedules with no inserted idle time have the same value for $\underset{j=1}{\overset{N}{Max}} F_j$. The justification to the following rule is left as an exercise.

FCFS Rule For the problem of minimizing the maximum job flow time under a single-machine dynamic system in which all jobs are equally important, the optimal solution is to order the jobs so that:

$$r_{[1]} \leq r_{[2]} \leq r_{[3]} \leq \cdots \leq r_{[N]} \tag{11.21}$$

Ties can be broken arbitrarily.

The rule orders jobs by release time, or equivalently first come first served, and is the simplest of all scheduling rules. Its performance, however, is only optimal in general for the $\underset{j=1}{\overset{N}{Max}} F_j$ criterion and can lead to poor system flow time performance when a long job arrives early in the scheduling horizon.

11.4.2 Lateness and Tardiness Related Criteria

When considering lateness and tardiness criteria, one must consider the job due dates as well as the processing times. The first result shows that minimizing total system lateness:

$$\sum_{j=1}^{N} L_j \tag{11.22}$$

is a simple problem that we have already solved.

SPT Rule For the problem of minimizing system lateness under a single machine static system in which all jobs are equally important, the optimal solution is to order the jobs so that:

$$t_{[1]} \leq t_{[2]} \leq t_{[3]} \leq \cdots \leq t_{[N]} \tag{11.23}$$

Ties can be broken arbitrarily.

This result is identical to the SPT rule for minimizing system flow time and follows from a small bit of algebra:

$$\sum_{j=1}^{N} L_j = \sum_{j=1}^{N} (C_j - d_j) = \sum_{j=1}^{N} C_j - \sum_{j=1}^{N} d_j \tag{11.24}$$

Minimizing the system lateness is equivalent to minimizing the sum of completion times minus a constant (the sum of the due dates over all of the jobs is a constant, independent of the schedule). Because the SPT rule minimizes the sum of completion times, it must also minimize the system lateness.

The above argument will not work for the system tardiness because we cannot make an easy substitution for job tardiness because of the nonlinear "max" operator in the definition. The system tardiness problem cannot be solved by a simple sequencing rule and will be solved using mathematical programming and heuristic techniques at the end of this subsection.

For a static system with equal weight jobs, the criteria of:

$$\textit{Minimize } \underset{j=1}{\overset{N}{Max}} L_j \quad \text{or} \quad \textit{Minimize } \underset{j=1}{\overset{N}{Max}} T_j \tag{11.25}$$

can be solved with a simple scheduling rule. The basic idea is that no job should be too late, thus we must schedule jobs so that lateness is somewhat balanced across all jobs. Note, that the maximum lateness may be negative, whereas 0 bounds the maximum tardiness from below.

EDD Rule   For the problem of minimizing the maximum job lateness or tardiness under a single-machine static system in which all jobs are equally important, the optimal solution is to order the jobs so that:

$$d_{[1]} \leq d_{[2]} \leq d_{[3]} \leq \cdots \leq d_{[N]} \tag{11.26}$$

Ties can be broken arbitrarily.

The rule orders jobs by due date, or equivalently **early due date** and requires only simple sorting to implement. It is interesting that the approach does not depend in any way on the specific job processing times (however, the specific value of the optimal objective is directly connected to the specific values of the due dates and the processing times). A small example follows.

EXAMPLE 11.7

Consider a static single-machine system with five jobs and processing times and due dates as follows (time in hours):

Job	1	2	3	4	5
t_j	10	20	8	12	18
d_j	15	45	10	70	45

How should the jobs be sequenced to minimize maximum tardiness, and what is the optimal objective value?

SOLUTION

Following Equation (11.26), job 3 should be first because it has the earliest due date of 30 hours. Job 1 should be second. There is a tie for third between jobs 2 and 5. We can break ties arbitrarily thus we place job 2 in the third position and job 5 in the fourth position. Job 4 has the latest due date and therefore is placed fifth. The completion times and tardiness values are given in the following Table:

Sequence	[1]	[2]	[3]	[4]	[5]
Job	3	1	2	5	4
t_j	8	10	20	18	12
C_j	8	18	38	56	68
d_j	10	15	45	45	70
T_j	0	3	0	11	0

The maximum tardiness time for these five jobs is 11 hours and it occurs on job 5. Note, that if job 2 had been placed after job 5, then job 2 would have been 11 hours late.

The proof that the EDD rule is indeed optimal is instructive in the general approach of proving that a particular rule is optimal for a particular scheduling scenario. The method is called "adjacent pairwise interchange" and can be used to prove that the SPT-based rules solve the system flow time problem, and the **First-Come, First-Served (FCFS)** rule solves the maximum flow time problem. The method assumes that a sequence that does not follow the rule is optimal and then proceeds to set up a contradiction when two adjacent jobs in the sequence are interchanged to get "closer" to the rule. The proof uses the maximum tardiness criterion, but the maximum lateness criterion follows similarly.

Proof that EDD is Optimal Consider the following schedule S, represented by a Gantt chart that is assumed to be optimal but is not in EDD order.

[1]	[2]	[3]	...	[k]	[k + 1]	...	[N]

$$0 \quad\quad C_{[1]} \quad\quad C_{[2]} \quad\quad C_{[3]} \quad\quad C_{[k-1]} \quad\quad C_{[k]} \quad\quad C_{[k+1]} \quad\quad C_{[N-1]} \quad\quad C_{[N]}$$

Because S is not in EDD order, there must exist some pair of adjacent jobs in the schedule where $d_{[j]} > d_{[j+1]}$ ([j] is due after [j + 1] yet [j] is clearly before [j + 1] in the schedule). There may be many such pairs, thus assume that the pair earliest in the schedule occurs with jobs [k] and [k + 1] ($d_{[k]} > d_{[k+1]}$). Create a new schedule SI by interchanging jobs [k] and [k + 1] in S and keeping all other jobs in the same positions. The schedule SI appears as:

[1]	[2]	[3]	...	[k + 1]	[k]	...	[N]

$$0 \quad\quad C_{[1]} \quad\quad C_{[2]} \quad\quad C_{[3]} \quad\quad C_{[k-1]} \quad\quad C_{[k+1]} \quad\quad C_{[k]} \quad\quad C_{[N-1]} \quad\quad C_{[N]}$$

and is "closer" to EDD because it has one less pair of jobs that is out of EDD order.

It is immediate from the interchange selection that the completion times for all jobs except jobs [k] and [k + 1] are identical in both S and SI. Therefore, the lateness and tardiness values for these jobs are the same in both schedules. Consider, job [k + 1] in schedule S and job [k] in schedule SI. Because these jobs have the identical set of preceding jobs in both schedules, their completion times must be identical. Therefore,

$$C_{[k]} \text{ in schedule } SI = C_{[k+1]} \text{ in schedule } S \tag{11.27}$$

Because $d_{[k]} > d_{[k+1]}$,

$$T_{[k+1]} \text{ in schedule } S \geq T_{[k]} \text{ in schedule } SI \tag{11.28}$$

Also, because [k + 1] is now done before [k] in schedule *SI*,

$$C_{[k+1]} \text{ in schedule } S \geq C_{[k+1]} \text{ in schedule } SI \tag{11.29}$$

Therefore,

$$T_{[k+1]} \text{ in schedule } S \geq T_{[k+1]} \text{ in schedule } SI \tag{11.30}$$

Because [k] is done before [k + 1] in S and $d_{[k]} > d_{[k+1]}$,

$$T_{[k+1]} \text{ in schedule } S > T_{[k]} \text{ in schedule } S \tag{11.31}$$

Summarizing the above three statements on job tardiness, $T_{[k+1]}$ in schedule *S* is no smaller than both $T_{[k]}$ and $T_{[k+1]}$ in schedule *SI* and larger than $T_{[k]}$ in schedule *S*.

Recall that the criterion is to minimize the maximum tardiness over all jobs. It is possible that the job with maximum tardiness in *S* occurs in jobs [1] through [k − 1] or in jobs $[k + 2]$ through $[N]$. In this case, after the interchange, the maximum tardiness value does not change. If the maximum tardiness value in S occurs in job $[k + 1]$ (it cannot be in job $[k]$ based on the above arguments), then after the interchange, the tardiness values all stay the same, or are changed to values that are no worse than $T_{[k+1]}$ in S. Therefore, the maximum tardiness value in SI is no worse than the maximum tardiness value in S.

Therefore, after the interchange, the maximum tardiness value may decrease. This is a contradiction with the first statement that the nonEDD sequence S is optimal. If the maximum tardiness remains the same after the interchange, then continue with the interchange process until arriving at an EDD sequence (finite number of interchanges because N is finite). Either the EDD sequence will be better, setting up the contradiction, or the EDD sequence is also optimal, justifying the result and completing the proof.

The EDD rule can be extended to the case in which the jobs have different weights. The rule is similar to the extension of the SPT rule for this case.

WEDD Rule For the problem of minimizing maximum job lateness or tardiness under a single-machine static system, the optimal solution is to order the jobs so that:

$$d_{[1]}/w_{[1]} \le d_{[2]}/w_{[2]} \le d_{[3]}/w_{[3]} \le \cdots \le d_{[N]}/w_{[N]} \tag{11.32}$$

Ties can be broken arbitrarily.

This is called the **"WEDD rule"** because the jobs are in **w**eighted **e**arly **d**ue **d**ate order in which each due date is adjusted by the job importance. The ratio has the desired bias in that low due dates and/or high weights can make the ratio low.

Both EDD and WEDD can be extended to dynamic systems as long as preempt resume processing holds. The basic idea is that the machine should always be busy processing the job with the nearest due date (equal weight case) or the smallest ratio of due date/job importance (unequal weight case). For each point in time, when a job completes processing or when a new job arrives to the system, the schedule must be reevaluated to ensure that the correct job is being processed. It is a simple matter to compare the due date of an arriving job with the due date of the job on the machine to see if preemption should occur. Similar to the case of dynamic SPT, one must keep track of the remaining processing time of all jobs to ensure that exactly the required amount of time is spent in processing each job.

A clever algorithm attributable to Moore and Hodgson [1968] uses properties of EDD sequences to find the minimum value of NT, the number of tardy jobs. This criterion is especially useful in cases in which the shop is over capacitated, and we must decide the jobs that will be late to do as little harm as possible to the tardiness of the remaining jobs. Alternatively, a firm may promise to complete jobs on time, and wants to decide on the maximum number of jobs that can be accepted. Define two sets of jobs, E, the set of early jobs, and T, the set of tardy jobs. The algorithm schedules E first so that no jobs in E are tardy and then schedules T. The key is to set T so that it has minimum cardinality. The formal algorithm (*NT rule*) is:

Step 0 (Initialization): Set E to the set of all jobs and T to the empty set.

Step 1 (Check for termination): Order the jobs in E in EDD order. If no job in E is tardy, then stop the optimal schedule is to do the jobs in E in EDD order, and then do the jobs in T in any order. The optimal value of NT is the cardinality of T. Otherwise, go to Step 2.

Step 2 (Finding a tardy job): Starting with the first job in the sequence of E, compute the completion times of each job and find the first job in the sequence that is tardy. Assume that the first tardy job in the sequence is $[k]$. The sequence appears as:

[1]	[2]	[3]	...	[k]	[k + 1]	...	[|E|]

$$0 \qquad C_{[1]} \qquad C_{[2]} \qquad C_{[3]} \qquad C_{[k-1]} \qquad C_{[k]} \qquad C_{[k+1]} \qquad\qquad C_{[|E|]}$$

and $C_{[j]} \le d_{[j]}$ for $j = 1 \ldots k - 1$ and $C_{[k]} > d_{[k]}$. Go to Step 3.

Step 3 (Selecting the tardy job): Find the longest job in the subsequence [1] through [k]. Assume that this is job $[j]$. Delete $[j]$ from E and add it to T. Return to Step 1.

The validity of the *NT* rule hinges on properties of EDD sequences. In step 2, when we find a tardy job, it can be shown that at least one job in the subsequence [1] through [k] must be tardy. The longest job contributes the most to tardiness because it causes the largest delay in the subsequence [1] through [k] and any future processing after job [k].

EXAMPLE 11.8

You have six homework assignments to complete. Each assignment has a due date and an expected time to completion to thoroughly learn the material and achieve an A grade (the only grade that is acceptable to you at this point in your career).

Job	1	2	3	4	5	6
t_j	10	20	8	12	18	15
d_j	15	45	10	70	45	55

If an assignment is late, you receive no credit; however, you must still hand in the assignment. How should you sequence the assignments to maximize the credit that you receive?

SOLUTION

$E = \{1, 2, 3, 4, 5, 6\}$ and $T = \Phi$. The EDD sequence of E is:

$$(3, 1, 2, 5, 6, 4) \text{ (ties are broken arbitrarily)}$$

The completion times for these jobs are:

Job	3	1	2	5	6	4
t_j	8	10	20	18	15	12
d_j	10	15	45	45	55	70
C_j	8	18				
T_j	0	3				

Job 1 is tardy, thus one of jobs 3 and 1 must be tardy. We move the longer of 1 and 3 to the tardy set T, and that is job 1. Therefore, $E = \{2, 3, 4, 5, 6\}$ and $T = \{1\}$. The EDD sequence of E is:

$$(3, 2, 5, 6, 4) \text{ (ties are broken arbitrarily)}$$

The completion times for these jobs are:

Job	3	2	5	6	4
t_j	8	20	18	15	12
d_j	10	45	45	55	70
C_j	8	28	46		
T_j	0	0	1		

Job 5 is tardy, so one of jobs 3, 2, and 5 must be tardy. We move the longest of 3, 2, and 5 to the tardy set T, and that is job 2. Therefore, $E = \{3, 4, 5, 6\}$ and $T = \{2, 1\}$. The EDD sequence of E is:

$$(3, 5, 6, 4) \text{ (ties are broken arbitrarily)}$$

The completion times for these jobs are:

Job	3	5	6	4
t_j	8	18	15	12
d_j	10	45	55	70
C_j	8	26	41	53
T_j	0	0	0	0

All jobs in E are on time, thus the algorithm terminates with two jobs tardy. E should be done first and scheduled in EDD order, and T is done afterward.

The *NT* rule cannot be easily extended to the case of weighted jobs and/or nonzero release times, regardless of the preemption structure. These cases generally require integer programming or dynamic programming techniques to find optimal solutions for specific problem instances. Also, the techniques are often complicated by the nonlinear nature of the tardiness function.

At this point, the system total tardiness criterion has been ignored in our discussion. This is not because of the lack of importance of the criterion and indeed, it could be argued that this criterion is the most important of all criteria discussed. However, this criterion does not admit a simple sequencing rule. An integer programming formulation of the system tardiness problem concludes the subsection.

There are many approaches to formulating scheduling problems. The approach here uses decisions of job starting times and indicator variables for sequencing pairs of jobs. Define the following additional notation:

Decision variables:

$$s_j = \text{starting time for job } j$$

$$\delta_{ij} = \begin{cases} 1 & \text{if job } i \text{ is before job } j \\ 0 & \text{otherwise} \end{cases}$$

$$M = \text{large constant}$$

The model is (denoted model T):

$$\text{Minimize} \quad \sum_{j=1}^{N} w_j T_j \tag{11.33}$$

Subject to: $\qquad T_j \geq C_j - d_j \qquad\qquad$ for $j = 1 \dots N \qquad\qquad$ (11.34)

$$s_j \geq r_j \qquad\qquad\qquad \text{for } j = 1 \dots N \qquad\qquad (11.35)$$

$$C_j = s_j + t_j \qquad\qquad \text{for } j = 1 \dots N \qquad\qquad (11.36)$$

$$s_i \geq C_j - M\delta_{ij} \qquad\qquad \text{for all job pairs } (i, j) \qquad\quad (11.37)$$

$$s_j \geq C_i - M(1 - \delta_{ij}) \qquad \text{for all job pairs } (i, j) \qquad\quad (11.38)$$

$$T_j \geq 0, s_j \geq 0, C_j \geq 0 \qquad \text{for } j = 1 \dots N \qquad\qquad (11.39)$$

The objective [Equation (11.33)] represents the total weighted tardiness of the system. Constraint set [Equation (11.34)] defines one of the bounds for tardiness. Because the objective is to minimize tardiness, constraints in this set will be tight at optimality whenever T_j is positive. Constraint set [Equation (11.35)] bounds the job starting times to be after the job release times. Constraint set [Equation (11.36)] defines the job completion time to be the job starting time plus the job processing time. Constraint sets [Equation (11.37)] and [Equation (11.38)] ensure that only one job is on the machine at any one time. When $\delta_{ij} = 1$, and job i is before job j, the constraint in Equation (11.37) is trivially satisfied. Equation (11.38) requires that the starting time for job j must be after the completion time for job i. When $\delta_{ij} = 0$, indicating job j is before job i, the constraint in [Equation (11.38)] is trivially satisfied and the starting time for job i must be after the completion time for job j to satisfy (11.37). Finally, constraint set [Equation (11.39)] ensures that the problem decisions are all non-negative.

The model is called a **"disjunctive programming formulation,"** because constraint sets [Equation (11.37)] and [Equation (11.38)] are disjunctions (either job i precedes job j or job j precedes job i). The formulation can be solved with integer programming techniques such as branch and bound; however, solution time can be extensive for large problems.

There is a large body of heuristics that have been proposed for the static system tardiness problem. Key insights in heuristic development come from two very simple results. First, if you consider an **EDD** sequence, and **only one job is tardy,** then that sequence minimizes system tardiness. Second, if you consider an **SPT** sequence and **all jobs are tardy,** then that sequence minimizes system tardiness. The insight is that if due dates are tight, making many jobs tardy, then an optimal sequence for system tardiness should be close to an SPT sequence. If due dates have slack, and few jobs are tardy, then an optimal sequence for system tardiness should be close to an EDD sequence. Potts and Wassenhove [1991] and Panwalkar, Smith, and Koulamas [1993] contain computational results for a wide variety of heuristics including construction approaches, improvement approaches, meta approaches that combine simple approaches depending on the specific situation, and heuristic search approaches such as simulated annealing.

To complete the discussion, it is important to understand the structure of heuristics that perform well in computational tests. The construction heuristics of Wilkinson and Irwin **(WI)** [1971], Morton, Rachamadugu, and Vepsalainen **(AU)** [1984], and Panwalkar, Smith, and Koulamas **(PSK)** heuristic [1993] are addressed next.

The **WI heuristic** is based on the idea that it is easy to sequence systems with only two jobs. If two jobs j and k are adjacent in the sequence and start at time $P(S)$, then you should place the job with the earlier due date first as long as

$$P(S) + max\ (t_i, t_j) \leq max\ (d_i, d_j). \qquad\qquad (11.40)$$

If the inequality does not hold, then the shorter of the two jobs should come first. The heuristic maintains two subsequences; U is in EDD order and contains jobs not yet scheduled, and S is in sequence order and can be revised during the procedure. Denote job i^* as the index of the last job on the scheduled list, job j^* as the "pivot job," and k^* as the first job in U. Finally, $P(S)$ denotes the finish time of jobs in S assuming that the subsequence starts at time 0. The specific procedure follows:

Step 1 (Initialization): Set $U = $ an EDD sequence of the jobs. Let a and b denote the first two jobs on the EDD list. If $max\ (t_a, t_b) \leq max\ (d_a, d_b)$, then assign the job with the earliest due date to S, otherwise assign the job with the shortest processing time to S. The assigned job is i^* and $P(S)$ is computed. The nonassigned job is the pivot job, j^*. k^* is the first job in U. Go to step 2.

Step 2 (Scheduling of original pivot job): If $P(S) + max\ (t_{j^*}, t_{k^*}) \leq max\ (d_{j^*}, d_{k^*})$ or if $t_{j^*} \leq t_{k^*}$, then add job j^* to S, recompute $P(S)$, i^* becomes j^*, j^* becomes k^*, and k^* becomes the next job in U after k^*. If k^* cannot be set because there are no jobs remaining in U, then add j^* to the end of S, and stop with S as the sequence. Otherwise, return to step 2.

If $P(S) + max\ (t_{j^*}, t_{k^*}) > max\ (d_{j^*}, d_{k^*})$ and if $t_{j^*} > t_{k^*}$, then consider job k^* before job j^* by making job k^* the pivot job. This is done by setting j^* to k^*. Return the original pivot job to U, and keep U in EDD order. k^* is not relevant at this point.

Step 3 (Scheduling of secondary pivot job): If $P(S) - t_{i^*} + max\ (t_{i^*}, t_{j^*}) \leq max\ (d_{i^*}, d_{j^*})$ or if $t_{i^*} \leq t_{j^*}$, then add job j^* to S, recompute $P(S)$, i^* becomes j^*, j^* becomes the first job in U, and k^* becomes the next job in U after j^*. Return to Step 2.

If $P(S) - t_{i^*} + max\ (t_{i^*}, t_{j^*}) > max\ (d_{i^*}, d_{j^*})$ and if $t_{i^*} > t_{j^*}$, then a jump condition should occur. Go to step 4.

Step 4 (Jump condition, change the pivot job): Remove job i^* from S and place it in U in EDD order. If there are still jobs in S, let i^* be the last job in S and return to step 3 (note, that j^* is the same from the previous iteration and k^* is not relevant). If S is empty, then place j^* in the first position of S and i^* becomes j^*. The first job in U now becomes j^*, and the next job in U becomes k^*. Go to step 2.

The WI heuristic looks at pairs of jobs (jobs j^* and k^*) and tries to schedule the appropriate job next based on Equation (11.40). Sometimes, the jump condition holds and neither job should be scheduled and in fact, an already scheduled job must be moved off of the scheduled list. The following example demonstrates the method.

EXAMPLE 11.9

Consider the following data set for a single-machine static scheduling problem in which system tardiness is the primary objective.

Job	1	2	3	4	5	6
t_j	6	7	8	12	5	7
d_j	10	12	16	34	29	38

SOLUTION

The EDD ordering of the jobs is:

$$(1, 2, 3, 5, 4, 6)$$

Assign $a = 1$ and $b = 2$. Because $max\ (t_a, t_b) = 7 \leq 12 = max\ (d_a, d_b)$, job 1 is assigned first because it has the earlier due date ($S = \{1\}$). Therefore, $P(S) = 6$, $i^* = 1$, $j^* = 2$, and $k = 3$.

Proceeding to step 2, both conditions $P(S) + max\ (t_{j^*}, t_{k^*}) = 14 \leq 16 = max\ (d_{j^*}, d_{k^*})$ and $t_{j^*} \leq t_{k^*}$ hold, therefore, j^* is added to the end of $S(S = \{1, 2\})$, $P(S) = 13$, $i^* = 2$, $j^* = 3$, and $k^* = 5$.

The method continues with step 2 again. The following Table details the sequence of steps and the results of each iteration of the method.

Iter	S	P(S)	U	i*	j*	k*	Step	Comparison
1	{1}	6	{3,4,5,6}	1	2	3	2	$P(S) + max\ (t_{j*}, t_{k*}) = 14 \leq 16 = max\ (d_{j*}, d_{k*})$
2	{1,2}	13	{4,5,6}	2	3	5	2	Both conditions fail
2	{1,2}	13	{3,4,6}	2	5	–	3	$P(S) - t_{i*} + max\ (t_{i*}, t_{j*}) = 8 \leq 29 = max\ (d_{i*}, d_{j*})$
3	{1,2,5}	18	{4,6}	5	3	4	2	$t_{j*} = 8 \leq 12 = t_{k*}$
4	{1,2,5,3}	26	{6}	3	4	6	2	$P(S) - t_{i*} + max\ (t_{j*}, t_{k*}) = 30 \leq 38 = max\ (d_{j*}, d_{k*})$
5	{1,2,5,3,4}	38	Φ	4	6	–	2	Finish since U empty

The final sequence from the WI heuristic is {1, 2, 5, 3, 4, 6}. The following Gantt chart details the job completion times and the job tardiness values.

	1	2	5	3	4	6
C_j	6	13	18	26	38	45
d_j	10	12	29	16	34	38
T_j	0	1	0	10	4	7

Total system tardiness is $1 + 10 + 4 + 7 = 22$.

The **AU or Apparent Urgency heuristic** is a construction heuristic that sequences jobs based on the urgency priority index for job j. In Example 11.9, the WI heuristic scheduled job 5 very early in the sequence although it was not due. Job 3 was delayed significantly causing excessive tardiness. The AU method tries to correct this shortcoming.

Assume that a partial sequence S has been determined, it starts at time 0 and completes processing at $P(S)$. To find the next job in the sequence, AU considers the following index for each job still to be scheduled:

$$AU_j = (w_j/t_j)\ e^{(-max\{0,\ d_j - P(S) - t_j\}/(k\ *\ \bar{t}))} \tag{11.41}$$

where k is the "look-ahead" parameter that is set based on the perceived tightness of the due dates, and $\bar{t}$ is the average of the job processing times for jobs not yet scheduled. The job with the largest AU_j value is scheduled next. Note, that this method can also be used for weighted total tardiness problems.

The term $max\ \{0, d_j - P(S) - t_j\}$ represents the slack remaining for job j. If the maximum is at 0, then j is already tardy. Jobs that are tardy have AU_j value of w_j/t_j. When all jobs are equal weight and many jobs are tardy, then the method is essentially an SPT scheduling rule as expected. In general, the higher the slack, the lower the AU_j value, all other things being equal. When k is set high, the effect of positive slack is reduced, and the index value approaches w_j/t_j.

The method starts with S empty and proceeds to schedule one job in each iteration. The following example demonstrates the approach.

EXAMPLE 11.10

Using the data from Example 11.9, schedule the jobs using the AU heuristic with $k = 1$, and compute the system tardiness.

SOLUTION

Initially $S = \Phi$. The following Table contains the AU_j values at each iteration and the selections for building S.

It.	S	$P(S)$	$\bar{t}$	1	2	3	4	5	6
						AU_j			
1	Φ	0	7.5	.098	.073	.043	.004	.008	.002
2	{1}	6	7.8	–	.143	.097	.011	.020	.006
3	{1,2}	13	8	–	–	.125	.023	.051	.015
4	{1,2,3}	21	8	–	–	–	.074	.137	.041
5	{1,2,3,5}	26	9.5	–	–	–	.083	–	.084
6	{1,2,3,5,6}	33	12	–	–	–	.083	–	–
7	{1,2,3,5,6,4}	45	–	–	–	–	–	–	–

In the first iteration, job 1 has the largest value, so it is placed first in the sequence. In each successive iteration, the method selects the job with the largest AU_j value and augments it to the existing sequence. The final Gantt chart for the schedule and the job tardiness values are:

	1	2	3	5	6	4
C_j	6	13	21	26	33	45
d_j	10	12	16	29	38	34
T_j	0	1	5	0	0	11

Total system tardiness is $1 + 5 + 11 = 17$.

The **PSK heuristic** is similar to the AU heuristic in that it starts with an empty scheduled set S and then adds jobs one at a time to the end of S. The method tries to keep short jobs early in the sequence, and generally, first considers the shortest job not yet scheduled to be included next.

Let S be the set of scheduled jobs in order, and let U be the set of unscheduled jobs. Let $P(S)$ be the sum of the processing times of jobs in S and hence represents the completion time of S if S starts at time 0. The heuristic process is:

Step 0 (Initialization): $S = \Phi$ and $U =$ the SPT sequence of all jobs with ties broken by EDD sequencing. $P(S) = 0$.

Step 1 (Setting active and candidate jobs): If U has only one job, then schedule that job in the last position in S and stop; S is the generated schedule. Otherwise, let job j^* be the first job in the set U, and we call this the "active job." The job directly next to j^* in U is denoted k^* and is called the "candidate job."

Step 2 (Schedule active if tardy): If $P(S) + t_{j^*} \geq d_{j^*}$, then go to step 7.

Step 3 (Schedule active if tardy when done after candidate): If $d_{j^*} \leq P(S) + t_{k^*}$, then go to step 7.

Step 4 (Move to the next candidate if active job due date is early): If $d_{j^*} \leq d_{k^*}$, then go to step 6.

Step 5 (Change active jobs): Job k^* now becomes the active job. Set $j^* = k^*$. If j^* is the last job in U, then go to step 7. Otherwise, let the job to the right of j^* in U be k^*, and return to step 2.

Step 6 (Change candidate job): If job k^* is the last job in U, go to step 7, otherwise reset k^* to be the job in U directly after the current job k^*, and return to step 3.

Step 7 (Update schedule): Remove job j^* from U, and put it last in S. Increase $P(S)$ by t_{j^*}, and return to step 1.

The PSK heuristic starts with an SPT schedule and then tries to schedule jobs when delaying those jobs causes lateness (steps 2 and 3). In steps 4 and 6, the active job, the job that is currently being scheduled, is compared to all jobs later in the SPT sequence. In step 5, the method changes the active job when there is a longer job that has a more pressing due date. The following example demonstrates computations and steps of the PSK heuristic.

EXAMPLE 11.11

Using the data from Example 11.9, schedule the jobs using the PSK heuristic and compute the system tardiness.

SOLUTION

The SPT sequence is $\{5, 1, 2, 6, 3, 4\}$. The method starts with $j^* = 5$ and $k^* = 1$. The following Table contains the progression of the U and S sequences, the changes in the j^* and k^* values, and the sequence of heuristic steps processed in each iteration.

U	S	$P(S)$	j^*	k^*	Key steps
$\{5, 1, 2, 6, 3, 4\}$	Φ	0	5	1	5
$\{5, 1, 2, 6, 3, 4\}$	Φ	0	1	2	$4 \to 6$
$\{5, 1, 2, 6, 3, 4\}$	Φ	0	1	6	$4 \to 6$
$\{5, 1, 2, 6, 3, 4\}$	Φ	0	1	3	$4 \to 6$
$\{5, 1, 2, 6, 3, 4\}$	Φ	0	1	4	$4 \to 6 \to 7$
$\{5, 2, 6, 3, 4\}$	$\{1\}$	6	5	2	5
$\{5, 2, 6, 3, 4\}$	$\{1\}$	6	2	6	$2 \to 7$
$\{5, 6, 3, 4\}$	$\{1, 2\}$	13	5	6	$4 \to 6$
$\{5, 6, 3, 4\}$	$\{1, 2\}$	13	5	3	5
$\{5, 6, 3, 4\}$	$\{1, 2\}$	13	3	4	$2 \to 7$
$\{5, 6, 4\}$	$\{1, 2, 3\}$	21	5	6	$4 \to 6 \to 7$
$\{6, 4\}$	$\{1, 2, 3, 5\}$	26	6	4	$3 \to 7$
$\{4\}$	$\{1, 2, 3, 5, 6\}$	33	4	–	1
Φ	$\{1, 2, 3, 5, 6, 4\}$	45	–	–	Done

The schedule computed is identical to that generated by the AU heuristic, hence the system tardiness is 17 days. Each time step 7 is executed, one more job is added to S. Also, we can see the progression of searching the entire set of jobs in the first five iterations before we place job 1. The method quickly moves off jobs that do not need immediate scheduling.

11.5   FLOW-SHOP SCHEDULING

A flow shop is a system in which all jobs have the same routing, and each job visits each machine only one time. One can picture the system as a linear sequence of machines like an assembly line. Each job is processed sequentially, moving from one machine to the next.

Even with this simplified structure, finding the optimal schedules can be extremely difficult. Schedules in which the same job sequence is used on each machine are called **"permutation schedules."** As long as the criterion is a regular measure, only active schedules without unnecessary idle time need be considered; however, it is not necessarily optimal to use a permutation schedule. Therefore, the number of schedules that must be considered when finding an optimal solution increases exponentially as the number of jobs and/or the number of machines increases. Typically, problems with ten machines and ten to 20 jobs are considered large and are difficult to solve optimally.

Two generalizations of the flow shop have received a great deal of recent study. Figure 11.2 depicts the underlying structures of these systems as compared to the standard flow shop. The first is called a **"flexible flow shop."** Here, each job in the system follows the same route through the departments in the shop; however, in each department there are many machines that can be used to process the jobs. The second is called a **"re-entrant flow shop."** Here, all jobs take the same route through the shop, however jobs may visit machines more than once. This type of system is typically found in wafer fabrication plants and is extensively addressed in Graves et al. [1983], and Wein [1988]. Both systems are addressed in the next chapter.

11.5.1 Scheduling Two Machines—Johnson's Rule

When the system is static, has only two machines, and the criterion is to minimize makespan, then optimal solutions are easily found using an algorithm attributable to Johnson [1954]. In this case, it can be shown that one needs only to consider permutation schedules on order to find an optimal schedule. The key concept in the algorithm is that minimizing makespan in this case is equivalent to minimizing the amount of idle time on the second machine. Because all jobs are released at time 0, there is no need to have idle time on the first machine. Idle time on the second machine is caused by having to wait for a job to complete processing on the first machine. Idleness occurs when a job has a long processing time on the first machine, and all jobs before it in the sequence have completed processing on the second machine. The sequencing rule orders the jobs so that this idle time is minimized.

To set up the algorithm, let t_{1j} be the processing time of job j on machine 1 and t_{2j} be the processing time of job j on machine 2. All jobs must be processed on both machines. The formal algorithm (**Johnson's rule**) is:

Step 0 (Initialization): Find the minimum of t_{1j} and t_{2j} for each job j. Denote this value as t_j, and note, for each job the machine in which the minimum occurs. Sort the t_j values from low to high. Initialize the sequence S to Φ; the pointers into the sequence, H to N and L to 1; and the index in the list of t_j values, k to 1. Go to step 1.

Step 1 (Insert in Sequence): Consider the kth value in the sorted list of t_j values. If this minimum occurs on the first machine, then put job j in the Lth position in S and increase L by 1. If this minimum occurs on the second machine, then put job j in the Hth position in S and decrease H by 1. Increase k by 1, and go to step 2.

Step 2 (Completion): If $k > N$, then stop, otherwise return to step 1.

The sequence is constructed from the outside to the middle, as opposed to first to last. Jobs that have a short processing time on the first machine tend to be early in S, whereas jobs that have short processing time on the second machine are late in the sequence. The basic idea is to get jobs to machine 2, have them wait there for processing, and incur no machine idle time at machine 2. The following example demonstrates Johnson's rule.

EXAMPLE 11.12

Consider a laundry operation with one washing machine and one dryer. Six loads of clothing must go through both operations in the sequence washing → drying. The following Table lists the processing time for each load on each machine.

Load	1	2	3	4	5	6
t_{1j}	15	40	5	20	25	20
t_{2j}	20	20	15	15	30	25

How should you sequence the loads so that the laundry is completed as soon as possible?

SOLUTION

First, t_j must be computed for each load and noted for machine occurrence:

t_j	$15(1^{st})$	$20(2^{nd})$	$5(1^{st})$	$15(2^{nd})$	$25(1^{st})$	$20(1^{st})$

The sorted list is (ties can be sorted arbitrarily):

t_j	$5(1^{st})$	$15(1^{st})$	$15(2^{nd})$	$20(2^{nd})$	$20(1^{st})$	$25(1^{st})$
Job	3	1	4	2	6	5

The algorithm starts with an empty sequence with L and H at the extremes. t_3 is the smallest t_j value and this occurs on the first machine. Therefore, it is placed in the Lth position in the sequence (first because $L = 1$). L is increased by 1, and the next job in the t_j list is selected. t_1 is the next smallest t_j value, and this also occurs on the first machine. Therefore, it is placed in the Lth position in the sequence (second because $L = 2$). L is increased by 1 and the next job in the t_j list is selected. t_4 is the next smallest t_j value, and this occurs on the second machine. Therefore, it is placed in the Hth position in the sequence (sixth because $H = 6$). H is decreased by 1, and the next job in the t_j list is selected. The method continues until all jobs have been sequenced. The solution process is summarized below.

K (iteration)	L (low index)	H (high index)	S (Sequence)
1–job 3	1	6	{3, –, –, –, –, –,}
2–job 1	2	6	{3, 1, –, –, –, –,}
3–job 4	3	6	{3, 1, –, –, –, 4}
4–job 2	3	5	{3, 1, –, –, 2, 4}
5–job 6	3	4	{3, 1, 6 –, –, 2, 4}
6–job 5	4	4	{3, 1, 6, 5, 2, 4}

The Gantt chart for the solution and the makespan is:

Washer	3	1	6	5	2	4

Time	5	20	40	65	105	125	140

Dryer	3	1	6	5	2	4

	0	5	20	40	65	105	125	140 time

Johnson's rule can be easily extended to other cases. A three-machine problem with processing times having one of the following two properties:

$$\operatorname*{Min}_{j=1}^{N} t_{1j} \geq \operatorname*{Max}_{j=1}^{N} t_{2j}$$
$$\text{or} \tag{11.42}$$
$$\operatorname*{Min}_{j=1}^{N} t_{3j} \geq \operatorname*{Max}_{j=1}^{N} t_{2j}$$

can be transformed into an equivalent two-machine problem. The transformation is:

$$t'_{1j} = t_{1j} + t_{2j} \tag{11.43}$$

$$t'_{2j} = t_{2j} + t_{3j} \tag{11.44}$$

where t'_{1j} and t'_{2j} are the processing times on machines 1 and 2 in the transformed problem. Johnson's rule is now used on the "prime" problem and the resulting permutation schedule is optimal in the three-machine problem. The transformation works because the processing times on machine 2 are dominated by the processing times on machine 1 or machine 3. Machine 2 does not force idle time in the system and is called a **"nonbottleneck processor."**

Another extension is to a two-machine static system in which the job set can be partitioned into the following four subsets:

Subset A: Jobs processed on machine 1 only

Subset B: Jobs processed on machine 2 only

Subset C: Jobs processed on machine 1 and then machine 2

Subset D: Jobs processed on machine 2 and then machine 1

and the criterion is to minimize makespan. The key insight here is that minimizing makespan is equivalent to minimizing idle time on both machines simultaneously. In Johnson's rule, idle time occurs when a job is running on the first machine, and the second machine completes the job's predecessor. To avoid this phenomenon, the jobs in subsets A and B are used to space out the jobs in subsets C and D so that as little idle time as possible results. The following scheduling strategy can be used:

Step 1: Order the jobs in A in any sequence denoted S_A.

Step 2: Order the jobs in B in any sequence denoted S_B.

Step 3: Order the jobs in C using Johnson's rule to obtain the sequence denoted S_C.

Step 4: Order the jobs in D using Johnson's rule to obtain the sequence denoted S_D (note, that the machines are ordered differently in D with machine 2 being the first machine, thus care must be taken to get the sequence in the correct order).

Step 5: Use the following sequences on each machine:

$$\text{Machine } 1 - S_C, S_A, S_D$$
$$\text{Machine } 2 - S_D, S_B, S_C$$

Note that this problem generally does not yield a permutation schedule. The following example demonstrates the method.

EXAMPLE 11.13

Find a schedule for the following two-machine system in which all jobs are released at time 0 and the criterion is to minimize makespan:

Job	First machine	Processing time	Second machine	Processing time
1	Machine 1	5	Machine 2	15
2	Machine 1	10	Machine 2	10
3	Machine 2	5	Machine 1	10
4	Machine 2	25	Machine 1	10
5	Machine 2	15	Machine 1	20
6	Machine 1	10		
7	Machine 1	20		
8	Machine 2	5		
9	Machine 2	15		
10	Machine 2	25		

SOLUTION

The subsets are easily defined from the above data:

$$A = \{6, 7\}$$
$$B = \{8, 9, 10\}$$
$$C = \{1, 2\}$$
$$D = \{3, 4, 5\}$$

Using Johnson's rule for C, we arrive at the schedule $S_C = (1, 2)$ (job 1 is first because it has the minimum processing time and that occurs on machine 1). Using Johnson's rule for D, we arrive at the schedule $S_D = (3, 5, 4)$ (job 3 is first when we recall that machine 2 is first and this job/machine has the smallest processing time; job 4 is last because it has the second smallest processing time and that occurs on machine 1). Using any order for A and B, we have $S_A = (6, 7)$ and $S_B = (8, 9, 10)$. Therefore, the optimal schedule for each machine is:

$$\text{Machine 1: } (1, 2, 6, 7, 3, 5, 4)$$
$$\text{Machine 2: } (3, 5, 4, 8, 9, 10, 1, 2)$$

The following Gantt chart depicts the results and the optimal makespan of 115.

11.5.2   *M*-Machine Flow Shops

The static *M*-machine flow shop problem with makespan as the objective has a long history. Solving the problem optimally is difficult even for problems as small as ten machines and ten jobs. Most heuristic approaches concentrate on permutation schedules, and this simplifies the problem somewhat (and generates schedules that are easier to execute on the production floor). The Palmer (P) heuristic [1965], the Campbell, Dudek, and Smith (CDS) heuristic [1970], and the Nawaz, Enscore, and Ham (NEH) heuristic [1983] tend to perform well [Taillard (1990)] and yield reasonable computation times. These three heuristics are presented next.

Heuristic P sequences jobs based on a "slope index" for each job. The slope index for job j, S_j, is defined as:

$$S_j = (M - 1)t_{Mj} + (M - 3)t_{(M-1)j} + (M - 5)t_{(M-3)j} + \cdots - (M - 5)t_{3j}$$
$$- (M - 3)t_{2j} - (M - 1)t_{1j} \qquad (11.45)$$

and the method sequences jobs in decreasing slope index order. The slope index gives a high positive weight to processing times on later machines and a high negative weight to processing times on machines early in the flow. Jobs with shorter processing times generally are scheduled early in the sequence. The sign flip in the slope index formula occurs at the mid-point of the flow-line.

EXAMPLE 11.14

Use heuristic P to find a sequence for the following 4-machine, 5-job flow-shop problem data:

	Machine			
Job	1	2	3	4
1	3	6	9	12
2	12	6	9	6
3	6	9	12	6
4	9	3	9	6
5	6	6	15	6

SOLUTION

The slope index for job j is:

$$3t_{4j} + t_{3j} - t_{2j} - 3t_{1j} \qquad (11.46)$$

Substituting in the processing time values, the slope index values are:

Job	Slope index
1	$3*12 + 9 - 6 - 3*3 = 30$
2	$3*6 + 9 - 6 - 3*12 = -15$
3	$3*6 + 12 - 9 - 3*6 = 3$
4	$3*6 + 9 - 3 - 3*9 = -3$
5	$3*6 + 15 - 6 - 3*6 = 9$

Sorting the slope values in descending order, the P heuristic yields the sequence:

$$(1, 5, 3, 4, 2)$$

The Gantt chart for this sequence is:

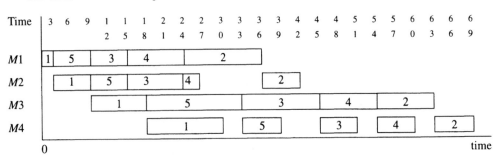

The makespan for the sequence is 69.

Heuristic CDS generates $M - 1$ sequences and chooses the one with the lowest makespan. Define l as the sequence index, $l = 1 \ldots M - 1$. The basic idea is to turn the M-machine problem into a two-machine problem, and then use Johnson's method to find a sequence. Similarly to the three-machine nonbottleneck problem, we define a two-machine problem by adding processing times. For the lth problem, define:

$$t_{1j}^l = \sum_{i=1}^{l} t_{ij} \tag{11.47}$$

$$t_{2j}^l = \sum_{i=l}^{M} t_{(M-i+1),j} \tag{11.48}$$

Now, using t_{1j}^l and t_{2j}^l, solve a two-machine problem to get the lth sequence. The first problem uses the processing time on the first machine for t_{1j}^l and the processing time on the Mth machine for t_{2j}^l. The lth problem uses processing times on the first l machines for t_{1j}^l and processing times on the last l machines for t_{2j}^l. The method stops at $l = M - 1$ because when $l = M$, t_{1j}^l and t_{2j}^l are identical.

EXAMPLE 11.15

Use the CDS heuristic to generate a sequence for the 4-machine, 5-job problem in Example 11.14.

SOLUTION

Since there are 4 machines, the CDS heuristic generates 3 sequences and the one with the best makespan is chosen. The processing times used in the 2-machine problems, the sequence for Johnson's algorithm, and the sequence makespans are:

Job	$l = 1$		$l = 2$		$l = 3$	
	t_{1j}^1	t_{2j}^1	t_{1j}^2	t_{2j}^2	t_{1j}^3	t_{2j}^3
1	3	12	9	21	18	27
2	12	6	18	15	27	21
3	6	6	15	18	27	27
4	9	6	12	15	21	18
5	6	6	12	21	27	27
Seq.	(1, 3, 5, 4, 2)		(1, 4, 5, 3, 2)		(1, 3, 5, 2, 4)	
Makespan	66		69		66	

Select sequence for $l = 1$ or 3 because these have the lowest makespan.

Heuristic NEH first adds the operation processing times for each job and orders these job times in decreasing order. The first two jobs, j_1 and j_2, are considered first. Evaluate the makespan for the two job sequences, (j_1, j_2) and (j_2, j_1), and select the sequence that yields the lowest makespan. Next, take the third job, j_3 and test it at three locations in the partial sequence by evaluating the makespan of the partial sequence. For example, if (j_2, j_1) was selected, then the three sequences to test are (j_3, j_2, j_1), (j_2, j_3, j_1), and (j_2, j_1, j_3). Once the three-job sequence is chosen, consider inserting j_4 in four possible positions. Continue the insertion and selection process until all jobs are in the sequence. The final selection is the sequence generated by the NEH heuristic.

EXAMPLE 11.16

Use the NEH heuristic to generate a sequence for the four-machine, five-job problem in Example 11.14.

SOLUTION   Summing the processing times for each job, we arrive at the following SPT sequence (breaking ties arbitrarily):

$$\{4, 1, 2, 3, 5\}$$

The heuristic starts by testing two subsequences, $\{4, 1\}$ and $\{1, 4\}$. When $\{4, 1\}$ is used, these jobs have a makespan of 42 hours, and when $\{1, 4\}$ is used the makespan is 36 hours. $\{1, 4\}$ is chosen because it has a lower makespan, and the heuristic proceeds to inserting job 2. The following Table summarizes the process of testing and inserting until the final sequence is selected.

Job	Sequence 1	Sequence 2	Sequence 3	Sequence 4	Sequence 5
Initial	$\{4, 1\} = 42$	$\{1, 4\} = 36$	–	–	–
2	$\{2, 1, 4\} = 54$	$\{1, 2, 4\} = 45$	$\{1, 4, 2\} = 45$	–	–
3	$\{3, 1, 2, 4\} = 60$	$\{1, 3, 2, 4\} = 54$	$\{1, 2, 3, 4\} = 57$	$\{1, 2, 4, 3\} = 57$	–
5	$\{5, 1, 3, 2, 4\} = 72$	$\{1, 5, 3, 2, 4\} = 69$	$\{1, 3, 5, 2, 4\} = 66$	$\{1, 3, 2, 5, 4\} = 69$	$\{1, 3, 2, 4, 5\} = 69$

We note that there is a tie in the selection process when job 2 was inserted and this tie was broken arbitrarily here. One, however, could use more structured methods of selection. The best sequence found has a makespan of 66 hours and is $\{1, 3, 5, 2, 4\}$.

11.6 JOB SHOP SCHEDULING

Job shop scheduling problems are notoriously difficult optimization problems, and there has been little general success at finding optimal solutions for problems with more than 20 jobs and ten machines. Hence, the focus of this section is heuristic approaches to the problem that generate high-quality solutions with reasonable computational efforts. Because the job shop is traditionally a highly dynamic environment, rescheduling of operations occurs often. Therefore, it is imperative that the scheduler be given tools that generate schedules in a timely manner.

The system considered is a job shop. All jobs have different routings through the system. The first situation considered is the traditional dynamic environment problem—a machine becomes idle, what operation is processed next? Next, a more static situation is addressed there is a set of jobs that must be processed; how should these be scheduled to completion? To conclude, rule-based expert systems and systems that use artificial intelligence techniques will be addressed briefly.

11.6.1 Real-Time Dispatching Rules for Multiple Machines

There has been a long history of the use of dispatching rules for the job shop [Panwalker and Iskander (1977) list more than 100 dispatching rules]. The dispatching rules listed in section 11.2.1 for determining priorities for releasing jobs to the shop can also be used in the job shop for determining operations to process next. Assume that machine m^* has just become idle, and denote J^* as the set of jobs in the system that require the use of m^*. Note, that J^* may contain jobs that are waiting in line at m^*, jobs elsewhere in the system in line or being processed that eventually use m^*, and jobs that will soon be released to the shop that use m^*. The basic algorithm for dispatching is:

Step 1: Compute the set J^* of jobs that will be under consideration for processing next on m^*

Step 2: For each job $j \in J^*$, compute a priority index

Step 3: The next job on J^* is the job with the best priority index value. If that job is immediately available, then start processing, otherwise keep m^* idle until the job arrives.

The process requires a specification of the priority index used in step 2, and this determines the effectiveness of the schedule. Some rules only look at processing time on m^* and other rules look at the total processing time remaining for job completion. Example rules are similar to those used to set release dates in section 11.2.1:

SPT – select the job with the shortest processing time on m^*;

LWR – select the job with the least work remaining summed over all operations;

FCFS – select the job that has been waiting at the machine the longest;

FISFS – select the job that has been in the shop the longest;

MOR – select the job with the most number of operations remaining;

LOR – select the job with the least number of operations remaining;

SLK – select the job with the least slack remaining where slack is defined as the difference between time until the due date and total remaining processing time;

EDD – select the job with the earliest due date;

RND – select a job at random.

The idea of keeping m^* idle while waiting for a job may seem a bit odd at first. However, examples in which it is best to wait a short amount of time for an important job to arrive at a machine rather than immediately tie up the machine for a long operation are easy to construct.

The following example uses the LWR rule where J^* consists of all jobs waiting in line at machine m^*.

EXAMPLE 11.15

Consider a three-machine four-job situation with the following data:

Job	Routing–machine number, processing time required			
1	M1, 10	M2, 6	M3, 10	M2, 8
2	M1, 6	M3, 10	M2, 8	
3	M3, 10	M2, 4	M1, 4	M3, 8
4	M2, 6	M1, 10		

All jobs are released at time 0 and preemption is not allowed. Develop the schedule generated using the least work remaining rule for selecting jobs to process. Use only jobs currently in line at the machine when making scheduling selections. Construct a Gantt chart and compute the makespan for the schedule.

SOLUTION The following Gantt chart depicts the schedule derived:

| 2 | 4 | 6 | 8 | 10 | 12 | 14 | 16 | 18 | 20 | 22 | 24 | 26 | 28 | 30 | 32 | 34 | 36 | 38 | 40 | 42 | 44 | 46 | 48 | 50 | 52 | 54 |

*M*1 — 2 | 4 | 3 | 1

*M*2 — 4 | 3 | 2 | 1 | 1

*M*3 — 3 | 2 | 3 | 1

0 .... time

At hour 0, job 4 starts on machine 2, job 3 starts on machine 3, and we select job 2 to start on machine 1 because its total work remaining is 24 hours while the work remaining on job 1 (both are ready in front of machine 1) is 34 hours. At hour 6, job 2 completes on machine 1 and goes to the queue for machine 3 while job 4 completes on machine 2 and goes to the queue for machine 1. Machine 2 is idle because no jobs remain in its queue. Jobs 1 and 4 are in the queue for machine 1, and we select job 4 because it has the least work remaining (10 versus 34). At hour 10, job 3 completes on machine 3 and immediately starts processing on machine 2. The next event occurs at hour 14 when job 3 completes on machine 2 and joins job 1 in the queue for machine 1. At hour 16, job 4 completes on machine 1 and we choose job 2 next because it has the least remaining work (12 versus 34). The process continues, and eventually job 1 is started. The schedule has a makespan of 54, and it is clear that delaying job 1 has caused a delay in the makespan (for example, consider interchanging job 1 and job 4 on machine 1).

Different dispatching rules perform well for different objectives. For example, LWR tends to get small jobs out of the shop quickly and hence does well when the criteria is to minimize total flow time or system WIP. Because LWR delays long jobs, it can perform poorly for makespan and tardiness objectives. Note, the similarity to SPT for the single-machine problem. Simply using the myopic single-machine SPT in the multiple-machine environment (pick the job with the shortest operation time) can lead to poor performance. Sule [1997], Pinedo [1995], and Panwalker [1977] address the performance of different dispatching rules and conclude that there is little that can be said in general. System rules tend to perform better than myopic, or single-machine rules. Often, combinations of rules work well, especially when the particular rule used depends on the state of the system. Due date rules perform better on tardiness related criteria particularly when due dates are neither so tight such that many jobs are tardy or so loose such that almost all jobs are easily completed on time. Processing time based rules perform better on flow time related criteria.

11.6.2 Mathematical Programming Approaches

Mathematical programming approaches for the *M*-machine job shop problem generally center around the disjunctive programming formulation. Recall that such a formulation was used to model the single-machine problem in which the objective was to minimize system tardiness (Model T). The adjustments to extend Model T to *M* machines are relatively simple and only add another dimension to the problem.

Each job j has a fixed routing and denote i as the ith operation on the route for j. There are $N(j)$ operations on route j. Each operation (i, j) requires t_{ij} units of time on machine m_{ij}. There are no options on machine selection. Operation i of job j completes at C_{ij} and job j completes at the completion time of operation $N(j)$ denoted $C_{N(J),j}$. It is convenient to recognize the starting time of each operation (i, j) as s_{ij}. Finally, the key decision on sequence on a machine is defined using a binary decision variable:

$\delta_{ij,kl} = 1$ if operation (i, j) is done before operation (k, l). Note, that m_{ij} and m_{kl} must be the same machine; however, j and l may or may not be the same job (a job could have multiple tasks on a single machine).

To simplify the model, assume that all job ready times are 0 (this is easily relaxed), that no preemption can occur, and that the objective is a regular measure. The model for the job shop (denoted **model JS**) is:

$$\text{Minimize} \sum_{j=1}^{N} f(C_{N(j),j}) \tag{11.49}$$

$$\text{Subject to: } s_{ij} \geq C_{i-1,j} \qquad \text{for each } (i, j) \text{ pair where } C_{0j} = 0 \tag{11.50}$$

$$C_{ij} = s_{ij} + t_{ij} \qquad \text{for each } (i, j) \text{ pair} \tag{11.51}$$

$$s_{ij} \geq C_{kl} - M\delta_{ij,kl} \qquad \begin{array}{l}\text{for each pair of operations } (i, j) \text{ and} \\ (k, l) \text{ that share the same machine}\end{array} \tag{11.52}$$

$$s_{kl} \geq C_{ij} - M(1 - \delta_{ij,kl}) \qquad \begin{array}{l}\text{for each pair of operations } (i, j) \text{ and} \\ (k, l) \text{ that share the same machine}\end{array} \tag{11.53}$$

$$\delta_{ij,kl} \in \{0, 1\}, s_{ij} \geq 0, C_{ij} \geq 0 \tag{11.54}$$

The objective [Equation (11.49)] simply states that the goal is to minimize some function of the completion times. The specified function could be flow time, makespan, tardiness, or even NT. Constraint set [Equation (11.50)] ensures that the routing is followed. The starting time for each operation for each job must be no less than the completion time of the predecessor operation. Constraint set [Equation (11.51)] sets the completion time for an operation to be the starting time plus the processing time (no intentionally inserted idle time). Constraint sets [Equation (11.52)] and [Equation (11.53)] make up the difficult portion of the formulation. A sequence of operations must be set for each machine. For any two operations, (i, j) and (k, l) on the same machine, either (i, j) must be before (k, l) or vice versa. If (i, j) is before (k, l) then $\delta_{ij,kl} = 1$. Now, the corresponding constraint in Equation (11.52) is:

$$s_{ij} \geq C_{kl} - M\delta_{ij,kl} = C_{kl} - M \tag{11.55}$$

The constraint is trivially satisfied by any s_{ij} and C_{kl} settings as long as M is sufficiently large. On the other hand, the corresponding constraint in Equation (11.53) is:

$$s_{kl} \geq C_{ij} - M(1 - \delta_{ij,kl}) = C_{ij} \tag{11.56}$$

Now, the start time for (k, l) must be no less than the completion time for (i, j). Recall that both (i, j) and (k, l) are on the same machine, therefore, if $\delta_{ij,kl} = 1$ Equation (11.52) and Equation (11.53) ensure that the start and completion times of the operations obey the constraint that each machine works only on a single operation at any point in time. A similar argument can be made when $\delta_{ij,kl} = 0$ and (k, l) precedes (i, j) on the machine. In this case, the corresponding constraint between (i, j) and (k, l) in Equation (11.53) is trivially satisfied, and the constraint in Equation (11.52) ensures that the start and completion times obey the correct relationship.

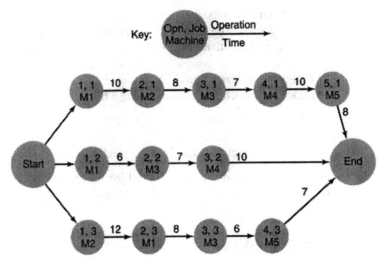

Figure 11.6 Sequence of Operations for Jobs, 3 Jobs, 5 Machines

Model JS has two basic requirements on each operation (i, j). First, all of the predecessor operations of (i, j) for job j, must be completed before (i, j) can start. This is depicted graphically in Figure 11.6 above. Each node in the Figure represents an operation labeled with (i, j) and m_{ij}, whereas each arc represents the processing time of the operation. The "start" and "end" nodes tie the system together and the arcs from the start to the first node in each job have no processing time. Note, that this figure shows the **critical path.** The top path in the figure (Job 1) requires 43 time units, the longest path from start to end. Operation (i, j) cannot start until all predecessors have completed.

It is a simple matter to recursively compute the operation starting times from the diagram using the relationships:

$$C_{0j} = 0 \text{ for all } j, \quad s_{ij} = C_{i-1,j} \quad \text{and} \quad C_{ij} = s_{ij} + t_{ij}. \tag{11.57}$$

These calculations ensure that no idle time is inserted. These would be the actual operation starting and completion times, if each job had its own set of machines, however the values must be adjusted because of the shared machine constraints.

EXAMPLE 11.18

Using the data in Figure 11.6, compute s_{ij} and C_{ij} for all operations under the assumption that each job has its own set of machines.

SOLUTION

Job 1 has five operations. Using Equation (11.57) for computing s_{ij} and C_{ij}, the computations are:

Operation	(1, 1)	(2, 1)	(3, 1)	(4, 1)	(5, 1)
t_{i1}	10	8	7	10	8
s_{i1}	0	10	18	25	35
C_{i1}	10	18	25	35	43

The starting time of any operation is the completion time of the predecessor operation. For jobs 2 and 3, the following results are similarly obtained:

Operation	(1, 2)	(2, 2)	(3, 2)
t_{i2}	6	7	10
s_{i2}	0	6	13
C_{i2}	6	13	23

Operation	(1, 3)	(2, 3)	(3, 3)	(4, 3)
t_{i3}	12	8	6	7
s_{i3}	0	12	20	26
C_{i3}	12	20	26	33

The second basic requirement deals with the sharing of machines. All operations (k, l) scheduled before operation (i, j) on m_{ij} must complete before (i, j) can start. Figure 11.7 shows how the operation sequence on a machine can be included in a graph similar to Figure 11.6. The dotted arcs represent the sequence on each machine. The arcs start at one node and move to the next operation in the sequence. The lengths of the dotted arcs are the operation times for the operation referenced at the originating node of the arc. In Figure 11.7, operations (1, 1), (1, 2), and (2, 3) are all done on machine 1. The dotted arcs show that (1, 1) is done first, (1, 2) is done second, and (2, 3) is done third. Similarly, for machine 2, there are two operations and (1, 3) is done first and (2, 1) is done second. Note, that for the last job in any machine sequence, there is no operation time given and this will be incorporated later.

If there were only constraints for machine sequence, and not job routing constraints, it is a simple matter to compute s_{ij} and C_{ij} for all operations on each machine. The relationships are:

$$C_{[0]} = 0 \text{ for all } j, \; s_{[l]} = C_{[l]-1} \quad \text{and} \quad C_{[l]} = s_{[l]} + t_{ij}^{[l]} \tag{11.58}$$

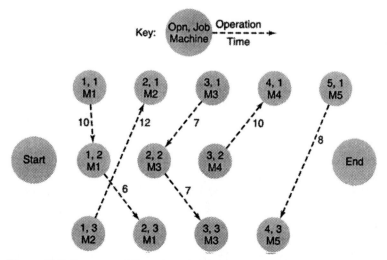

Figure 11.7 Sequence of Operations for Jobs and Machines, 3 Jobs, 5 Machines

where [*l*] is the lth operation in the sequence for the machine and $t_{ij}^{[l]}$ is the processing time for the lth operation in the sequence. The following example demonstrates the computations.

EXAMPLE 11.19

For the data in Figure 11.7, compute the operation starting and completion times for operations on each machine individually.

SOLUTION

Machine 1 has three operations. Using Equation (11.58) for computing s_{ij} and C_{ij} the computations are:

Operation	(1, 1)	(1, 2)	(2, 3)
t_{ij}	10	6	–
s_{ij}	0	10	16
C_{ij}	10	16	–

Similarly for the other machines (in order of machine number):

Operation	(1, 3)	(2, 1)
t_{ij}	12	–
s_{ij}	0	12
C_{ij}	12	–

Operation	(3, 1)	(2, 2)	(3, 3)
t_{ij}	7	7	–
s_{ij}	0	7	14
C_{ij}	7	14	–

Operation	(3, 2)	(4, 1)
t_{ij}	10	–
s_{ij}	0	10
C_{ij}	10	–

Operation	(5, 1)	(4, 3)
t_{ij}	8	–
s_{ij}	0	8
C_{ij}	8	–

One can combine Figures 11.6 and 11.7 to compute the operation starting times for a particular job shop sequence. The key idea is that operations cannot start until all predecessor operations for the job are completed and all operations earlier in the sequence for a machine are completed. Figure 11.8 depicts the combined graph for the routing and schedules used previously. Each operation node can have at most two input arcs, one solid and one dotted for a given set of sequences (if an operation is first in the sequence of a machine, then it has only one solid input arc). **The operation cannot start until both of these predecessors have been completed.** Therefore, the start time for any operation must be **equal to the maximum of the completion times of the two operations that come directly before it in the diagram.** Beginning with the "start" node, it is a simple matter to set $s_{ij} = 0$ for all operations that are first in their routing and first on a machine (at least one such operation must exist, otherwise, the sequence does not obey the routings and is not feasible). Once the starting time of the operation is known, the completion time is easily computed by:

$$C_{ij} = s_{ij} + t_{ij} \tag{11.59}$$

The network has no directed cycles, hence there is an ordering of the nodes so that when computing the start time for each node, both of the completion times for the predecessor nodes have already been completed. The next example demonstrates the method to compute the starting times and completion times of all operations.

EXAMPLE 11.20

For the data and sequences depicted in Figure 11.8, compute the completion times for each job and the makespan for the system.

SOLUTION

Begin at "start" and set its start and completion times to 0. Next, go to any nodes that have only "start" as a predecessor (operations $(1, 1)$ and $(1, 3)$). The start time is 0 (completion time of single predecessor node) and the finish time is $0 + 10 = 10$ for operation $(1, 1)$ and $0 + 12 = 12$ for operation $(1, 3)$. Next, look for any operations in which both of its predecessors on the network have been completed. Both $(1, 2)$ and $(2, 1)$ satisfy this requirement. For operation $(1, 2)$, it follows $(1, 1)$ on machine 1 and "start" on the routing. Operation $(1, 1)$ completes at time 10 and "start" completes at time 0, therefore, the starting time for $(1, 2)$ is:

$$C_{12} = \max (10, 0) = 10 \tag{11.60}$$

For operation $(2, 1)$, operations $(1, 1)$ and $(1, 3)$ precede, therefore:

$$C_{21} = \max (10, 12) = 12 \tag{11.61}$$

The calculations for the other nodes proceed similarly, and the following Table lists the computations in the order they were made:

Opn.	t_{ij}	S_{ij}	C_{ij}
Start	0	0	0
$(1, 1)$	10	$\max (C_{start}) = 0$	10
$(1, 3)$	12	$\max (C_{start}) = 0$	12
$(1, 2)$	6	$\max (C_{start}, C_{11}) = 10$	16
$(2, 1)$	8	$\max (C_{11}, C_{13}) = 12$	20
$(3, 1)$	7	$\max (C_{21}) = 20$	27
$(2, 2)$	7	$\max (C_{12}, C_{31}) = 27$	34
$(2, 3)$	8	$\max (C_{13}, C_{12}) = 16$	24
$(3, 2)$	10	$\max (C_{22}) = 34$	44
$(4, 1)$	10	$\max (C_{31}, C_{32}) = 44$	54
$(5, 1)$	8	$\max (C_{41}) = 54$	62
$(3, 3)$	6	$\max (C_{23}, C_{22}) = 34$	40
$(4, 3)$	7	$\max (C_{33}, C_{51}) = 62$	69
End	0	$\max (C_{51}, C_{32}, C_{43}) = 69$	69

The makespan for the schedule is 69. Job 1 completes at 62, job 2 at 44 and job 3 at 69.

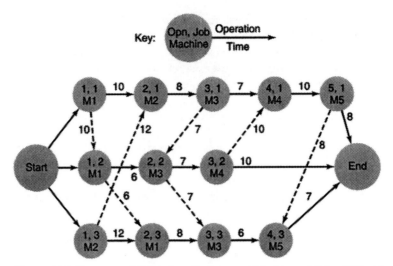

Figure 11.8 Sequence of Operations for Jobs and Machines, 3 Jobs, 5 Machines

Two points are in order. First, the calculations in all of the examples are precisely those used in the forward pass to compute early start and early completion task times in the **critical path method (CPM).** The early start time for operation (i, j) is S_{ij}, and the early finish time is C_{ij}. Here, because the machine sequences are given, the networks are more specialized than general CPM networks and therefore the calculations can be done efficiently. **The makespan for the schedule is exactly the length of the critical path in the network.**

Extending the analogy to the CPM method, a backward pass can be used to compute late start and late finish times for all operations. Here, the makespan is fixed at the value computed in the forward pass. The late finish time for any operation is defined to be the latest time that the operation can be completed without increasing the makespan. The late finish time for the operation "end" is defined to be the makespan value. The recursive computational formulas for the remaining operations (i, j) are:

$$LF_{ij} = \text{late finish time of } (i, j) = \text{minimum } \{LS_{lk}\} \text{ over all operations } (l, k)$$
$$\text{that follow operation } (i, j) \tag{11.62}$$

$$LS_{ij} = \text{late start of } (i, j) = LF_{ij} - t_{ij} \tag{11.63}$$

The operations that follow (i, j) are simply any remaining operations in job j and any operations that are done after (i, j) on the same machine. Meeting late start values ensures that all operations will be completed by the makespan date. Therefore, the late start values can be used as pseudo due dates during the actual completion of the operations. The following example demonstrates the computation process on the data from Example 11.20.

EXAMPLE 11.21

For the data and sequences depicted in Figure 11.8, compute the late start and late finish times for all operations.

SOLUTION

Begin with operation "end" and set its late finish and late start times to 69, the makespan value computed in Example 11.20. Next, go to operations that only have "end" as a successor—operation $(4, 3)$. The late finish time for $(4, 3)$ is 69, the late start time of "end," and the late start time is $69 - 7 = 62$ (using Equation (11.63) and t_{43}). Next, consider any operations in which the late start times of all of its successors have been computed. Operations (5, 1) and (3, 3) satisfy this requirement. Operation (5, 1) precedes "end" and (4, 3), therefore, its late finish and start times are:

$$LF_{51} = \text{Minimum } (LS_{\text{End}}, LS_{43}) = \text{Minimum } (69, 62) = 62 \tag{11.64}$$

$$LS_{51} = LF_{51} - t_{51} = 62 - 8 = 54 \tag{11.65}$$

For operation (3, 3), the calculations are:

$$LF_{33} = \text{Minimum } (LS_{43}) = \text{Minimum } (62) = 62 \tag{11.66}$$

$$LS_{33} = LF_{33} - t_{33} = 62 - 6 = 56 \tag{11.67}$$

The calculations for the remaining operations proceed similarly, and the following Table contains the results in the order computed (note, that the s_{ij} value is listed simply for comparison):

Opn.	t_{ij}	LF_{ij}	LS_{ij}	s_{ij}
End	**0**	**69**	**69**	**69**
(4, 3)	**7**	**Min $(LS_{\text{End}}) = 69$**	**62**	**62**
(5, 1)	**8**	**Min $(LS_{\text{End}}, LS_{43}) = 62$**	**54**	**54**
(3, 3)	6	Min $(LS_{\text{End}}, LS_{43}) = 62$	56	34
(4, 1)	**10**	**Min $(LS_{51}) = 54$**	**44**	**44**
(3, 2)	**10**	**Min $(LS_{\text{End}}, LS_{41}) = 44$**	**34**	**34**
(2, 3)	8	Min $(LS_{33}) = 56$	48	16
(2, 2)	**7**	**Min $(LS_{32}, LS_{33}) = 34$**	**27**	**27**
(3, 1)	**7**	**Min $(LS_{41}, LS_{22}) = 27$**	**20**	**20**
(2, 1)	**8**	**Min $(LS_{31}) = 20$**	**12**	**12**
(1, 2)	6	Min $(LS_{23}, LS_{22}) = 27$	21	10
(1, 3)	**12**	**Min $(LS_{21}, LS_{23}) = 12$**	**0**	**0**
(1, 1)	10	Min $(LS_{21}, LS_{12}) = 12$	2	0
Start	**0**	**Min $(LS_{11}, LS_{12}, LS_{13}) = 62$**	**0**	**0**

Any operations where $LS_{ij} = s_{ij}$ are critical in the sense that they must start precisely as soon as the machine is available for processing, otherwise the makespan will increase. These critical operations are bolded in the solution Table. Figure 11.9 contains the arcs from Figure 11.8 for the crit-

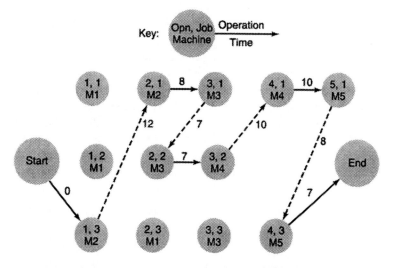

Path Length of 12 + 8 + 7 + 7 + 10 + 10 + 8 + 7 = 69

Figure 11.9 Critical Path of Operations for Jobs and Machines, 3 Jobs, 5 Machines

ical operations that connect to form the critical start to end path. The resulting is the longest path in the network. Operations with $LS_{ij} > s_{ij}$ have slack and can be delayed up to $LS_{ij} - s_{ij}$ time units without affecting the makespan.

The second point is that the calculations here are only for one possible choice of operation sequences on machines. To solve the job shop problem, one must determine the optimal sequence for each machine so that the criterion is optimized. The examples presented are descriptive of what will happen in the system once the decisions are made; however, it remains to set individual machine sequences.

A generic approach (denoted **GE**) for job shop scheduling can be developed based on the preceding computational method. Let MA be the set of machines that have been scheduled.

Step 1 (Initialization): Set $MA = \Phi$. Set up a network diagram with only the routing constraints (example in Figure 11.6).

Step 2 (Machine selection): Select a machine $i^* \notin MA$ to sequence. Find the sequence of jobs for i^* that optimizes the objective, given that the sequences for $i \in MA$ are fixed. Add i^* to MA.

Step 3 (Adjustment): For each machine $i \in MA$, $i \neq i^*$, individually check its schedule and see if the schedule can be changed to improve the objective. If $|MA| = M$, then stop the current set of machine schedules is the heuristic solution. Otherwise, return to step 2.

GE decomposes the M-machine problem into a set of single-machine problems. The approach gains in computational efficiency; however, this method is not guaranteed to find the optimal set of sequences.

One can vary the generic approach by specifying the method of solving for the appropriate schedule for each machine and selecting i^* in step 2, and the method of evaluating potential schedule adjustments in step 3. Two related approaches are addressed next.

Adams, Balas, and Zawack [1988] approach the problem of minimizing makespan for a job shop by using a sequential single-machine scheduling heuristic. Their approach, "The Shifting Bottleneck Heuristic," denoted **SB,** requires the solution of an integer programming problem for each machine to determine i^* in step 2 of GE and another integer programming problem for each machine when evaluating adjustments in step 3. The selection of i^* is based on the operation (and hence the machine) that has the largest tardiness value over all operations. Sule [1997] instead considers selecting i^* by finding the machine with the largest total tardiness. This "Modified Shifted Bottleneck Approach," denoted **MSB,** requires the solution of simpler integer programming problems (an implicit enumeration procedure is described) and performs reasonably well when compared to the original approach. The approaches are similar and are described below.

Step 1 (Initialization): Set $MA = \Phi$. Set up a network diagram with only the routing constraints (example in Figure 11.6). Compute s_{ij} and LS_{ij} for all operations (i, j)

Step 2 (Machine selection): For each machine $i \in M - MA$ formulate a single-machine problem with release times and due dates, where the operations to be done on that machine constitute the jobs. Assume that there are K operations that require processing on machine i. For each job/operation $k = 1 \ldots K$, let $r_k = s_{ij}$ and let d_k be the smallest LS_{lk} value over all operations (l, k) that are immediate successors of operation (i, j). No preemption is permitted. Solve each single-machine nonzero release problem using the L_{max} objective (SB) or the system tardiness objective (MSB). Select as i^*, the bottleneck machine, the machine that yields the largest L_{max} value (SB) or the largest system tardiness value (MSB). For machine i^*, use the sequence of operations determined from the single-machine problem, and add the schedule arcs into the graph. Add i^* to MA.

Step 3 (Adjustment): For each machine $i \in MA$, $i \neq i^*$, individually check its schedule and see if the schedule can be changed to improve the objective. This is done by deleting the scheduling arcs associated with machine i (the dotted arcs in Figure 11.8), recomputing s_{ij} and LS_{ij}, forming the appropriate release and due times ($r_k = s_{ij}$ and $d_k = LS_{ij}$), and solving the resulting problem using either L_{max} (SB) or system tardiness (MSB) as the objective. If a new schedule results, then test the schedule in the current network. If the makespan improves, then accept the new schedule, otherwise revert back to the existing machine schedule, and go to the next machine until all machines are tested for adjustment. If $|MA| = M$, then stop the current set of machine schedules is the heuristic solution. Otherwise, return to step 2.

The methods are called "bottleneck heuristics" because the machine selected in every iteration is the machine that has the worst tardiness or lateness value, and hence puts a constraint on the makespan. Therefore, the lateness criterion is a surrogate for the true criterion, makespan. The use of s_{ij} and LS_{kl} as release and due times is an intuitive way to model the competition of the operations for the machines. If an operation has successors that must be done quickly to ensure the makespan is met, then that operation has a sense of urgency and should be scheduled. If lateness occurs, then the due dates cannot be met and the makespan will increase. The goal is to select the machine that increases makespan as much as necessary and then fix that schedule so that little further delay is required.

Ties when selecting machines in step 2 can be broken arbitrarily and if the optimal sequence for a single machine is difficult to determine, then heuristics can be used (performance may decline, however, for this meta-heuristic). Also, if there are machine sequences that yield identical tardiness values, then these can be individually tested for the sequence that minimizes makespan. The following example uses system tardiness as the single-machine criteria, and hence is an application of MSB.

EXAMPLE 11.22

Using the data in Figure 11.6, use the MSB heuristic to find a low makespan schedule.

SOLUTION

Each iteration of the algorithm is addressed below. Initially, $MA = \Phi$ and machines will be tested for improvement in the order that they enter MA.

Iteration 1 Using the routing arcs in Figure 11.6, s_{ij} and LS_{ij} are computed for all (i, j). The makespan is 43. The release time for each operation is s_{ij}, and the due date is the minimum LS_{lk} value over all successor operations of (i, j). Because there are no machine sequencing arcs in the network, the successor operations are simply the next opertion in the routing after (i, j). The following Table lists the operations assigned to each machine. For each operation, the ID, the processing time, the release time, and the due time are given. The fifth column gives the optimal sequence for the objective of minimizing system tardiness subject to the given release times. The final column contains the value of the optimal objective.

Machine	OpnID. $-t_{ij}$ (r_k, d_k)	OpnID. $-t_{ij}$ (r_k, d_k)	OpnID. $-t_{ij}$ (r_k, d_k)	Sequence completion times	System tardiness
M1	(1, 1) − 10 (0, 10)	(1, 2) − 6 (0, 26)	(2, 3) − 8 (12, 30)	(1, 1) → (1, 2) → (2, 3) 10 16 24	0
M2	(2, 1) − 8 (10, 18)	(1, 3) − 12 (0, 22)		(1, 3) → (2, 1) 12 20	2
M3	(3, 1) − 7 (18, 25)	(2, 2) − 7 (6, 33)	(3, 3) − 6 (20, 36)	(2, 2) → (3, 1) → (3, 3) 6 25 31	0
M4	(4, 1) − 10 (25, 35)	(3, 2) − 10 (13, 43)		(3, 2) → (4, 1) 23 35	0
M5	(4, 3) − 7 (26, 43)	(5, 1) − 8 (35, 43)		(4, 3) → (5, 1) 33 43	0

$M2$ is the bottleneck machine because it has the largest system tardiness value. $M2$ is added to MA ($MA = \{M2\}$) and the sequence arc (1, 3) to (2, 1) is inserted. The resulting network is depicted in Figure 11.10 and has a makespan of 45. There are no machines to adjust in step 3.

Iteration 2 Using Figure 11.10, s_{ij} and LS_{ij} are computed for all (i, j). The release time for each operation is s_{ij}, and the due date is the minimum LS_{lk} value over all successor operations of (i, j). The successor operations are the next operation in the routing for job j, operation $(i + 1, j)$ and the next job in the sequence for the machine of (i, j) if it has been scheduled. The following Table has a structure identical to that of the corresponding Table for iteration 1 except $M2$ is not completed because it is in MA already. Instead, for $M2$, the optimal sequence is listed.

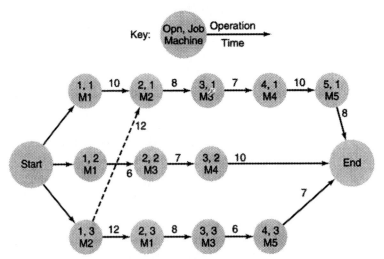

Figure 11.10 MSB Example—Iteration 1 M2 Sequence Set

Machine	OpnID. $- t_{ij}$ (r_k, d_k)	OpnID. $- t_{ij}$ (r_k, d_k)	OpnID. $- t_{ij}$ (r_k, d_k)	Sequence completion times	System tardiness
M1	(1, 1) − 10 (0, 12)	(1, 2) − 6 (0, 28)	(2, 3) − 8 (12, 32)	(1, 1) → (1, 2) → (2, 3) 10 16 24	0
M2	(2, 1) − 8	(1, 3) − 12		(1, 3) → (2, 1)	
M3	(3, 1) − 7 (20, 27)	(2, 2) − 7 (6, 35)	(3, 3) − 6 (20, 38)	(2, 2) → (3, 1) → (3, 3) 6 27 33	0
M4	(4, 1) − 10 (27, 37)	(3, 2) − 10 (13, 45)		(3, 2) → (4, 1) 23 37	0
M5	(4, 3) − 7 (26, 45)	(5, 1) − 8 (37, 45)		(4, 3) → (5, 1) 33 45	0

No machines have any tardiness, so arbitrarily $M1$ is added to MA ($MA = \{M2, M1\}$) and the sequence arcs (1, 1) to (1, 2) to (2, 3) are inserted. The resulting network is depicted in Figure 11.11. Now, step 3 is executed to see if changing the sequence on $M2$ will result in a lower makespan. Arc (1, 3) to (2, 1) is deleted from Figure 11.11, early and late start times are computed, and the resulting information for jobs on $M2$ and the optimal system tardiness sequence is:

M2	(2, 1) − 8 (10, 18)	(1, 3) − 12 (0, 22)	(1, 3) → (2, 1) 2

There is no change from the existing ordering, so the arc (1, 3) to (2, 1) is re-inserted into the network.

Iteration 3   Similarly to iteration 2, using Figure 11.11, s_{ij} and LS_{ij} are computed for all (i, j). The release times and due dates for each operation are computed. The resulting makespan remains 45. The following Table contains the relevant information and the optimal sequences for $M1$ and $M2$.

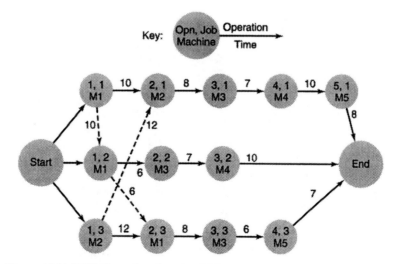

Figure 11.11 MSB Example—Iteration 2 M1 Sequence Set

Machine	OpnID. $- t_{ij}$ (r_k, d_k)	OpnID. $- t_{ij}$ (r_k, d_k)	OpnID. $- t_{ij}$ (r_k, d_k)	Sequence completion times	System tardiness
$M1$	$(1, 1) - 10$	$(1, 2) - 6$	$(2, 3) - 8$	$(1, 1) \rightarrow (1, 2) \rightarrow (2, 3)$	
$M2$	$(2, 1) - 8$	$(1, 3) - 12$		$(1, 3) \rightarrow (2, 1)$	
$M3$	$(3, 1) - 7$	$(2, 2) - 7$	$(3, 3) - 6$	$(3, 1) \rightarrow (2, 2) \rightarrow (3, 3)$	2
	$(20, 27)$	$(16, 35)$	$(24, 38)$	27 $\quad$ 34 $\quad$ 40	
$M4$	$(4, 1) - 10$	$(3, 2) - 10$		$(4, 1) \rightarrow (3, 2)$	2
	$(27, 37)$	$(23, 45)$		37 $\quad$ 47	
$M5$	$(4, 3) - 7$	$(5, 1) - 8$		$(4, 3) \rightarrow (5, 1)$	0
	$(30, 45)$	$(37, 45)$		37 $\quad$ 45	

$M3$ and $M4$ tie for the largest system tardiness, so $M3$ is added to MA ($MA = \{M2, M1, M3\}$) and the sequence arcs $(3, 1)$ to $(2, 2)$ to $(3, 3)$ are inserted. The resulting network is depicted in Figure 11.12 and the makespan is 47. Now, step 3 is executed to see if changing the sequence on $M2$ or $M1$ will result in a lower makespan. Arc $(1, 3)$ to $(2, 1)$ is deleted from Figure 11.12, early and late start times are computed, and the resulting information for jobs on $M2$ and the optimal system tardiness sequence is:

$M2$	$(2, 1) - 8$	$(1, 3) - 12$	$(1, 3) \rightarrow (2, 1)$	2
	$(10, 18)$	$(0, 24)$	12 $\quad$ 20	

There is no change from the existing ordering, so the arc $(1, 3)$ to $(2, 1)$ is re-inserted into the network. Next, $M1$ is tested and arcs $(1, 1)$ to $(1, 2)$ to $(2, 3)$ are deleted from Figure 11.12, early and late start times are computed, and the resulting information for jobs on $M1$ and the optimal system tardiness sequence is:

$M1$	$(1, 1) - 10$	$(1, 2) - 6$	$(2, 3) - 8$	$(1, 1) \rightarrow (1, 2) \rightarrow (2, 3)$	0
	$(0, 12)$	$(0, 27)$	$(12, 34)$	10 $\quad$ 16 $\quad$ 24	

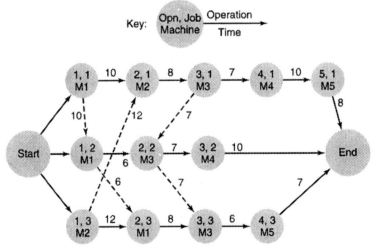

Figure 11.12 MSB Example—Iteration 3 M3 Sequence Set

There is no change from the existing ordering, so the arcs (1, 1) to (1, 2) to (2, 3) are reinserted into the network.

Iteration 4 Similarly to iteration 3, using Figure 11.12, s_{ij} and LS_{ij} are computed for all (i, j). The release times and due dates for each operation are computed. The resulting makespan is 47. The following Table contains the relevant information and the optimal sequences for $M1$, $M2$, and $M3$.

Machine	OpnID. $- t_{ij}$ (r_k, d_k)	OpnID. $- t_{ij}$ (r_k, d_k)	OpnID. $- t_{ij}$ (r_k, d_k)	Sequence completion times	System tardiness
$M1$	(1, 1) $-$ 10	(1, 2) $-$ 6	(2, 3) $-$ 8	(1, 1) $\to$ (1, 2) $\to$ (2, 3)	
$M2$	(2, 1) $-$ 8	(1, 3) $-$ 12		(1, 3) $\to$ (2, 1)	
$M3$	(3, 1) $-$ 7	(2, 2) $-$ 7	(3, 3) $-$ 6	(3, 1) $\to$ (2, 2) $\to$ (3, 3)	
$M4$	(4, 1) $-$ 10 (27, 39)	(3, 2) $-$ 10 (34, 47)		(4, 1) $\to$ (3, 2) 37 47	0
$M5$	(4, 3) $-$ 7 (40, 47)	(5, 1) $-$ 8 (37, 47)		(5, 1) $\to$ (4, 3) 45 52	5

$M5$ has the largest system tardiness, so $M5$ is added to MA ($MA = \{M2, M1, M3, M5\}$) and the sequence arc from (5, 1) to (4, 3) is inserted. The resulting network is depicted in Figure 11.13 and the makespan is 52. Now, step 3 is executed to see if changing the sequence on $M2$, $M1$, and $M3$ will result in a lower makespan. $M2$ is tested first, so arc (1, 3) to (2, 1) is deleted from Figure 11.13, early and late start times are computed, and the resulting information for jobs on $M2$ and the optimal system tardiness sequence is:

$M2$	(2, 1) $-$ 8 (10, 18)	(1, 3) $-$ 12 (0, 29)	(2, 1) $\to$ (1, 3) 18 30	1

This is a change from the existing ordering, so the arc (2, 1) to (1, 3) is re-inserted into the network and the makespan is recomputed. The resulting makespan is 51, and this is better than 52. Therefore, the ordering of jobs on $M2$ is changed to (2, 1) to (1, 3). The resulting network is depicted in Figure 11.14. Next, $M1$ is tested and arcs (1, 1) to (1, 2) to (2, 3) are deleted from Figure 11.14,

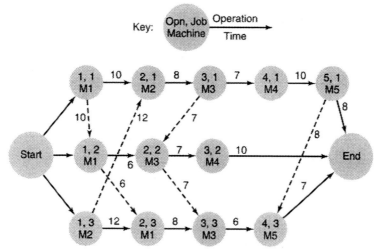

Figure 11.13 MSB Example—Iteration 4 M5 Sequence Set

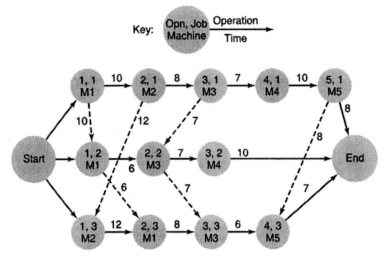

Figure 11.14 MSB Example—Iteration 4 M2 Sequence Changed

early and late start times are computed, and the resulting information for jobs on $M1$ and the optimal system tardiness sequence is:

$M1$	$(1, 1) - 10$	$(1, 2) - 6$	$(2, 3) - 8$	$(1, 1) \to (1, 2) \to (2, 3)$	0
	$(0, 10)$	$(0, 31)$	$(30, 38)$	$\quad 10 \qquad 16 \qquad 38$	

There is no change from the existing ordering, thus the arcs $(1, 1)$ to $(1, 2)$ to $(2, 3)$ are re-inserted into the network. Next, $M3$ is tested and arcs $(3, 1)$ to $(2, 2)$ to $(3, 3)$ are deleted from Figure 11.14, early and late start times are computed, and the resulting information for jobs on $M3$ and the optimal system tardiness sequence is:

$M3$	$(3, 1) - 7$	$(2, 2) - 7$	$(3, 3) - 6$	$(3, 1) \to (2, 2) \to (3, 3)$	0
	$(18, 26)$	$(16, 41)$	$(38, 44)$	$\quad 25 \qquad 32 \qquad 44$	

There is no change from the existing ordering, thus the arcs $(3, 1)$ to $(2, 2)$ to $(3, 3)$ are re-inserted into the network.

Iteration 5 Similarly to iteration 4, using Figure 11.14, s_{ij} and LS_{ij} are computed for all (i, j). The release times and due dates for each operation are computed. The resulting makespan is 51. The following Table contains the relevant information and the optimal sequences for $M1$, $M2$, $M3$, and $M5$.

Machine	OpnID. $- t_{ij}$ (r_k, d_k)	OpnID. $- t_{ij}$ (r_k, d_k)	OpnID. $- t_{ij}$ (r_k, d_k)	Sequence completion times	System tardiness
$M1$	$(1, 1) - 10$	$(1, 2) - 6$	$(2, 3) - 8$	$(1, 1) \to (1, 2) \to (2, 3)$	
$M2$	$(2, 1) - 8$	$(1, 3) - 12$		$(2, 1) \to (1, 3)$	
$M3$	$(3, 1) - 7$	$(2, 2) - 7$	$(3, 3) - 6$	$(3, 1) \to (2, 2) \to (3, 3)$	
$M4$	$(4, 1) - 10$	$(3, 2) - 10$		$(4, 1) \to (3, 2)$	0
	$(25, 36)$	$(32, 51)$		$\quad 35 \qquad 45$	
$M5$	$(4, 3) - 7$	$(5, 1) - 8$		$(5, 1) \to (4, 3)$	

*M*4 is the only remaining machine and is added to *MA* (*MA* = {*M*2, *M*1, *M*3, *M*5, *M*4}) and the sequence arc from (4, 1) to (3, 2) is inserted. The resulting network is depicted in Figure 11.15, and the makespan is 51. Now, step 3 is executed to see if changing the sequence on *M*2, *M*1, *M*3, and *M*5 will result in a lower makespan. Starting with *M*2, Arc (1, 3) to (2, 1) is deleted from Figure 11.15, early and late start times are computed, and the resulting information for jobs on *M*2 and the optimal system tardiness sequence is:

*M*2	(2, 1) − 8	(1, 3) − 12	(2, 1) → (1, 3)	1
	(10, 18)	(0, 29)	18 30	

This is no change from the existing ordering, thus the arc (2, 1) to (1, 3) is re-inserted into the network. Next, *M*1 is tested and arcs (1, 1) to (1, 2) to (2, 3) are deleted from Figure 11.15, early and late start times are computed, and the resulting information for jobs on *M*1 and the optimal system tardiness sequence is:

*M*1	(1, 1) − 10	(1, 2) − 6	(2, 3) − 8	(1, 1) →(1, 2) →(2, 3)	0
	(0, 10)	(0, 31)	(30, 38)	10 16 38	

There is no change from the existing ordering, thus the arcs (1, 1) to (1, 2) to (2, 3) are re-inserted into the network. Next, *M*3 is tested and arcs (3, 1) to (2, 2) to (3, 3) are deleted from Figure 11.15, early and late start times are computed, and the resulting information for jobs on *M*3 and the optimal system tardiness sequence is:

*M*3	(3, 1) − 7	(2, 2) − 7	(3, 3) − 6	(3, 1) →(2, 2) →(3, 3)	0
	(18, 26)	(16, 41)	(38, 44)	25 32 44	

There is no change from the existing ordering, so the arcs (3, 1) to (2, 2) to (3, 3) are re-inserted into the network. Finally, *M*5 is tested and arc (5, 1) to (4, 3) is deleted from Figure 11.15, early and late start times are computed, and the resulting information for jobs on *M*5 and the optimal system tardiness sequence is:

*M*5	(5, 1) − 8	(4, 3) − 7	(5, 1) →(4, 3)	0
	(35, 50)	(44, 51)	43 51	

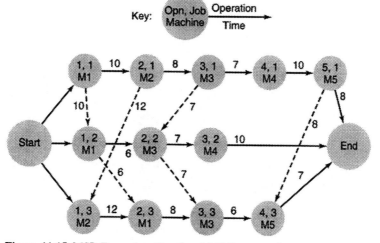

Figure 11.15 MSB Example—Iteration 5 M4 Sequence Set

There is no change from the existing ordering, thus the arc (5, 1) to (4, 3) is re-inserted into the network. The heuristic terminates because $|MA| = M = 5$. The resulting makespan is 51, and the final machine sequences are depicted in Figure 11.15.

11.6.3 Intelligent Scheduling Systems

The material discussed previously in the chapter has centered on mathematical programming, optimal sequencing, and heuristic approaches. As we have seen, optimal solutions are often hard to obtain and not available in real-time. Actual problems are often even more complex than the problems described. Processing times are random, machines break down, due dates change, and there may be multiple objectives. Real-time scheduling systems often use a combination of algorithms, heuristics, and expert knowledge. Expert knowledge is defined to be heuristic rules and strategies that a firm has found to work well on related (previously solved) scheduling problems.

Kusiak and Chen [1988] provide a general structure for classifying intelligent scheduling systems according to the way that the system represents knowledge about the scheduling problem. Scheduling systems should be able to perform the following tasks:

provide on-line decision support,

enable the dynamic scheduling of operations,

synchronize processes for different jobs simultaneously,

coordinate the use of machines, labor, and material handling resources, and

facilitate monitoring the execution of the schedule.

Typically, the term **Expert System** is used as a "catch-all" for computer systems that use expert knowledge in conjunction with models and algorithms. To build an expert system, one must consider the issues of:

knowledge representation,

the inference engine, and

knowledge acquisition.

Each of these are briefly addressed next.

The two main structures for knowledge representation are **production rules** and **frames.** A production rule has the structures:

IF <conditions>

THEN <actions>

or

IF <conditions>

THEN <conclusions>

An example of the first structure might be:

IF due date for job 1 < due date for job 2

THEN schedule operations of job 1 before those of job 2

while an example of the second structure is:

IF all jobs have positive slack values

THEN use an SPT rule for dispatching.

A frame is simply defined as a structured piece of knowledge and resembles a standard data structure found in a programming language like C. The frame contains the identity of the item of information and all ancillary data about the item that are required to make scheduling decisions. For example, the following frame descriptions contain information about a job and its initial operation:

Frame: Job 3254

Is-A: Job

Customer: Big Industries (Importance-value 6)

Release-Date: 7 September 1999

Due-Date: 10 October 1999

Operations-Sequence: OP1, OP6, OP9

Frame: OP1

Is-A: Operation

Work center: Edge Saw (BARMAC machine)

Processing-Time: 6 minutes (Range: 3, 15)

Setup Time: 3 minutes (Range: 2, 4)

Possible-Employees: EM23, EM27, EM30 (If-Modified: update)

The **inference engine** is the control structure of the expert system and is used to process the knowledge representation to develop the actual schedule or to give aid to a human scheduler. In rule-based systems, there must be an ordering of the rules and a processor for evaluating each rule and combining evaluations into a recommendation. In frame-based systems, the state of the system must be evaluated and the future states must be estimated. The system will make decisions to move toward more advantageous states, and the logic of testing and evaluating different states is the major part of the inference engine. Pattern matching may be used in both rule- and frame-based systems to match rule conditions with the current state of the system.

Knowledge acquisition enables the system to acquire the knowledge and data necessary to make decisions and usually closely follows the data representation scheme. Both static and dynamic knowledge must be included. For example, one can often use surveys of human schedulers to develop scheduling rules or one can use mathematical programming model solutions to formulate rules. Both methods require information on the current state of the system and that data must be updated continuously.

Intelligent scheduling systems have been in use since the early 1980s with the advent of the affordable computing. A few of the systems are addressed next, and additional material and references can be found in Noronha and Sarma [1991], Pinedo [1999], and Fraser [1995].

The OPAL system [Bensana et al. (1988)] is a rule-based system developed at Université Paul Sabatier. OPAL uses three kinds of knowledge, theoretical scheduling results, empirical knowledge about setting priorities for dispatching rules, and practical knowledge provided by shop floor managers. The system provides schedules for a job shop environment with multiple release times and due date related objectives. Batch transportation and setup times are considered. The system has three main modules:

A database that contains information such as a description of the shop, the data required for the scheduling problem, and the current schedule,

A constraint-based analysis tool that computes feasible time windows for the processing of each operation based on partial sequences and due dates (similar to the method used in the shifted bottleneck heuristic discussed previously), and

A decision support module that provides advice based on the practical knowledge.

A supervisor coordinates the passing of information between the modules and constructs the schedule based on the output from the modules.

PROVISIA (Lanner Group), the RHYTHM system (i2 Technology) and the MIMI system (Aspen Technology) are generic scheduling systems used in industry. All three have developed from relatively simple expert systems to comprehensive supply chain optimization systems. They consider constraint-based solution techniques as well as current optimization techniques such as simulated annealing and genetic search algorithms. They have built-in simulation features and links to cost data so that schedules can be evaluated and tested in an environment similar to that encountered in operation. The packages have advanced user interface features to facilitate data input and output as well as features to include company specific expert information into the scheduling processes.

11.7   SUMMARY

This chapter has covered the issues that typically arise in the scheduling of production systems. First, the theory of constraints was used to find the bottleneck processes that must be efficiently utilized to ensure that the system achieves maximum output. Scheduling involves the allocation of machines and other resources to specific job operations at precise times to meet some objective. Typical objectives for scheduling problem formulations are to minimize the system flow time, minimize system tardiness, minimize maximum tardiness, and to minimize makespan. Typical constraints are that operations for each job must be processed in their specified routing, and each machine may work on only one operation at a time.

Mathematical programming models were developed for problems on a single machine, problems on flow shops, and problems on job shops. Optimal sequencing rules exist for single-machine problems and two and three-machine flow shop problems. However, as the number of machines and the number of jobs increase, or when the objective is nonlinear, or when the jobs are ready for processing at different times, finding optimal solutions becomes computationally difficult. In these cases, heuristic procedures can be used to obtain high quality feasible solutions with significantly less computational effort than complete or partial enumeration of schedules. Also rule-based or frame-based intelligent scheduling systems can be used when past expertise in scheduling has proven to be effective.

The topics in this chapter are a small portion of the research on scheduling problems that has been developed during the past 45 years. The interested student is encouraged to look at the more theoretical operations research literature as well as the more applied production planning and scheduling literature. To effectively solve complex scheduling problems, it is necessary to understand the key objectives and constraints of the actual system, as well as the theoretical topics of mathematical modeling, discrete optimization, and heuristic development.

11.8   REFERENCES

ADAMS, J., BALAS, E., & ZAWACK, D. The shifting bottleneck procedure for job shop scheduling, *Management Science,* 1988, 34, 391–401.

ASKIN, R., & STANDRIDGE, C. *Modeling and analysis of manufacturing systems.* New York: John Wiley & Sons, Inc., 1993.

BAKKE, N., & HELLBERG, R. Relevance lost? A critical discussion of different cost accounting principles in connection with decision making for both short and long term production scheduling. *International Journal of Production Economics,* 1991, 24, 1–18.

CAMPBELL, H., DUDEK, R., & SMITH, M. A heuristic algorithm for the N job M machine sequencing problem, *Management Science,* 1970, 16, B630–B637.

CARLIER, J. The one-machine sequencing problem. *European Journal of Operations Research,* 1982, 11, 42–47.

CONWAY, R., MAXWELL, W., & MILLER, L. *Theory of Scheduling.* Reading MA: Addison-Wesley, 1967.

FRASER, J. Finite scheduling and manufacturing synchronization: Tools for real plant productivity. *IIE Solutions,* 1995, 27, 44–53.

FRENCH, S. *Sequencing and scheduling: An introduction to the mathematics of the job shop.* NY: John Wiley & Sons, Inc., 1982.

GOLDRATT, E. *The haystack syndrome.* Croton-on-Hudson, NY: North River Press, 1990.

GOLDRATT, E. *The theory of constraints.* Croton-on-Hudson, NY: North River Press, 1990.

GOLDRATT, E., & COX, J. *The goal: A process of ongoing improvement, revised edition.* Croton-on-Hudson, NY: North River Press, 1987.

GOLDRATT, E., & FOX, R. *The race.* Croton-on-Hudson, NY: North River Press, 1986.

GRAVES, S., MEAL, H., STEFEK, D., & ZEGHMI, A. Scheduling of reentrant flow shops. *Journal of Operations Management,* 1983, 3, 197–207.

JOHNSON, S. Optimal two- and three stage production schedules with setup times included. *Naval Research Logistics Quarterly,* 1954, 1, 61–67.

KUSIAK, A., & CHEN, M. Expert systems for planning and scheduling manufacturing systems. *European Journal of Operational Research,* 1988, 34, 113–130.

MOORE, J.M. An N job, one machine sequencing algorithm for minimizing the number of late jobs. *Management Science,* 1968, 15, 102–109.

NAWAZ, M., ENSCORE Jr, E., & HAM, I. A heuristic algorithm for the m-machine, n-job flow-shop sequencing problem. *OMEGA, The International Journal of Management Science,* 1983, 11, 91–95.

NORONHA, S., & SARMA, V. Knowledge-based approaches for scheduling problems: A survey. *IEEE Transactions on Knowledge and Data Engineering,* 1991, 3, 160–171.

PALMER, D. Sequencing jobs through a multistage process in the minimum total time—A quick method of obtaining a near optimum. *Operations Research Quarterly,* 1965, 16, 101–107.

PANWALKER, S., & ISKANDER, W. A survey of scheduling rules. *Operations Research,* 1977, 25, 45–61.

PINEDO, M. *Scheduling theory, algorithms and systems.* Englewood Cliffs, NJ: Prentice-Hall Inc., 1995.

PINEDO, M. & CHAO, X. *Operations scheduling with applications in manufacturing and services,* New York, NY: McGraw-Hill Inc., 1999.

RONEN, B., & STARR, M. Synchronized manufacturing as in OPT: From practice to theory. *Computers and Industrial Engineering,* 1990, 18, 585–600.

SMITH, J. *Theory of constraints and MRP II: From theory to results,* [Technical Report]. Bradley University, 1994.

STECKE, K. Nonlinear integer production planning problems for flexible manufacturing systems. *Management Science,* 1983, 29, 273–288.

SULE, D. *Industrial scheduling.* Boston, MA: PWS Publishing Company, 1997.

TAILLARD, E. Some efficient heuristic methods for the flow shop sequencing problem. *European Journal of Operational Research,* 1990, 47, 65–74.

WEIN, L. Scheduling semi-conductor wafer fabrication. *IEEE Transactions on Semiconductor Manufacturing,* 1988, 1, 115–129.

11.9   PROBLEMS

11.1.   A regular objective is defined to be an objective that remains constant or improves as job completion times decrease. Explain why it is never optimal to insert unnecessary idle time into a schedule when you are optimizing regular measures.

11.2.   Give an example of a nonregular measure, and devise problem data that show it is better to have machine idle time, although none is necessary.

11.3.   Consider, a single-machine scheduling problem in which jobs have nonzero release times and preempt resume is the method of operation. Show that:

a.   If the objective is to minimize the maximum F_j value, then the optimal sequence processes the jobs in a first come first served order.

b.   There exists an optimal sequence that has no preemption.

11.4.   For a single-machine problem, the VIP sequence is to order jobs so that:

$$w_{[1]} \geq w_{[2]} \geq \cdots \geq w_{[N-1]} \geq w_{[N]}$$

a.   If job weights are proportional to job processing time ($w_j = \alpha t_j$) then show that all sequences are equivalent for minimizing total system weighted flow time.

b.   If all jobs have equal processing time, then show that VIP sequencing is optimal for minimizing total system weighted flow time.

11.5.   Consider a single-machine problem in which all jobs are released at time 0, all jobs are equally weighted, and the objective is to minimize total system flow time. Use the adjacent pairwise interchange method to show that the optimal sequence follows the SPT rule.

11.6.   Consider a single-machine problem in which all jobs are released at time 0. Show that if an EDD sequence has only one tardy job, then that sequence must be optimal for the NT objective.

11.7.   Consider a single-machine problem in which all jobs are released at time 0. Show that if an SPT sequence has all jobs tardy, then this sequence is optimal when the objective is total system tardiness.

11.8.   Consider a single-machine problem where all jobs are released at time 0. Show that if all jobs have a common due date d, then the SPT rule is optimal when the objective is total system tardiness.

11.9.   A shop consisting of three machines has seven jobs waiting to be released. Jobs are released in daily batches. Batches may have at most eight hours of work per machine. All jobs are considered to be of equal importance.

Job j	d_j	t_{1j}	t_{2j}	t_{3j}
1	30	6	4	1
2	8	3	4	4
3	54	4	3	1
4	16	2	3	6
5	36	5	4	2
6	24	3	2	3
7	40	4	2	5

a. Construct release batches using Earliest Due Date.

b. Construct release batches using Shortest Processing Time.

c. Construct release batches using Minimum Slack Time.

d. Construct release batches using Critical Ratio.

11.10. Suppose in the previous problem that the work release rules are applied on an aggregate basis. Each job is assumed to have a work content equal to the sum of its three machine times. Each daily batch is allowed to contain up to 22 hours of total work.

a. Construct release batches using Earliest Due Date.

b. Construct release batches using Shortest Processing Time.

c. Construct release batches using Minimum Slack Time.

d. Construct release batches using Critical Ratio.

11.11. Using the data in problem 11.9, formulate and solve model R1 to determine release dates. Assume that the jobs have weights (1, 3, 2, 2, 3, 1, 4) respectively. (You may ignore the due dates).

11.12. Using the data in problem 11.9, write out the formulation for enhanced model R2 to determine release dates. Assume that the jobs have weights (1, 3, 2, 2, 3, 1, 4) respectfully.

11.13. Machine three has been identified as the bottleneck machine in a four machine serial system. Average flow time from release until reaching the buffer at machine three is five hours. To avoid shortages, releases are scheduled to maintain 15 hours of machine three workload upstream of machine three. Jobs are released in due date order. Given the following set of jobs in the system and waiting to be released, what action should be taken? When do you expect each of the waiting jobs to be released? All processing times and due dates are in hours.

Job j	d_j	t_{j1}	t_{j2}	t_{j3}	t_{j4}	Current status
1	12	1	2	4	2	Machine 3 buffer
2	24	1	3	3	1	At machine 2
3	24	1	1	2	1	At machine 1
4	40	2	1	3	3	At machine 1
5	40	1	2	5	2	Waiting for release
6	56	1	1	2	1	Waiting for release

11.14. The following eight jobs are waiting in queue to be scheduled on a machining center.

Job j	1	2	3	4	5	6	7	8
t_j	12	5	8	21	6	9	11	4
d_j	100	35	86	19	12	45	75	64

a. Sequence the jobs to minimize total flow time.

b. Sequence the jobs to minimize maximum tardiness.

c. Sequence the jobs to minimize the number of tardy jobs.

11.15. In Example 11.8, a homework assignment scheduling problem is used to demonstrate the Moore and Hodgson Algorithm for optimizing NT. Do you use such an approach when scheduling your own completion of homework assignments? What are the drawbacks of using such an approach for your own homework scheduling problem. What approach do you use for scheduling events in your daily life?

11.16. Consider the following data for a single-machine problem:

Job	Processing time	Due time
1	6	41
2	8	34
3	7	12
4	6	16
5	10	24
6	5	17

Find the optimal sequence that minimizes NT using the Moore and Hodgson Algorithm.

11.17. Consider the following data for a single-machine problem:

Job	Processing time	Due time
1	6	41
2	8	34
3	7	12
4	6	16
5	10	24
6	5	17

Assume that we run in a preempt resume operation.

a. Find the optimal sequence that minimizes total system flow time.

b. Find the optimal sequence that minimizes maximum tardiness.

11.18. Construct a numerical example that shows that an EDD sequence does not minimize total system tardiness.

11.19. Consider the following data for a single-machine problem:

Job	Processing time	Due time
1	6	41
2	8	34
3	7	12
4	6	16
5	10	24
6	5	17

a. Use the WI heuristic to find a sequence that gives low total system tardiness.

b. Use the AU heuristic to find a sequence that gives low total system tardiness.

c. Use the PSK heuristic to find a sequence that gives low total system tardiness.

11.20. Consider the following data for a single-machine problem:

Job	Processing time	Due time
1	10	19
2	9	47
3	5	7
4	4	6
5	8	24
6	11	28

a. Use the WI heuristic to find a sequence that gives low total system tardiness.

b. Use the AU heuristic to find a sequence that gives low total system tardiness.

c. Use the PSK heuristic to find a sequence that gives low total system tardiness.

11.21. Construct the optimal sequence and its Gantt chart for the following data for a two-machine flow shop problem.

Job	t_{1j}	t_{2j}
1	4	6
2	5	3
3	2	7
4	3	4
5	8	4
6	6	5

11.22. Construct the optimal sequence and Gantt chart for the following data for a three-machine flow shop problem.

Job	t_{1j}	t_{2j}	t_{3j}
1	4	3	6
2	3	3	5
3	6	2	6
4	7	2	3
5	4	1	4

11.23. Consider the following data for a two-machine shop problem: (Define the subsets A through D. Construct an optimal sequence and Gantt chart.)

Job	t_{1j}	t_{2j}	Machine order
1	8	–	M1 Only
2	–	6	M2 Only
3	3	6	M1 then M2
4	4	8	M1 then M2
5	6	2	M2 then M1
6	5	3	M2 then M1
7	1	7	M2 then M1

11.24. Consider the following data for a four-machine flow shop problem:

Job	t_{1j}	t_{2j}	t_{3j}	t_{4j}
1	6	8	3	5
2	4	4	2	10
3	7	3	5	8
4	3	8	7	4
5	6	8	2	5

a. Use the P heuristic to determine a permutation schedule with low makespan.

b. Use the CDS heuristic to determine a permutation schedule with low makespan.

c. Use the NEH heuristic to determine a permutation schedule with low makespan.

11.25. Consider the following data for a four-machine flow shop problem:

Job	t_{1j}	t_{2j}	t_{3j}	t_{4j}
1	3	6	5	2
2	7	6	2	6
3	4	7	8	4
4	6	3	2	8
5	3	5	3	8

a. Use the P heuristic to determine a permutation schedule with low makespan.

b. Use the CDS heuristic to determine a permutation schedule with low makespan.

c. Use the NEH heuristic to determine a permutation schedule with low makespan.

11.26. Consider the following data for a five-machine job shop:

Job	Machine routing—processing time				
1	M1–4	M2–5	M3–8	M4–5	M5–5
2	M2–5	M1–3	M4–6	M3–8	M5–6
3	M2–3	M3–6	M4–3	M1–7	
4	M3–6	M1–7	M2–5	M5–9	M4–7
5	M2–4	M3–4	M4–3	M1–3	
6	M1–7	M5–5	M4–7	M3–6	M2–3

a. Use the SPT rule to develop a schedule and Gantt chart for the set of jobs and compute the makespan.

b. Use the LWR rule to develop a schedule and Gantt chart for the set of jobs and compute the makespan.

c. Use the SLK rule to develop a schedule and Gantt chart for the set of jobs and compute the makespan.

d. Use the LOR rule to develop a schedule and Gantt chart for the set of jobs and compute the makespan.

11.27. Consider the following data for a five-machine job shop:

Job	Machine routing—processing time				
1	M1–4	M2–5	M3–8	M4–5	M5–5
2	M2–5	M1–3	M4–6	M3–8	M5–6
3	M2–3	M3–6	M4–3	M1–7	
4	M3–6	M1–7	M2–5	M5–9	M4–7
5	M2–4	M3–4	M4–3	M1–3	
6	M1–7	M5–5	M4–7	M3–6	M2–3

a. Set up model JS for this data set assuming that the objective is to minimize makespan, and all jobs are released at time 0.

b. How many variables does the formulation have? How many constraints?

c. If we double the number of jobs in the problem, then how does the number of variables and constraints change?

d. If we double the number of machines in the problem, then how does the number of variables and constraints change?

11.28. Consider the data for problem 26.

a. Draw a graph similar to Figure 11.6 and compute the makespan when each job has its own set of machines.

b. Assume that the sequence of operations on each machine is lowest job number first. For example, all jobs require processing on machine 1, thus the sequence is (1, 2, 3, 4, 5, 6). Only four jobs require processing on machine 5, thus its sequence is (1, 2, 4, 6). Draw a graph similar to Figure 11.8 and compute the makespan for the jobs under this sequencing rule.

11.29. Use the generic GE approach heuristic for job shop scheduling to find a low makespan schedule for the following job shop data set:

Job	Machine routing—processing time			
1	M1–4	M2–5	M3–8	M4–5
2	M2–5	M1–3	M4–6	
3	M2–3	M3–6	M4–3	M1–7
4	M3–6	M1–7	M2–5	
5	M2–4	M3–4	M4–3	M1–3
6	M1–7	M4–7	M3–6	

In step 2 of the approach, select the machines in numerical order. Note, that because there are at most four jobs to be sequenced on each machine, you can use enumerative techniques in step 2 and step 3 to find locally optimal sequences.

11.30. Use the modified shifted bottleneck heuristic for job shop scheduling to find a low makespan schedule for the following data set:

Job	Machine routing—processing time		
1	M1–4	M2–5	M3–8
2	M2–5	M1–3	
3	M2–3	M3–6	
4	M3–6	M1–7	
5	M2–4	M3–4	M1–3
6	M1–7	M3–7	M2–6

CASE STUDY 11.9 *Scheduling Systems*

Elecs, Inc. (EI) is a small company that subcontracts small electronic components manufacturing. Customer companies provide EI with components, assembly specifications, and equipment for producing their particular products, and EI provides labor and space for the assembly operations. The labor pool is largely low-skilled and therefore, the assembly operations must be simple and easy to teach.

EI receives forecasts for each week's demand some three months in advance. The components are shipped to a warehouse and then are trucked to EI. The components are inspected for quality and the appropriate quantity. Once all components for a particular order are available, the order is scheduled for production. The scheduler divides the order into a set of equal sized batches (each batch is called a kit) and components for each kit are collected and grouped together. The schedule for the next week is planned on the Friday before and includes orders due in the next week and any backorders. The schedule consists of a time table of operations for each employee and each machine. The primary objective is a feasible schedule that meets due dates.

Often, the customer changes the forecast, and these changes can occur as late as during the week of production. EI is in the business of satisfying demand, so it is critical that these changes be handled effectively. If the forecast increases significantly, then components can be shipped overnight to facilitate assembly. The scheduling system must be able to react quickly to changes in the forecast. Both employees and machines can be impacted by schedule changes.

The production of each kit follows a serial routing with between five and ten operations, and different employees can work sequentially on the kit. Between operations, the kits are stored in a storage area at the end of the plant. Each operation must be performed on a specific machine type and specific set of skills is needed. Operation times are known and generally take about one hour per kit. Each employee has a set of skills and these are matched to the requirements for each operation when assigning employees to operations. Because the operations are simple, there is little learning curve effect and all employees trained in an operation tend to work at the same speed

for that operation. Machines and employees can work on only one kit at a time and there is no pre-emption of operations. Each kit need not be completed by the end of the day, and work can continue to the next day.

The plant layout is a typical process layout — machines of similar type are located together. Each machine and employee must be scheduled over the day and the standard time period size for planning is one hour. Machine fixtures are not a scarce resource and are not critical in the scheduling process. Material handling is done manually, and the entire kit fits into a small plastic container. Upon completion of a kit, there is a 100% inspection of the parts, and parts may be sent for rework (rework must now be included in the schedule). On completion of all kits for an order, the parts are packaged and shipped to the customer.

The current database system is an 1987 MS-DOS application that reports inventory levels of all components and all products. The tool has been in use since 1987, requires input from a single person trained in operation, and is considered obsolete.

EI has enlisted your team to help improve its production planning system. Specifically, they need help with the following tasks:

1. Design and implementation of a database system that tracks order progress, provides input to the scheduling process, provides data for a method to determine employee compensation, and provides reports necessary for customer relations.

2. Design and implementation of a scheduling system that provides a weekly schedule for each machine and each employee.

3. Design and implementation of an incentive system for rewarding employees for obtaining higher skills and for completing more parts.

Chapter 12

Shop Floor Control: Systems and Extensions

$\mathbf{E}$nterprise information systems require data from the shop floor. Purchasing and short-term planning (MRP) require knowledge of inventory levels. Marketing will frequently need to locate open shop orders to respond to customer inquiries. Accounting requires input on worker activity to run payroll and industrial engineering needs these data to develop standards. Schedulers need complete shop status of orders, workers, machines, and tools to produce short-term schedules. The list goes on and on. The shop floor control (SFC) system accumulates these data for higher level administrative and planning functions and controls the execution of detailed tasks that define the short-term production schedule.

The logical functions of SFC consist of data acquisition, planning, scheduling, and execution control. Embedded in these is the communication protocol and network for informing and coordinating the shop's resources. The schedule converts the production plan of what to produce by time period into a detailed plan of when each operation will be performed and the equipment, tooling, instructions, and personnel required to perform those operations. Execution control enacts the schedule. This includes dispatching and monitoring subfunctions. The dispatcher informs the resources (equipment and personnel) that it is time to start an operation using the specified tools and instructions. The dispatcher takes this action based on the schedule and current status of the shop resources and open orders. The monitor lies in the background continually (or frequently) observing the status of the system and comparing that status to that expected by the schedule. Violations are fed back to the scheduler. The monitor must contain the capability to capture data, analyze that data for discrepancy from the schedule, and provide decision support by defining the appropriate message to feedback to the schedule. For example, if a job is behind schedule because the machine it requires has broken down, the scheduler should know the condition of the machine, the location and status of the job, the expected renewal time for this machine, the other jobs waiting for this machine, and the status of any other jobs that depend on the delayed job. The monitor must integrate order status, machine status, and maintenance schedule data. Quality data is also relevant. The data capture function should record machine performance in regard to time and quality standards as well as current status.

In this chapter, we address the basic mechanisms of shop floor control and several commonly encountered situations that call for special-purpose scheduling and resource management methods. We begin with a discussion of general control architectures for manufacturing shops. This is followed by a description of the functions that should be included in a SFC system. From there, we look at several special manufacturing environments—high volume assembly lines, flexible flow systems, re-entrant flow systems, and flexible manufacturing systems.

12.1   CONTROL SYSTEM ARCHITECTURE

The control system architecture provides the framework within which the shop floor control system (SFCS) must operate. Architecture defines the linkages between components, how they will interact and the allowable ways to communicate. Traditionally, hierarchical structures have been used, but heterarchical approaches have been contemplated in recent years. In the following sections, we describe these two conceptual frameworks.

12.1.1   Hierarchical Control Systems

Historically, shop floor control systems (SFCSs) have been implemented through an hierarchical architecture. In the hierarchical paradigm, communication occurs "vertically." Each resource entity, such as a machine, transporter, or worker, has a single supervisor that provides instructions. The resource may have one or more subordinates under its authority. Instructions are passed down to subordinates. Subordinates acknowledge the receipt of instructions, attempt to execute the instruction, and then report back to the supervisor. Each entity may include a task manager that breaks up its instructions into more detailed tasks, schedules these tasks, and controls their execution. Figure 12.1 defines a four-level model encompassing plant, department, workstation, and equipment levels. Arcs indicate allowable communication paths. A spot welding robot, component insertion machine, or lathe would be equipment entities. Equipment entities connected by an integrated material handling system and usually assigned to a single shop order at any time constitute a workstation. The shop, also referred to as a department as shown in the Figure, is composed of workstations. A simple example will illustrate the concept. The shop will send a message to a workstation to produce a part that is now located in its input queue. The workstation's task manager will break this up into instructions to load the part on a machine, perform the required operations, and unload the part. In so doing, these tasks

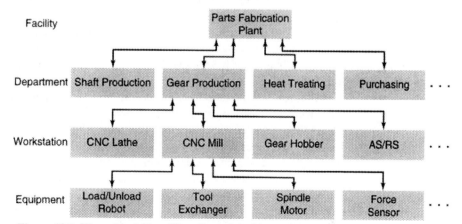

Figure 12.1 Hierarchical Shop Floor Control Architecture

are subdivided into more specific tasks and instructions for lower-level entities such as actuators on the pick and place robot, the position locator for the tool magazine, and the spindles on the machine. These lower level devices could be considered as a fifth level.

12.1.2 Heterarchical Control Systems

Large systems are difficult to coordinate. Simplification offers one approach to overcome this difficulty. In production scheduling, we saw how kanban control can be used to simplify the information structure and distribute decision-making authority to individual workstations. As system complexity and information requirements increase, distributed processing becomes more common. Each object such as a machine can be responsible for monitoring, scheduling, and controlling itself, as well as passing along only the information necessary to coordinate with the system. For example, a machine may track its usage and contact the maintenance department directly when it is time to schedule maintenance. Heterarchical control systems carry this paradigm a step further. Taking advantage of modern communication and decision making technologies, parts, transporters, and machines are autonomous entities that negotiate. Each job entity carries attributes detailing the required operations and due date. Each machine understands its current processing capabilities and current scheduled workload. Jobs are allowed to broadcast requests for services and the machines place bids. The job then selects the best offer. The job will have to coordinate its activities such as the completion time at one machine, obtaining a transporter, and arriving at the next machine. When conflicts exist such as two jobs waiting for a machine at the same time, control algorithms can be used to find solutions with the least deviation from the individually optimal schedules. Resource requirements are generally conditional on multiple events. As the requests are interrelated, planning processes may require iterative schemes for broadcasting needs, responding with offers, and making agreements. Consider a specific machine. It would need to possess a process plan evaluator to determine if it can fulfill the service requirements broadcast by a part. The machine must then determine if necessary tools are already available on the machine or if they can be obtained by broadcasting a request for the tools to other machines and the tool crib. If the responses are affirmative, the machine would respond with a time commitment for these tasks. The offer is kept on the machine's blackboard until confirmation is received or the deadline for the offer expires. Once the bid is accepted, transportation must be arranged by sending out requests for transport devices. In turn, devices respond with estimated delivery times.

Large heterarchical systems experience significant communication demands. Advantages, however, include eliminating the need for continuous centralized replanning, the robustness of the system to breakdowns in individual entities, and the use of current status data for decision making. The modular nature of the system allows new objects to be brought on-line quickly. Veeramani et al. [1992] describe a complete information system architecture for such systems.

Although heterarchical systems have received a fair amount of attention in the literature, their introduction in practice has been slow. Smooth production and small queues require that machines coordinate their schedules. Conceptually, it is easier to see how this can be accomplished explicitly using a centralized scheduling system. We know, however, that these "best laid plans" seldom work out. Indeed, kanban systems represent one recognition that centralized control may not always be best in practice. Moreover, the production stages do cooperate in pull systems. Information is passed between adjacent stages instead of centrally. The issue then becomes how to force autonomous objects (machines and jobs) to cooperate without compromising their freedom. An important heterarchical system design principle decomposes the system into nearly-independent entities connected by a few, well-defined inter-relationships that enable communication and cooperation. These interactions should be of sufficiently low frequency and impact that the entities are robust in the short run to external changes. Minimization of linkages also makes it easier to add and subtract entities over time as systems evolve.

12.2   MANUFACTURING EXECUTION SYSTEMS

After creating the production plan, we execute the plan. Plan execution requires forming a sequence of tasks to be performed, communicating that sequence to the resources, co-ordinating the authorization to initiate the tasks, monitoring the completion of those tasks, and updating the sequence over time as planned and unexpected events occur. In addition, it is helpful to use the data collected during the execution process to support other functions including the preparation of reports for higher management, scheduling preventive maintenance, and feeding administrative functions such as purchasing, accounts payable, payroll, and personnel. Integrated software-based manufacturing execution systems, or MES, have become popular in recent years for controlling production operations and sharing operations data with related business and engineering functions. The MES assumes responsibility for converting the production plan into a set of tasks and monitoring those tasks. The core activities are as follows:

1. **Interface to production planning system.** The production planning system determines what is to be produced. The MES must convert this plan into a required set of activities and adjust those activities as the higher level plan changes each period. Planned batch quantities and due dates are combined with process plans to create the list of activities. In turn, the production planning system will use the shop status information supplied by the MES when updating production plans for the next period.

2. **Work order management.** The progress of each shop order must be tracked as it goes through the plant. The MES should coordinate the scheduling of all operations for an order to ensure it is completed on time. These operations must be tracked and adjustments made to the schedule for remaining operations as unexpected deviations in the outcome of individual operations occur. The priority of each work order is maintained in the system as a function of its due date and remaining operations. The due date may be set as a function of related work orders that ultimately match up at final assembly. Order tracking serves external uses as well. Billing may issue invoices for partial payments as orders progress. Customers may periodically request updates on order status. This has become an expected service as companies from Federal Express to Dell Computer allow customers to track the progress of their order on-line over the Internet. We may also need to check order status in response to customer or engineering initiated change orders. To speed flow times, orders may be broken into sublots. In this case, the SFCS should track these sublots to coordinate their progression and allow for recombining these sublots at a later step.

3. **Workstation management.** The system should maintain a schedule for each workstation and monitor the status of the workstations. The set of orders available for processing at the workstation is kept and used to create a schedule by applying the desired scheduling rules. The system therefore knows the tasks and setup planned for each workstation at all times. This can be used to plan the delivery of tooling and processing instructions. It also allows determination of the workload on each workstation and can anticipate long queues and short-term capacity problems that may require a change in planned order releases or rerouting of some jobs. A list of orders available for processing at each

workstation is maintained along with links to the location of setup and processing plans for each order destined for the workstation. The status of the workstation should be monitored with that information being available in the MES. If breakdowns or other problems occur at the workstation, the system should be able to automatically replan the routing of affected jobs.

4. Tooling management. Each operation of each work order requires specific tooling. The location of all tools should be kept current in the MES. Tool life should be monitored and any retirement or regrinding of tools should be automatically scheduled by the MES when appropriate. As the start time for an operation approaches, the MES should be able to ensure that tooling will be available and have it sent to the workstation so that it will be available when needed without delaying setup.

5. Labor management and effort reporting. Each worker must be tracked with respect to availability and capability. The system will use this information along with skill requirements noted in the process plan of each part to assign workers to tasks. Monitoring of worker performance and status serves personnel management, accounting, and industrial engineering functions. Historically, workers used time cards to report their activities. More typically today, workers scan in their ID number from a badge along with the bar code from a job whenever they begin or complete an operation. Data collected by an automatic identification system on the shop floor can be fed directly to personnel/wage administration to prepare performance reports on individual workers and to automatically compute hours for each pay period. In addition, quality problems can be traced back to specific task times, workers, and machines. When the operator inputs his ID and order number, the SFCS should check to ensure that the operator has the appropriate certification for this operation. In addition, the combination of tooling and order can be compared as one check that the operator is about to perform the operation properly.

6. Inventory tracking and traceability. Specifics on the location, quantity, and operational history of all materials, WIP and finished goods still in the system should be maintained by the MES. MRP systems and purchasing rely on accurate inventory records. As available parts are assigned to orders and then removed from stores, we need to track the available quantities. In many industries, we will need to trace each item from the time it is received until it is combined into a finished product and shipped. This is necessary for health reasons in the pharmaceuticals and food industries, for safety and service reasons in the consumer goods industries, and for accounting purposes on government contracts. This information can also be helpful in identifying quality problems by tracing commonalities among returned products. Imagine the value of being able to identify that the majority of products that fail early in the field are all produced on the same machine or use a component from the same supplier.

7. Material handling. The movement of unit loads of material through the system is controlled by the MES. The MES schedules requests for movement and identifies the appropriate transport device. This device is then located and given the necessary coordinates and details to pick up and deliver this load.

8. Automated data collection and management. The MES should control general data collection activities and manage these data. As we have seen, these data are used for management reports and also for tracking the state of the system and making production schedules. Although data can be collected and input manually, automatic identification devices such as bar code scanners and radio frequency (RF) transmitters are often used. With the large quantity and rapid turnover of order and resource status data, it is imperative that the MES have a well-designed database structure and management system. The MES must exchange data with the centralized engineering, business, and marketing databases to obtain bills of materials, process plans (including current routings and time standards), due dates, and report on job status and labor activity.

9. Production control and exception management. The actual prioritized schedule of tasks should be determined by the MES. In doing so, the MES will normally make use of resource and order data. Knowledge of organizational procedures and constraints is also stored and becomes important in setting schedules. Once planned, schedules need to be communicated and executed. The SFCS dispatches commands to machines, workers, and material handlers to execute tasks. As time passes, shop status is updated as responses are received from these resources announcing the completion of their assigned tasks. This would likely be the case in which simple dispatching rules are used to set priorities for processing resources. Allowing the SFCS to make decisions such as which order to begin processing when a machine becomes available reduces the extensive need for communication and continuous replanning that accompanies centralized planning. The MES may use simple dispatching rules for setting schedules, logical reasoning, or more complex decision algorithms. These approaches were addressed in the previous chapter.

10. Maintenance management. Planned maintenance can be crucial to removing variability in product quality and resource availability. Automated data collection can be fed to a computerized system for scheduling machine maintenance. The system can track the time that has transpired since the last overhaul for parts and tooling and, therefore, schedule routine maintenance. This schedule can in turn be fed back to the resource scheduling module to ensure feasible production schedules. Basic procedures such as calling for regrinding of a tool when its planned tool life is reached can significantly reduce quality and downtime problems. If quality data is recorded, the system can also track deterioration in machine (and operator) performance and schedule extraordinary maintenance checks if they appear to be warranted.

11. Quality assurance. It is useful to have the MES directly interfaced to the quality assurance system. If a problem develops with a workstation that affects its availability or performance, schedules for all shop resources should be updated. If fully integrated, the MES can track status of gages and other test equipment, and schedule calibration activities much in the same manner as preventive maintenance is scheduled on machines. SPC results can also be tied to maintenance.

12. Supplier interface. The need for materials derives from inventory levels, bills of materials, and operation schedules. Thus, linking material purchases into the MES seems natural. The system can monitor stocks of purchased parts and order replenishments automatically using the rules developed in this text. Moreover, the MES can automatically report on the on-time delivery of each supplier. With a tie-in to the quality assurance system, the results of incoming inspections and internal failures can be linked to vendor claims and material specifications.

12.3 DESIGN AND CONTROL OF FLOW SHOP SYSTEMS

Serial production systems are used to produce a wide variety of products. The simplified flow pattern provides an opportunity for minimizing material handling distances and coordinating schedules between adjacent processors. In Chapter 11, we provided basic methods for scheduling jobs in a flow shop. We now address several special topics related to controlling the flow of jobs in serial systems. In particular, we discuss three important topics—the allocation of tasks to workstations (assembly line balancing), selecting the proper number of units to transfer between workstations at a time (lot-streaming), and scheduling of flow shops with parallel processors (flexible flow lines). In subsequent sections, we cover re-entrant flow shops and an introduction to the operational problems associated with computer-controlled flexible manufacturing systems with general flow patterns.

12.3.1   Assembly Line Balancing

Assembly line balancing (ALB) is the term commonly used to refer to the decision process of assigning tasks to workstations in a serial production system. The tasks consist of the elemental operations required to convert raw materials into finished products. The basic ALB problem addresses the case of a single product being produced in high volume with a labor-intensive process. Workstations represent individual workers who are viewed as identical and capable of performing or learning all tasks. Whereas it may be possible to let each workstation produce entire products from start to finish, there are advantages to splitting the total production process into a series of stages with a different worker used for each stage. The serial and parallel system design extremes are shown in Figure 12.2. Advantages of the single serial assembly line include:

the ability to use a synchronous part entry and transfer mechanism to pace the production rate;

reduced training requirements as each worker must only learn a subset of tasks;

shorter cycles usually have a faster learning curve rate (greater reduction per cycle);

shorter cycles imply less chronological time until workers exceed the 100% standard pace resulting from learning (less time until the nth cycle);

reduced capital cost because each task is performed at a single workstation thus avoiding the need to duplicate tooling;

elimination of setup time that might otherwise be required if workers constantly switch back and forth between tasks.

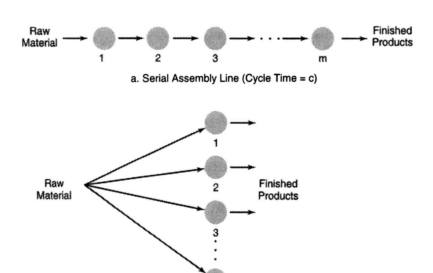

a. Serial Assembly Line (Cycle Time = c)

b. Parallel Assembly Line (Cycle Time = mc)

Figure 12.2 Serial vs. Parallel Workstation Design

The parallel system of Figure 12.2b on the other hand has its own advantages, notably:

job enrichment for employees;

longer cycle times reduce fatigue and repetitive stress illnesses;

direct accountability for tracing and correcting quality problems;

lower idle times caused by imbalance in total task times between workstations;

multiple products or models can be assigned to different workers without forcing idle time (balancing task times across workstations in a serial line is more difficult with multiple products);

minimization of task reassignment problems and confusion because of absenteeism;

independence of workers whereby an individual's output is not affected by the absences of others or differences in work rates.

Other factors can favor either or neither line design depending on the situation. In serial lines, changes in production requirements can be accomplished by adjusting the number of workers on the line and their task assignments, or changing the number of hours the line is scheduled. Changing the number of parallel lines or their scheduled hours will likewise vary output.

If we want to design a line with more than one serial workstation, the primary decision problem involves assigning tasks to workstations. Process planning defines the set of tasks. Normally, we know or can estimate the time required for each task. We label this time t_i for task i. Physical limitations may partially restrict the order in which tasks are performed. We refer to these as **precedence constraints.** It is necessary, for example, to mount a component on a circuit board before it can be soldered. The **cycle time, c,** determines the production rate for the line. A new product unit enters the first workstation once every c time units. Every c time units thereafter, it moves to the next workstation until it reaches the end of the line. If the line has k stages, and there are no part buffers between workstations, then each unit will spend kc time in process. To achieve a desired production rate of D units/time, we must set $c = D^{-1}$. We can adjust this cycle time by replicating the process and using m lines instead of one. In this case, we set the cycle time to $c_m = \dfrac{m}{D}$. For a single line, the objective is to **minimize k** subject to satisfying the precedence constraints, assigning all tasks, and ensuring that no workstation must work more than c time per unit. If multiple lines are being considered, we can attempt to **minimize k · m** with cycle time c_m. In the following section, we describe a basic heuristic for the ALB problem. Thereafter, we generalize the basic problem to allow for the production of multiple products on the same line.

12.3.1.1   Ranked Positional Weight

The ALB problem requires packing n objects of different sizes into as few ordered bins as possible in which each bin has the same capacity and the packing order is partially constrained. The finiteness of the task sizes makes this a difficult problem to solve. The motivation for the problem is the assignment of tasks (objects) to operators (bins) along an assembly line. Task size corresponds to its operation time (t_i), and bin size represents the cycle time for the line (c). The partially-constrained ordering results from technological constraints that require certain assembly tasks to be performed before others.

To solve the problem, we begin with the well-known ranked positional weight heuristic. **Positional weight** defines a useful attribute of a task. The positional weight of task i,

PW_i, is defined as the sum of task i's processing time plus that of all its necessary successor tasks. Notationally, we can write this as $PW_i = \sum_{j \in S_i} t_j$ where S_i is the set of tasks that must come after task i in the assembly sequence to satisfy precedence constraints. You can think of positional weight as a lower bound on the amount of work remaining once task i is started. In this sense, positional weight is one measure of the degree to which task i is a bottleneck, and we should do tasks with high PW_i values as soon as possible. The ranked positional weight technique simply orders tasks from high to low PW_i, and then sequentially assigns tasks to the first workstation in which they will "fit." Fit requires enough free time and the satisfaction of all precedence constraints.

EXAMPLE 12.1

Consider the product described in Table 12.1. Demand calls for producing 450 units in each 7.5-hour shift. Precedence constraints are recorded in the table in the form of "immediate predecessors." Thus, task e can begin as soon as both c and d are complete. The fact that e also requires task b to be complete is implied through the requirement that d cannot begin until b is completed. Find the ranked positional weight line balance and the idle time of the workers for that assignment.

SOLUTION

To produce 450 units in 7.5 hours with a single line requires a cycle time of:

$$c = \frac{(7.5 \ hr/day)(3600 \ seconds/hr)}{450 \ units/day} = 60 \ seconds/unit.$$

(This assumes an ideal environment with no down time during the shift nor any defective items). Figure 12.3 illustrates the data from the Table. Task times are included in each task node. Precedence constraints are indicated by directed arcs between task nodes. It is only necessary to draw arcs for immediate predecessors. Task c must precede f, but this is indicated by the path $c \to e \to f$.

The first step involves labeling each task with its positional weight. In general, we would need to process these data to determine the set of successors for each task. Immediate successors (ISs) are found by inverting the immediate predecessor (IP) relationships. Because d is an IP of e, e is an IS of d. For any task, we can trace through its set of ISs to find the entire set of successors. To find PW values, we start with tasks without successors and work backward always selecting a task with ISs that have been visited. For each task, we can retain its list of successors if desired. For any task, its list of successors is the union of its immediate successors and their successors:

$$S_i = \{IS_i \cup_{j \in IS_i} S_j\} \tag{12.1}$$

To begin, h has no ISs and we set $PW_h = t_h = 10$. Because the only IS of g is h, we can compute PW_g. $PW_g = t_g + \sum_{j \in S_g} t_j = 20 + t_h = 30$. Next, we can select either d or f because in each case, all

Table 12.1 Assembly Tasks for a Speaker Module

Task	Description	Task time (seconds)	Immediate predecessor tasks
a	Mount Bracket to Baseboard	10	—
b	Mount Baseboard to Frame	45	a
c	Insert 2-way Speaker	30	a
d	Insert and Test Circuit Card	25	b
e	Connect Power Cables	25	c, d
f	Test Assembly	20	e
g	Enclose and Seal	20	d, f
h	Pack	10	g

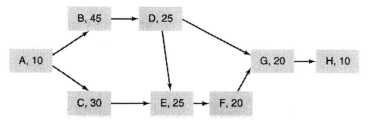

Figure 12.3 Illustration of Precedence Structure for Example 12.1

their *IS*s have been "labeled" with their set of successors. For task f, $PW_f = t_f + \sum_{j \in S_f} t_j = 20 + t_g + t_h = 50$. Similarly, $PW_e = 75$. Task d's *IS*s are e and g. Both of these have been labeled. We, therefore, know its successors and compute: $PW_d = t_d + \sum_{j \in S_d} t_j = 25 + t_e + t_f + t_g + t_h = 100$. Continuing in this manner: $PW_c = 30 + t_e + t_f + t_g + t_h = 105$, $PW_b = 145$, and $PW_a = 185$. Placing the task in nonincreasing order of PW yields the ranking a, b, c, d, e, f, g, h.

Next, we sequentially assign each task to the first workstation in which it will fit. Task a is first thus we assign it to the first workstation. This consumes ten of the 60 seconds available. Next, we examine task b. By using PW order, we can be assured that all of task b's predecessors are assigned. There are 50 seconds of available time remaining in workstation 1, and b only requires 45 seconds thus we add it to workstation 1. Task c is next and requires 30 seconds. It will not fit in workstation 1 thus we start a new workstation. Task d requires 25 seconds. It will not fit in workstation 1, but it will fit in workstation 2 and, thus, we assign it there leaving $30 - 25 = 5$ seconds free. Task e takes 25 seconds. This exceeds the time available in the opened workstations and so we start a new workstation 3. Of course, even if workstation 1 had enough idle time, we could not assign task e because e requires that c is completed first, and c has been assigned to workstation 2. Task f finds workstation 3 as the first fitable location leaving $60 - t_e - t_f = 15$ seconds of idle time there. Task g will not fit in workstation 3 and, thus, we start a new workstation 4. Finally, task h must be assigned to workstation 4. There is sufficient time remaining in workstation 3, but the precedence constraint of task g before h must be satisfied. The RPW steps are summarized in the following Table.

Iteration	Task assigned	Workstation selected	Time remaining
1	a	1	$60 - 10 = 50$
2	b	1	$50 - 5 = 45$
3	c	2	$60 - 30 = 30$
4	d	2	$30 - 25 = 5$
5	e	3	$60 - 25 = 35$
6	f	3	$35 - 20 = 15$
7	g	4	$60 - 20 = 40$
8	h	4	$40 - 10 = 30$

The problem statement asked us to find worker idle times. If we add the idle times per cycle for the four workstations, we obtain $5 + 5 + 15 + 30 = 55$ seconds. Thus, the proportion of idle labor time for workers is $\dfrac{55}{4 \cdot (60)} = 0.229$. This varies between $\dfrac{5}{60} = 0.083$ for workers 1 and 2 and $\dfrac{30}{60} = 0.50$ for the last workstation.

We observed in Example 12.1 that 23% of the human resources in the line were idle on average. This seems high but for the given set of tasks, shift length, and production quota, the idle time is unavoidable. This can be seen by noting that a lower bound on the number of workstations needed is given by:

$$LB_k = \left\lceil \frac{\sum_i t_i}{c} \right\rceil = \left\lceil \frac{185}{60} \right\rceil = 4 \tag{12.2}$$

Three workers will never be able to produce enough of these products because even a perfectly balanced three workstation line would require $185/3 = 62$ seconds per cycle. Because our solution required four workstations, it is "optimal" for the conditions stated in the problem description. Of course, the solution does not look very good, and we may want to redefine those conditions.

In general, several alternatives can be considered to improve line efficiency. Four major strategies present themselves. Options include:

1. Using optimization tools and more sophisticated heuristics. The RPW technique may yield a solution that utilizes more workstations than the lower bound. In these instances, it may be possible to find better solutions by examining more task assignment alternatives. A variety of techniques have been proposed. Implicit enumeration schemes [see Johnson (1988) for example] have been used to solve problems optimally with 1,000 or more tasks. We have characterized a solution to the ALB problem as an assignment of tasks to workstations. Define a **feasible** task sequence as any ordering of tasks that satisfies the precedence constraints. The sequence a, c, b, e, d, f, g, h is not feasible because e is not allowed to precede d. The recognition that for a given cycle time, **specifying a feasible task sequence also specifies a solution** provides a useful result. The statement follows because we would always perform as many tasks as possible in a workstation before starting another. Passing a product unit to the next workstation when enough time remains to perform one or more tasks is never advantageous. Doing so will always require at least as many downstream workstations and may require more. Suppose that for example the cycle time was 85, and we specified a feasible sequence of tasks with times 19, 37, 28, 36, 25, 53, and 32. If we want to use as few workstations as possible, we would assign tasks [1], [2], and [3] to the first workstation. This would take $19 + 37 + 28 = 84$ time units leaving one time unit idle. We then assign as many tasks as possible to workstation 2. We can fit tasks [4] and [5] for a cumulative load of $36 + 25 = 61$ time units, but there is not enough free time to include task [6]. Tasks [6] and [7] would, however, fit in workstation 3. This yields an optimal solution because the lower bound is

$$LB = \left\lceil \frac{\sum_i t_i}{c} \right\rceil = \left\lceil \frac{26 + 37 + \cdots + 32}{85} \right\rceil = 3.$$ Suppose that we decided to assign only

tasks [1] and [2] to the first worker leaving 29 units of idle time per cycle. Worker 2 could then be assigned tasks [3] and [4] leaving 24 units of idle time. Worker 3 only has enough time to do [5] and [6]. We would need a fourth worker for the final task. The moral is to pack the tasks as tightly as possible for any specific task sequence.

Implicit enumeration attempts to explore every possible task sequence to ensure that we find an optimal solution. This can be a formidable computational problem if there are many tasks. The trick is to find ways to quickly exclude many solutions (sequences) without explicitly testing their cost. Precedence constraints may help reduce the number of feasible sequences. Consider the problem in Example 12.1. At first glance, there appears

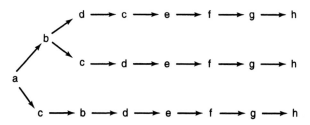

Figure 12.4 Tree of Feasible Task Sequences for Example 12.1

to be 8! = 40,320 possible ways to sequence the eight tasks. But, precedence restrictions actually reduce this to the three sequences shown in Figure 12.4. Task a must come first, and tasks a, b, c, and d must be completed before any of e through h are started. A number of useful rules can also be applied to eliminate sequences. For example, suppose that $\sum_{i=1}^{n} t_i = T$, and RPW provided a solution of k workstations. We either want to find a sequence requiring at most $k - 1$ workstations or prove that none exist. The maximum total idle time allowed in a solution with $k - 1$ or fewer workstations is $I^{max} = (k - 1) \cdot c - T$. If $k = 8$, $c = 100$, and $T = 645$, then we are interested in solutions with at most $7(100) - 645 = 55$ total units of idle workstation time. If in examining a task sequence the first three workstations have idle times of 17, 21, and 19 time units respectively, then we know **all** sequences that start this way can be excluded because we already have 57 units of idle time. Idle time cannot be reduced by subsequent workstations unless we somehow overlap station boundaries and allow workers to share task responsibilities.[1] We refer to this process of eliminating all sequences that start this way as **fathoming** this partial sequence. A variety of other rules for implicitly examining and excluding sets of sequences are discussed in Johnson [1988].

2. Varying worker schedules. Cycle time was selected based on a specific shop schedule in order to meet demand. We could try running a second shift or overtime. In a shift with seven hours of normal work time, a demand rate of 140 units requires a cycle time of: $c = \dfrac{(7\ hours) \cdot (60\ min/hr)}{140\ units} = 3.0\ min/unit$. If we add two hours of overtime per day, the required cycle time becomes $c = \dfrac{(9\ hours) \cdot (60\ min/hr)}{140\ units} = 3.86\ min/unit$.

If the line is rebalanced with this new time, a more efficient configuration may be found.

3. Building parallel lines. Building parallel lines generalizes the approach of varying worker schedules. Instead of a second temporal shift, we could construct two lines side by side, or three, or four, and so on. If we have m lines, we only need to produce $1/m$ as many units per line. Cycle time for each line changes from c to $c \cdot m$. We are not forced to construct identical lines. If we need 140 units per day, two lines could each produce 70 per day or one line could produce 100 and the second line could produce 40 if that happens to maximize efficiency. A variety of options can be examined.

---

[1]Such sharing may be a good idea, and indeed cross-training and flexible movement of workers along the line can be very valuable. However, these actions require more sophisticated operations planning and violate the assumed protocol of the ALB problem.

4. Redesigning the product or assembly process. The production engineer may have the luxury of redesigning the product configuration. Design for manufacturability discussions should include the possible efficiencies associated with a particular design. Perhaps replacing several metal parts by a single injection molded part will change the set of production tasks and simplify the material flow path or reduce idle time. A different design arising from breaking up the product into different modules to accomplish its required functions may also present better fabrication and assembly solution options.

The methods addressed thus far assume well-structured problems. Real life can be more complex. Some precedence constraints may be desirable but not essential. Some workers may already have certain task knowledge and, thus, it would be preferable to assign those tasks accordingly. Some task times may exceed the desired cycle time and, thus, multiple workers are assigned (either to work on alternating items or to work jointly on each item). Roy and Allchurch [1996] describe a knowledge-based system that incorporates such factors and can be used to modify line balances as requirements change.

12.3.1.2 Random Processing Times and Buffers

When workstations are reliable and processing times are consistent from unit to unit, there is no need for buffers to store items between workstations. Such lines are usually paced either by a continuously moving conveyor or a material transfer mechanism that discretely indexes parts between workstations at fixed cycle time intervals. On the other hand, when processing times can vary, either because of random operation times or because different versions of the main product are produced on the line, then the line is typically run asynchronously. Parts stay in workstations as long as necessary to complete the assigned operations before being passed on. As soon as the part completes at a workstation, the worker attempts to pass it on to the next workstation. After passing the current unit downstream, the worker looks upstream to see if the next unit is available. If a unit of product is available, the worker immediately begins working.

Adjacent workers do not necessarily complete their operations at the same time nor always within the average cycle time. For such asynchronous operation, inventory buffers may be added to the line to reduce blocking and starvation of workstations.[2] The asynchronous mode of operation allows workers to keep their current unit of product until all assigned tasks are completed. Buffers increase the level of independence between workstations because of short-term, random processing time variation. If downstream buffer space is available, the workstation can pass on the completed unit and start working on the next unit even if the downstream station is not ready. Likewise, upon passing along a unit, a workstation can withdraw a unit from its upstream buffer even if the previous workstation has not completed its assigned tasks. Blumenfeld [1990] developed an empirical model for estimating the production rate of a line with random processing times. The model assumes there are k identical workstations with mean processing time of c per unit and coefficient of variation C_V (the coefficient of variation is the ratio of standard deviation

---

[2]Blocking occurs when the workstation completes an item, but the next workstation is not prepared to receive it, nor is there space in the buffer to store it. Starvation occurs when the workstation passes on its completed unit, but there is no new unit available to work on. The workstation remains idle until the blocking or starvation condition changes.

of processing time to the mean processing time, i.e. $C_V = \dfrac{\sigma}{c}$). Between each pair of adjacent workstations there exists a buffer that can hold up to B units. The summary finding is:

$$X = c^{-1}\left[1 + \frac{1.67 \cdot (k-1) \cdot C_V}{1 + k + 0.31 \cdot C_V + \dfrac{1.67 \cdot k \cdot B}{2 \cdot C_V}}\right]^{-1} \tag{12.3}$$

where X is the production rate of the line in units/time. With infinitely large buffers to eliminate blocking and starvation we would have $X = c^{-1}$. We approach this as B increases.

EXAMPLE 12.2

A ten-station line has an average cycle time of six minutes, but actual processing time at each workstation is normally distributed with mean of six minutes and a standard deviation of 1.2 minutes. A buffer with capacity of two units will be placed between each workstation. Estimate the maximum production rate of the line and determine the loss resulting from blocking and starvation.

SOLUTION

We know that $c = 6$ minutes, $k = 10$ workstations, $B = 2$ spaces, and $\sigma = 1.2$ minutes. We first compute the coefficient of variation for processing time as $C_V = \dfrac{1.2}{6} = 0.2$. We can then use Equation (12.3) and find:

$$X = c^{-1}\left[1 + \frac{1.67 \cdot (k-1) \cdot C_V}{1 + k + 0.31 \cdot C_V + \dfrac{1.67 \cdot k \cdot B}{2 \cdot C_V}}\right]^{-1}$$

$$= \left(\frac{1}{6}\right) \cdot \left[1 + \frac{1.67(9)(0.2)}{1 + 10 + (0.31)(0.2) + \dfrac{1.67(10)(2)}{2(0.2)}}\right]^{-1} = 0.1615$$

This line design will produce 0.1615 units per minute or 9.69 units per hour. Note that as B approaches infinity, the term in brackets disappears, and we produce $1/6$ unit per minute or 10 per hour. The loss caused by variability is therefore 0.31 units per hour. However, without buffers (i.e. $B = 0$), Equation (12.3) would yield:

$$X = c^{-1}\left[1 + \frac{1.67 \cdot (k-1) \cdot C_V}{1 + k + 0.31 \cdot C_V + \dfrac{1.67 \cdot k \cdot B}{2 \cdot C_V}}\right]^{-1}$$

$$= \left(\frac{1}{6}\right) \cdot \left[1 + \frac{1.67(9)(0.2)}{1 + 10 + (0.31)(0.2) + \dfrac{1.67(10)(0)}{2(0.2)}}\right]^{-1} = 0.1311$$

or 7.87 units per hour. The relatively small buffers increase output by over 20%!

We have made the assumption that workstation time is valuable relative to buffer capacity and inventory. Another approach to managing under uncertainty would be to increase the allowable cycle time. We choose to use c_{min} as our cycle time. Alternatively, we could add slack time to the system. If we can describe the distribution of processing times for each station, then we can set the cycle time whereby $100(1 - \alpha)\%$ of units are completed with c for some small α. For example, if the most active workstation has a uniformly distributed processing time between 15.00 and 18.00 minutes, then setting cycle time to 17.94 minutes will ensure that the workstation completes the part in 98% of the regular cycles.

12.3.1.3 Mixed-Model Releases

Customers expect options. They want to be able to choose among features, price, and styles. As such, many modern serial production systems permit multiple versions of a single product, or multiple similar products to be made at the same facilities. If you tour an automobile assembly plant, you will see two-, three-, and four-door models, and four- and six-cylinder engines coming down the line in seemingly random order. The line must be able to produce a variety of products each with its own option profile. We call this a mixed-model line. Mixed-model lines generally run asynchronously. Each workstation is allowed to start on the next unit coming down the line as soon as the unit enters their work sphere and the workstation has completed its operations on the previous unit.

Line balancing is performed using average task times $\bar{t}_i = \sum_j w_j \cdot t_{ij}$ where w_j is the proportion of units that are of type j, and t_{ij} is the task time for task i on model j.

Mixed-model lines pose additional design and operating requirements. The order in which models enter the line should be planned to smooth the time demands on workstations. As some models may require more than c time units in some stations, we should offset long task time models with short task time models to avoid short-term overloading of workstations. As we have seen, if we batch (or sequence) similar units, the variances between task times from model to model have a larger impact on feasible cycle time and inventory accumulation. Instead, we smooth the entry rate for each model. Smoothing the entry rate also smoothes the demand for the components produced by the feeder lines. This allows for effective kanban control of the multistage fabrication and assembly process.

We will present a simple procedure for sequencing the entry of units to a mixed-model line. Our approach smoothes the production rate of each model and guarantees that we produce each in the desired quantity. More sophisticated methods can be found in the references listed at the end of the chapter. See, for example, Miltenburg [1989] and Bard et al. [1994].

Let d_j be the demand as dictated by the production plan for model j this period. Total production for the period will be $D = \sum_j d_j$. We assume that the minimal feasible average cycle time is such that this demand can be met during the current period, i.e. $c_{\min} \cdot D \leq T$ where $c_{\min} = \max_k \sum_{i \in S_k} \bar{t}_i$, T is the length of the period, and S_k is the set of tasks assigned to station k. Then the ideal smoothed start time for the nth unit of model j is $s_{nj} = \dfrac{(n-1) \cdot T}{d_j}$. Each unit therefore has a desired start time. By merging these sequences of start times for each model into a single nondecreasing time sequence, we create the sequence for model starts. Ties can be broken by giving preference to the most demanded model or to the model that will minimize blocking and starvation if it is released next in sequence. To spread production over the period we release units once every $c = \dfrac{T}{D}$ time units. Thus, we release units at times $(0, c, 2c, 3c, ..., [D-1] \cdot c)$.

EXAMPLE 12.3

An automobile assembly line is scheduled to produce 10 four-door sedans, 6 three-door hatchbacks, and 4 station wagons during the upcoming 4-hour period. Sequence these starts to balance the production rate.

SOLUTION

Total production for the next four hours is $10 + 6 + 4 = 20$ vehicles. For a 240 minute shift, this means starting a unit every $240/20 = 12$ minutes. The sedans would like to start every 24 minutes, the hatchbacks every 40 minutes, and the station wagons every 60 minutes producing ideal time vectors of (0, 24, 48, 72, 96, 120, 144, 168, 192, 216), (0, 40, 80, 120, 160, 200) and (0, 60, 120, 180) respectively. Using "H" for hatchback, "S" for sedan, and "W" for wagon if we combine these ideal vectors breaking ties on the basis of largest demand, we obtain the entry sequence (S, H, W, S, H, S, W, S, H, S, S, H, W, S, H, S, W, S, H, S). These models enter the line at times (0, 12, 24, 36, 48, 60, 72, 84, 96, 108, 120, 132, 144, 156, 168, 180, 192, 204, 216, 228) minutes into the shift.

12.3.1.4　Determining Workstation Spacing

Variability in task times between products creates a design issue when constructing the line. In the purely deterministic, single-model case, with synchronous movement, workstations are simply placed along the line at a separation distance equal to the distance the line travels in c time units. The asynchronous case requires more careful design. Assume that the line is moving continuously and workers stay with each unit until they have finished their assigned tasks. The distance between workers should be set to avoid interference among workers recognizing that some workers will stay with some models more than c time units. Workstations must be separated by enough space to avoid the conflict of two adjacent workstations wanting to work on the same unit in the same space at the same time. Given a prospective order entry sequence and the time each unit requires at each workstation, we can quickly determine the necessary spread between workstations. We have a sequence of D units to be produced where the nth unit corresponds to model $j(n)$ and requires time $T_{nk} = \sum_{j \in S_k} t_{j(n),k}$ at station k. Envision the line as moving continuously at a speed of v ft/min. For the first unit, the first worker will start at location 0 and stay with the item until the line reaches point $v \cdot T_{11}$. The worker then moves back up the line until he/she encounters the next unit which is c minutes (or $c \cdot v$ ft.) behind. If the worker moves at the rate v_1 ft./min. back up the line as measured relative to the ground, then the worker can move upstream for $\dfrac{c \cdot v}{v + v_1}$ min. before encountering the second unit. The potential distance covered in this time is the return length $l_r = v_1 \cdot \left(\dfrac{c \cdot v}{v + v_1} \right)$. If the worker keeps moving until encountering the unit, they meet at point $v \cdot T_{11} - l_r$ along the line. Note that if the return speed v_1 is large relative to the line speed v, then $l_r \cong c \cdot v$.

We can continue this model through all D units to find the location of the first worker at all times. A similar approach can be used for the workers at stations 2, 3, ..., K if we measure distance relative to their starting points. Figure 12.5 illustrates the movement of workers along the line over time. The space required by worker k is the difference between their earliest starting and latest ending point over all D units. The minimum line length is found by moving the spans for each worker in Figure 12.5 leftward until the span for worker k just touches that of worker $k - 1$. Algebraically, we can define $B(n, k)$ to be the point at which workstation k begins operating on the nth unit and $E(n, k)$ as its ending point along the line. A simple recursion will generate the locations. We initialize with:

$$B(1, k) = 0. \tag{12.4}$$

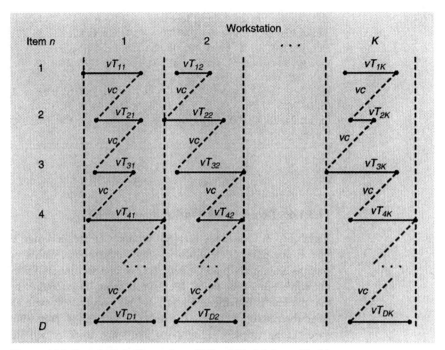

Figure 12.5 Sample Path Illustration of Mixed Model Assembly Line

Then,

$$E(n, k) = B(n, k) + v \cdot T_{n,k} \tag{12.5}$$

$$B(n, k) = E(n - 1, k) - l_r, \quad \text{for } n = 2, ..., D \tag{12.6}$$

The length of workstation k should be at least:

$$L(k) = \max_n E(n, k) - \min_n B(n, k) \tag{12.7}$$

EXAMPLE 12.4

A four-workstation assembly line will produce five different versions of a product. Model A has twice the demand of the other four. Products will be released to the line in the repeating sequence A, B, C, A, D, E. The assembly times for each model are shown in the Table. The line will be paced to produce ten products per hour, and the conveyor will move at a maximum speed of 1 ft. per min. The time for the worker to move back upstream from one unit to the next is negligible. Determine the dedicated space needed for each workstation.

Assembly Times in Minutes

	Workstation			
Product	1	2	3	4
A	5.6	7.1	6.2	4.0
B	3.3	5.3	4.5	5.7
C	6.2	4.0	8.3	3.6
D	5.0	4.0	5.6	8.7
E	5.0	5.7	5.2	7.9

Table 12.2   Beginning and Ending Positions for Example 12.4

Item no.	Product		Workstation k			
			1	2	3	4
1	A	$B(1, k)$	0	0	0	0
		$E(1, k)$	5.6	7.1	6.2	4.0
2	B	$B(2, k)$	−0.4	1.1	0.2	−2.0
		$E(2, k)$	2.9	6.4	4.7	3.7
3	C	$B(3, k)$	−3.1	0.4	−1.3	−2.3
		$E(3, k)$	3.1	4.4	7.0	1.3
4	A	$B(4, k)$	−2.9	−1.6	1.0	−4.7
		$E(4, k)$	2.7	5.5	7.2	−0.7
5	D	$B(5, k)$	−3.3	−0.5	1.2	−6.7
		$E(5, k)$	1.7	3.5	6.8	2.0
6	E	$B(6, k)$	−4.3	−2.5	0.8	−4.0
		$E(6, k)$	0.7	3.2	6.0	3.9

SOLUTION

The beginning and ending times are found using Equations (12.4) through (12.6) with $v = 1$ ft./min. and $c = 6$ min. The values of $n = (1, 2, 3, 4, 5, 6)$ correspond to products (A, B, C, A, D, E). Because $v_1 \rightarrow \infty$, $l_r = c \cdot v = 6$ ft. Using this data,

$$B(1, 1) = 0.$$
$$E(1, 1) = B(1, 1) + vT_{A1} = 0 + (1)(5.6) = 5.6 \text{ ft.}$$
$$B(2, 1) = E(1, 1) - vc = 5.6 - (6) = -0.4.$$
$$E(2, 1) = B(2, 1) + vT_{B1} = -0.4 + (1)(3.3) = 2.9.$$
$$B(3, 1) = E(2, 1) - vc = 2.9 - 6 = -3.1.$$
$$E(3, 1) = B(3, 1) + vT_{c1} = -3.1 + 6.2 = 3.1.$$

Continuing in this manner, we obtain the values in the second column of Table 12.2. A similar recursive process is used to obtain the values in the columns for workstations 2, 3, and 4.

If we note the ending positions after the six-unit cycle, each worker can return to their original relative starting position of 0.0 prior to the start of the next six-unit cycle. In fact, workers 1, 2, and 4 will return early and have a short wait. Worker 3 will return to his/her starting point just in time for the next cycle of parts to begin. The required length of each workstation, $L(k)$, is given in Equation (12.7) as the difference between the worker's most upstream and downstream locations. For workstation 1, this is $L_1 = \max_n E(n,1) - \min_n B(n,1) = 5.6 - (-4.3) = 9.9$ ft. Similarly, $L_2 = \max_n E(n,2) - \min_n B(n,2) = 7.1 - (-2.5) = 9.6$ ft., $L_3 = 7.2 - (-1.3) = 8.5$ ft. and $L_4 = 10.7$ ft.

Two procedural assumptions drove this analysis. First, we assumed workstations were "closed" in the sense that all space was dedicated to a unique worker. In practice, it may only be necessary that two workers never occupy the same space at the same time. In the end-of-chapter exercises, we ask the reader to adapt the analysis to this "open" model. Second, we assumed workers will continue moving backward until encountering the next unit. Alternatively, assuming the average cycle time at a workstation is less than c, we can use the available idle time to restrict the movement of the worker and allow the unit to come to the workstation. Suppose that we instruct workers not to cross the "0" point on their return trips. The worker begins the first unit at location $B'(1, k) = 0$ and completes the assigned tasks at location $E'(1, k) = vT_{1k}$. The worker then returns upstream until either encountering the next unit or reaching the starting point. For this protocol,

$$B'(n, k) = \max\{E'(n - 1, k) - l_r, 0\} \tag{12.8}$$

$$E'(n, k) = B'(n, k) + v \cdot T_{nk} \tag{12.9}$$

For workstation k, the required length is:

$$L'(k) = \max_n E'(n, k) \tag{12.10}$$

If the cycle of models is repeating, and we end in a position beyond the starting point, i.e., $E'(D, k) > l_r$ then we would need to continue the recursion until returning to a beginning point of 0 to ensure the workstation length is sufficient.

EXAMPLE 12.5

Determine the space requirements for workstations in Example 12.4 assuming the workers do not return past their starting point.

SOLUTION

Worker 1 begins at $B'(1, 1) = 0$ and completes the first unit at $E'(1, 1) = 5.6$. The worker then returns upstream. Because $l_r = 6 > E'(1, 1)$, the worker stops at location 0 and waits for the second unit. The path of each worker is summarized in Table 12.3.

Workers 1 and 2 can return to their starting points after the cycle. However, workers 3 and 4 are beyond the l_r point, and we should continue examining their paths. If the cycle of products A, B, C, A, D, E continues, worker 3 will start A at 1.3 ft., B at 1.5 ft., and the C at 0. Thereafter, the pattern will resemble that shown in Table 12.3. Worker 4 will start A at 4.6 ft., B at 2.6 ft., C at 2.3 ft., and then A at 0 and will thereafter match the Table. Using the largest ending values as the required lengths, we have $L'(1) = 6.2$ ft., $L'(2) = 7.1$ ft., $L'(3) = 8.5$ ft., and $L'(4) = 10.6$ ft. Comparing this solution with the previous example's, we note that restricting workers' travel resulted in

Table 12.3 Beginning and Ending Times with Restricted Travel

Item no.	Product		Workstation k			
			1	2	3	4
1	A	$B(1, k)$	0	0	0	0
		$E(1, k)$	5.6	7.1	6.2	4.0
2	B	$B(2, k)$	0	1.1	0.2	0
		$E(2, k)$	3.3	6.4	4.7	5.7
3	C	$B(3, k)$	0	0.4	0	0
		$E(3, k)$	6.2	4.4	8.3	3.6
4	A	$B(4, k)$	0.2	0	2.3	0
		$E(4, k)$	5.8	7.1	8.5	4.0
5	D	$B(5, k)$	0	1.1	2.5	0
		$E(5, k)$	5.0	5.1	8.1	8.7
6	E	$B(6, k)$	0	0	2.1	2.7
		$E(6, k)$	5.0	5.7	7.3	10.6

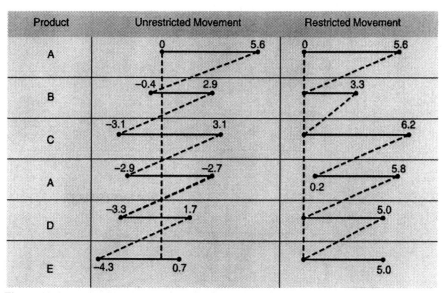

Product	Unrestricted Movement	Restricted Movement
A	0 5.6	0 5.6
B	−0.4 2.9	3.3
C	−3.1 3.1	6.2
A	−2.9 −2.7	0.2 5.8
D	−3.3 1.7	5.0
E	−4.3 0.7	5.0

Figure 12.6 Worker One Path Trajectories with and without Restricted Flow

lower resource (space) requirements. Figure 12.6 shows the trajectories of worker 1 under both movement protocols.

12.3.2 Lot-Streaming

We have spent considerable effort in this text in addressing models for setting the size of production batches. In kanban systems, this may be a single container. However, in many instances, the production batch size will exceed the natural material handling container size. In such cases, movement of parts in smaller **transfer batches** can have significant benefits in reducing work-in-process (WIP) inventory and flow times. By splitting the production batch into multiple sublots for intermachine transfer, the production batch can be in process at multiple operations simultaneously. This practice of using small transfer batches, also known as **lot-streaming,** can significantly reduce throughput time for the production batch. Each transfer batch is moved to the next operation as soon as all its units are completed. In addition to having parts ready for processing sooner at the next workstation, transfer batches effectively increase the service rate of processors. From queuing theory, we know that for systems with the same level of overall utilization, average waiting times are smaller in systems that produce many small jobs than those that produce a few large jobs. The disadvantage is the increase in the frequency of material handling moves. For a material handling system with low variable move cost such as a conveyor, the impact of transfer batches is minimal until the conveyor capacity is reached. However, for movement with significant variable cost, such as manually-assisted cranes and push trucks, the impact of transfer batches can be substantial if the load size is reduced. This impact becomes even more significant if production capacity is lost in each transport. Such would be the case, for example, if production line workers were responsible for movement of parts between workstations or if a small setup was needed for each transfer batch.

The primary issues with sublots are when to use them and how many to use. First, in a serial line with no loss of capacity, the more sublots the better for reducing through-

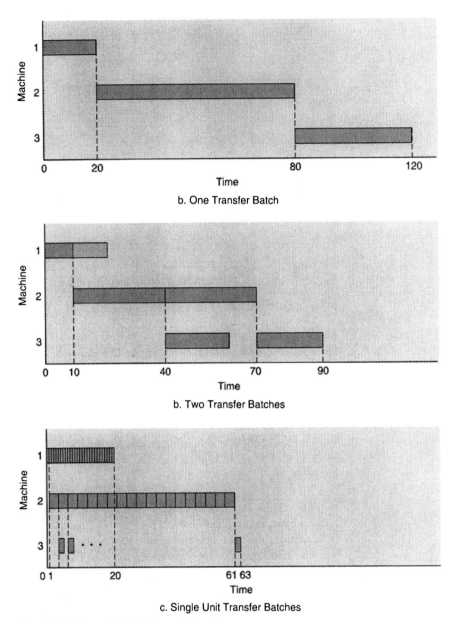

b. One Transfer Batch

b. Two Transfer Batches

c. Single Unit Transfer Batches

Figure 12.7 Impact of Sublots on Flow Time

put time. Figure 12.7 illustrates this phenomenon for a simple system without queuing or setup. We have a process batch of twenty units to be manufactured in a three-stage system. Unit processing times at the three stages are one minute, three minutes, and two minutes, respectively. With a single-transfer batch, the batch spends 20 minutes at the first stage, 60 minutes at the second stage, and 40 minutes at the third stage for a total throughput time of 120 minutes (Figure 12.7a). If we divide the batch into two equal-sized sublots of ten units each, then after ten minutes, the first sublot can begin processing at the second stage. After 30 minutes of processing, that sublot is ready for the third stage and thus it begins at stage 3 at time 40 minutes. It requires 20 minutes of processing and thus finishes at time 60. The second sublot finishes at machine 1 at time 20, begins processing at time 40 at machine 2, and time 70 at machine 3. It then finishes at time 90 (Figure 12.7b). As we add more sublots, the total throughput time for the batch reduces even further. With 20 single-unit sublots, total throughput time is 63 minutes (Figure 12.7c). This example indicates that more sublots are better, but there is a definite diminishing return as we add additional sublots. In general, two sublots will produce about 50% of the maximum possible savings in throughput time. Additional sublots significantly increase material handling with smaller decreases in throughput time. The trade-off between material handling cost and flow time for this example is shown in Figure 12.8.

We can easily show why two sublots provide almost half the advantage of single-unit sublots. Consider an isolated production batch of Q units going through a serial line of M workstations. The unit processing time at workstation i is p_i. We let L denote the number of sublots. The busiest machine is the bottleneck, denoted as B and identified by $p_B \geq p_i \ \forall i$. For equal-sized sublots of size $q_l = \dfrac{Q}{L}$; $l = 1, \ldots, L$, the makespan for the production batch in the system is:

$$Makespan = Q \cdot p_B + q \cdot \left(\sum_{i=1}^{B-1} p_i + \sum_{i=B+1}^{m} p_i \right) \tag{12.11}$$

This result can be seen by considering the flow of sublots. The first sublot will flow through the system spending $q \cdot p_i$ at workstation i. The first term in parentheses gives the time until this sublot reaches the bottleneck. Other sublots will follow, and sublots will queue in front of the bottleneck machine. The bottleneck will work continuously for Qp_B time once the first sublot arrives. This yields the first term in Equation (12.11). The last sublot will never have to wait after leaving the bottleneck because the bottleneck paces the line and sublots ahead of it will always have time to clear the machine before the next arrives. With no waiting, its remaining processing time will be $q \sum_{i=B+1}^{m} p_i$. Now, using Equation (12.11), as we go from one to two to Q sublots, makespan reduces by:

$$Makespan(L = 1) = Q \cdot p_B + Q \cdot \left(\sum_{i=1}^{B-1} p_i + \sum_{i=B+1}^{m} p_i \right) \tag{12.12}$$

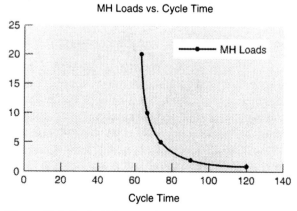

Figure 12.8 Tradeoff Between Material Handling Loads and Cycle Time

$$Makespan(L = 2) = Q \cdot p_B + \frac{Q}{2} \cdot \left(\sum_{i=1}^{B-1} p_i + \sum_{i=B+1}^{m} p_i \right) \tag{12.13}$$

$$Makespan(L = Q) = Q \cdot p_B + 1 \cdot \left(\sum_{i=1}^{B-1} p_i + \sum_{i=B+1}^{m} p_i \right) \tag{12.14}$$

The maximum improvement is found from the difference between Equations (12.12) and (12.14) with $\Delta Makespan = (Q - 1) \cdot \left(\sum_{i=1}^{B-1} p_i + \sum_{i=B+1}^{m} p_i \right)$. But, with just two sublots, we have a reduction of $\Delta Makespan = \frac{Q}{2} \cdot \left(\sum_{i=1}^{B-1} p_i + \sum_{i=B+1}^{m} p_i \right)$ between Equations (12.12) and (12.13).

Several authors have examined the impact of lot-streaming. The emphasis has been on how to apportion the units in the process batch among a fixed number of sublots. Potts and Baker [1989] show that for the case of two machines with varying sublot sizes q_l, $l = 1, \dots, L$, the makespan is minimized by using the geometric rule $q_{l+1} = \frac{p_2}{p_1} \cdot q_l$. This rule equates the processing time for the $l + 1^{st}$ sublot at machine 1 with that of sublot l at machine 2 to eliminate blocking and starvation. Kropp and Smunt [1990] present a linear programming formulation for determining the optimal size of each sublot if the number of sublots and batch size are known. Simple extensions to the model to accommodate setup and transport times are described. The authors conclude that the value of lot-splitting is small when setup times are large. Baker and Pyke [1990] develop heuristics for sizing sublots. Among their conclusions are that equal transfer batches work well with an average loss of only about 5%. And, in practice, equal-sized sublots may be considerably easier to implement. Equal transfer batches work particularly well with small setup times, and in general, a smaller first sublot is best. Potts and Baker [1989] show that using equal sublots has a worst case makespan ratio of 1.53 relative to the best solution with consistent sublots. A general survey of results can be found in Baker [1995].

Each sublot may have a minor associated setup. This may result from the movement and loading of parts or just the general tendency for workers to lose time when switching between activities—the miscellaneous break. We can build a simplistic model to provide a first approximation to the desirability of using sublots in this environment. We borrow the result from the theory of exponential queues that the expected waiting time in a queue is given by $E(W_q) = \dfrac{\rho}{\mu \cdot (1 - \rho)}$ where μ is the service rate of jobs and ρ is the average utilization for the workstation. Utilization is composed of processing time and setup time. The total variable processing time plus the basic setup for a production batch, measured as a proportion of time available at workstations, will be denoted ρ_0. We use s as the sublot setup factor indicating the proportion of capacity nominally used by any additional sublot setup. This is the capacity loss attributed to each sublot. If we use production batches, utilization is $\rho_0 + s$, if we use two sublots per production batch, utilization is $\rho_0 + 2s$. The service rate μ is a function of L resulting from sublot setup. Because the bottleneck limits service, we will use Qp_b as the mean service time per order (any required internal batch setup time should be added as well). Beyond this, the service rate is slowed down by sublot setups. Putting this together, as a function of the number of sublots L, the expected time in system (TIS) for a production batch can be approximated by:

$$TIS = p_b \cdot Q + \frac{Q}{L} \cdot \left(\sum_{i \neq B} p_i \right) + \frac{\rho_0 + s \cdot L}{\left(\dfrac{1}{Q \cdot p_b + s \cdot L} \right) \cdot (1 - \rho_0 - s \cdot L)}$$

$$= p_b \cdot Q + \frac{Q}{L} \cdot \left(\sum_{i \neq B} p_i \right) + \frac{(Q \cdot p_b + s \cdot L) \cdot (\rho_0 + s \cdot L)}{(1 - \rho_0 - s \cdot L)} \tag{12.15}$$

and for feasibility we add the constraint $\rho_0 + s \cdot L < 1$. The first observation is that if $s = 0$, we should use as many sublots as possible, i.e. $L = Q$. Utilization is unaffected and TIS [Equation (12.15)] becomes a decreasing function of L. For positive s, TIS can be searched or plotted as function of L to find the optimal number of sublots.

EXAMPLE 12.6

A production facility assembles production orders in batches averaging 100 items. Using full production batches, shop utilization is about 50%. As such, an engineer has proposed implementing transfer batches. Shorter throughput times will make the facility more competitive and increase sales. Processing time at the bottleneck workstation averages 0.2 weeks per batch. The other eight processing stages account for an average total of 1.1 weeks of processing time per batch. The engineer estimates that each additional transfer batch per order will increase utilization about 1% at the bottleneck. The increase is in external setup. How many transfer batches should be used per order?

SOLUTION

First, note that processing times are given per batch instead of per unit. Thus, $0.2 = Q \cdot p_b$. Using Equation (12.15) we find the relationship:

$$TIS = (0.2) + \frac{(1.1)}{L} + \frac{(0.2 + 0.01 \cdot L) \cdot (0.5 + 0.01 \cdot L)}{(1 - 0.5 - 0.01 \cdot L)} \quad \text{for } 1 \leq L < 50.$$

A plot of throughput time versus the number of sublots is shown in Figure 12.9. Adding sublots reduces throughput time from 1.4 weeks using a single-process batch to 0.52 weeks for 9 sublots. We also see that throughput time is somewhat insensitive to the number of sublots near the optimum. Using anywhere from two to 34 sublots per batch makes average flow time less than one week.

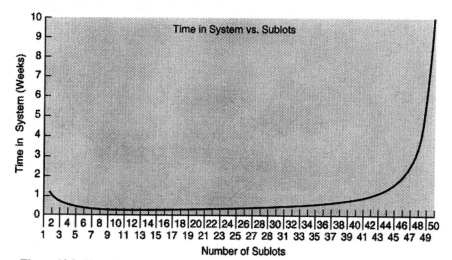

Figure 12.9 Time in System vs. Number of Sublots

Our model assumed that sublots would be used at the bottleneck as well as elsewhere. If we can manage tasks to avoid the loss of capacity for each sublot at the bottleneck, then the dominant system utilization measure would not vary with L. In that case, more sublots would be better. (This observation was also made when discussing the theory of constraints in the previous chapter.) Minimizing the loss of capacity may be achieved by having other workers assist in setup and material handling or by simply aggregating all sublots into a single batch at the bottleneck before processing. Of course, the incremental improvement would still be small beyond the first few sublots.

12.3.3   Flexible Flow Lines

The assembly line balancing problem assigns repetitive tasks to workstations. The model views processors as identical, flexible machines that are capable of performing any task. In many plants, the processors are specialized. The product progresses through a sequence of different machine types. Each machine has its own operating capabilities. In producing a variety of circuit cards, each card may go through stages in which DIPS, then SIPS, and then axial components are placed on the card. If the number of components of one type exceeds the feeder capacity of its machine type, then there may even be sequential stages required for the same class of component. In automobile assembly, workstations are not identical. They are characterized by specialized tooling for lifting, positioning, attaching and testing components at each assembly stage. Whether it is circuit cards or automobiles, multiple models are produced on the same line with each product unit passing through most, if not all, of the sequential stages. Workstations are specially equipped to perform one or more specific tasks and processing times at a stage may vary substantially from unit to unit. In order to provide sufficient capacity to meet demand, parallel workstations may be required at one or more stages. These parallel workstations may be identical or tuned to specific models. A schematic of a four-stage flow line with a total of ten workstations appears in Figure 12.10. The material handling system must be capable of transporting a product unit from any workstation at stage i to any workstation at stage $i + 1$ (or $j > i$ if some stages are not required for all models). We refer to such arrangements as **flexible flow lines.**

With parallel workstations and significant processing time differences between models, the operations planning for flexible flow lines involves scheduling production on a daily or continuous basis. Scheduling entails selecting the specific processor and time for

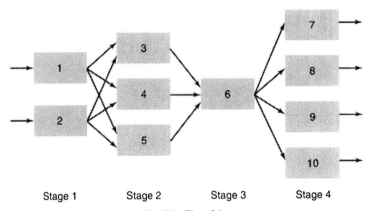

Figure 12.10 A Four Stage Flexible Flow Line

each operation of each unit of product. The objectives of interest are likely to be some combination of minimizing makespan, minimizing average flow time, and satisfying due dates. Because the parallel processor scheduling problem for an individual stage is difficult by itself, it is clear that we will not be able to guarantee finding optimal schedules for the entire line for realistically-sized problems. We present a scheduling heuristic for the static approach. In the static problem, we are given a set (batch) of product models to make this period. The schedule details the planned start time and specific machine for each operation. Creating a schedule allows production control to plan release times and coordinate the delivery of parts. With only modest product variety and inexpensive parts, we can maintain bins of all parts used by one or more models at each of the parallel workstations in a stage. However, if subassemblies and detailing are expensive and specific to product units, a schedule is necessary for coordinating the arrival of the main product body and its specific components at workstations.

The procedure described below draws its motivation from the procedure described by Wittrock [1988]. We simplify, however, the process of sequencing jobs and timing releases. The objective of minimizing makespan motivates the procedure. Our approach uses the suboptimal but time-honored tradition of solving difficult problems by breaking them into pieces, solving each piece separately, and then linking the pieces together. We do this to make the difficult problem more manageable. We first solve a set of parallel scheduling problems to assign jobs to specific processors at each stage and then determine the sequence for releasing jobs to the system. The weakness of the approach is that it does not permit use of due dates for specific product units.

Step 1. Assigning units to workstations. At each stage, we have a parallel processor scheduling problem. Although stages are actually linked by product flow, for now we treat them independently. Our goal is to equalize the workload on processors. Assuming all workstations can begin at the same time, this would potentially minimize the makespan for this stage. To heuristically equalize the workload, we use the Longest Processing Time (**LPT**) assignment rule. Order the units in descending order of processing time for the stage. Sequentially assign units to the workstation that will be able to complete it soonest.

EXAMPLE 12.7

A flexible flow line produces three styles of vents for commercial heating and air conditioning systems. Each order specifies a style and number of units. The orders pass through the stages of stamping, welding, and assembly. Operation times for today's orders are shown in Table 12.4. As indi-

Table 12.4   Operation Times for Example 12.7

Order	Operation times		
	Stamping (hrs).	Welding (hrs.)	Assembly (hrs.)
1	1.2	2.0	4.8
2	0.9	1.6	3.6
3	1.4	3.2	6.1
4	0.6		1.3
5	3.2	5.3	10.4
6	0.4		1.2
7	0.3		1.1
8	1.5	2.2	4.4
9	2.7	3.1	7.8

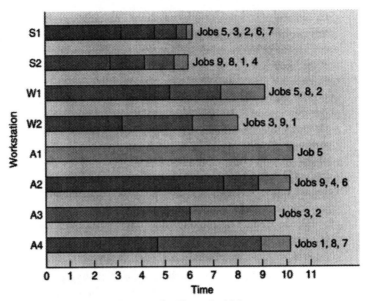

Figure 12.11 LPT Assignment for Example 12.8

cated, orders for the budget model do not require welding. There are two stamping workstations, two welding workstations, and four assembly workstations. Assign operations to workstations, and find the workload for each workstation.

SOLUTION

Consider stamping first. Ranking orders from largest to smallest operation time produces the sequence 5, 9, 8, 3, 1, 2, 4, 6, 7. Without implying a production sequence, we assign these to the two parallel stamping workstations. Order 5 goes to workstation 1. Order 9 then goes to workstation 2. Because workstation 2 only has 2.7 hours assigned in comparison to the 3.2 hours assigned to workstation 1, order 8 is assigned to workstation 2. Continuing in this manner, we obtain the assignment shown in Figure 12.11.

Using the same procedure, for welding we make assignments using the order sequence 5, 3, 9, 8, 1, 2. The result has orders 5, 8, and 2 on the first workstation and 3, 9, and 1 on the second. For assembly, the sequence for assigning orders is 5, 9, 3, 1, 8, 2, 4, 6, 7. Workstation 1 will assemble order 5. Workstation 2 will assemble orders 9, 4, and 6. Workstation 3 will assemble orders 3 and 2. Workstation 4 will assemble orders 1, 8, and 7. The assignments appear in Figure 12.11.

The task assignments determine workload. The first stamping workstation will do orders 5, 3, 2, 6, and 7. This adds to a workload of 3.2 + 1.4 + 0.9 + 0.4 + 0.3 = 6.2 hours. Repeating this computation for each workstation, we obtain the workloads shown in Table 12.5.

Table 12.5 Workloads Assigned in Example 12.7

	Stamping		Welding		Assembly			
Workstation	1	2	1	2	1	2	3	4
Time (Hrs).	6.2	6.0	9.1	8.3	10.4	10.3	9.7	10.3

For use in the next step, we will need to identify the bottleneck machine. Let t_{jm} be the processing time order j requires at stage m. $V(k)$ will be the set of jobs that have been assigned to machine k (k identifies both a stage and specific workstation). The total workload for machine k is $W(k) = \sum_{j \in V(k)} t_{j,m(k)}$, and the bottleneck machine is $k^* = \arg \max_k W(k)$. These values are reported in Table 12.5. We see that the bottleneck workstation is the first assembly workstation with 10.4 hours of work.

Step 2.   Sequencing order releases. It would be possible to use any of the dispatching rules discussed previously in the text to control the flow of orders. One only needs a rule such as SPT, LWKR, or EDD to decide which order to produce first at a machine when the machine becomes available, and there are multiple orders in queue. Wittrock suggested a more sophisticated rule for the case of minimizing makespan with a secondary concern for minimizing WIP. The rule builds a sequence of orders one at a time. At each step, we consider the impact of assigning each available order and select the order that appears to keep closest to the ideal processing rate. To keep computations tractable, the method overlooks potential competition for resources among orders not yet scheduled. Borrowing the concept of an ideal activation time for processors, we present a greedy procedure for constructing a job release sequence.

Orders should be sequenced and released to minimize starvation at the near-bottleneck workstations while avoiding excessive WIP. We can control this by planning the arrival of jobs at workstations. Given a partial sequence and shop schedule, the "ideal" arrival time for the next job at the bottleneck workstation is immediately following completion of its previous job. We gradually construct a sequence. To the partial sequence, we add the job that in some sense best conforms to the ideal arrival time. At each step, we identify the processor most desirous of receiving a job and then add the job that can arrive soonest.

Recognizing that flow must be balanced, and that bottleneck machines will typically carry a heavier load, ideal arrival time at nonbottlenecks is set later than completion time of the previous job. For nonbottleneck processor k, an average planned idle time of:

$$I_k^* = \frac{W(k^*) - W(k)}{N(k)} \qquad (12.16)$$

should precede the start of each job where $N(k)$ is the number of orders that will visit machine k. This lag evenly distributes the required idle time for nonbottleneck workstations over the jobs visiting that workstation. The even distribution serves to make the schedule robust to random events at all times but admittedly will require greater care in labor management.

At any point in constructing a schedule, we have a partial schedule containing a set P of sequenced jobs. These jobs place $N_{P,k}$ jobs on processor k with a cumulative workload of $w_{P,k}$. The ideal start time for the next operation on k is therefore:

$$S_{P,k}^* = w_{P,k} + (N_{P,k} + 1) \cdot I_k^* \qquad (12.17)$$

The critical machine will be k_P^* such that:

$$k_P^* = \arg \min_k \{S_{P,k}^*\}. \qquad (12.18)$$

The next question concerns when each currently unsequenced job could start on k_P^*. Let $R(j)$ be the set of machines visited by job j. We consider only those jobs that visit k_P^*, that is j with k_P^* an element of their route $R(j)$. In each case, we attempt to add j to the schedule starting each operation as early as possible without delaying an already scheduled job. We let $A_j^{k_p}$ be the earliest arrival time of j at k_P^*. Then, select the job that can start first, i.e.,

$$j^* = \arg \min_{j \notin P \cap k_P^* \in R(j)} \{A_j^{k_p}\} \tag{12.19}$$

More formally,

Sequencing Heuristic

1. Set $P = \varnothing$, $w_{P,k} = N_{P,k} = 0$.
2. Compute $S_{P,k}^*$ for all k using Equation (12.17). Select k_P^* by Equation (12.18). For all jobs $j \notin P \cap k_P^* \in R(j)$ find $A_j^{k_p}$ and the desired job j^* using Equation (12.19).
3. Add j^* to P. Schedule j^* and update $N_{P,k}$ and $w_{P,k}$. If all jobs that visit k_P^* are sequenced, remove k_P^* from further consideration. If unsequenced jobs remain, go to 2, otherwise stop.

Jobs currently in process can be easily accommodated by including this information in the initial $w_{P,k}$ values.

EXAMPLE 12.8

Find the desired idle time per order for each machine and then determine a job release sequence for Example 12.7. Assume move times are 0.

SOLUTION

Desired idle times can be found by inserting the computed workloads from Table 12.5 into Equation (12.12). Consider for instance the second welding station. From the Table we know that workstation A1 is the bottleneck. For workstation W2 we obtain $I_{W2}^* = \dfrac{W(A1) - W(W2)}{N(W2)} = \dfrac{10.4 - 8.3}{3} =$

0.7. The denominator indicates that there are three orders routed to this workstation, namely, orders 1, 3 and 9. Similarly, we can compute the other desired idle times and obtain

Workstation	S1	S2	W1	W2	A1	A2	A3	A4
Desired Idle Time/Order	0.84	1.1	0.43	0.70	0	0.033	0.35	0.033

To select the first order to release, we first select the key machine. Because all $w_{P,k}$ and $N_{P,k}$ are 0, $S_{\varnothing,k}^* = I_k^*$. The smallest value is for the bottleneck machine A1. Because only job 5 uses this machine, job 5 is sequenced first. The corresponding schedule appears in Figure 12.12a. We update $w_{\{5\},S1} = 3.2$, $w_{\{5\},W1} = 5.3$, $w_{\{5\},A1} = 10.4$ and $N_{\{5\},S1} = N_{\{5\},W1} = N_{\{5\},A1} = 1$.

Next, select the second job. For the processors used by job 5, we update $S_{\{5\},S1}^* = 3.2 + 2 \cdot (0.84) = 4.88$, and $S_{\{5\},W1}^* = 5.3 + 2 \cdot (0.43) = 6.16$. Processor A1 has no remaining jobs and is removed from consideration. With the updated values, processor A2 is selected as critical (A4 could also have been selected as they are tied with $S_{\{5\},k}^* = 0.033$). We, therefore, choose between jobs 9, 4, and 6 (Figure 12.11). Projecting operations, job 9 could finish at S1 at time $3.2 + 2.7 = 5.9$ and at W2 at time $5.9 + 3.1 = 9$. Job 4 could arrive at A2 at time 0.6. Job 6 could arrive at A2 at time $3.2 + 0.6 = 3.8$. Because job 4 can arrive first, we schedule it next. We continue in

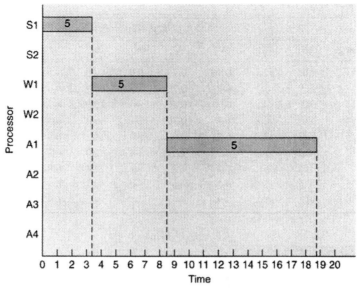

a. Loading of First Job (Job No. 5)

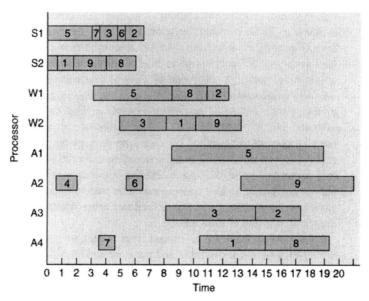

b. Complete Schedule

Figure 12.12 Flexible Flow Line Schedule for Example 12.8

this manner until obtaining the schedule in Figure 12.12b. Details of the process are shown in Table 12.6.

The procedure is fairly quick to execute. In practice, we could maintain a schedule but also update it whenever jobs arrive or complete an operation to include the latest random events.

Table 12.6 Job Sequencing for Example 12.8

Stage	$S^*_{P,k}$ Values for processors								k_P^*	Job choices	j^*
	S1	S2	W1	W2	A1	A2	A3	A4			
1	0.84	1.1	0.43	0.70	0.0	0.03	0.35	0.03	A1	5	5
2	4.9	1.1	6.2	0.70	—	0.03	0.35	0.03	A2	9,4,6	4
3	4.9	2.8	6.2	0.70	—	1.4	0.35	0.03	A4	1,8,7	7
4	6.0	2.8	6.2	0.70	—	1.4	0.35	1.2	A3	3,2	3
5	8.3	2.8	6.2	4.6	—	1.4	6.8	1.2	A4	1,8	1
6	8.3	5.1	6.2	7.3	—	1.4	6.8	6.0	A2	9,6	6
7	9.5	5.1	6.2	7.3	—	2.60	6.8	6.0	A2	9	9
8	9.5	8.9	6.2	—	—	—	6.8	6.0	A4	8	8
9	9.5	—	8.8	—	—	—	6.8	—	A3	2	2

12.4 RE-ENTRANT FLOW LINES

Cycles in product routes pose a challenge for schedulers. Semiconductor manufacturing in particular experiences cycling or **re-entrant flow** patterns as wafers containing hundreds of integrated circuit dies flow repeatedly through the same work centers for etching, photolithography, ion implant, and other processes. An integrated circuit may have as many as 20 layers, and the wafer must flow through the work centers once for each layer. Figure 12.13 illustrates a simple re-entrant flow pattern. The action of jobs cycling back for repeated visits to one or more work centers during its process route[3] defines re-entrant flow. Note that each visit by a job to a workstation is assigned a unique buffer label. This allows us to track where a job is in its overall route. Although semiconductor manufacturing and other re-entrant systems exhibit some of the characteristics of flow systems, they are also characterized by different routings for each job and the fact that not all processes are required for each loop through the system. The variability in process routes and operation times make scheduling difficult. At the same time, the large capital investment tied up in WIP, the marketing imperative of meeting tight delivery schedules, and the increasing risk of contamination as wafers sit idle, make it essential to minimize cycle time for wafers. Likewise, to facilitate order-release planning, it is important to minimize the variance of flow time.

The key decision for re-entrant lines relates to selecting the job to process when a workstation becomes available and there are orders waiting at different stages (cycles) of their route. Suppose that machine 3 in Figure 12.13 becomes available, and there are jobs waiting in buffers 3, 6, and 9. Which buffer receives priority? In general, we would like to produce the jobs with the fewest remaining operations to both create output for the system and minimize WIP. This would suggest priority to buffer 9. However, if workstations 4 or 5 happen to be long-term bottlenecks that are currently idle, it would seem prudent to try to force work toward them by selecting buffer 3 or 6. We need general rules for making such decisions. Scheduling in re-entrant systems has been widely studied in recent years, both from a theoretical perspective using stochastic queuing models and an empirical perspective by simulating the performance of dispatching rules. We describe the fluctuation smoothing policy for mean cycle time (FSMCT). This dispatching technique has been observed to perform well for minimizing the mean and standard deviation of cycle time in re-entrant systems [Lu et al. (1994)]. Queuing theory relates that the variabil-

---

[3]In semiconductor manufacturing, the process route is referred to as the **recipe** for the job.

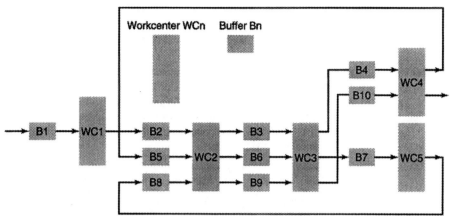

Figure 12.13 A Re-Entrant Flow Line

ity of interarrival times affects mean time at a work center. FSMCT attempts to minimize mean flow time by reducing the interarrival time variability. When a processor becomes available and more than one of its input buffers has jobs waiting, the FSMCT rule gives priority to the buffer with the maximum downstream shortfall between the desired and actual number of jobs in process. The rule balances the trade-off between trying to complete jobs and avoiding later starvation at connected work centers. Relative to FISFS, savings of 20% in mean cycle time and 40% in standard deviation of cycle time have been reported.

Operationally, the FSMCT rule selects the next job to process using a modified measure of slack. FSMCT checks all buffers at the processor and selects the job that has the least slack. Recall that slack is measured by **due date − remaining processing time − current time.** FSMCT sets job due dates at equal intervals with the motivation that this will yield a steady production rate. A steady completion rate of jobs should in turn help produce steady arrival rates at each processor. If jobs arrive to the system at the average rate λ, jobs should exit the system once every $\dfrac{1}{\lambda}$ time units. The jth job should therefore be given a due date of $\dfrac{j}{\lambda}$ plus some constant. The constant accounts for the mean flow time in the system. (Aiming for a constant flow time will help minimize the variance of flow time which in turn enhances the fluidity of the system and simplifies order release decisions.) In another update to the slack expression, expected remaining flow time substitutes for remaining processing time. For any job at buffer i let $\overline{F}_i^R$ be the expected remaining flow time once the job leaves this buffer. $\overline{F}_i^R$ can be estimated from historical data or simple mean value procedures such as those described in Chapter 7. Because the current time and any constants added to all due dates do not affect the relative ordering of job slacks, they can be ignored. Thus, we can use the measure:

$$S_j = \frac{j}{\lambda} - \overline{F}_i^R \tag{12.20}$$

where S_j is the relative slack for job j.

EXAMPLE 12.9

Consider the re-entrant system shown in Figure 12.13. The system produces three jobs per hour on average. Historically, it takes about two hours to pass through work centers 1, 2, and 4. It takes four hours to pass through work centers 3 and 5. A processor at work center 2 has just become available for processing. Job numbers 19 and 22 are in buffer 8, job 37 is in buffer 5 and jobs 47 and 48 are in buffer 2. Which job should be processed?

SOLUTION

For each job, we find the slack measure from Equation (12.20). Remaining processing time is estimated by historical workstation flow times and process routes. A job in buffer 8 must still go to workstations 3 and 4. This yields an expected remaining flow time of $4 + 2 = 6$ hours. In buffer 5, a job must proceed through workstations 3, 5, 2, 3, and 4 with a corresponding expected time of $4 + 4 + 2 + 4 + 2 = 16$ hours. For buffer 2, the expected remaining flow time is 24 hours. The relative slacks are therefore:

Job	Due date	Remaining flow time	Slack
19	$19/3 = 6.3$	6	0.3
22	$22/3 = 7.3$	6	1.3
37	$37/3 = 12.3$	16	-3.7
47	$47/3 = 15.7$	24	-8.3
48	$48/3 = 16$	24	-8

Job 47 has the minimum slack and should be scheduled next at work center 2.

12.5   TOOL MANAGEMENT SYSTEMS

Production operations require materials, machines, operators, instructions, and tools. We have generally assumed that the tools would be available at a machine when needed. In practice, we need a system for controlling tool availability. Tools must be acquired, stored, maintained, and transported to support the production plan. In so doing, tool location and condition must be tracked. Gray et al. [1993] provide an overview of tool management strategies and issues. The tool management system should operate as an integrated part of the overall production control system. By combining the output of the MRP system or the smoothed production schedule for a kanban system with the process plans for individual parts, weekly tooling requirements can be computed. This information forms the input for tool purchasing and maintenance. Disposable tooling can be treated as any other raw material with basic inventory control policies being applied. Tool movement from storage areas to machines, should be coordinated with detailed production schedules to avoid delaying the start of operations but also to avoid the need for excess tooling on the shop floor.

Fixtures are one aspect of tooling. For many operations, the part will need to be securely and accurately positioned in a fixture. The fixture is loaded onto the machine with precise positioning. The fixture must also hold the part firmly and support it against the forces imparted during the operation without allowing linear slippage, bending, or rotation of the part and also without damaging the part. In some cases, the part remains in the fixture for transporting between operations and even the entire process flow. For machining operations, cutting tools or inserts are required in addition to fixturing. Depending on the operation type, either the tool or part will spin at a high cutting speed. The force of the tool against the work piece removes unwanted material from the part but also causes wear on the tool. This wear must be monitored and the tool replaced when its useful life expires. Depending on the cutting conditions, tools may wear out from gradual wear or from sudden breakage under high stress. General wear of tools was one of the first subjects for scientific management. Frederick Taylor published his tool life studies in 1907 noting the logarithmic relationship $V \cdot T^n = C$ between tool life (T) and cutting speed (V). The constants n and C depend on the part material, tool material, and other conditions. More advanced models take into account the effect of feed rate and depth of cut. Modern carbide and ceramic tool inserts have significantly longer life times, and thus parameter values are different, but the general model forms are still useful.

The tool management system must be prepared to replace worn or broken tools and to continue with the operation. Thus, a system to replace tools at the workstation is required. Rather than leave tools on machines, in some cases tools are routed with parts and loaded at the same time. This approach makes sense when tools are unique to parts, but otherwise it suffers the disadvantage of frequent tool changeovers and premature removal of tools before their useful life has expired. In addition to being able to replace worn tools, the tool management system must have the capability to move tools as well as parts from one machine to another when machines break down.

Modern automated manufacturing systems pose new problems for scheduling. The necessity to integrate tooling considerations into machine scheduling and setup accompanies many automated systems. In the next section, we discuss the problems that develop when planning and scheduling flexible manufacturing systems.

12.6   FLEXIBLE MANUFACTURING SYSTEMS

Flexible manufacturing systems (FMS) generally consist of a set of computer numerically controlled (CNC) machines and an integrated material-handling system that can be used to produce a wide variety of part types. These systems are economically attractive in situations in which a variety of part types are produced on the same equipment. A part's volume should be sufficient to merit investment in part specific fixtures and computerization of the processing instructions, but not sufficient to warrant its own dedicated processing lines. The job shop is best suited for high-variety, low-volume parts. Flexible or hard automated transfer lines are used for low-variety, high-volume parts. The strength of the FMS is that one can attain the efficiency of dedicated serial lines while retaining the flexibility of the job shop. Computerized control permits flexible machines to stay active with minimal down time for part and tooling changeovers. Figure 12.14 shows an FMS setup.

When scheduling an FMS system, the following interrelated problems must be solved:

1. Part type selection problem. Given a set of part types, and their production quota for the mid-term planning horizon, determine a subset of part types for immediate, simultaneous processing.

2. Production ratio problem. For the part types selected in problem 1, determine their relative production ratios or the quantity of each to be produced during this short-term period.

3. Resource allocation problem. Determine the allocation of pallets and fixtures to the selected parts from problem 1. The mix of parts kept in the system at any one time should be determined to achieve the production ratios or rates selected in the previous problem. Note, that because of differences in routings and processing times, the ratio of WIP may not equal the ratio of parts produced.

4. Loading problem. Allocate the operations and tools for the selected parts to the machines in the cells. All required tools for the parts being produced must be available on one or more machines. It is advantageous to provide tooling on as many machines as possible to increase routing flexibility. We may also choose to allocate tools whereby parts can have the maximum number of sequential operations performed on a single machine

Figure 12.14 Flexible manufacturing system. (Photo courtesy of Giddings and Lewis Machine Tools.)

when the part is loaded. Stecke [1983] suggests that the machines of each type in the system be divided into groups, with all the machines in a group being loaded with the same tools. This provides routing flexibility while keeping routing relatively simple. The grouping of machines of each type could be viewed as a fifth problem in this process.

A variety of approaches have been proposed for solving these problems. It is important to keep in mind that these are operational problems and must be solved repeatedly either on a daily, shift, or real-time basis. We will propose both mathematical programming and heuristic approaches. The choice between the two depends on problem size, computational resources available, and the length and number of planning periods in the planning horizon. One approach would be to plan production batches and quantities for each shift during the week but only commit to executing the resultant plan for the first shift. The problem is then resolved each period. The tentative plan for future periods ensures a feasible path for meeting MRP goals for the week, but the setup worker only needs to know what tools to load this morning. Using this **rolling schedule** approach allows scheduling plans to adapt quickly to changes in production plans and randomness in shop outcomes.

12.6.1   Formation of Part Batches

Part batches are groups of parts that are produced together on the FMS. Our batch definition indicates a specific number of parts of each type and therefore answers both the part selection and production ratio subproblems. Each batch uses a set of machine tools, requires machine processing time, operator time, tool magazine capacity, and has a customer due date. The system may have a limit on the processing time for the batch (one shift or one day typically), a limit on the number of tools in each machine tool magazine, and a limit on WIP because of material-handling constraints. The formation of effective batches requires considering all of the issues simultaneously.

Heuristic methods similar to release time assignment methods in section 11.2.1 can be devised and executed. Denote B_k as the set of parts assigned to batch k, $k = 1 \dots K$ (note that at the start of the process, we do not know the value K because it depends on the batching strategy and the specific capacities of the system). Consider, the situation in which jobs arrive at the FMS for processing. Each job contains one or more parts of a single part type. We assume that jobs cannot be further divided. (If this is not the case, it is a simple matter to break up the arriving jobs into smaller jobs, perhaps as small as a single part each.) A simple heuristic process for formulating batches where meeting due dates is the primary objective is:

Step 1:   Initialize each B_k to the empty set, and order the jobs by early due date. Set j, the index into the part list, to 0.

Step 2:   If no jobs remain to be batched, then stop with the current B_k sets. Otherwise, increment j by 1 and go to the next part. Set $k = 1$.

Step 3:   Include job j in batch k if there is sufficient capacity in the system. That is, there must be sufficient machine time remaining in the batch production period for producing j, sufficient tool magazine capacity for tools required by j (not already included in the batch), and sufficient space in the material-handling system. If j can be added to batch k, then set $B_k = B_k \cup \{j\}$, update the remaining batch production time, the available tool magazine capacity, and the material-handling system capacity for batch k, and return to step 2. Otherwise, increment k by one and return to step 3 to try part j in the next batch.

The method starts with an EDD sequence, therefore, jobs with the closest due dates will be placed in the lowest indexed batches and can be scheduled early. Also, the method does not allow the splitting of a job's demand between two batches. This is easily remedied in step 3 by simply filling up the remaining capacity and adjusting the demand requirements by the amount partially satisfied. The following example demonstrates the approach.

EXAMPLE 12.10   Consider the following part data set for an FMS batching problem:

Part (job)	Tools used (type, slots/tool, no. req'd)	Machine time (hours)	Due date
1	(A, 2, 0.7), (B, 3, 0.4), (C, 1, 0.8), (D, 2, 0.1)	6	2
2	(A, 2, 0.6), (B, 3, 0.9)	3	1
3	(A, 2, 0.5), (C, 1, 1.7)	2	3
4	(A, 2, 0.5), (D, 2, 0.4)	8	1
5	(A, 2, 1.4), (C, 1, 0.5)	7	2
6	(B, 3, 0.1), (C, 1, 0.7)	3	2

For example, part 1 uses 70% of the tool life of a type A tool, 40% of a type B tool, 80% of a type C tool, and 10% of a type D tool. Tool A requires two tool spaces in the magazine, B requires three, C requires one, and D requires two. Part 1 requires six hours to process all operations, and is due in two days. Assume that the tool magazine has 12 slots, there are ten hours per day available for processing a batch, and the material-handling system can accommodate all parts simultaneously. Formulate a small set of batches that meet all constraints.

SOLUTION

Using the part batching heuristic, the parts are placed in EDD order, and the batches are initially empty. The following Table contains the step-by-step decisions made for each part.

Part	Batch 1	Batch 2	Batch 3	Reason
0	Φ	Φ	Φ	Initialize
2	{2}	Φ	Φ	Place part 2 in batch 1. The batch has seven slots and seven hours remaining. One A and one B are loaded.
4	{2}	{4}	Φ	Does not fit in batch 1 because eight hours are required. Start batch 2. Batch 2 has eight slots and two hours remaining.
1	{2}	{4}	{1}	Does not fit into batch 1 because of tool space (another A would have to be loaded in addition to a B, C, and D). Does not fit into batch 2 because of hours. Start batch 3. Batch 3 has four slots and four hours remaining.
5	{2, 5}	{4}	{1}	Fits in batch 1. Batch 1 has one slot (we added two A's and a C) and 0 hours remaining. Batch 1 is complete because there is no time left.
6	{2, 5}	{4}	{1, 6}	Does not fit in batch 2 because of hours. Fits in batch 3. We must add a tool C but B can be shared. Batch 3 has three slots and one hour remaining.
3	{2, 5}	{4, 3}	{1, 6}	Part 3 fits in batch 2. We add a tool D and share tool A. Finished.

The final batches are {2, 5} for day 1, {4, 3} for day 2, and {1, 6} for day 3. Part 4 finishes one day late. Also, parts 1 and 6 finish one day late.

The heuristic can be improved by trying different initial orderings and different decision rules; however, an alternate approach is to use mathematical programming. The approach presented next, models the objective of scheduling parts so that they are done as close as possible to their due dates. Each batch is associated with a different time period, thus batches and time periods are synonymous. Also, the model considers jobs requiring multiple machines and limits on the number of each tool available. The system has M machine types indexed $i = 1 \ldots M$. For each machine type i, P_i is the available production time and S_i is the capacity of the tool magazine in each period. The P_i and S_i are proportional to the number of identical type i machines. The N jobs are indexed $j = 1 \ldots N$. Each job j, has a due date d_j, production time requirements on each machine p_{ji}, and the tools required on each machine $l(j, i)$. Here, the approach assumes that each job uses a unique tool set (this can be relaxed). Finally, there are L tools indexed $l = 1 \ldots L$. For each tool l, s_l represents the number of slots required in the tool magazine, and N_l denotes the number of tools available in the system for any single period. Define the following additional notation concerning the decision variables:

$\delta_{jk} = 1$ if part j is produced in batch k, 0 otherwise;

$y_{lik} = 1$ if tool l is on machine i for batch k, 0 otherwise.

The model denoted **PS** is:

$$Minimize \sum_{j=1}^{N} \sum_{k=1}^{K} \left| k \cdot \delta_{jk} - d_j \right| \tag{12.21}$$

Subject to:

$$\sum_{j=1}^{N} \delta_{jk} \cdot p_{ji} \leq P_i \qquad \text{for each pair } (k, i) \tag{12.22}$$

$$\sum_{l=1}^{L} s_l \cdot y_{lik} \leq S_i \qquad \text{for each pair } (k, i) \tag{12.23}$$

$$\sum_{i=1}^{M} y_{lik} \leq N_l \qquad \text{for each pair } (k, l) \tag{12.24}$$

$$\delta_{jk} \leq y_{l(j,i),i,k} \qquad \text{for each triple } (i, j, k) \tag{12.25}$$

$$\sum_{k=1}^{K} \delta_{jk} = 1 \qquad \text{for } j = 1, ..., N \tag{12.26}$$

$$\delta_{jk} \in \{0, 1\}, \, y_{lik} \in \{0, 1\} \tag{12.27}$$

The objective function [Equation (12.21)] totals the part earliness and lateness. The term $k\delta_{jk}$ picks off the production time of job j and the function compares this with the due time. Note, that early and late penalties are assumed to be equal; however, this is easily relaxed when the absolute value term is linearized into its positive and negative components. Constraint sets [Equation (12.22)] and [Equation (12.23)] ensure that machine production limits and tool magazine limits are not violated. Constraint set [Equation (12.24)] ensures that no more than the available number of each tool is allocated for any one batch. Constraint set [Equation (12.25)] is a logical constraint that ensures that tool assignments match job assignments. If a job is assigned to a batch, then the tools required must also be assigned. Note, that if two jobs require the same tool, then their $l(j, i)$ values will be the same, and the tool will be assigned only one time. Constraint set [Equation (12.26)] ensures that each job is assigned to one batch and [Equation (12.27)] constrains the variables to the $\{0, 1\}$ set.

Model PS is a large integer programming model. In contrast to the heuristic batching method that front loads the schedule, PS tends to delay parts until their due dates. The model has flexibility in that it can easily be changed to consider situations other than those stated in the assumptions. When the model is solved, the scheduler will obtain a set of parts that should be included in each batch. The next problem to solve is to allocate specific parts to specific machines. Whereas in PS, the scheduler can aggregate machines and consider capacity rather coarsely; now the scheduler must get down to a finer set of details.

12.6.2 Tool Loading

The tool loading and operation assignment problem assumes that the part batching problem has been solved, and we must now assign tools to the locations of the tool magazines for each machine. In some cases, we may have a tool delivery system that can automatically deliver tools to machines much in the same way parts are delivered. In other cases, the tool magazines will be set up by technicians prior to running each batch. The requirement is that the batch be completed on time. Several objectives may be relevant. We may want to minimize the cost of tools required or the number of times parts are moved. To guard against random break downs, we may want to duplicate tools on several machines so that parts can be rerouted when necessary. This routing flexibility may also help reduce congestion and waiting time in the system. Assigning tools to machines whereby multiple operations can be performed consecutively on the same machine will also reduce material handling congestion and quality problems stemming from setup errors.

The problem setup consists of N operations that must be assigned to M machines. L tools are used to process the operations. Each operation j must be performed D_j times in the system. The tool that is required for operation j is $l(j)$. The machines are flexible in that if tooling is assigned to the machine, then the operation can be performed on the machine. There are N_l copies available of each tool. There are S_i tooling slots in the magazine of machine i, and each tool l takes up c_l slots in the magazine.

The decision variables are:

x_{ij} = number of operations j assigned to machine i;

y_{li} = 1 if tool l is assigned to machine i, 0 otherwise.

Additional model parameters are:

c_{ij} = cost per unit of operation j assigned to machine i;

s_{li} = tool magazine space required by tool l on machine i;

p_{ij} = processing time of operation j performed on machine i;

P_i = total processing time available on machine i (excluding normal idle time);

S_i = space available in the magazine for machine i.

The model, denoted **TO**, is:

$$Minimize \sum_{i=1}^{M} \sum_{j=1}^{N} c_{ij} \cdot x_{ij} \tag{12.28}$$

subject to:

$$\sum_{i=1}^{M} x_{ij} = D_j, \qquad \text{for } j = 1, ..., N \tag{12.29}$$

$$\sum_{j=1}^{N} p_{ij} \cdot x_{ij} \leq P_i, \qquad \text{for } i = 1, ..., M \tag{12.30}$$

$$\sum_{l=1}^{L} s_{li} \cdot y_{li} \leq S_i, \qquad \text{for } i = 1, ..., M \tag{12.31}$$

$$x_{ij} - D_j \cdot y_{li} \leq 0, \qquad \text{for all pairs } (i, j) \tag{12.32}$$

$$\sum_{l=1}^{L} y_{li} \leq N_l, \qquad \text{for } l = 1, ..., L \tag{12.33}$$

$$y_{li} \in \{0, 1\}, x_{ij} \geq 0 \tag{12.34}$$

The objective [Equation (12.28)] minimizes the total cost of production. This allows for the case in which some operations may be performed on different machines, possibly different machine types, but at different efficiencies. For example, we can drill a pattern of holes one at a time on one machine or by using a multispindle tool head on another. Constraint set [Equation (12.29)] ensures that the total production meets the required demand whereas set [Equation (12.30)] ensures that there is sufficient production time available to implement the solution. Constraint set [Equation (12.31)] limits the number of tool magazine slots used on each machine to those that are available. Constraint set [Equation

(12.32)] ensures that the correct tools are assigned to the machine if production takes place. Constraint set [Equation (12.33)] ensures that the number of available tools is not exceeded, and constraint set [Equation (12.34)] limits the decision variables to binary integer values or non-negative values. The model adapts readily to many situations. Suppose that, for example, we wanted to enforce routing flexibility in case of break downs. Adding the constraint:

$$\sum_{i=1}^{M} y_{li} \geq 2, \qquad \forall l \tag{12.35}$$

will guarantee that flexibility. If we want to reduce the number of material handling moves, we could externally aggregate clusters of adjacent operations that can be performed on the same machine. Each cluster becomes a single operation with the associated tool set and processing time. Assigning these clusters to machines both reduces the problem size and the number of part transfers. If the machine type for each operation is fixed whereby processing cost is constant, we can modify the objective function to minimize the makespan. Simply replace P_i by P in Equation (12.30) and use the objective *Minimize P*.

The solution to Model TO is an assignment of tools to machines and operations to machines. Model solution for small to mid-sized problems (10 to 15 operations, 4 to 5 machines) can be obtained using integer programming approaches. Heuristic approaches exist for larger problem instances. Several approaches are described in the references Sarin and Chen [1987], Chen and Askin [1990], Askin and Standridge [1993] listed at the end of the chapter.

12.6.3   Resource Allocation

The production goals for the period are now determined, and the machines are configured. The remaining issue concerns the level of WIP for the system. Parts will be placed on fixtures and loaded into the system. Fixtures may hold one or more parts. A job of ten parts may be broken up whereby each part is in its own fixture and transported through the system individually on separate pallets.

We could use the simple order release and dispatching rules of Chapter 11 to control the flow of parts. The difference here is that the control will be automated and implemented through the system's computer controller. In addition to deciding which parts to move when and where, the controller must integrate material handling decisions and control the devices that physically move parts between machines. Because fixtures are typically part specific, the resource allocation problem also entails determining the number of parts of each type in the batch that should be actively in process at all times. The remaining parts will either be at the input station waiting to be loaded or in the output buffer having finished processing and waiting to be used. The active number of parts of each type indicates the number of fixtures needed. Modeling tools such as mean value analysis (see Chapter 7) can be used to determine the necessary WIP levels to achieve the desired production rate of each part type in the batch. If space permits, the entire batch can be loaded into the system with an overflow storage area being used to hold those not currently in use. Alternatively, we can control the input process to release the parts that comprise the batch at smooth intervals over the period. This open approach will yield the desired production but WIP levels for each part type will seek their own stationary level.

12.7   SUMMARY

Shop floor control systems execute the production plan. Hierarchical structures are the most typical architecture for command, communication, and control. Each entity receives commands from its supervisory parent, manages the execution of those commands by assigning and coordinating the tasks of its subordinates, and then reports back to the supervisor. Heterarchical structures of autonomous entities have also been discussed in recent years.

By serving as a central repository for operations data and authorization, manufacturing execution systems provide an effective mechanism for controlling production operations and sharing production data with engineering and administrative functions. MESs are directly responsible for work order management, workstation management, tooling management, labor management, effort reporting, inventory tracking, material handling control, shop floor data collection, and schedule construction and control. The MES has significant interfaces with the production planning, maintenance, quality assurance, accounting, and supplier management systems.

A variety of manufacturing systems merit special consideration. Serial systems such as assembly lines can be very effective at producing similar products quickly with minimal cost. The key issue concerns the allocation of tasks to workers. Short repetitive cycles allow rapid learning, lower capital costs, and lower skill requirements but complicate the task assignment problem potentially leading to increased idle time and repetitive motion injuries. For a given cycle time, the ranked positional weight technique provides a quick heuristic for balancing the workload at workstations when solving the task assignment problem. With multiple product models, the randomness in processing times favors the use of asynchronous operation and small buffers in between workstations. The various product models should be sequenced to match the relative demand ratios and to smooth the workload at workstations over time. In general, flow systems, lot-streaming should be considered to reduce the throughput time. In many cases, simply dividing a production batch into two sublots will provide substantial time reduction. Scheduling in flexible flow lines with parallel processors at each stage presents a challenge. One approach utilizes a combination of the longest processing time heuristic to assign jobs to processors at each stage and then attempts to find a job release sequence that gives priority to feeding critical machines. Re-entrant flow lines, commonly found in semiconductor manufacturing and elsewhere, present the additional challenge of trying to select jobs for processing that will maintain short-term throughput without causing starvation in the longer term. The FSMCT dispatching rule combines the desire for a steady arrival rate at processors with expected remaining flow times to produce effective real-time control.

Tool management forms a critical aspect of shop floor control. Tools must be acquired, maintained, transported, and loaded as required at work centers. The performance of flexible manufacturing systems (FMSs) in particular, is driven by tooling issues. FMS scheduling requires selecting a set of part types to make this period, determining their production ratio, allocating tools to machines, and setting WIP levels accordingly. Mixed integer programming approaches can be used for small problems but heuristics may be required for finding solutions for mid- and large-sized systems.

12.8   REFERENCES

ASKIN, R.G., & STANDRIDGE, C. *Modeling and analysis of manufacturing systems.* New York: John Wiley & Sons, Inc., 1993.

BARD, J.F., DAR-EL, E., & SHTUB, A. An analytical framework for sequencing mixed model assembly lines. *International Journal of Production Research,* 1992, 30(1), 2431–2454.

BARD, J.F., SHTUB, A., & JOSHI, S.B. Sequencing mixed-model assembly lines to level parts usage and minimize line length. *International Journal of Production Research,* 1994, 32(10), 2431–2454.

BAUER, A., BOWDEN, J., BROWNE, J., DUGGAN, J., & LYONS, G. *Shop floor control systems.* New York: Chapman and Hall, 1994.

BERRY, G.L., JUNG, D., BEDWORTH, D., & YOUNG, H.H. Design of a semiautomated system for capturing and processing shop floor information. *Journal of Manufacturing Systems,* 1983, 2(1), 39–51.

BLUMENFELD, D.E. A simple formula for estimating throughput of serial production lines. *International Journal of Production Research,* 1990, 28(6), 1163–1182.

CHEN, Y.-J., & ASKIN, R.G. A multiobjective evaluation of FMS loading heuristics. *International Journal of Production Research,* 1990, 28(5), 895–911.

CONWAY, R., MAXWELL, W., McCLAIN, J.O., & THOMAS, L.J. The role of WIP inventory in serial production lines. *Operations Research,* 1988, 36(2), pp. 229–241.

DAR-EL, E.M., & COTHER, R.F. Assembly line sequencing for model mix. *International Journal of Production Research,* 1975, 13(5), 463–477.

DUFFIE, N.A., & PIPER, R.S. Non-hierarchical control of a flexible manufacturing cell. *Robotics and Computer Integrated Manufacturing,* 1987, 3(2), 175–179.

DUFFIE, N.A., & PRABHU, V.V. Heterarchical control of highly distributed manufacturing systems. *International Journal of Computer Integrated Manufacturing,* 1996, 9(4), 270–281.

FOX, M.S., ALLEN, B.P., SMITH, S.F., & STROHM, G.A. ISIS: A constraint-directed reasoning approach to job shop scheduling. *Proceedings of IEEE Conference on Trends and Applications,* Gaithersburg, MD: National Bureau of Standards; 1983.

GERTSBAKH, I.B. *Statistical reliability theory.* New York: Marcel Dekker, Inc., 1989.

GRAY, A.E., SEIDMANN, A., & STECKE, K.E. A synthesis of decision models for tool management in automated manufacturing. *Management Science,* 1993, 39(5), 549–567.

KUMAR, P.R. Re-entrant lines. *Queuing systems: Theory and applications*, 1993, 13, 87–110.

LIN, G.Y., & SOLBERG, J.J. Integrated shop floor control using autonomous agents. *IIE Transactions*, 1992, 24(3), 57–71.

LU, S.C.H., Ramaswamy, D., & KUMAR, P.R. Efficient scheduling policies to reduce mean and variance of cycle-time in semiconductor manufacturing plants. *IEEE Transactions on Semiconductor Manufacturing*, 1994, 7(3), 374–388.

McCLELLAN, M. *Manufacturing execution systems*, Falls Church, VA: APICS, 1997.

MILTENBURG, J. Level schedules for mixed model assembly lines in JIT production systems. *Management Science*, 1989, 35(2), 192–207.

MORRISON, J.R., & KUMAR, P.R. On the guaranteed throughput and efficiency of closed re-entrant lines. *Queuing Systems*, 1998, 28, 33–54.

O'GRADY, P., & LEE, K.H. An intelligent cell control system for automated manufacturing. *International Journal of Production Research*, 1988, 26(5), 845–861.

PRENTING, T.O., & THOMOPOULOS, N.T. *Humanism and technology in assembly line systems*. New Rochelle, N.J.: Hayden Book Co., Inc., 1974.

RACHAMADUGU, R., & SHANTHIKUMAR, J.G. Layout considerations in assembly line design. *International Journal of Production Research*, 1991, 29(4), 755–768.

ROY, R., & ALLCHURCH, M.J. Development of a knowledge based system for balancing complex mixed model assembly lines. *International Journal of Computer Integrated Manufacturing*. 1996, 9(3), 205–216.

SARIN, S.C., & CHEN, C.S. The machine loading and tool allocation problem in an FMS. *International Journal of Production Research*, 1987, 27(7), 1081–1094.

STECKE, K. Formulation and solution of nonlinear integer production planning problems for flexible manufacturing systems. *Management Science*, 1983, 29(3), 273–288.

STECKE, K., & KIM, I. A study of FMS part type selection approaches for short-term production planning. *International Journal of Flexible Manufacturing Systems*, 1987, 1(1), 7–29.

TAYLOR, F.W. On the art of cutting metals. *ASME Transactions*, 1907, 28, 310–350.

UZSOY, R., LEE, C.-Y., & MARTIN-VEGA, L. A review of production planning and scheduling models in the semiconductor industry part II: Shop floor control. *IIE Transactions on Scheduling and Logistics*, 1994, 26(5), 44–55.

VEERAMANI, D., BHARGAVA, B., & BARASH, M.M. Information system architecture for heterarchical control of large FMSs. *Computer Integrated Manufacturing Systems*, 1993, 6(2), 76–92.

WITTROCK, R. An adaptable scheduling algorithm for flexible flow lines. *Operations Research*, 1988, 36(3), 445–453.

YANO, C.A., & BOLAT, A. Survey, development, and application of algorithms for sequencing paced assembly lines. *Journal of Manufacturing and Operations Management*, 1989, 2(3), 172–198.

ZACKS, S. *Introduction to reliability analysis*. New York: Springer-Verlag, 1992.

12.9   PROBLEMS

12.1. List the principal functions of a shop floor control system.

12.2. List the advantages and disadvantages of a single assembly line relative to having each worker perform all tasks in a parallel manner.

12.3. Explain how a re-entrant flow shop differs from a standard flow shop and the additional issues this presents for scheduling.

12.4. Describe the advantages of an automated flexible manufacturing system compared with a traditional shop of special-purpose, manually controlled machine tools.

12.5. List the basic operational scheduling problems that must be solved for a flexible manufacturing system. Describe how the inputs and decisions for each of these problems interacts with the other problems.

12.6. Go to the World Wide Web and look up commercially available manufacturing execution systems. Choose one system and list its features. Does this system appear to address all the responsibilities of a shop floor control system?

12.7. Consider the assembly line balancing problem described in Table 12.7. Use the ranked positional weight technique to balance the line. Demand will be met if the cycle time is at most 40.

12.8. Table 12.8 contains the set of tasks required to assemble a portable CD radio player.

a. Determine the cycle time necessary to produce 18 units per hour.

b. Find a lower bound on the number of workstations required.

c. Balance the assembly line using the RPW technique.

Table 12.7   Assembly Tasks

Task	Task time (min.)	Immediate predecessors
1	10	—
2	14	1
3	8	1
4	5	3
5	26	2
6	9	2
7	11	4, 5
8	13	6, 7
9	22	8
10	21	4
11	19	9, 10
12	17	4, 11

Table 12.8   Portable Radio CD Player Assembly Tasks

Task number	Description	Task time (min.)	Immediate predecessors
1	Inspect Boards	1.1	—
2	Insert Boards	0.85	1
3	Insert Power Module	0.45	—
4	Insert CD Player	0.68	—
5	Wire Speakers	2.35	2, 3
6	Test Speakers	2.25	5
7	Insert Speakers	0.75	6
8	Insert Tape Door	0.25	—
9	Close Frame	0.40	4, 7, 8
10	Visual Inspect	0.40	9
11	Pack	1.00	10

d. List the set of possible sequences that must be considered in order to find an optimal line balance.

12.9. Suppose in the previous problem that the 18 units per hour represent four different models of portable players. Model A accounts for 50% of demand. Model B represents one-third of demand. Model C sells two units per hour, and model D accounts for one unit per hour. Determine a cyclical release pattern for the four models.

12.10. A three-stage production system produces a cyclical pattern of four product models. The models are A, B, C, and D and the pattern is A, B, C, B, D, B, D, B. The line moves at the rate of 2 ft./min. Find the space needed for each workstation if workers may not overlap. The time per model per workstation is as follows:

	Time per stage (min.)		
Model	1	2	3
A	3.0	4.0	7.0
B	4.6	4.3	5.0
C	5.2	4.3	3.2
D	3.6	5.0	3.6

12.11. A five-stage production line has a mean cycle time of 12 minutes for each stage. However, the processing times at each stage are random with a standard deviation of two minutes. Estimate the number of units produced per hour if the line has no buffers. How much will output increase if buffer space for two units is placed between each pair of workstations. The line operates asynchronously.

12.12. A product requires ten production operations. A system is to be designed to produce 50 units per eight-hour shift.

Operations two and four use the same tooling as do six and eight. Each workstation has a fixed cost of $300 per day plus tooling cost. Tooling capability has a depreciation cost of $50 per tool per workstation per day. The first three operations can be performed in any order. All other operations cannot start until the first three operations are completed. The only other precedence constraints are that operation 8 requires that operation 7 be completed first and operation 10 must be last. Operation times are 5.5, 3.3, 8.9, 7.1, 8.5, 2.3, 5.0, 4.8, 2.7, and 4.3 minutes, respectively. Propose an assembly system design. Estimate the manufacturing cost per part (exclusive of material cost).

12.13. Batches containing 50 units are released to the shop floor every 200 minutes. Each batch flows through four processing stages with unit processing times of 2.7 minutes, 4.0 minutes, 1.9 minutes, and 3.8 minutes, respectively. Compute the throughput time of the batch if entire process batches are moved at the same time between machines. How would the throughput time change if each batch was divided into two sublots?

12.14. A balanced six-stage flow line produces batches of 25 units. Batch processing time at each stage is one hour. Plot the throughput time of batches as a function of the number of sublots used.

12.15. Batches arrive to a flow line for processing at the rate of 15 per hour. Mean batch processing time at each workstation is nine minutes including setup. An engineer believes about one minute of capacity would be lost per sublot at each stage if batches were subdivided. Estimate the optimal number of sublots that should be used per batch to minimize flow time. What is the maximum feasible number of sublots to use per batch?

12.16. Extend the results in Equation (12.15) to the case in which there are m_b parallel servers at the bottleneck. Let ρ_0 still represent the average utilization per server without sublots.

Table 12.9 Flexible Flow Line Processing Times (Hours) for Problem 12.17

Job	Processing stage			
	1	2	3	4
1	1.2	2.0	0.5	2.6
2	2.0	3.0	1.1	4.2
3	0.6	0.7	0.2	0.9
4	0.3	0.4	0.5	0.7
5	2.2	3.1	1.7	4.0
6	1.7	2.4	0.3	2.6
7	0.4	0.7	0.7	0.9
8	0.7	1.2	0.2	1.9
9	0.8	0.9	0.9	2.1
10	1.0	2.1	1.0	2.8
11	0.9	3.3	1.3	3.9

12.17. Consider the four stage flexible flow line system (Figure 12.10). Eleven jobs need to be produced during the next shift. Processing times are provided in Table 12.9. Schedule these jobs in an attempt to minimize the makespan so that all jobs can be finished as soon as possible.

12.18. Suppose that your objective is to minimize flow times in a multistage process. Would you prefer two identical parallel workstations that can process μ jobs per hour each or one workstation that can process 2μ jobs per hour? Show that the single fast processor is preferable. Now, suppose that the system produces multiple part types and changeover time is re-

quired when switching between part types. Is your conclusion still always valid?

12.19. A three-stage flow line has three processors at stage 1, four processors at stage 2, and two processors at stage 3. Five jobs of type A and seven jobs of type B are to be produced today. Type A jobs require two hours at stage 1, three hours at stage 2, and one hour at stage 3. Type B jobs require 1.5 hours at stage 1, four hours at stage 2, and 1.5 hours at stage 3. Operators are necessary for loading and unloading jobs and to make occasional repairs. As such, workers are required to be available while one or more processors are operating and are paid for the number of hours present. Schedule the jobs.

12.20. Consider the re-entrant flow system (Figure 12.13). Throughput for the system is 40 jobs per eight-hour day. Average time at each workstation is two hours. Workstation 4 just completed the first job of the day. Jobs 2, 3, and 5 are waiting in buffer 10. Jobs 16 and 19 are waiting in buffer 4. Which job should be produced next according to the FMSCT rule?

12.21. An engineer has collected data on job flow for a re-entrant shop. Records for ten jobs traveling through the facility described in Figure 12.13 are shown in Table 12.10. The records record the time each job was logged into the workstation. Jobs are supposed to be logged in as soon as they arrive.

a. Estimate the average time spent at each workstation.

b. Estimate the average remaining flow times for each buffer.

c. What statistical claims can you make about your estimates? (Hint: Are the recorded times independent?)

12.22. Suppose that the system shown in Figure 12.13 produces 50 jobs per day on average. Average times at workstations are:

	Average hours per visit									
Buffer	1	2	3	4	5	6	7	8	9	10
Hours	1.0	4.3	2.1	0.5	3.7	1.8	0.2	1.1	1.3	0.8

Table 12.10 Shop Records for Job Arrivals (Hour.Minute of Arrival)

Job	Operation										
	1	2	3	4	5	6	7	8	9	10	End
A26	1.19	2.31	4.59	5.13	6.09	9.31	11.02	12.56	15.34	18.11	20.34
A35	1.40	3.02	5.31	6.45	6.59	10.21	10.59	12.33	14.58	18.45	20.55
B01	2.04	3.37	5.49	7.06	7.55	10.49	11.56	13.06	15.58	19.11	21.23
B22	2.36	4.02	6.31	7.54	8.32	11.28	12.35	13.32	16.44	19.46	21.51
C19	3.00	4.24	6.56	8.22	8.53	11.53	12.59	13.52	17.33	20.41	22.37
A43	3.16	4.56	7.22	8.50	9.26	12.17	12.59	14.18	18.15	21.32	23.13
B16	3.45	5.29	7.51	9.42	10.29	12.52	13.48	14.37	18.40	21.58	23.56
A41	4.02	5.59	8.30	9.57	10.43	13.20	15.06	15.41	19.03	22.20	24.29
C29	4.43	6.31	8.54	10.41	11.22	13.52	15.31	16.07	19.50	22.56	25.03
B31	4.41	6.55	9.22	11.31	11.56	15.18	15.49	16.41	20.32	23.21	25.33

Workstation 2 just became available. Develop an expression for the relative slack for an arbitrary job j sitting in buffers 2, 5, and 8.

12.23. Consider the model for workstation space allocation in mixed model assembly lines that was described in section 12.3.1.4. This model assumed a "closed" line spacing such that space was dedicated to a single workstation. Suppose that workers can share space in an "open" manner provided the workers never occupy the same space at the same time. Modify the model for this case.

12.24. Modify the batch formation model presented in Equations (12.21) to (12.27) to allow for some tools to be used by multiple jobs. Assume that jobs that require a common tool may share that tool.

12.25. Consider a flexible manufacturing system with four machines—A, B, C, and D. The following jobs are waiting to be produced:

	Operation 1	Operation 2	
Job	Mach., # tool slots, prod. hrs.	Mach., # tool slots, prod. hrs.	Due date
1	A, 3, 3.0	B, 2, 2.4	1
2	A, 3, 1.5	C, 1, 1.5	2
3	B, 2, 1.5	D, 3, 1.5	2
4	A, 3, 3.0	C, 1, 3.0	3
5	A, 3, 7.0	—	3
6	B, 2, 2.5	D, 3, 3.0	2
7	B, 2, 5.0	D, 3, 5.0	1

Each operation has its own tools. Each tool magazine has five slots. There are eight hours of production time available for each batch (each batch must be done in one day). Use the part batch heuristic to find the batches to load into the system. If each batch is produced daily, compute the total job tardiness for the batches.

12.26. Suppose that in the previous problem there were two type A machines available. Construct the daily part batches and also solve the tool loading problem to indicate the specific type A machine to which each part or tool is assigned.

12.27. Consider a FMS with the following part data:

	Operation 1	Operation 2	
Job	Mach., # tool slots, prod. hrs.	Mach., # tool slots, prod. hrs.	Due date
1	A, 2, 5.0	B, 2, 4.5	1
2	A, 4, 3.0	C, 1, 3.0	2
3	B, 2, 3.0	D, 3, 2.5	2
4	A, 2, 2.8	C, 2, 3.0	3
5	A, 2, 6.5	D, 3, 7.0	3
6	B, 2, 3.0	D, 3, 3.0	2
7	B, 2, 5.5	D, 3, 4.5	1

There is one machine of each type, A, B, C, and D. Each tool magazine has six slots. Part batches 3 and 6 use the same tools, all other tools are specific to the part and operation. There are eight hours of production time available for each batch (each batch must be done in one day). Use the part batch heuristic to find the batches to load into the system. If each batch is produced daily, compute the total job tardiness for the batches.

12.28. For the data in problem 12.25, write out the detailed batch formation model PS. Assume that there are at most four batches and that the number of each tool available is large enough to ensure that the corresponding constraints are never tight. If you have access to integer programming software, solve the model for the optimal solution and compare the results to those of problem 12.25.

12.29. Extend batch formation model PS to the case in which you can split a part's demand over multiple batches. Hint: Using the tardy/early objective is difficult because it is difficult to determine when the production for the part is actually finished. How might tardiness be defined when a portion of the demand is on time? Consider different objectives in the extension. Can you simply let δ_{jk} be a continuous decision variable?

12.30. Extend batch formation model PS to the case in which part j earliness is penalized at a rate of $\$E_j$ per day, and tardiness is penalized at a rate of $\$T_j$ per day.

Appendix I

Appendix IA Standard Normal Distribution Function

Z	0	0.01	0.02	0.03	0.04	0.05	0.06	0.07	0.08	0.09
0	0.50000	0.50399	0.50798	0.51197	0.51595	0.51994	0.52392	0.52790	0.53188	0.53586
0.1	0.53983	0.54380	0.54776	0.55172	0.55567	0.55962	0.56356	0.56749	0.57142	0.57535
0.2	0.57926	0.58317	0.58706	0.59095	0.59483	0.59871	0.60257	0.60642	0.61026	0.61409
0.3	0.61791	0.62172	0.62552	0.62930	0.63307	0.63683	0.64058	0.64431	0.64803	0.65173
0.4	0.65542	0.65910	0.66276	0.66640	0.67003	0.67364	0.67724	0.68082	0.68439	0.68793
0.5	0.69146	0.69497	0.69847	0.70194	0.70540	0.70884	0.71226	0.71566	0.71904	0.72240
0.6	0.72575	0.72907	0.73237	0.73565	0.73891	0.74215	0.74537	0.74857	0.75175	0.75490
0.7	0.75804	0.76115	0.76424	0.76730	0.77035	0.77337	0.77637	0.77935	0.78230	0.78524
0.8	0.78814	0.79103	0.79389	0.79673	0.79955	0.80234	0.80511	0.80785	0.81057	0.81327
0.9	0.81594	0.81859	0.82121	0.82381	0.82639	0.82894	0.83147	0.83398	0.83646	0.83891
1	0.84134	0.84375	0.84614	0.84849	0.85083	0.85314	0.85543	0.85769	0.85993	0.86214
1.1	0.86433	0.86650	0.86864	0.87076	0.87286	0.87493	0.87698	0.87900	0.88100	0.88298
1.2	0.88493	0.88686	0.88877	0.89065	0.89251	0.89435	0.89617	0.89796	0.89973	0.90147
1.3	0.90320	0.90490	0.90658	0.90824	0.90988	0.91149	0.91308	0.91466	0.91621	0.91774
1.4	0.91924	0.92073	0.92220	0.92364	0.92507	0.92647	0.92785	0.92922	0.93056	0.93189
1.5	0.93319	0.93448	0.93574	0.93699	0.93822	0.93943	0.94062	0.94179	0.94295	0.94408
1.6	0.94520	0.94630	0.94738	0.94845	0.94950	0.95053	0.95154	0.95254	0.95352	0.95449
1.7	0.95543	0.95637	0.95728	0.95818	0.95907	0.95994	0.96080	0.96164	0.96246	0.96327
1.8	0.96407	0.96485	0.96562	0.96638	0.96712	0.96784	0.96856	0.96926	0.96995	0.97062
1.9	0.97128	0.97193	0.97257	0.97320	0.97381	0.97441	0.97500	0.97558	0.97615	0.97670
2	0.97725	0.97778	0.97831	0.97882	0.97932	0.97982	0.98030	0.98077	0.98124	0.98169
2.1	0.98214	0.98257	0.98300	0.98341	0.98382	0.98422	0.98461	0.98500	0.98537	0.98574
2.2	0.98610	0.98645	0.98679	0.98713	0.98745	0.98778	0.98809	0.98840	0.98870	0.98899
2.3	0.98928	0.98956	0.98983	0.99010	0.99036	0.99061	0.99086	0.99111	0.99134	0.99158
2.4	0.99180	0.99202	0.99224	0.99245	0.99266	0.99286	0.99305	0.99324	0.99343	0.99361
2.5	0.99379	0.99396	0.99413	0.99430	0.99446	0.99461	0.99477	0.99492	0.99506	0.99520
2.6	0.99534	0.99547	0.99560	0.99573	0.99585	0.99598	0.99609	0.99621	0.99632	0.99643
2.7	0.99653	0.99664	0.99674	0.99683	0.99693	0.99702	0.99711	0.99720	0.99728	0.99736
2.8	0.99744	0.99752	0.99760	0.99767	0.99774	0.99781	0.99788	0.99795	0.99801	0.99807
2.9	0.99813	0.99819	0.99825	0.99831	0.99836	0.99841	0.99846	0.99851	0.99856	0.99861
3	0.99865	0.99869	0.99874	0.99878	0.99882	0.99886	0.99889	0.99893	0.99896	0.99900
3.1	0.99903	0.99906	0.99910	0.99913	0.99916	0.99918	0.99921	0.99924	0.99926	0.99929
3.2	0.99931	0.99934	0.99936	0.99938	0.99940	0.99942	0.99944	0.99946	0.99948	0.99950
3.3	0.99952	0.99953	0.99955	0.99957	0.99958	0.99960	0.99961	0.99962	0.99964	0.99965
3.4	0.99966	0.99968	0.99969	0.99970	0.99971	0.99972	0.99973	0.99974	0.99975	0.99976
3.5	0.99977	0.99978	0.99978	0.99979	0.99980	0.99981	0.99981	0.99982	0.99983	0.99983
3.6	0.99984	0.99985	0.99985	0.99986	0.99986	0.99987	0.99987	0.99988	0.99988	0.99989
3.7	0.99989	0.99990	0.99990	0.99990	0.99991	0.99991	0.99992	0.99992	0.99992	0.99992
3.8	0.99993	0.99993	0.99993	0.99994	0.99994	0.99994	0.99994	0.99995	0.99995	0.99995
3.9	0.99995	0.99995	0.99996	0.99996	0.99996	0.99996	0.99996	0.99996	0.99997	0.99997
4	0.99997	0.99997	0.99997	0.99997	0.99997	0.99997	0.99998	0.99998	0.99998	0.99998

Appendix IB Unit Normal Linear Loss (L(u))

u	0	0.01	0.02	0.03	0.04	0.05	0.06	0.07	0.08	0.09
0	0.39894	0.39396	0.38902	0.38412	0.37926	0.37444	0.36966	0.36492	0.36022	0.35556
0.1	0.35094	0.34635	0.34181	0.33731	0.33285	0.32842	0.32404	0.31969	0.31539	0.31112
0.2	0.30689	0.30271	0.29856	0.29445	0.29038	0.28634	0.28235	0.27840	0.27448	0.27060
0.3	0.26676	0.26296	0.25920	0.25547	0.25178	0.24813	0.24452	0.24094	0.23740	0.23390
0.4	0.23044	0.22701	0.22362	0.22027	0.21695	0.21367	0.21042	0.20721	0.20404	0.20090
0.5	0.19780	0.19473	0.19170	0.18870	0.18573	0.18281	0.17991	0.17705	0.17422	0.17143
0.6	0.16867	0.16595	0.16325	0.16059	0.15797	0.15537	0.15281	0.15028	0.14778	0.14531
0.7	0.14288	0.14048	0.13810	0.13576	0.13345	0.13117	0.12892	0.12669	0.12450	0.12234
0.8	0.12021	0.11810	0.11603	0.11398	0.11196	0.10997	0.10801	0.10607	0.10417	0.10229
0.9	0.10043	0.09860	0.09680	0.09503	0.09328	0.09156	0.08986	0.08819	0.08654	0.08491
1	0.08332	0.08174	0.08019	0.07866	0.07716	0.07568	0.07422	0.07279	0.07138	0.06999
1.1	0.06862	0.06727	0.06595	0.06465	0.06336	0.06210	0.06086	0.05964	0.05844	0.05726
1.2	0.05610	0.05496	0.05384	0.05274	0.05165	0.05059	0.04954	0.04851	0.04750	0.04650
1.3	0.04553	0.04457	0.04363	0.04270	0.04179	0.04090	0.04002	0.03916	0.03831	0.03748
1.4	0.03667	0.03587	0.03508	0.03431	0.03356	0.03281	0.03208	0.03137	0.03067	0.02998
1.5	0.02931	0.02865	0.02800	0.02736	0.02674	0.02612	0.02552	0.02494	0.02436	0.02380
1.6	0.02324	0.02270	0.02217	0.02165	0.02114	0.02064	0.02015	0.01967	0.01920	0.01874
1.7	0.01829	0.01785	0.01742	0.01699	0.01658	0.01617	0.01578	0.01539	0.01501	0.01464
1.8	0.01428	0.01392	0.01357	0.01323	0.01290	0.01257	0.01226	0.01195	0.01164	0.01134
1.9	0.01105	0.01077	0.01049	0.01022	0.00996	0.00970	0.00945	0.00920	0.00896	0.00872
2	0.00849	0.00827	0.00805	0.00783	0.00762	0.00742	0.00722	0.00702	0.00683	0.00665
2.1	0.00647	0.00629	0.00612	0.00595	0.00579	0.00563	0.00547	0.00532	0.00517	0.00503
2.2	0.00489	0.00475	0.00462	0.00449	0.00436	0.00423	0.00411	0.00400	0.00388	0.00377
2.3	0.00366	0.00356	0.00345	0.00335	0.00325	0.00316	0.00307	0.00298	0.00289	0.00280
2.4	0.00272	0.00264	0.00256	0.00248	0.00241	0.00234	0.00227	0.00220	0.00213	0.00207
2.5	0.00200	0.00194	0.00188	0.00183	0.00177	0.00171	0.00166	0.00161	0.00156	0.00151
2.6	0.00146	0.00142	0.00137	0.00133	0.00129	0.00125	0.00121	0.00117	0.00113	0.00110
2.7	0.00106	0.00103	0.00099	0.00096	0.00093	0.00090	0.00087	0.00084	0.00081	0.00079
2.8	0.00076	0.00074	0.00071	0.00069	0.00066	0.00064	0.00062	0.00060	0.00058	0.00056
2.9	0.00054	0.00052	0.00051	0.00049	0.00047	0.00046	0.00044	0.00042	0.00041	0.00040
3	0.00038	0.00037	0.00036	0.00034	0.00033	0.00032	0.00031	0.00030	0.00029	0.00028
3.1	0.00027	0.00026	0.00025	0.00024	0.00023	0.00022	0.00021	0.00021	0.00020	0.00019
3.2	0.00019	0.00018	0.00017	0.00017	0.00016	0.00015	0.00015	0.00014	0.00014	0.00013
3.3	0.00013	0.00012	0.00012	0.00011	0.00011	0.00010	0.00010	0.00010	0.00009	0.00009
3.4	0.00009	0.00008	0.00008	0.00008	0.00007	0.00007	0.00007	0.00007	0.00006	0.00006
3.5	0.00006	0.00006	0.00005	0.00005	0.00005	0.00005	0.00005	0.00004	0.00004	0.00004
3.6	0.00004	0.00004	0.00004	0.00003	0.00003	0.00003	0.00003	0.00003	0.00003	0.00003
3.7	0.00003	0.00002	0.00002	0.00002	0.00002	0.00002	0.00002	0.00002	0.00002	0.00002
3.8	0.00002	0.00002	0.00002	0.00001	0.00001	0.00001	0.00001	0.00001	0.00001	0.00001
3.9	0.00001	0.00001	0.00001	0.00001	0.00001	0.00001	0.00001	0.00001	0.00001	0.00001
4	0.00001	0.00001	0.00001	0.00001	0.00001	0.00001	0.00001	0.00001	0.00000	0.00000

Appendix II

Linear Programming

A.1 MODELING AND FORMULATION

Linear Programming (LP) is a modeling technique that is used to develop insight into the solutions of problems that are composed predominantly of linear functions of decision variables. To specify the model, you must precisely specify all of the following pieces of information:

Decision variables—notation, meaning, and sign (non-negative, nonpositive, or unrestricted),

Objective function to optimize—function of the decision variables and the direction of improvement (maximize or minimize),

Constraint functions that the solution must satisfy—each constraint consists of a function of the decision variables, a relationship $(=, \leq, \geq)$, and a right side value (scalar).

LP can be used to model many decision problems. The following requirements on the problem are necessary to ensure that LP is a valid modeling technique for the problem.

Deterministic data (function coefficients from cost data and constraint data)

Linear functions (both objective and ALL constraints)

Single objective

Continuous decision variables (discrete decisions are not permitted)

Before considering how LP fits into the general engineering design process, we provide a small example that will be used throughout the discussion.

EXAMPLE A.1

Consider a manufacturing facility that produces two products—sofas and chairs. Each sofa requires 20 feet of framing wood, whereas each chair requires only ten feet. 5,000 feet of wood are available per month. Also, each sofa requires 25 minutes to assemble, and each chair requires ten minutes. During the month, there are 6,000 minutes of labor time available. Finally, demand for each product is limited. You can only sell 200 sofas and 300 chairs per month. The company makes $250 per chair produced and sold and $400 per sofa. How should we plan production over the next month?

SOLUTION

The first issue in LP modeling is to determine if this problem fits the LP assumptions. First, consider deterministic data. Clearly, data such as demand limits and production materials are NOT deterministic, and estimates must be used. We will consider "sensitivity analysis" to consider the impact of poor estimates; however, we must realize that LP has a weakness here. The assumption of linear functions is easier to satisfy here. We construct linear functions for the objective and the constraints. The assumption concerning a single objective is also easily satisfied here because the key

517

objective is to maximize profit. Finally, the assumption concerning continuous decisions is problematic. If our decisions are the production level of each product, then clearly, we cannot actually produce a fraction of a chair or sofa. Here, we try to bound the effects of the assumption (or consider an alternate approach called "integer programming" that can consider discrete decisions).

Once we believe that LP modeling is an appropriate solution tool, we have to construct the model. Here is one approach to the problem:

$$\text{Define } S = \text{number of sofas produced in a month}$$
$$C = \text{number of chairs produced in a month}$$

The model to solve is:

Maximize	$Z = \$400\, S + \$250\, C$	(A.1)
Subject to:	$S \leq 200$ sofas	(A.2)
	$C \leq 300$ chairs	(A.3)
	$20S + 10C \leq 5000$ feet	(A.4)
	$25S + 10C \leq 6000$ minutes	(A.5)
	$S \geq 0,\ C \geq 0.$	(A.6)

The objective A.1 seeks to maximize profit by multiplying the per unit profit by the production units and summing over both products. Constraints A.2 and A.3 represent the demand limits per month. Constraint A.4 limits the wood usage. The left side is a function quantifying the wood used whereas the right side is the wood available. A similar strategy is used in A.5 to model the labor time limit. Finally, constraint set A.6 ensures that we don't produce a negative amount of either product (we cannot subcontract out production in this situation).

The general LP modeling process requires the following steps:

Problem identification (identification of organizational objective, constraints, and range of decision alternatives)

Model formulation (converting problem characteristics into decision variables and appropriate linear functional relationships)

Data collection (determination of cost, productivity, and resource availability coefficients)

Model validation and verification (ensuring that the model represents what is actually happening in the system)

Model solution (finding the optimal solution and validating)

Sensitivity analysis (attacking the "determinism" assumption)

Economic evaluation and interpretation ("what iffing")

Recommendation and communication

If you fail on any part of the process, you are likely to have difficulty when you try to implement the recommendation. The specific techniques of building formulations are often covered in operations research courses and hence are omitted here (other than the formulation in Example A.1). The focus at this point is on solution method and economic evaluation and interpretation through duality.

A.2   SOLUTION METHOD

The general linear programming model can be written as follows:

$$\text{Maximize } Z = \mathbf{cx} \qquad (A.7)$$

$$\text{Subject to: } Ax \leq b \tag{A.8}$$

$$x \geq 0 \tag{A.9}$$

Where A is an m $\times$ n matrix of constraint coefficients, x is a n $\times$ 1 vector of decision variables, c is a 1 $\times$ n vector of cost coefficients, and b is an m $\times$ 1 vector of constraint right hand side values. A **feasible solution** to the model is a set of decision variable values that satisfies ALL constraints. An **optimal solution** is a feasible solution with the "best" objective function value.

The model is general in that simple transformations can be made to turn maximization problems into minimization problems and vice versa (multiply each objective coefficient by -1) and "$\leq$" into "$\geq$" and vice versa (multiply each constraint coefficient and the right side value by -1). Problems can be formulated as "$=$" constraints by adding slack or subtracting surplus variables (these variables have a "0" objective function coefficient and are used to take up the "space" remaining in an inequality constraint). In short, the modeling construction is extremely general.

When solving an LP model one of the following situations must occur:

No feasible solutions (the model is said to be **infeasible**),

Many feasible solutions with no limit on the optimal value of the objective (the model is said to be **unbounded**),

Feasible solutions with a unique optimal solution, or

Feasible solutions with an infinite number of optimal solutions.

The set of feasible points must be a **convex set.** The constraints that form the boundaries are called **faces of the convex set,** and the corner points where constraints cross are called **extreme points.** We demonstrate a solution technique that uses a graphical representation of the feasible set when problems have two decisions (continuation of Example A.1).

EXAMPLE A.2   Find the optimal solution of the model in Example A.1, if it exists.

SOLUTION   Recall the formulation:

$$\text{Maximize} \quad Z = \$400\, S + \$250\, C \tag{A.1}$$
$$\text{Subject to:} \quad S \leq 200 \text{ sofas} \tag{A.2}$$
$$C \leq 300 \text{ chairs} \tag{A.3}$$
$$20S + 10C \leq 5000 \text{ feet} \tag{A.4}$$
$$25S + 10C \leq 6000 \text{ minutes} \tag{A.5}$$
$$S \geq 0,\ C \geq 0. \tag{A.6}$$

Because the model has only two variables, the set of feasible solutions can be graphed in Figure A.1. Here, each constraint is a line where one side is feasible (direction of the arrows). Note, that there are four lines, one for each constraint, and we are only concerned with the non-negative quadrant because of constraint set A.6. The feasible set for the entire problem is the **intersection** of the feasible sets of all of the constraints. The convex feasible set is highlighted on Figure A.1. The faces of the set are the constraint segments that make up the intersection area. In this example, each constraint contributes a segment. However, in general, this need not happen and constraints may be redundant.

Figure A.2 depicts the **isocost** or **isoquant** method of solving LP models. Here, we take the feasible set (the extreme points are highlighted) and plot the objective function line for various

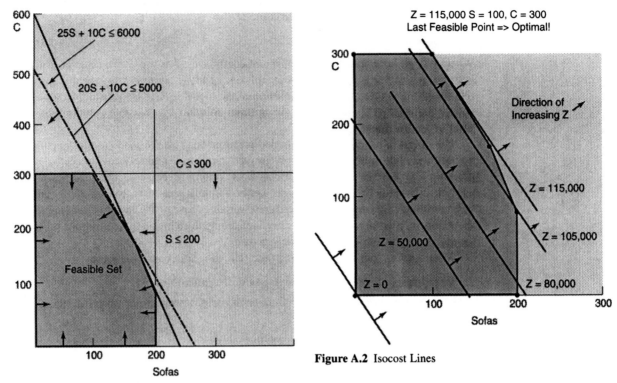

Figure A.1 Feasible Region

Figure A.2 Isocost Lines

values of Z. Once we fix Z, the line represents all pairs of S and C that yield the particular Z value. For example, when $Z = 0$, the line goes through the origin ($S = 0$, $C = 0$). As we increase Z the line moves to the northeast and starts to move through the feasible set. As long as the objective line is in the interior of the feasible set, then we can keep increasing Z. The process stops when the objective **last touches the feasible set.** This must occur at an extreme point—as depicted in Figure A.2—or occurs along a face of the feasible set. For the example formulation, the optimal Z value occurs at the point $C = 300$, $S = 100$ and has an objective value of 115,000.

The following results are true for all LP models (not just those with two variables):

Theorem A.1: If an LP model has an optimal solution, then it must have an optimal solution at an extreme point.

Theorem A.2: If an LP model has multiple optimal solutions, then the objective line must be parallel to one of the faces of the feasible set, and there are at least two optimal extreme points.

The simplex method is an organized fashion of implicitly checking all of the extreme points for finding the optimal solution. Assume that we have transformed the model into one with **all equality constraints** (we have added slack and surplus variables if they are necessary). Assume that a feasible solution vector $x = (x_1, x_2, \ldots, x_n)$ is given. Define the index set $B = (j : x_j > 0)$. The set x^B is called the set of **basic variables.** These are the variables that are strictly positive in x. The set x^N is the remainder of the variables and these are called the **nonbasic variables.** All nonbasic variables take on a value of 0 in the solution. We use B and N to represent the matrices composed of the constraint coefficient columns for the basic and nonbasic variables respectively (order is important, so the column order of B and N must be the same as the order of the variables in the vectors x^B and x^N. Finally, let c^B be the vector of cost coefficients for the basic variables, and let c^N be the vector of cost coefficients for the nonbasic variables.

Theorem A.3: A feasible solution x to an LP is a **Basic Feasible Solution (BFS)** if the solution is feasible and the columns in the matrix B are linearly independent. A solution x corresponds to an extreme point, if and only if, x is a basic feasible solution.

If x is a BFS, then it is simple to solve for the values for the variables so that the constraints are satisfied at equality. Use the partition x^B and x^N, plug into the constraint equations, realize that the matrix B is invertible and that the values for x^N equal 0:

$$Bx^B + Nx^N = b \qquad (A.10)$$

$$x^B = B^{-1}b, \, x^N = 0 \qquad (A.11)$$

Once an extreme point is found, optimality is evaluated using **reduced costs.** If a nonbasic variable at 0 is increased, then some of the variables in x^B that are currently positive change and the objective increases or decreases. The reduced cost for a variable tabulates the total change in the objective when we increase the variable from 0 to 1. For any nonbasic variable, the formula for computing reduced cost is:

$$c_j' = c_j - c^B B^{-1} N_j \qquad (A.12)$$

where N_j is the column of N associated with the nonbasic variable.

Theorem A.4: For a maximization (minimization) problem, if $c_j' \le 0$ ($c_j' \ge 0$) for all x_j that are in x^N, then, x is an optimal solution to the LP.

Because the equations are linear, if changing a variable value from 0 to 1 improves the objective, then changes to values larger than 1 improve the objective more. We would like to increase the variable value until another constraint says that we must stop.

The simplex method picks a variable with an objective improving reduced cost and switches it from x^N to x^B. The variable that leaves x^B is determined by the **minimum ratio rule** and is the first variable that goes to value 0 as you increase the entering variable above 0. The process repeats until Theorem A.4 is satisfied. The following algorithm finds an optimal solution to an LP with a "minimization" objective:

Step 1: Find an initial BFS. Define x^N, x^B, B, N, c^N, and c^B.

Step 2: Compute the vector $c' = (c^N - c^B B^{-1} N)$. If $c_j' \ge 0$ for all j, then stop, the solution $x^B = B^{-1}b$ is optimal and has objective $c^B B^{-1} b$. Otherwise, select the variable in x^N with the minimum value of c_j' and denote this a variable x_p.

Step 3: Compute $A_p' = B^{-1}A_p$. If $A_{ip}' \le 0$ for all i, then stop, the problem is unbounded and the objective can decrease to $-\infty$. Otherwise, compute $b' = B^{-1}b$ and find

$$\text{Min } b_i'/a_{ij}'$$

where the minimum is taken over i whereby $a_{ij}' > 0$. Assume that the minimum ratio occurs in row r. Insert x_p into the rth position of x^B and take the variable that was in this position and move it to x^N. Update B, N, c^N, c^B and return to Step 2.

A.3 DUALITY AND ECONOMIC INTERPRETATION

The problem that we have been formulating and solving is called the **"primal"** problem. The decision is to set production levels so that costs are minimized and resources are not overused. An alternate idea is to sell resources. The decisions are the per unit prices to charge for each of the resources (termed the **"shadow price for a resource"**). This second problem is called the **"dual"** problem and is formulated based on the primal problem. The objective is to minimize total cost from buying resources. The prices are set so that it is at least as desirable to sell resources as to produce product. This idea carries over from microeconomics in that in a production venture, marginal revenue of production must be greater than or equal to the marginal cost of the resources used in production. The dual formulation for a primal maximization problem with "$\leq$ constraints" and "≥ 0 variables" is:

$$\text{Minimize } yb \tag{A.13}$$

$$\text{Subject to: } yA \geq c \tag{A.14}$$

$$y \geq 0 \tag{A.15}$$

where b, A, and c take dimensions as defined before and y is a $1 \times m$ vector. y_i is the decision variable for the price that you should pay for a single unit of resource i.

The following results are all useful in determining the value of resources and in relating the primal and dual problems:

Theorem A.5 (weak duality): Assume that you are given a solution x' that is feasible to the primal problem and a solution y' that is feasible to the dual problem (as defined above) then: $y'b \geq cx'$.

Theorem A.6 (strong duality): Assume that the primal problem has a finite optimal solution x^*. Then, dual also has an optimal solution y^* and $y^*b = cx^*$.

Theorem A.7 (complementary slackness): Assume that y' and x' are feasible to the dual and primal problems respectively. Then, y' and x' are optimal to their respective problems, if and only if:

$$(b_i - A_i'x') * y_i' = 0 \qquad \text{for all } i = 1 \ldots m \tag{A.16}$$

$$(c_j - y'A_j) * x_j' = 0 \qquad \text{for all } j = 1 \ldots n. \tag{A.17}$$

where A_j is the constraint coefficient column associated with x_j and A_i' is the ith row of the constraint coefficient matrix.

The complementary slackness conditions, A.16 show that in an optimal solution, if you have a positive price for a resource, then you must be using **ALL** of that resource (there is no slack in the corresponding primal constraint). The second condition A.17 states that if you decide to produce an item (make x_j positive), then for that item, you must have the marginal cost of the item equal to the marginal revenue obtained from selling resources.

The shadow price of a resource represents the maximum that you would be willing to pay for an extra unit of resource because this is the amount that the optimal objective value would increase if you obtained the extra unit. This price is only valid for small increases in the resource availability because you may move to a new optimal extreme point if you are given sufficient resources. Duality, and shadow prices in particular, provide valuable information concerning the limiting resources for a decision problem and help a planner make decisions in situations in which it is possible to invest in buying additional resources.

Index

CPSIA information can be obtained at www.ICGtesting.com
Printed in the USA
BVOW05s0424090813

328163BV00002B/9/A

9 780471 115939